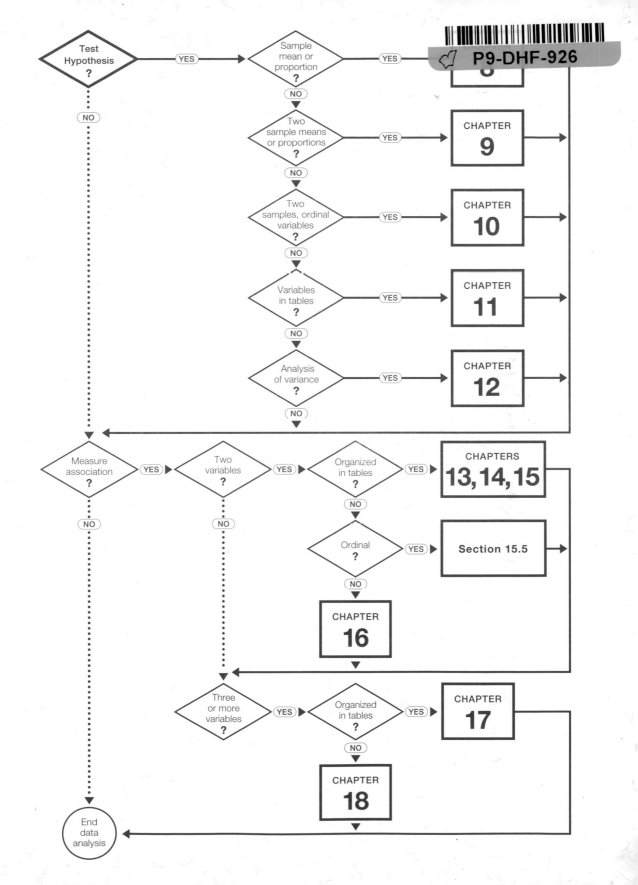

Statistics: A Tool for Social Research

Fourth Edition

Statistics: A Tool for Social Research

Fourth Edition

Joseph F. Healey
Christopher Newport University

 Wadsworth Publishing Company

I(T)P™ An International Thomson Publishing Company

Belmont • Albany • Bonn • Boston • Cincinnati • Detroit • London • Madrid • Melbourne
Mexico City • New York • Paris • San Francisco • Singapore • Tokyo • Toronto • Washington

Editor: Eve Howard
Assistant Editor: Susan Shook
Editorial Assistant: Julie McDonald
Production: The Wheetley Company
Print Buyer: Karen Hunt
Cover Design: Sandy Drooker
Cover Illustration: Giovanni Battista Piranesi (1720–1778), *Piazza del Campidoglio and S. Maria in Aracoeli,* from *Vedute di Roma;* reproduced by permission of the Huntington Library, San Marino, California.
Compositor: G&S Typesetters, Inc.
Printer: Quebecor Printing Group/Fairfield

Printed in the United States of America
2 3 4 5 6 7 8 9 10—02 01 00 99 98 97 96

For more information, contact Wadsworth Publishing Company:

Wadsworth Publishing Company
10 Davis Drive
Belmont, California 94002, USA

International Thomson Publishing Europe
Berkshire House 168-173
High Holborn
London, WC1V 7AA, England

Thomas Nelson Australia
102 Dodds Street
South Melbourne 3205
Victoria, Australia

Nelson Canada
1120 Birchmount Road
Scarborough, Ontario
Canada M1K 5G4

International Thomson Editores
Campos Eliseos 385, Piso 7
Col. Polanco
11560 México D.F. México

International Thomson Publishing GmbH
Königswinterer Strasse 418
53227 Bonn, Germany

International Thomson Publishing Asia
221 Henderson Road
#05-10 Henderson Building
Singapore 0315

International Thomson Publishing Japan
Hirakawacho Kyowa Building, 3F
2-2-1 Hirakawacho
Chiyoda-ku Tokyo 102, Japan

Library of Congress Cataloging-in-Publication Data

Healey, Joseph F., 1945–
    Statistics : a tool for social research / Joseph F. Healey.—4th ed.
    p. cm.
    Includes index.
    ISBN 0-534-25152-8
    1. Social sciences—Statistical methods. 2. Statistics. I. Title.
HA29.H39   1995
519.5—dc20   95-11723
                                                                   CIP

Brief Contents

Detailed Contents

Preface

Sociology and the other social sciences are research-based disciplines. Social scientists test and refine their ideas against the highest quality evidence available. The research capabilities of the social sciences have grown enormously over the past several decades and, as the end of the twentieth century approaches, the essential "tools of the trade" for the social scientist now routinely include calculators and computers.

Statistics and research methodology are essential components in the education of the social scientist. However, learning (and teaching) these subjects can present challenges that are not present in other areas of the curriculum. Sociology and other social science majors typically begin a statistics course with a wide range of mathematical backgrounds and an equally diverse set of career goals. Students are often puzzled about how statistics relates to sociology in general and, not infrequently, there is some math anxiety to overcome.

This text was written to meet the challenge of introducing statistical analysis to social science majors while addressing the instructional challenges that are inherent in the course. The text makes minimal assumptions about the mathematical background of the students (the ability to read a simple formula is sufficient preparation for virtually all of the material in the text), and a variety of special features have been integrated into the text to assist students to successfully analyze data. The text has been written especially for sociology and social work programs but is sufficiently flexible to be used in statistics courses in political science or in any program with a social science base and an applied focus (for example, public administration, criminal justice, urban studies, and gerontology).

The text is written at a level intermediate between the more rigorous and sophisticated texts on one hand and the mere "cookbook" on the other. That is, while I have not sacrificed comprehensive coverage or statistical correctness, the theoretical and mathematical explanations of why statistics "do what they do" are kept at an elementary level, as is appropriate in a first exposure to social statistics. For example, I do not treat formal probability theory per se. Rather, the background necessary for an understanding of inferential statistics is introduced, informally and intuitively, in Chapters 5 and 6 while considering the concepts of the normal curve and the sampling distribution. The text makes no claim that statistics are "fun" or that the material can be mastered without considerable effort. At the same time, students are not overwhelmed with abstract proofs and mathematical theory, which at this level needlessly frustrate the learning experience.

My major goal is basic statistical literacy. The text is designed to provide a solid foundation in statistical analysis and to prepare students to be intelligent consumers of social research. More specifically, I believe that basic statistical literacy can be defined in terms of three interrelated qualities and, as a way of further describing the nature of this text, I would like to list each of these qualities and briefly summarize how the text is designed to develop them.

Computational Competence. At a minimum, students should emerge from their first course in statistics with the ability to perform elementary forms of data analysis—to execute a series of calculations and arrive at the correct answer. Since students in social science statistics courses frequently do not have strong quantitative backgrounds, I have included a number of features to help students cope with computations:

- *Step-by-step computational algorithms* are provided for each statistic.

- *Extensive problem sets* are provided at the end of each chapter. For the most part, these problems use fictitious data and are designed for relative ease of computation.

- *Cumulative exercises* are included at the end of each part to provide practice in choosing statistics. These exercises present only data sets and research questions. Students must choose appropriate statistics as part of the exercise.

- *Solutions* to odd-numbered problems are provided so that students may check their answers.

- *SPSS/PC+,* the leading social science statistical software, is incorporated to give students access to the computational power of the computer. This feature is explained in more detail below.

An Appreciation of Statistics. A statistically literate person can do much more, of course, than merely calculate correct answers. Such a person understands the relevance of statistics for social research, can select an appropriate statistic for a given purpose and a given set of data, and can analyze and interpret the meaning of that statistic. This textbook begins to develop these qualities, within the constraints imposed by the introductory nature of the course, in the following ways:

- *The relevance of statistics.* Chapter 1 includes a discussion of the role of statistics in social research and stresses the usefulness of these techniques as ways of analyzing and manipulating data and answering research questions. Each example problem is framed in the context of a research problem. A question is posed and then, with the aid of a statistic, answered. The relevance of statistics for answering questions is thus stressed throughout the text. This central theme of usefulness is further reinforced by a series of boxes labeled "Applications," each of which illustrates some specific way statistics can be used to answer questions.

    The great majority of the end-of-chapter problems are labeled by the social science discipline or subdiscipline from which they are drawn. The following abbreviations are used as labels: *SOC* for *sociology, SW* for *social work, PS* for *political science, CJ* for *criminal justice, PA* for *public administration*, and *GER* for *gerontology*. By identifying problems with specific disciplines, students can more easily see the relevance of statistics to their own academic interests. (Not incidentally, they will also see that the disciplines have a large subject matter in common.)

- *Selecting appropriate statistics*. A series of flowcharts are included to help students select appropriate statistics. These flowcharts have two components. Decision points are represented by diamonds and information by rectangles. The selection process is represented in general terms on the inside front cover and at the beginning of each part. Chapters begin with detailed flowcharts that, based on a consideration of the purpose of the analysis, the format of the data, and the level of measurement criterion, lead students to specific formulas or sections of the chapter.

- *Interpreting statistics*. After selecting and computing a statistic, students still face difficulties in understanding what the statistic means. The ability to interpret statistics can be developed only by exposure and experience. To provide exposure, I have been careful, in the example problems, always to express the meaning of the statistic in terms of the original research question. To provide experience, the end-of-chapter problems almost always call for an interpretation of the statistic calculated. To provide examples, many of the Answers to Odd-Numbered Problems in the back of the text are expressed in words as well as numbers. In particular, the answers to the cumulative exercises include extensive discussions on the process of selecting statistics.

The Ability to Read the Professional Social Science Literature. The statistically literate person can comprehend and critically appreciate research reports written by others. The development of this quality is a particular problem at the introductory level because of the marked disparity between the concise language of the professional researcher and the rather wordy vocabulary of the classroom. To help bridge this gap, I have included a series of boxes labeled "Reading Statistics." These begin in Chapter 1 and appear every two or three chapters. In each box, I briefly describe the reporting style typically used for the statistic in question and try to alert students about what to expect when they approach the professional literature.

These inserts are supplemented by excerpts from the professional literature. The excerpts illustrate how statistics are actually applied and interpreted by social science researchers.

Additional Features. A number of other features make the text more meaningful for students and more useful for instructors:

- *Readability.* The writing style is informal and accessible to students without ignoring the traditional vocabulary of statistics. Problems and examples have been written to maximize student interest and to focus on issues of concern and significance. For the more difficult material (such as hypothesis testing), students are first walked through an example problem before being confronted by formal terminology and concepts. Each chapter ends with a summary of major points and formulas and a glossary of important concepts. A glossary of symbols inside the back cover can be used for quick reference.

- *Organization and coverage.* The text is divided into four parts, with most of the coverage devoted to univariate descriptive statistics, inferential statistics, and bivariate measures of association. The distinction between description and inference is introduced in the first chapter and maintained throughout the text.

  In selecting statistics for inclusion, I have tried to strike a balance between the essential concepts with which students must be familiar and the amount of material students can reasonably be expected to learn in their first (and perhaps only) statistics course, while bearing in mind that different instructors will naturally wish to stress different aspects of the subject. Thus, the text covers a full gamut of the usual statistics, with each chapter broken into subsections so that instructors may choose the particular statistics they wish to include.

- *Review of mathematical skills.* Appendix H provides a comprehensive review of all of the mathematical skills that will be used in this text. Students who are inexperienced or out of practice with mathematics may want to study this section early in the course and/or refer to it as needed. A self-test is included in Appendix H so students may check their level of preparation for the course.

- *Computer applications and realistic data.* In order to help students learn to take advantage of the power of the computer, this text integrates SPSS/PC+, the most widely used statistical software in the social sciences. Appendix F is an introduction to statistical packages in general and SPSS/PC+ in particular. There are demonstrations at the end of each chapter that explain how to use SPSS/PC+ to produce the statistics presented in the chapter. Student exercises in analyzing data with SPSS/PC+ are also included.

  The main data base for computer applications is taken from the General Social Survey. This data base will give students the opportunity to practice their statistical skills on "real-life" survey data. The data base is described in Appendix G, and instructors may request a data diskette from Wadsworth. The data can be used with many statistical programs other than SPSS/PC+ so, even if students do not have access to SPSS/PC+, they can still do the exercises and compare their results with those presented in the demonstrations.

- *Instructor's Manual.* The Instructor's Manual includes learning objectives, chapter summaries, a test item file of multiple-choice questions, answers to even-numbered computational problems, and step-by-step solutions to selected problems. In addition, the Instructor's Manual includes cumulative exercises (with answers) similar to those in the text but with different variables or scores. These may be used for testing purposes.

- *Study Guide.* The Study Guide, written by Rebecca Davis, University of Maryland–College Park, contains additional examples to illuminate basic principles, review problems with detailed answers, SPSS work, and multiple-choice questions and answers that complement but do not duplicate the test item file.

Changes in this edition.   There are two major changes in this edition, both related to the increasing importance of computers in the research process. First, the data base available for this text has been updated to the 1993 version of the General Social Survey. The number of variables and the sample size have both been increased, and the data base now includes 77 variables and almost 800 cases. Second, the format of bivariate tables with ordinal level variables has been changed to be consistent with computerized statistics packages. Analysis and interpretation of the direction of relationships no longer has to be "translated" from the text to computer output.

The text has been thoroughly reviewed for clarity and readability, and some minor problems with terminology have been resolved. As with previous editions, my goal is to provide a comprehensive, flexible, and student-oriented text that will provide a challenging first exposure to social statistics.

Joseph F. Healey

Acknowledgments

This text has been in development, in one form or another, for more than ten years. An enormous number of people have made contributions, both great and small, to this project and, at the risk of inadvertently omitting someone, I am bound at least to attempt to acknowledge my many debts. This edition and the third have benefitted greatly from the expert editorial guidance of Serina Beauparlant. Although she is, sadly, no longer with Wadsworth, I thank her for her hard work, her countless contributions, and her unflagging support. I would also like to acknowledge the contributions of Bob Podstepny of Wadsworth during the early stages of the project and Susan Shook during the late stages.

Much of whatever integrity and quality this book has is a direct result of the contributions of colleagues who reviewed the manuscript during the various stages of writing. I have been consistently impressed by their sensitivity to the needs of the students, and I would like to thank Barbara Coventry, University of Toledo; David Fasenfest, Purdue University; Gerard Grzyb, University of Wisconsin–Oshkosh; Dennis Palumbo, Arizona State University; Paul Raffoul, University of Houston; Lawrence Rosen, Temple University; and Bruce Wade, Spelman College. Whatever failings are contained in the text are, of course, my responsibility and are probably the results of my occasional decisions not to follow the advice of the reviewers.

I would like to thank the instructors who made statistics understandable to me (Professors Satoshi Ito, Noelie Herzog, and Ed Erikson), all of my colleagues at Christopher Newport University for their support and encouragement (especially Professors F. Samuel Bauer, Robert Durel, James Forte, Ruth Kernodle, Cheryl Mathews, Lea Pellet, Virginia Purtle, and William Winter), and, in particular, all of my students for their patience and support. Also, I am grateful to the Literary Executor of the late Sir Ronald A. Fisher, F.R.S., to Dr. Frank Yates, F.R.S., and to Longman Group Ltd., London, for permission to reprint Appendixes B, C, and D, from their book *Statistical Tables for Biological, Agricultural and Medical Research* (6th edition, 1974).

Finally, I want to acknowledge the support of my family and rededicate this work to them. I am fortunate to be a member of an extended family that is remarkable in many ways. Although I cannot list everyone (my family is also remarkably large), I would like to especially thank my mother, Alice T. Healey, and my sons, Kevin and Christopher.

J. F. H.

1 Introduction

1.1 WHY STUDY
STATISTICS?

Students sometimes approach their first course in statistics with questions about the value of the subject matter. What, after all, do numbers and statistics have to do with understanding people and society? In a sense, this entire book will attempt to answer this question, and the value of statistics will become clear as we move from chapter to chapter. For now, the importance of statistics can be demonstrated, in a preliminary way, by briefly reviewing the research process as it operates in the social sciences. These disciplines are scientific in the sense that social scientists attempt to verify their ideas and theories through research. Broadly conceived, research is any process by which information is systematically and carefully gathered for the purpose of answering questions, examining ideas, or testing theories. Research is a disciplined inquiry that can take numerous forms. Statistical analysis is relevant only for those research projects where the information collected is represented by numbers. Numerical information of this sort is called **data**, and the sole purpose of statistics is to manipulate and analyze data. **Statistics**, then, are a set of mathematical techniques used by social scientists to organize and manipulate data for the purpose of answering questions and testing theories.

What is so important about learning how to manipulate data? On one hand, some of the most important and enlightening works in the social sciences do not utilize any statistical techniques. There is nothing magical about data and statistics. The mere presence of numbers guarantees nothing about the quality of a scientific inquiry. On the other hand, data can be the most trustworthy kind of information available to the researcher and, consequently, deserve special attention. Data that have been carefully collected and thoughtfully analyzed are the strongest, most objective foundations for building theory and enhancing understanding. Without a firm base in data, the social sciences would lose the right to the name *science* and would be of far less value to humanity.

Thus, the social sciences rely heavily on data-gathering for the advancement of knowledge. Let me be very clear about one point: it is never enough merely to gather data (or, for that matter, any kind of information). Even the most objective and carefully collected numerical information does not and cannot speak for itself. The researcher must be able to use statistics effectively. To be useful, the data must be organized, evaluated, and analyzed. Without a good understanding of the principles of statistical analysis, the researcher will be unable to make sense of the data. Without the appropriate application of statistical techniques, the data will remain mute and useless.

Statistics are an indispensable tool for the social sciences. They provide the scientist with some of the most useful techniques for evaluating ideas, testing theory, and discovering the truth. In the next section, I'll describe the relationships between theory, research, and statistics in more detail.

1.2 THE ROLE OF STATISTICS IN SCIENTIFIC INQUIRY

Figure 1.1 graphically represents the role of statistics in the research process. The diagram is based on the thinking of Walter Wallace and illustrates how the knowledge base of any scientific enterprise grows and develops. One point the diagram makes is that scientific knowledge accumulates by a circular process in which theory and research continually shape each other. Statistics are one of the most important means by which research and theory interact. Let's take a closer look at the wheel.

Since the figure is circular, it has no beginning or end and we could begin our discussion at any point. For the sake of convenience, let's begin at the top with the box labeled theory and follow the arrows around the circle. A **theory** is an explanation of the relationships between phenomena. This definition might seem academic and formal, but let me hasten to point out that we all "theorize" quite naturally. We wonder about, for example, what causes some people to be prejudiced, what causes child abuse, poverty, war, or success. We also develop explanations for these phenomena in an attempt to understand them. A major difference between our informal, everyday explanations and scientific theory is that the latter is subject to a rigorous testing process. Let's take the problem of racial prejudice as an example to illustrate how the research process works.

If we spent some time thinking about what causes prejudice, it might occur to us that the region of the country in which a person was socialized might be one important causal factor. The South is often identified as the most traditional region of the nation, especially with respect to matters of race and ethnicity. Perhaps white Southerners are more prejudiced as a result of growing up in an area where prejudice has a stronger hold on the ways people think about each other.

These thoughts are not a complete explanation of prejudice, but they will serve to illustrate the role of theory. We are trying to explain the relationship between two social phenomena: (1) prejudice, and (2) region of birth. The proposed connection between the two is that people socialized in the South will be more prejudiced as a result of acquiring and internalizing a more traditional and prejudiced culture. Thus, we have a coherent and, at least on its face, a logical theory—an attempt to describe and explain the relationship between two phenomena.

Before moving to the next area of Figure 1.1, let's take a moment to further examine theory. Our "theory" is very simplified but, like most theories, it is stated in terms of causal relationships between variables. A **variable** is

FIGURE 1.1   THE WHEEL OF SCIENCE

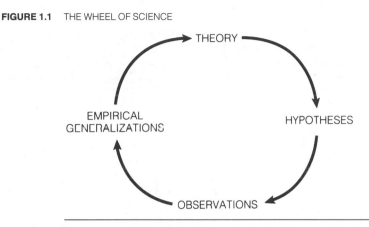

Source: Adapted from Walter Wallace, *The Logic of Science in Sociology* (Chicago: Aldine-Atherton, 1971).

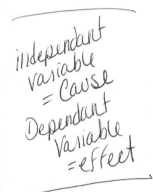

any trait that can change values from case to case. Examples of variables would be gender, age, income, or political party affiliation. In any specific theory, some variables will be identified as causes and others will be identified as effects or results. In the language of science, the causes are called **independent variables** and the "effects" or result variables are called **dependent variables**. In our theory of prejudice, region would be the independent variable (or the cause) and prejudice would be the dependent variable (the result or effect). In other words, we are arguing that region of birth is a cause of level of prejudice or that an individual's level of prejudice depends on the region of birth.

So far, we have a theory of prejudice complete with an explanation and an independent and a dependent variable. What we don't know yet is whether the theory is true or false. To find out, we need to compare our theory with the facts: we need to do some research. The next steps in the process would be to define our terms and ideas more specifically and exactly. One problem we often face in doing research is that, unlike our simplified example, scientific theories are too complex and abstract to be fully tested in a single research project. To conduct research, one or more hypotheses must be derived from the theory. A **hypothesis** is a statement about the relationship between variables which, while logically derived from the theory, is much more specific and exact.

For example, if we wished to test our theory of prejudice and region, we would have to say exactly what we mean by these terms. While it might not be much of a challenge to define "region of birth," our other variable, prejudice, will require considerable thought to define with clarity. It happens that prejudice in all its forms has been a major research focus in American sociology, so there are both a considerable literature and a well-established research tradition that will help us clarify what we will mean by this term.

As our definitions develop and the hypotheses take shape, we enter the next phase of the research process. Now, we will design the data-gathering phase of the project. We must decide how cases will be tested, how these cases will be selected, how exactly the variables will be measured, and a host of related matters. Ultimately, these plans will lead to the observation phase (the bottom of the wheel), where we actually measure social reality. Before we can do this, we must have a very clear idea of what we are looking for and a well-defined strategy for conducting the search.

If we decided to study racial prejudice, for example, we might decide to interview 200 individuals selected from our local community. As part of the interview, we might ask: "In which region of the country were you born?" or "In which region of the country did you grow up?" as a way of measuring our independent variable. (Note that the second question is more consistent with what we want to know, but it is also more ambiguous than the first. We might need to clarify what we mean by "grow up" so as to reduce misunderstandings.) We would also ask questions designed to measure levels of prejudice. For example, we might ask the respondents to agree or disagree with a series of statements such as "It is very important that every effort be made to racially integrate the public school system." This interviewing process would generate "observations" that we could then use to assess our theory. We will be able to see from the completed interviews if Southerners really do respond in more prejudiced ways to our survey items.

Now, finally, we come to statistics. As the observation phase of our research project comes to an end, we will be confronted with a large collection of numerical information or data. Try to imagine dealing with 200 completed interviews. In our example, if we had asked each of the respondents just five questions, we would have a total of 1000 separate pieces of information to deal with. What do we do? We have to have some systematic way to organize and analyze this information and, at this point, statistics will become very valuable. Statistics will supply us with many ideas about "what to do" with the data; we will begin to look at some of the options in the next chapter. For now, let me stress two points about statistics.

First, statistics are crucial. Simply put, without statistics, quantitative research is impossible. Without quantitative research, the development of the social sciences would be severely impaired and perhaps even arrested. Only by the application of statistical techniques can mere data help us shape and refine our theories and understand the social world better. Second, and somewhat paradoxically, the role of statistics is rather limited. As Figure 1.1 makes clear, scientific research proceeds through several mutually interdependent stages, and statistics become directly relevant only at the end of the observation stage. Before any statistical analysis can be legitimately applied, the preceding phases of the process must have been successfully completed. If the researcher has asked poorly conceived questions or has made serious errors of design or method, then even the most sophisticated statistical analysis is valueless. As useful as they can be, statistics cannot substitute for rigorous conceptualization, detailed and careful planning, or creative use of theory. Statistics cannot salvage a poorly conceived or designed research project. They cannot make sense out of garbage.

On the other hand, inappropriate statistical applications can limit the usefulness of an otherwise carefully done project. Only by successfully completing all phases of the process can a quantitative research project hope to contribute to understanding. A reasonable knowledge of the uses and limitations of statistics is as essential to the education of the social scientist as is training in theory and methodology.

As the statistical analysis comes to an end, we would move on to the next stage of the process. In this phase, we would primarily be concerned with assessing our theory but we would also look for other trends in the data. Assuming that we found that Southerners are more prejudiced, we might go on to ask if the pattern applies to males as well as females, to the well educated as well as the poorly educated, to older respondents as well as to the younger. As we probed the data, we might begin to develop some generalizations based on the empirical patterns we observe. For example, what if we found that regional differences existed for older respondents but not for the younger? We might suppose that regional differences in prejudice had diminished over the past generations. We might wonder why regional differences had decayed in recent years and, as we developed tentative explanations, we might revise our theory.

If we change the theory to take account of these findings, however, a new research project designed to test the revised theory is called for, and the wheel of science would begin to turn again. We, or some other researchers, would go through the entire process once again with this new—and, hopefully, improved—theory. This second project might result in further revisions and elaborations that would (you guessed it) require still more research projects, and the wheel of science would continue turning as long as scientists were able to suggest additional revisions or develop new insights. Every time the wheel turned, our understandings of the phenomena under consideration would (at least hopefully) improve.

This characterization of the research process does not include white-coated, clipboard-carrying scientists who, in a blinding flash of inspiration, discover some fundamental truth about reality and shout "Eureka!" The truth is that, in the normal course of science, there are only rarely times when we can say with absolute certainty that a given theory or idea is definitely true or false. Rather, evidence for (or against) a theory will gradually accumulate over time, and ultimate judgments of truth will likely be the result of many years of hard work, research, and debate.

Let's briefly review our imaginary research project. We began with an idea or theory about regional differences in racial prejudice. We imagined some of the steps we would have to take to test the theory and took a quick look at the various stages of the research project. We wound up back at the level of theory, ready to begin a new project guided by a revised theory. We saw how theory can motivate a research project and how our observations might cause us to revise the theory and, thus, motivate a new research project. One of the great values of Wallace's wheel of science is that it clearly illustrates how theory stimulates research and how research shapes theory. This constant interaction between theory and research is the lifeblood of science and the key to enhancing our understandings of the social world.

The dialog between theory and research occurs at many levels and in multiple forms. Statistics are one of the most important links between these two realms. Statistics permit us to analyze data, to identify and probe trends and relationships, to develop generalizations, and to revise and improve our theories. As you will see throughout this text, statistics are limited in many ways. They are also an indispensable part of the research enterprise. Without statistics, the interaction between theory and research would become extremely difficult, and the progress of our disciplines would be severely retarded.

1.3 THE GOALS OF THIS TEXT

In the preceding section, I argued that statistics are a crucial part of the process by which scientific investigations are carried out and that, therefore, some training in statistical analysis is a crucial component in the education of every social scientist. In this section, we will address the questions of how much training is necessary and what the purposes of that training are. First, this textbook takes the point of view that statistics are tools. They can be a very useful means of increasing our knowledge of the social world, but they are not ends in themselves. Thus, we will not take a "mathematical" approach to the subject. The techniques will be presented as a set of tools that can be used to answer important questions. To avoid misunderstandings, let me stress that we will be concerned with matters of mathematics. This text includes enough mathematical material so that you can develop a basic understanding of why statistics "do what they do." Our focus, however, will be on how these techniques are applied in the social sciences.

Second, all of you will soon become involved in advanced coursework in your major fields of study, and you will find that much of the literature used in these courses assumes at least basic statistical literacy. Furthermore, many of you, after graduation, will find yourselves in positions—either in a career or graduate school—where some understanding of statistics will be very helpful or perhaps even required. Very few of you will become statisticians per se (and this text is not intended for the preprofessional statistician), but you must have a grasp of statistics in order to read and critically appreciate your own professional literature. As a student in the social sciences and in many careers related to the social sciences, you simply cannot realize your full potential without a background in statistics.

Within these constraints, this textbook is an introduction to statistics as they are utilized in the social sciences. The general goal of the text is to develop an appreciation—a "healthy respect"—for statistics and their place in the research process. You should emerge from this experience with the ability to use statistics intelligently and to know when other people have done so. You should be familiar with the advantages and limitations of the more commonly used statistical techniques, and you should know which techniques are appropriate for a given set of data and a given purpose. Lastly, you should develop sufficient statistical and computational skills and enough experience in the interpretation of statistics to be able to carry out some elementary forms of data analysis by yourself.

1.4 DESCRIPTIVE AND INFERENTIAL STATISTICS

Descriptive Statistics.   As noted earlier, the general function of statistics is to manipulate data so that the original research question(s) can be answered. The researcher can call upon two general classes of statistical techniques that, depending on the research situation, are available to accomplish this task. The first class of techniques is called **descriptive statistics** and is relevant (1) when the researcher needs to summarize or describe the distribution of a single variable and (2) when the researcher wishes to understand the relationship between two or more variables. If we are concerned with describing a single variable, then our goal will be to arrange the values or scores of that variable so that the relevant information can be quickly understood and appreciated. Many of the statistics that might be appropriate for this summarizing task are probably familiar to you. For example, percentages, graphs, and charts can all be used as single-variable descriptive statistics.

To illustrate briefly the usefulness of these kinds of statistics, consider the following problem: suppose you wanted to summarize the distribution of the variable "family income" for a community of 10,000 families. How would you do it? Obviously, you couldn't simply list all incomes in the community and let it go at that. Presumably you would want to develop some summary measures of the overall income distributions—perhaps an arithmetic average or the proportions of incomes that fall in various ranges (such as low, middle, and high). Or perhaps a graph or a chart would be more useful. Whatever specific method you choose, its function is the same: to reduce these thousands of individual items of information into a few easily understood numbers. The process of allowing a few numbers to summarize many numbers is called **data reduction** and is the basic goal of single-variable descriptive statistical procedures. The first part of this text is devoted to these statistics.

The second type of descriptive statistics is designed to help the investigator understand the relationship between two or more variables. These statistics, called **measures of association**, allow the researcher to quantify the strength and direction of a relationship. These statistics are very useful because they enable us to investigate two matters of central theoretical and practical importance to any science: causation and prediction. These techniques help us disentangle and uncover the connections between variables. They help us trace the ways in which some variables might have causal influences on others; and, depending on the strength of the relationship, they enable us to predict scores on one variable from the score on another. Note that measures of association cannot, by themselves, prove that two variables are causally related. However, these techniques can provide valuable clues about causation and are therefore extremely important for theory testing and theory construction.

For example, suppose you were interested in the relationship between "time spent studying statistics" and "final grade in statistics" and had gathered the appropriate data from a group of college students. By calculating the appropriate measure of association, you could determine the strength of the relationship and its direction. Suppose you found a strong, positive relationship between these variables. You would infer that "study time" and "grade" were closely related (strength of the relationship) and that as one increased in value, the other also increased (direction of the relationship). You could make predictions from one variable to the other (for example, "the longer the study time, the higher the grade").

Now, as a result of finding this strong, positive relationship, you might be tempted to make causal inferences. That is, you might jump to such conclusions as "longer study time leads to (causes) higher grades." Such a conclusion might make a good deal of common sense and would certainly be supported by your statistical analysis. However, the causal nature of the relationship is in no way proven by the statistical analysis. Measures of association can be taken as important clues about causation, but the mere existence of a relationship should never be taken as conclusive proof of causation.

In fact, other variables might have an effect on the relationship. In the example above, we probably would not find a perfect relationship between "study time" and "final grade." That is, we will probably find some individuals who spend a great deal of time studying but receive low grades and some individuals who fit the opposite pattern. We know intuitively that other variables besides study time affect grades (such as efficiency of study techniques, amount of background in mathematics, and even random chance). Fortunately, researchers can incorporate these other variables into the analysis and measure their effects. Part III of this text is devoted to bivariate (two-variable) and Part IV to multivariate (more than two variables) descriptive statistics.

Inferential Statistics.   This second class of statistical techniques becomes relevant when we wish to generalize our findings from a **sample** to a **population**. A population is the total collection of all cases in which the researcher is interested and is thus the entity that the researcher wishes to understand. Examples of possible populations would be adult male voters in the United States, all parliamentary democracies, unemployed Puerto Ricans in Atlanta, or sophomore college football players in the Midwest.

Populations can theoretically range from inconceivable in size ("all humanity") to quite small (all 35-year-old red-haired belly dancers currently residing in downtown Cleveland) but are usually fairly large. In fact, they are almost always too large to be measured. To put the problem another way, social scientists almost never have the resources or time to test every case in a population. Hence the need for **inferential statistics**, which involve using information from samples (carefully chosen subsets of the defined populations) to make inferences about populations. Samples are, of course, much cheaper to assemble and, if the proper techniques are followed, generalizations based on these samples can be very accurate representations of the population.

Many of the concepts and procedures involved in inferential statistics may be unfamiliar. However, most of us are experienced consumers of inferential statistics—most familiarly, perhaps, in the form of public-opinion polls and election projections. When a public-opinion poll reports that 42% of the American electorate plans to vote for a certain presidential candidate, it is essentially reporting a generalization to a population ("the American electorate"—which numbers about 100 million people) from a carefully drawn sample (usually about 1500 respondents). Matters of inferential statistics will occupy our attention in Part II of this book.

1.5 DISCRETE AND CONTINUOUS VARIABLES

In the next chapter, you will begin to encounter some of the broad array of statistics available to the social scientist. One of the more puzzling aspects of studying statistics is learning when to use which statistic. You will learn specific guidelines as you go along, but let me introduce some basic and general guidelines at this point. The first of these concerns discrete and continuous variables; the second, covered in the next section, concerns level of measurement.

A variable is said to be **discrete** if it has a basic unit of measurement that cannot be subdivided. The measurement process for discrete variables involves accurate counting of the number of units per case. For example, number of people per household is a discrete variable. The basic unit is people, and the fewest you can have is one. Note that the score of a given household on this variable will always be a whole number (you'll never find 2.7 people living in a specific household), and as long as we are counting accurately, the scores we report will always be exact.

A variable is **continuous** if the measurement of it can be subdivided infinitely—at least in a theoretical sense. A good example of such a variable would be time, which can be measured in nanoseconds (billionths of a second) or even smaller units. In a sense, when we measure a continuous variable, we are always approximating and rounding off the scores. We could report somebody's time in the 100-yard dash as 10.7 seconds or 10.732451 seconds, but, since time is infinitely subdividable (if we have the technology to make the precise measurements), we will never be able to report the exact time elapsed. Since we cannot cite or work with infinitely long numbers, we must report the scores on continuous variables as if they were discrete. The distinction between the two types of variables relates more to measuring and processing the information than to the appearance of the data. This distinction between discrete and continuous variables is one of the most basic in statistics and will constitute one of the criteria by which we will choose among various statistics and graphic devices.

1.6 LEVEL OF MEASUREMENT

A second basic and general guideline for the selection of statistics is the level of measurement. Every statistical technique involves performing some mathematical operation such as adding scores or ranking cases. Before you can properly use a technique, you must measure the variable being processed in a way that justifies the required mathematical operations. For example, many statistical techniques require that scores be added together. These techniques could be legitimately used only when the variable is measured in a way that permits addition as a mathematical operation. Thus, the researcher's choice of statistical techniques is heavily dependent on the level at which the variables have been measured.

The three levels of measurement are, in order of increasing sophistication, nominal, ordinal, and interval-ratio. Since the concept of level of measurement is so central to statistics, let us consider the nature of the different levels at some length. I will make it a practice throughout this text to introduce level-of-measurement considerations for each statistical technique.

The Nominal Level of Measurement.   The most basic and the only universal measurement procedure is to classify cases into the preestablished categories of a variable. All measurement involves classification as a minimum. In nominal measurement, classification into categories is the only measurement procedure permitted. The categories themselves are not numerical and can be compared to each other only in terms of the number of cases classified in them. In no sense can the categories be thought of as "higher" or "lower" than each other along some numerical scale.

Although these measurement procedures are rudimentary, we do have criteria to identify adequately measured nominal variables. In brief:

1. The categories of the variable should be mutually exclusive of each other so that no ambiguity exists concerning classification of any given case.

2. The categories should be exhaustive; a category should exist for every manifestation of the variable being measured (at least an "other" or miscellaneous category).

3. The categories should be homogeneous in terms of the specific research project being conducted. Homogeneity is, of course, a relative thing and must be evaluated in terms of the specific purpose of the research. Cate-

gories that are too broad for some purposes may be perfectly adequate for others. Thus, the categories should incorporate elements that are sufficiently similar to facilitate descriptions and generalizations that are accurate and meaningful in the context of the research goals.

Tables 1.1 and 1.2 display four different schemes for measuring the variable "religious preference" and demonstrate some errors of measurement. In Table 1.1, Scale A violates the criterion of mutual exclusivity because of overlap between the categories Protestant and Episcopalian. Scale B does not provide a category for people with no religious preference (None), and Scale C uses a category (Non-Protestant) that would be too broad for many research purposes. Scale D in Table 1.2 represents the way religious preference is typically measured in North America.

There is a fourth criterion for judging the adequacy of a nominal variable: the categories should make theoretical sense. We could classify everything in the universe into tables and nontables, but such a scale is of no conceivable use to a social scientist. Variables should be categorized in ways that are relevant to theory and understanding.

As a final note, numerical labels are sometimes used to identify the categories of a variable measured at the nominal level. This practice is especially

TABLE 1.1 THREE INADEQUATE SCALES FOR MEASURING RELIGIOUS PREFERENCE

Scale A (not mutually exclusive)	Scale B (not exhaustive)	Scale C (not homogeneous)
Protestant	Protestant	Protestant
Episcopalian	Catholic	Non-Protestant
Catholic	Jew	
Jew		
None		
Other		

TABLE 1.2 AN ADEQUATE SCALE FOR MEASURING RELIGIOUS PREFERENCE

Scale D
Protestant
Catholic
Jew
None
Other

common when the data are being prepared for computer analysis. For example, the various religions might be labeled with a 1 indicating Protestant, a 2 signifying Catholic, and so on. You should understand that these numbers are merely labels or names and have no numerical quality to them. They cannot be added, subtracted, multiplied, or divided. The only mathematical operation permissible with nominal variables is counting the number of occurrences that have been classified into the various categories of the variable.

The Ordinal Level of Measurement.   In addition to classifying cases into categories, variables measured at the ordinal level allow the categories to be ranked with respect to how much of the trait being measured they possess. The categories form a kind of numerical scale that can be ordered from "high" to "low." Thus, variables measured at the ordinal level are more sophisticated than nominal-level variables because, in addition to counting the number of cases in a category, we can rank the cases with respect to each other. Not only can we say that one case is different from another; we can also say that one case is higher or lower, more or less than another.

For example, the variable socioeconomic status (SES) is usually measured at the ordinal level in the social sciences. The categories of the variable are often ordered according to the following scheme:

4. Upper class
3. Middle class
2. Working class
1. Lower class

Individual cases can be compared in terms of the categories into which they are classified. Thus, an individual classified as a 4 (upper class) would be ranked higher than an individual classified as a 2 (working class). Besides SES, examples of variables measured at the ordinal level would include virtually all attitude and opinion scales such as those that measure prejudice, alienation, or political conservatism.

The major limitation of the ordinal level of measurement is that a particular score represents only position with respect to some other score. We can distinguish between high and low scores, but the distance between the scores cannot be described in precise terms. Although we know that a score of 4 is more than a score of 2, we do not know if it is twice as much as 2.

By the same token, our inability to describe distances along ordinal scales in precise terms means we cannot assume that the distances between scores are always equal. Since the operations of addition, subtraction, multiplication, and division assume equal intervals between scores, they are not permitted with ordinally measured variables. For variables measured at the ordinal level, the most sophisticated mathematical operation permitted is ranking categories and cases.

TABLE 1.3   BASIC CHARACTERISTICS OF THE THREE LEVELS OF MEASUREMENT

Levels	Examples	Measurement Procedures	Mathematical Operations Permitted
Nominal	Sex, race, religion, marital status	Classification into categories	Counting number of cases in each category of the variable; comparing sizes of categories
Ordinal	Social class (SES), attitude and opinion scales	Classification into categories plus ranking of categories with respect to each other	All above plus judgments of "greater than" and "less than"
Interval-ratio	Age, number of children, income	All above plus description of distances between scores in terms of equal units	All above plus all other mathematical operations (addition, subtraction, multiplication, division, square roots, etc.)

The Interval-Ratio Level of Measurement.*   The categories of nominal-level variables have no numerical quality to them. Ordinal-level variables have categories that can be arrayed along a scale from high to low, but the exact distances between categories are unknown. Variables measured at the interval-ratio level not only permit classification and ranking but also allow the distance from category to category (or score to score) to be exactly defined. Interval-ratio variables are measured in units that have equal intervals and a true zero point. For example, recording the ages of your respondents is a measurement procedure that would produce interval-ratio data because the unit of measurement (years) has equal intervals (the distance from year to year is 365 days) and a true zero point (it is possible to be zero years old).

Other examples of interval-ratio variables would be income, number of children, weight, and years married. All mathematical operations are permitted for data measured at this level. Table 1.3 summarizes the basic characteristics of the three levels of measurement.

To summarize, different statistics require the use of different mathematical operations and, therefore, level-of-measurement considerations are the logical first guideline to use in selecting a statistic. For example, computation of a mean (or arithmetic average) requires that all observations be added together and then divided by the number of observations. Thus, computation

---

*Many statisticians distinguish between the interval level (equal intervals) and the ratio level (true zero point). I find the distinction unnecessarily cumbersome at the introductory level and will treat these two levels as one.

READING STATISTICS 1: INTRODUCTION

By this point in your education you have developed an impressive array of skills for reading words. Although you may sometimes struggle with a difficult idea or stumble over an obscure meaning, you can comprehend virtually any written work that you are likely to encounter.

As you continue your education in the social sciences, you must develop an analogous set of skills for reading numbers and statistics. To help you reach a reasonable level of literacy in statistics, I have included in this text a series of boxed inserts labeled "Reading Statistics." These will appear every two or three chapters and will discuss how statistical results are typically presented in the professional literature. Each installment will include an extract or quotation from the professional literature so that we can analyze a realistic example.

As you will see, professional researchers use a reporting style that is quite different from the statistical language you will find in this text. Space in research journals and other media is expensive, and the typical research project requires the analysis of many variables. Thus, a large volume of information must be summarized in very few words. Researchers may express in a word or two a result or an interpretation that will take us a paragraph or more to state.

Because this is an introductory textbook, I have been careful to break down the computational and logical processes that underlie each statistic and to identify, even to the point of redundancy, what we are doing when we use statistics. In this text we will never be concerned with more than a few variables at a time. We will have the luxury of analysis in detail and of being able to take pages or even entire chapters to develop a statistical idea or analyze a variable. Thus, a major theme of these boxed inserts will be to summarize how our comparatively long-winded (but more careful) vocabulary is translated into the concise language of the professional researcher.

When you have difficulty reading words, your tendency is (or, at least, should be) to consult reference books (especially dictionaries) to help you identify and analyze the elements (words) of the passage. When you have difficulty reading statistics, you should do exactly the same thing. I hope you will find this text a valuable reference book, but if you learn enough from this text to be able to use any statistics book to help you read statistics, this text will have fulfilled one of its major goals.

of the mean is fully justified only when a variable is measured at the interval-ratio level.

Ideally, the researcher would utilize only those statistics that were fully justified by the level-of-measurement criteria. In this imperfect world, however, the most powerful and useful statistics (such as the mean) require interval-ratio variables, while most of the variables of interest to the social sciences are only nominal (race, sex, marital status) or at best ordinal (attitude scales). Relatively few concepts of interest to the social sciences are so precisely defined that they can be measured at the interval-ratio level. This disparity creates some very real difficulties in the research process. On one hand, the researcher should use the most sophisticated statistical procedures fully justified by the level-of-measurement criteria. To treat interval-ratio data as if they were only ordinal, for example, results in a significant loss of information and precision. Treated as interval-ratio data, the variable "age" can supply us with exact information regarding the differences between the cases (for example, "Case A is three years and two months older than Case B"). If this trait was treated only as an ordinal variable, however, the precision of our comparisons would suffer and we could say only that "Case A is older (or greater than) Case B."

On the other hand, given the nature of the disparity, researchers are more likely to treat variables as if they were higher in level of measurement than they actually are. For example, ordinally measured variables might be treated as if they were interval-ratio because the statistical procedures available at the higher level are, in general, more powerful, flexible, and interesting than those appropriate for variables which are only ordinal. This practice is common but may lead to errors of logic and interpretation and, ultimately, to incorrect conclusions. In situations where level-of-measurement guidelines might have been violated, the researcher should be especially cautious and careful in assessing statistical results and developing interpretations.

At any rate, level-of-measurement considerations are important criteria and we will always consider them when presenting statistical procedures. Level of measurement is a major organizing principle for the material that follows, and you should make sure that you are familiar with these guidelines.

SUMMARY

1. Within the context of social research, the purpose of statistics is to organize, manipulate, and analyze data so that the researcher can more easily answer his or her original question. Along with theory and methodology, statistics are a basic tool by which social scientists attempt to enhance their understanding of the social world.

2. There are two general classes of statistics. Descriptive statistics are used to summarize the distribution of a single variable and the relationships between two or more variables. Inferential statistics provide us with techniques by which we can generalize to populations from random samples.

3. Two basic guidelines for selecting statistical techniques were presented. Variables may be either discrete or continuous and may be measured at any of three different levels. At the nominal level, we can classify cases into categories of the variable and compare category sizes. At the ordinal level, categories and cases can be ranked with respect to each other. At the interval-ratio level, all mathematical operations are permitted.

GLOSSARY

Continuous variable. A variable with a unit of measurement that can be subdivided infinitely.
Data. Any information collected as part of a research project and expressed as numbers.
Data reduction. Summarizing many scores with a few statistics. A major goal of descriptive statistics.
Dependent variable. A variable that is identified as an effect, result, or outcome variable. The dependent variable is thought to be caused by the independent variable.

Descriptive statistics. The branch of statistics concerned with (1) summarizing the distribution of a single variable or (2) measuring the relationship between two or more variables.

Discrete variable. A variable with a basic unit of measurement that cannot be subdivided.

Hypothesis. A statement about the relationship between variables that is derived from a theory. Hypotheses are more specific than theories, and all terms and concepts are fully defined.

Independent variable. A variable that is identified as a causal variable. The independent variable is thought to cause the dependent variable.

Inferential statistics. The branch of statistics concerned with making generalizations from samples to populations.

Level of measurement. The mathematical characteristics of a variable as determined by the measurement process. Variables can be measured at any of three levels, each permitting certain mathematical operations and statistical techniques. A major criterion for selecting statistical techniques. The characteristics of the three levels are summarized in Table 1.3.

Measures of association. Statistics that summarize the strength and direction of the relationship between variables.

Population. The total collection of all cases in which the researcher is interested.

Sample. A carefully chosen subset of a population. In inferential statistics, information is gathered from samples and then generalized to populations.

Statistics. A set of mathematical techniques for organizing and analyzing data.

Theory. A generalized explanation of the relationship between two or more variables.

Variable. Any trait that can change values from case to case.

PROBLEMS

1.1 In your own words, describe the role of statistics in the research process. Using the "wheel of science" as a framework, explain how statistics link theory with research.

1.2 Distinguish between descriptive and inferential statistics. Describe a research situation in which each would be useful.

1.3 Find a research article in any social science journal. Choose an article on a subject of interest to you and don't worry about being able to understand all of the statistics that are reported.
  a. How much of the article is devoted to statistics per se (as opposed to theory, ideas, discussion, and so on)?
  b. Was the research based on a sample from some population? How large is the sample? How were subjects or cases selected? Can the findings be generalized to some population?
  c. What variables are used? Which are independent and which are dependent? For each variable, determine level of measurement and whether the variable is discrete or continuous.
  d. What statistical techniques are used? Try to follow the statistical analysis and see how much you can understand. Save the article and read it again after you finish this course and see if you do any better.

1.4 Below are some items from a public opinion survey. For each item, indicate the level of measurement and whether the variable will be discrete or continuous.
  a. What is your occupation? _____
  b. How many years of school have you completed? _____
  c. If you were asked to use one of these four names for your social class, which would you say you belonged in?
    Upper ___Middle ___Working ___Lower ___
  d. What is your age? _____
  e. In what country were you born? _____
  f. What is your grade point average? _____
  g. What is your major? _____
  h. The only way to deal with the drug problem is to legalize all drugs.
    Strongly agree ___Agree ___Undecided ___
    Disagree ___Strongly disagree ___
  i. What is your astrological sign? _____
  j. How many brothers and sisters do you have?
    _____

1.5 Briefly describe a process for measuring each of the following variables. What level of measurement would be produced by your measurement technique? Are there other ways to measure the variable that would produce different levels of measurement? If so, indicate what these alternative measurement techniques would be.

Race	Honesty
Social class	Gross national product
Prejudice	Number of children
Height	Physicians per capita
Attractiveness	Distance from home to school

1.6 For each of the first 20 items in the 1993 General Social Survey (see Appendix G), indicate the level of measurement and whether the variable is continuous or discrete.

1.7 For each research situation summarized below, identify the level of measurement of all variables and indicate whether they are discrete or continuous. Also, decide which statistical applications are used: descriptive statistics (single variable), descriptive statistics (two or more variables), or inferential statistics. Remember that it is quite common for a given situation to require more than one type of application.

a. The administration of your university is proposing a change in parking policy. You select a random sample of students and ask each one if they favor or oppose the change.

b. You ask everyone in your social research class to tell you the highest grade they ever received in a math course and their grade on a recent statistics test. You then compare the two sets of scores to see if there is any relationship.

c. Your aunt is running for mayor and hires you (for a huge fee, incidentally) to question a sample of voters about their concerns in local politics. In particular, for each respondent, she wants to know party affiliation (Republican, Democrat, Independent), age, sex, and whether they favor or oppose a bond referendum.

d. Several years ago, a state reinstituted the death penalty for first degree homicide. Supporters of capital punishment argued that this change would reduce the homicide rate. To investigate this claim, a researcher has gathered information on number of homicides in the state for the two year periods before and after the change.

e. A local automobile dealer is concerned about customer satisfaction. He wants to mail a survey form to every customer for the past year and ask them if they are satisfied, very satisfied, or not satisfied with their purchase.

1.8 For each research situation below, identify the independent and dependent variables. Classify each in terms of level of measurement and whether or not the variable is discrete or continuous.

a. A graduate student is studying sexual harassment on college campuses and asks 500 female students if they personally have experienced any such incidents. Each student is asked to estimate the frequency of these incidents as either "often, sometimes, rarely, or never." The researcher also gathers data on age and major to see if there is any connection between these variables and frequency of sexual harassment.

b. A supervisor in the Solid Waste Management Division of a city government is attempting to assess two different methods of trash collection. One area of the city is served by trucks with two man crews who do "backyard" pickups and the rest of the city is served by "hi-tech" single person trucks with curbside pick-up. The assessment measures include the number of complaints received from the two different areas over a six-month period, the amount of time per day required to service each area, and the cost per ton of trash collected.

c. The adult book store near campus has been raided and closed by the police. Your social research class has decided to poll the student body about their reactions and opinions. The class decides to ask each student if they support or oppose the closing of the store, how many times they have visited the store, and if they strongly agree, agree, disagree, or strongly disagree with the statement that "pornography is a direct cause of sexual assaults on women." The class also collects information on the sex, age, and major of each student to see if opinions are related to these characteristics.

d. For a research project in a Political Science course, a student has collected information about the infant mortality rates and number of physicians per capita for fifty nations. The student wonders if these measures of quality of life vary by degree of economic modernization (measured by the percentage of the labor force employed in industrial occupations) and political democratization (measured by the percent-

age of all adults who are permitted to vote in national elections).

e. A highway engineer wonders if a planned increase in speed limit on a heavily traveled local avenue will result in any change in number of accidents. He plans to collect information on traffic volume, number of accidents, and number of fatalities for the six-month periods before and after the change.

INTRODUCTION TO SPSS/PC+ AND THE GENERAL SOCIAL SURVEY

The problems and exercises in this text have been written so that they can be solved with just a simple hand calculator. I've purposely kept the number of cases involved unrealistically low so that the tedium of mere calculation would not interfere unduly with the learning process. To provide a more realistic experience in the analysis of social science data, a shortened version of the 1993 General Social Survey (GSS) has been provided to your instructor. The GSS is a public-opinion poll conducted on a nationally representative sample of some 1400 to 1500 repondents every year. The full survey includes several hundred questions covering a broad range of social and political issues. The version supplied with this text has a limited number of variables and cases but is still actual, "real-life" data, so you have the opportunity to practice your statistical skills in a more realistic context.

One of the problems with reality, of course, is that it is often cumbersome and confusing. It's hard enough to do your homework with simplified problems, and you should be a little leery, in terms of your own time and effort, of promises of relevance and realism. This brings us to the second purpose of this section: computers and statistical packages. A statistical package is a set of computer programs for the analysis of data. The advantage of these packages is that, since the programs are already written, you can take advantage of the power of the computer with minimal computer literacy and virtually no programming experience.

One commonly used statistical package is called Statistical Package for the Social Sciences (SPSS). In these brief sections at the ends of chapters, I will explain how to use SPSS to manipulate and analyze the GSS data, and I will illustrate and interpret the results. My focus will be on SPSS as it is implemented on the personal computer (this version is called SPSS/PC+), but the translation to mainframe computers is straightforward. If you do not have access to SPSS at all, you can use whatever package is available (there are many others) to analyze the GSS data and at least compare results. Please read Appendix F before attempting any data analysis.

Part I   Descriptive Statistics

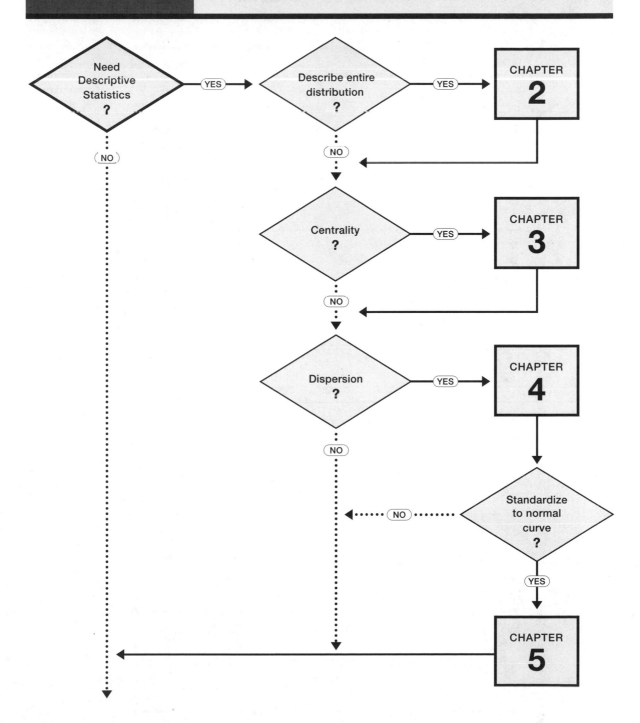

Basic Descriptive Statistics

Percentages, Ratios and Rates, Tables, Charts, and Graphs

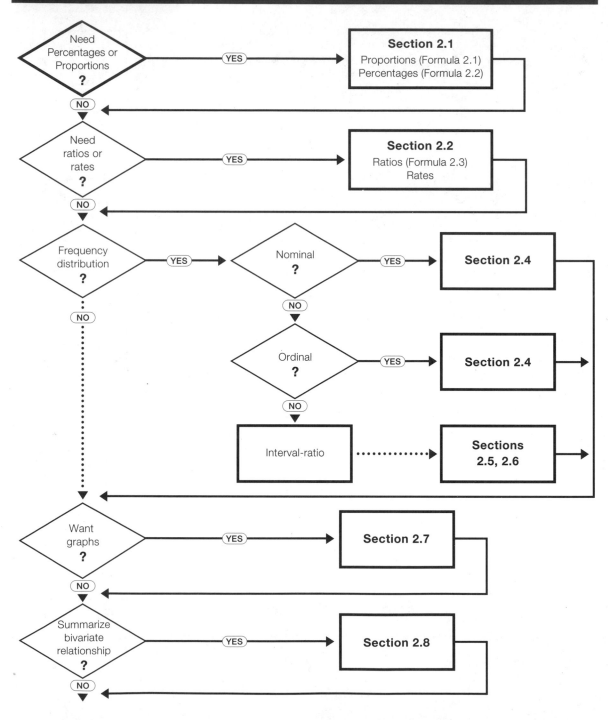

Research results do not speak for themselves. They must be arranged in ways that allow the researcher (and his or her readers) to comprehend their meaning quickly. The primary function of descriptive statistics is to present research results clearly and concisely. Researchers use a process called *data reduction* to organize data into presentable form. Data reduction involves using a few numbers, a table, or a graphic device to summarize or stand for a larger array of data.

Little about this process is mysterious or difficult, but, as a researcher or a consumer of research, you should be aware of one problem. Data reduction inevitably loses information (precision and detail); and, therefore, summarizing statistics might present a misleading picture of research results. The issue is not whether to use descriptive statistics—they are always necessary. The issue is which descriptive statistics to use. The various summarizing techniques present results in different ways, are based on different mathematical operations, and characteristically lose information in different ways. Thus, in choosing among the various summarizers, several decisions must be made: how to present the data, what kind of information to lose, and how much detail can safely be obscured.

In this chapter, we will consider several commonly used techniques for presenting research results: percentages and proportions, ratios and rates, tables, charts, and graphs.

2.1 PERCENTAGES AND PROPORTIONS

Consider the following statement: "Of the 269 cases handled by the court, 167 resulted in a prison sentence of five years or more." While there is nothing wrong with this statement, the same fact could have been more clearly conveyed if it had been reported as a percentage: "62.08% of all cases resulted in prison sentences of five or more years."

Percentages and proportions supply a frame of reference for reporting research results in the sense that they standardize the raw data: percentages to the base 100 and proportions to the base 1.00. The mathematical definitions of **proportions** and **percentages** are

FORMULA 2.1

$$\text{Proportion } (p) = \frac{f}{N}$$

FORMULA 2.2

$$\text{Percentage } (\%) = \left(\frac{f}{N}\right) \times 100$$

where f = frequency, the number of cases in any category
N = the number of cases in all categories

To illustrate the computation of percentages, consider the data presented in Table 2.1. To find the percentage of cases in the first category (sentences

TABLE 2.1 DISPOSITION OF 269 CRIMINAL CASES (fictitious data)*

Sentence	Frequency (f)	Proportion (p)	Percentage (%)
Five years or more	167	0.6208	62.08
Less than five years	72	0.2677	26.77
Suspended	20	0.0744	7.44
Acquitted	10	0.0372	3.72
	N = 269	1.0001	100.01%

*The slight discrepancies in the totals of the proportion and percentage columns are due to rounding error.

of five years or more), note that there are 167 cases in the category ($f = 167$) and a total of 269 cases in all ($N = 269$). So,

$$\text{Percentage (\%)} = \left(\frac{f}{N}\right) \times 100 = \left(\frac{167}{269}\right) \times 100 = (0.6208) \times 100 = 62.08\%$$

Using the same procedures, we can also find the percentage of cases in the second category:

$$\text{Percentage (\%)} = \left(\frac{f}{N}\right) \times 100 = \left(\frac{72}{269}\right) \times 100 = (0.2677) \times 100 = 26.77\%$$

Both results could have been expressed as proportions. For example, the proportion of cases in the third category is 0.0744.

$$\text{Proportion} = \left(\frac{f}{N}\right) = \frac{20}{269} = 0.0744$$

Percentages and proportions are easier to read and comprehend than frequencies. This advantage is particularly obvious when attempting to compare groups of different sizes. For example, based on the information presented in Table 2.2, which college has the higher relative number of social science majors? Because the total enrollments are so different, such comparisons are difficult to conceptualize from the raw frequencies. To make comparisons easier, the difference in size can be effectively eliminated by standardizing both distributions to the common base of 100 (or, in other words, by computing percentages for both distributions). The same data are presented in percentages in Table 2.3.

The percentages in Table 2.3 make it easier to identify both differences and similarities between the two colleges. College A has a much higher percentage of social science majors (even though the absolute number of social science majors is less than at College B) and about the same percentage of humanities majors. How would you describe the differences in the remaining two major fields?

TABLE 2.2 DECLARED MAJOR FIELDS OF STUDY ON TWO COLLEGE
CAMPUSES (fictitious data)

Major	College A	College B
Business	103	312
Natural sciences	82	279
Social sciences	137	188
Humanities	93	217
	$N = 415$	$N = 996$

TABLE 2.3 DECLARED MAJOR FIELDS OF STUDY ON TWO COLLEGE
CAMPUSES (fictitious data)

Major	College A (%)	College B (%)
Business	24.82	31.33
Natural sciences	19.76	28.01
Social sciences	33.01	18.88
Humanities	22.41	21.79
	100.00%	100.01%
	(415)	(996)

Some further rules on the use of percentages and proportions:

1. When working with a small number of cases (say, fewer than 20), it is usually preferable to report the actual frequencies rather than percentages or proportions. With a small number of cases, the percentages can change drastically with relatively minor changes in the data. For example, if you begin with a data set that includes 10 males and 10 females (that is, 50% of each gender) and then add another female, the percentage distributions will change noticeably to 52.38% female and 47.62% male. Of course, as the number of observations increases, each additional case will have a smaller impact. If we started with 500 males and females and then added one more female, the percentage of females would change by only a tenth of a percent (from 50% to 50.10%).

2. Always report the number of observations along with proportions and percentages. This will permit the reader to judge the adequacy of the sample size and, conversely, helps to prevent the researcher from lying with statistics. You might be impressed by statements like "two out of three people questioned prefer courses in statistics to any other course," but the claim would lose its gloss if you learned that only three people

APPLICATION 2.1

Not long ago, in a large social service agency, the following conversation took place between the executive director of the agency and a supervisor of one of the divisions.

Executive director: Well, I don't want to seem abrupt, but I've only got a few minutes. Tell me, as briefly as you can, about this staffing problem you claim to be having.

Supervisor: Ma'am, we just don't have enough people to handle our workload. Of the 177 full-time employees of the agency, only 50 are in my division. Yet, 6231 of the 16,722 cases handled by the agency last year were handled by my division.

Executive director (smothering a yawn): Very interesting. I'll certainly get back to you on this matter.

How could the supervisor have presented his case more effectively? Because he wants to com-

pare two sets of numbers (his staff versus the total staff and the workload of his division versus the total workload of the agency), proportions or percentages would be a more forceful way of presenting results. What if the supervisor had said, "Only 28.25% of the staff is assigned to my division, but we handle 37.26% of the total workload of the agency." Is this a clearer message?

The first percentage is found by

$$\% = \left(\frac{f}{N}\right) \times 100 = \left(\frac{50}{177}\right) \times 100$$
$$= (.2825) \times 100 = 28.25\%$$

and the second percentage is found by

$$\% = \left(\frac{f}{N}\right) \times 100 = \left(\frac{6231}{16,722}\right) \times 100$$
$$= (.3726) \times 100 = 37.26\%$$

were tested. You should be extremely suspicious of reports that fail to report the number of cases that were tested.

2.2 RATIOS AND RATES

Ratios and **rates** are two additional ways to simply and dramatically summarize the distribution of a variable. Ratios are especially useful for comparing categories in terms of relative frequency. Instead of standardizing the distribution of a variable to the base 100 or 1.00, as we did in computing percentages and proportions, we determine ratios by dividing the frequency of one category by the frequency in another. Mathematically, a ratio can be defined as

FORMULA 2.3

$$\text{Ratio} = \frac{f_1}{f_2}$$

where f_1 = the number of cases in the first category
f_2 = the number of cases in the second category

To illustrate the use of ratios, suppose that you were interested in the relative sizes of the various religious denominations and found that a particu-

APPLICATION 2.2

In Table 2.2, how many natural science majors are there compared to social science majors at College B? This question could be answered with frequencies, but a more easily understood way of expressing the answer would be with a ratio. The ratio of natural science to social science majors would be

$$\text{Ratio} = \frac{f_1}{f_2} = \frac{279}{188} = 1.48$$

For every social science major, there are 1.48 natural science majors at College B.

lar community included 1370 Protestant families and 930 Catholic families. To find the ratio of Protestants (f_1) to Catholics (f_2), divide 1370 by 930. The resultant ratio is 1.47. This number would mean that for every Catholic family, there are 1.47 Protestant families.

Note that ratios can be very economical ways of expressing the relative predominance of two categories. That Protestants outnumber Catholics in our example is obvious from the raw data. Percentages or proportions could have been used to summarize the overall distribution (that is, 59.56% of the families were Protestant, 40.44% were Catholic). Compared to these other methods, ratios are a precise measure of the relative frequency of one category per unit of the other category. They tell us in an exact way the extent to which one category outnumbers the other.

Ratios are often multiplied by some power of 10 to eliminate decimal points. For example, the ratio computed above might be reported as 147 instead of 1.47. This would mean that, for every 100 Catholic families, there are 147 Protestant families in the community. To ensure clarity, the comparison units for the ratio are often expressed as well. Based on a unit of ones, the ratio of Protestants to Catholics would be expressed as 1.47:1. Based on hundreds, the same statistic might be expressed as 147:100.

Rates provide still another way of summarizing the distribution of a single variable. Rates are defined as the number of actual occurrences of some phenomenon divided by the number of possible occurrences per some unit of time. Rates are usually multiplied by some power of 10 to eliminate decimal points. For example, the crude death rate for a population is defined as the number of deaths in that population (actual occurrences) divided by the number of people in the population (possible occurrences) per year. This quantity is then multiplied by 1000. The formula for the crude death rate can be expressed as:

$$\text{Crude death rate} = \frac{\text{number of deaths in year}}{\text{total population}} \times 1000$$

APPLICATION 2.3

In 1990, in a city of 167,000, there were 2500 births. In 1980, when the population of the city was only 133,000, there were 2700 births. Is the birthrate rising or falling? Although this question can be answered from the preceding information, the trend in birthrates will be much more obvious if we compute birthrates for both years. Like crude death rates, crude birthrates are usually multiplied by 1000 to eliminate decimal points. For 1980:

$$\text{Crude birthrate} = \left(\frac{2700}{133,000}\right) \times 1000$$
$$= 20.30$$

In 1980, there were 20.30 births for every 1000 people in the city.

For 1990:

$$\text{Crude birthrate} = \left(\frac{2500}{167,000}\right) \times 1000$$
$$= 14.97$$

In 1990, there were 14.97 births for every 1000 people in the city. With the help of these statistics, the decline in the birthrate is clearly expressed.

If there were 100 deaths during a given year in a town of 7000, the crude death rate for that year would be

$$\text{Crude death rate} = \frac{100}{7000} \times 1000 = (.01429) \times 1000 = 14.29$$

Or, for every 1000 people, there were 14.29 deaths during this particular year.

By the same token, if a city of 237,000 people experienced 120 auto thefts during a particular year, the auto theft rate would be

$$\text{Auto theft rate} = \frac{120}{237,000} \times 100,000 = (0.0005063) \times 100,000 = 50.63$$

Or, for every 100,000 people, there were 50.63 auto thefts during the year in question.

Up to now, we have considered three techniques (proportions and percentages, ratios, and rates) for describing and summarizing data. All three techniques express, clearly and concisely, the distribution of a single variable. They represent different ways of expressing information so that it can be quickly appreciated. All three techniques are also quite useful for purposes of comparison. For instance, the auto theft rate of 50.63 in our example might be difficult to interpret by itself (is this a high or a low rate?) but would take on much more meaning if you found that the rate for the preceding year had been 32.70. These statistics provide alternative ways of expressing relative frequency and are thus especially useful in making comparisons between different groups and/or different times.

2.3 FREQUENCY DISTRIBUTIONS: INTRODUCTION

Frequency distributions are tables that summarize the distribution of a variable by reporting the number of cases contained in each category of the variable. They are very helpful and commonly used ways of organizing and working with data. In fact, the construction of frequency distributions is almost always the first step in any statistical analysis.

To illustrate the usefulness of frequency distributions and to provide some data for examples, assume that a university counseling center is attempting to assess the effectiveness of its services. While any realistic evaluation research would collect a variety of information from a large group of students, we will, for the sake of convenience, confine our attention to four variables and 20 students. The data are reported in Table 2.4.

Note that, even though the data in Table 2.4 represent an unrealistically low number of cases, discerning patterns and trends is difficult. For example, try to ascertain the general level of satisfaction of the students from Table 2.4. You may be able to do so with just 20 cases, but it will take some time and effort. Imagine the difficulty with 50 cases or 100 cases presented in this fashion. Clearly the data need to be organized in a format that allows the researcher (and his or her audience) to understand easily the distribution of the variables.

Before we turn to specific techniques, let me state a general rule about the construction of frequency distributions: *The categories contained in the*

TABLE 2.4   DATA FROM COUNSELING CENTER SURVEY

Student	Sex	Marital Status	Satisfaction with Services*	Age
A	Male	Single	4	18
B	Male	Married	2	19
C	Female	Single	4	18
D	Female	Single	2	19
E	Male	Married	1	20
F	Male	Single	3	20
G	Female	Married	4	18
H	Female	Single	3	21
I	Male	Single	3	19
J	Female	Divorced	3	23
K	Female	Single	3	24
L	Male	Married	3	18
M	Female	Single	1	22
N	Female	Married	3	26
O	Male	Single	3	18
P	Male	Married	4	19
Q	Female	Married	2	19
R	Male	Divorced	1	19
S	Female	Divorced	3	21
T	Male	Single	2	20

*Key: (4) Very satisfied   (2) Dissatisfied
     (3) Satisfied   (1) Very dissatisfied

frequency distribution must be exhaustive and mutually exclusive. In other words, the categories must be stated in a way that permits each case to be counted in one and only one category. This basic principle applies to the construction of frequency distributions for variables measured at all three levels of measurement.

Beyond this general rule, there are only guidelines to help you construct useful frequency distributions. As you will see, the researcher has a fair amount of discretion in stating the categories of the frequency distribution (especially with variables measured at the interval-ratio level). I will identify the issues to consider as you make decisions about the nature of any particular frequency distribution. Ultimately, however, the guidelines I state are aids for decision making, nothing more than helpful suggestions. As always, the researcher has the final responsibility for making sensible decisions and presenting his or her data in a meaningful way.

2.4 FREQUENCY DISTRIBUTIONS FOR VARIABLES MEASURED AT THE NOMINAL AND ORDINAL LEVELS

Nominal-Level Variables. For nominal-level variables, construction of the frequency distribution is typically very straightforward. For each category of the variable being displayed, the occurrences are counted and the subtotals, along with the total number of cases (N), are reported. Table 2.5 displays a frequency distribution for the variable "sex" from the counseling center survey. For purposes of illustration, a column for tallies has been included in this table. Needless to say, this column would not be included in the final form of the frequency distribution. Take a moment to notice several other features of the table. Specifically, the table has a descriptive title, clearly labeled categories (male and female), and a report of the total number of cases at the bottom of the frequency column. *These items must be included in all tables regardless of the variable or level of measurement.*

The meaning of the table is quite clear. There are 10 males and 10 females in the sample. While not exactly startling, this information is much easier to comprehend when the data have been organized into a frequency distribution.

For some nominal variables, the researcher might have to make some choices about the number of categories he or she wishes to report. For example, the distribution of the variable "marital status" could be reported using the categories listed in Table 2.4. The resultant frequency distribution is

TABLE 2.5 SEX OF RESPONDENTS, COUNSELING CENTER SURVEY

Sex	Tallies	Frequency (*f*)
Male	LHT LHT	10
Female	LHT LHT	10
		$N = 20$

presented in Table 2.6. Although this is a perfectly fine frequency distribution, it may be needlessly detailed for some purposes. For example, the researcher might want to focus analytical attention solely on "nonmarried" as distinct from "married" students. That is, the researcher might not be concerned with the difference between single and divorced respondents but may want to treat both as simply "not married." In that case, these categories could be grouped together and treated as a single entity, as in Table 2.7. Notice that, by this collapsing, information and detail have been lost. This latter version of the table would not allow the researcher to discriminate between the various unmarried states.

Ordinal-Level Variables.   Frequency distributions for ordinal-level variables are constructed following the same routines used for nominal-level variables. Table 2.8 reports the frequency distribution of the "satisfaction" variable from the counseling center survey. Note that a column of percentages by category has been added to this table. Such columns heighten the

TABLE 2.6   MARITAL STATUS OF RESPONDENTS, COUNSELING CENTER SURVEY

Status	Frequency
Single	10
Married	7
Divorced	3
	$N = 20$

TABLE 2.7   MARITAL STATUS OF RESPONDENTS, COUNSELING CENTER SURVEY

Status	Frequency
Married	7
Not married	13
	$N = 20$

TABLE 2.8   SATISFACTION WITH SERVICES, COUNSELING CENTER SURVEY

Satisfaction	Frequency (f)	Percentage (%)
(4) Very satisfied	4	20
(3) Satisfied	9	45
(2) Dissatisfied	4	20
(1) Very dissatisfied	3	15
	$N = 20$	100%

TABLE 2.9 SATISFACTION WITH SERVICES, COUNSELING CENTER SURVEY

Satisfaction	Frequency (f)	Percentage (%)
Satisfied	13	65
Dissatisfied	7	35
	N = 20	100%

clarity of the table (especially with larger samples) and are common adjuncts to the basic frequency distribution for variables measured at all levels.

This table reports that most students were either satisfied or very satisfied with the services of the counseling center. The most common response (nearly half the sample) was "satisfied." If the researcher wanted to emphasize this major trend, the categories could be collapsed as in Table 2.9. Again, the price paid for this increased compactness is that some information (in this case, the exact composition of satisfied and dissatisfied students) is lost.

2.5 FREQUENCY DISTRIBUTIONS FOR VARIABLES MEASURED AT THE INTERVAL-RATIO LEVEL

Basic Considerations. In general, the construction of frequency distributions for variables measured at the interval-ratio level is more complex than for nominal and ordinal variables. Interval-ratio variables usually have a large number of possible scores (that is, a wide range from the lowest to the highest score). The large number of scores requires some collapsing or grouping of categories to produce reasonably compact frequency distributions. To construct frequency distributions for interval-ratio level variables, you must decide how many categories to use and how wide these categories should be. For example, suppose you wished to report the distribution of the variable "age" for a sample drawn from a community. Unlike the college data reported in Table 2.4, a community sample would have a very broad range of ages. If you simply reported the number of times each year of age (or score) occurred, you could easily wind up with a frequency distribution that contained 70, 80, or even more categories. Such a large frequency distribution would not present a concise picture. The scores (years) must be grouped into larger categories to heighten clarity and ease of comprehension. How large should these categories be? How many categories should be included in the table?

The exact way to make these decisions is somewhat arbitrary and a matter that the researcher must decide anew each time he or she confronts the problem. Even though there are no hard-and-fast guidelines, the nature of the decision always involves a trade-off between more detail (a greater number of narrow categories) or more compactness (a smaller number of wide categories). The researcher must keep in mind the purpose of the research and what he or she intends to do with the data (if anything) other than merely report it.

Constructing the Frequency Distribution. To introduce the mechanics and decision-making processes involved, we will construct a frequency distribution to display the ages of the students in the counseling center survey. Because of the narrow age range of a group of college students, we can use categories of only one year (these categories are often called **class intervals** when working with interval-ratio data). List the ages from youngest to oldest, count the number of times each score (year of age) occurs, and then total the number of scores for each category. Note that a concentration or clustering of scores in the 18 and 19 class intervals is immediately revealed after the categories are organized (Table 2.10).

Even though the picture presented in this table is fairly clear, assume for the sake of illustration that you desire a more compact (less detailed) summary. To do this, you will have to group scores into wider class intervals. By increasing the interval width (say to two years), you can reduce the number of intervals and achieve the desired compactness of expression. The grouping of scores in Table 2.11 clearly emphasizes the relative predominance of young respondents. This trend in the data can be stressed even more by the addition of a column displaying the percentage of cases in each category.

TABLE 2.10 AGE OF RESPONDENTS, COUNSELING CENTER SURVEY
(interval width = one year of age)

Class Intervals	Frequency (f)
18	5
19	6
20	3
21	2
22	1
23	1
24	1
25	0
26	1
	$N = 20$

TABLE 2.11 AGE OF RESPONDENTS, COUNSELING CENTER SURVEY
(interval width = two years)

Class Intervals	Frequency (f)	Percentage (%)
18–19	11	55
20–21	5	25
22–23	2	10
24–25	1	5
26–27	1	5
	$N = 20$	100%

Note that the class intervals in Table 2.11 have been stated with an apparent gap between them (that is, the class intervals are separated by a distance of one unit). At first glance, these gaps may appear to violate the principle of exhaustiveness; but, since age has been measured in whole units, the gaps actually pose no problem. Given the level of precision of the measurement (in whole years as opposed to 10ths or 100ths of a year), no case could have a score falling between these class intervals. In fact, for these data, the set of class intervals contained in Table 2.11 constitutes a scale that is exhaustive and mutually exclusive. Each of the 20 respondents in the sample can be sorted into one and only one age category.

However, consider the difficulties of sorting cases that might have been encountered if age had been measured with greater precision. If age had been measured in 10ths of a year, into which class interval in Table 2.11 would a 19.4-year-old subject be placed? You can avoid this potential difficulty by always stating the limits of the class intervals at the same level of precision at which you measure the data. Thus, if age were being measured in 10ths of a year, the limits of the class intervals in Table 2.11 would be stated in 10ths of a year. For example:

17.5–19.4

19.5–21.4

21.5–23.4

23.5–25.4

25.5–27.4

To maintain mutual exclusivity between categories, no overlapping of class intervals is allowed. If you state the limits of the class intervals at the same level of precision as the data and maintain a gap between intervals, you will always obtain a frequency distribution where each case can be assigned to one and only one category.

Real Class Limits and Midpoints.   If the sole purpose of the frequency distribution is to display the distribution of a variable, then no additional complexities need be introduced. However, for certain purposes (such as constructing the histograms described in Section 2.7), you must conceptualize the distribution not as a series of discrete, nonoverlapping class intervals but rather as a continuous series of categories. To work with the distribution of a variable as if it were continuous, statisticians use **real class limits**. To find the real limits of any class interval, begin with the limits as stated in the frequency distribution (or, the **stated limits**). Divide the distance between the stated class intervals in half and add the result to all upper stated limits and subtract it from all lower stated limits. This process is illustrated below with the class intervals stated in Table 2.11. The distance between intervals is one, so the real limits can be found by adding 0.5 to all upper limits and subtracting 0.5 from all lower limits.

Stated limits	Real limits
18–19	17.5–19.5
20–21	19.5–21.5
22–23	21.5–23.5
24–25	23.5–25.5
26–27	25.5–27.5

Note that, when conceptualized with real limits, the class intervals overlap with each other and the distribution can be seen as continuous.

In addition to real limits, you will also frequently need to work with the **midpoints** of the class intervals. Midpoints are defined as the points exactly halfway between the upper and lower real limits and can be found for any interval by dividing the sum of the upper and lower real limits by two. Table 2.12 displays the real limits and midpoints of two different sets of class intervals.

Cumulative Frequency and Cumulative Percentage. Two commonly used adjuncts to the basic frequency distribution for interval-ratio data are the **cumulative frequency** and **cumulative percentage** columns. Their primary purpose is to allow the researcher (and his or her audience) to tell at a glance how many cases fall below a given score or class interval in the distribution.

To construct a cumulative frequency column, begin with the lowest class interval in the distribution. The entry in the cumulative frequency columns for that interval will be the same as the number of cases in the interval. For

TABLE 2.12   REAL LIMITS AND MIDPOINTS

Interval Width of Three Units		
Stated Limits	Real Limits	Midpoints
0–2	−0.5–2.5	1
3–5	2.5–5.5	4
6–8	5.5–8.5	7
9–11	8.5–11.5	10
Interval Width of Six Units		
Stated Limits	Real Limits	Midpoints
100–105	99.5–105.5	102.5
106–111	105.5–111.5	108.5
112–117	111.5–117.5	114.5
118–123	117.5–123.5	120.5

the next higher interval, the cumulative frequency will be all cases in the interval plus all the cases in the first interval. For the third interval, the cumulative frequency will be all cases in the interval plus all cases in the first two intervals. Continue adding (or accumulating) cases until you reach the highest class interval, which will have a cumulative frequency of all the cases in the interval plus all cases in all other intervals (for the highest interval, cumulative frequency equals the total number of cases). Table 2.13 shows a cumulative frequency column added to Table 2.11. The arrows show the direction of addition.

The cumulative percentage column is quite similar to the cumulative frequency column. Begin by adding a column to the basic frequency distribution for percentages as in Table 2.11. This column shows the percentage of all cases in each class interval. To find cumulative percentages, follow the same addition pattern explained above for cumulative frequency. That is, the cumulative percentage for the lowest class interval will be the same as the percentage of cases in the interval. For the next higher interval, the cumulative percentage is the percentage of cases in the interval plus the percentage of cases in the first interval, and so on. Table 2.14 shows the age data with a cumulative percentage column added. Again, the arrows show the direction of addition.

TABLE 2.13   AGE OF RESPONDENTS, COUNSELING CENTER SURVEY

Class Interval	Frequency (f)	Cumulative Frequency
18–19	11	11
20–21	5	16
22–23	2	18
24–25	1	19
26–27	1	20
	N = 20	

TABLE 2.14   AGE OF RESPONDENTS, COUNSELING CENTER SURVEY

Class Interval	Frequency (f)	Cumulative Frequency	Percentage (%)	Cumulative Percentage (%)
18–19	11	11	55	55
20–21	5	16	25	80
22–23	2	18	10	90
24–25	1	19	5	95
26–27	1	20	5	100%
	N = 20		100%	

These cumulative columns are quite useful in situations where the researcher wants to make a point about how cases are spread across the range of scores. For example, Tables 2.13 and 2.14 show quite clearly that the great majority of students in the counseling center survey are less than 21 years of age. If the researcher wishes to impress this feature of the age distribution on his or her audience, then these cumulative columns are quite handy. Most realistic research situations will be concerned with many more than 20 cases and/or many more categories than our tables have. Since the cumulative percentage column is clearer and easier to interpret in such cases, it is normally preferred to the cumulative frequencies column.

Summary.   The basic guidelines for constructing frequency distributions for interval-ratio variables can now be stated:

1. Decide how many class intervals you wish to use. One reasonable convention suggests that the number of intervals should be about 10. Many research situations may require fewer than 10 intervals, and it is common to find frequency distributions with as many as 15 intervals. Only rarely will more than 15 intervals be used, since the resultant frequency distribution would not be very concise.

2. Find the size of the class interval. Once you have decided how many intervals you will use, interval size can be found by dividing the range of the scores by the number of intervals and rounding to a convenient whole number.

3. State the lowest interval so that its lower limit is equal to or below the lowest score. By the same token, your highest interval will be the one that contains the highest score. All intervals must be equal in size.

4. State the limits of the class intervals at the same level of precision as you have used to measure the data. Do not overlap intervals. You will then have defined the class intervals in such a way that each case can be sorted into one and only one category. If you anticipate using the frequency distribution for additional statistical operations, you might need to determine the real limits and/or the midpoints of the intervals.

5. Count the number of cases in each class interval and report these subtotals in a column labeled "frequency." Report the total number of cases (N) at the bottom of this column. The table may also include a column for percentages, cumulative frequencies, and cumulative percentages.

6. Inspect the frequency distribution carefully. Has too much detail been lost? If so, reconstruct the table with a greater number of class intervals (or smaller interval size). Is the table too detailed? If so, reconstruct the

APPLICATION 2.4

The following list shows the ages of 50 prisoners enrolled in a work-release program. Is this group young or old? A frequency distribution will provide an accurate picture of the overall age structure.

18	60	57	27	19
20	32	62	26	20
25	35	75	25	21
30	45	67	41	30
37	47	65	42	25
18	51	22	52	30
22	18	27	53	38
27	23	32	35	42
32	37	32	40	45
55	42	45	50	47

We will use about 10 intervals to display these data. By inspection we see that the youngest prisoner is 18 and the oldest is 75. The range is thus 57. Interval size will be 57/10, or 5.7, which we can round off to either 5 or 6. Let's use a six-year interval beginning at 18. The stated limits of the lowest interval will be 18–23. Now we must state the limits of all other intervals, count the number of cases in each interval, and display these counts in a frequency distribution. Columns may be added for percentages, cumulative percentages, and/or cumulative frequency. The complete distribution, with a column added for percentages is

Ages	Frequency	Percentages
18–23	10	20
24–29	7	14
30–35	9	18
36–41	5	10
42–47	8	16
48–53	4	8
54–59	2	4
60–65	3	6
66–71	1	2
72–77	1	2
	$N = 50$	100

The prisoners seem to be fairly evenly spread across the age groups up to the 48–53 interval. There is a noticeable lack of prisoners in the oldest age groups and a concentration of prisoners in their 20s and 30s.

table with fewer class intervals (or use wider intervals). Remember that the frequency distribution results from a number of decisions you make in a rather arbitrary manner. If the appearance of the table seems less than optimal given the purpose of the research, redo the table until you are satisfied that you have struck the best balance between detail and conciseness.

7. Remember to give your table a clear, concise title, and number the table if your report contains more than one. All categories and columns must also be clearly labeled.

2.6 CONSTRUCTING FREQUENCY DISTRIBUTIONS FOR INTERVAL-RATIO LEVEL VARIABLES: A REVIEW

We covered a lot of ground in the preceding section, so let's pause and review these principles by considering a specific research situation. Below are the numbers of visits received over the past year by 90 residents of a retirement community.

0	52	21	20	21	24	1	12	16	12
16	50	40	28	36	12	47	1	20	7
9	26	46	52	27	10	3	0	24	50
24	19	22	26	26	50	23	12	22	26
23	51	18	22	17	24	17	8	28	52
20	50	25	50	18	52	46	47	27	0
32	0	24	12	0	35	48	50	27	12
28	20	30	0	16	49	42	6	28	2
16	24	33	12	15	23	18	6	16	50

Listed in this format, the data are a hopeless jumble from which no one could derive much meaning. The function of the frequency distribution is to arrange and organize these data so that their meanings will be made obvious.

As a first step, we must decide how many class intervals to use in the frequency distribution. Following the guidelines established in the previous section, let's use about 10 intervals. By inspecting the data, we can see that the lowest score is 0 and the highest is 52. The range of these scores is 52 − 0, or 52. To find the approximate interval size, divide the range (52) by the number of intervals (10). Since $52/10 = 5.2$, we can set the interval size at 5.

The lowest score is 0, so the lowest class interval will be 0–4. The highest class interval will be 50–54, which will include the high score of 52. All that remains is to state the intervals in table format, count the number of scores that fall in each interval, and report the subtotals in a frequency column. These steps have been taken in Table 2.15, which also includes columns for the percentages and cumulative percentages. Note that this table is the product of several relatively arbitrary decisions. The researcher should remain aware of this fact and inspect the frequency distribution carefully. Remember that if the table is unsatisfactory for any reason, you can construct it anew with a different number of categories and interval sizes.

Now, with the aid of the frequency distribution, some patterns in the data can be discerned. There are three distinct clusterings of scores in the table. Ten residents were visited rarely, if at all (the 0–4 visits per year interval). The single largest interval, with 18 cases, is 20–24. Combined with the intervals immediately above and below, this represents quite a sizable grouping of

TABLE 2.15 NUMBER OF VISITS PER YEAR, 90 RETIREMENT-COMMUNITY RESIDENTS

Class Interval	Frequency (f)	Cumulative Frequency	Percentage (%)	Cumulative Percentage (%)
0–4	10	10	11.11	11.11
5–9	5	15	5.56	16.67
10–14	8	23	8.89	25.56
15–19	12	35	13.33	38.89
20–24	18	53	20.00	58.89
25–29	12	65	13.33	72.22
30–34	3	68	3.33	75.55
35–39	2	70	2.22	77.77
40–44	2	72	2.22	79.99
45–49	6	78	6.67	86.66
50–54	12	90	13.33	99.99%
	$N = 90$		99.99%*	

*Percentage columns will occasionally fail to total to 100% because of rounding error. If the total is between 99.90% and 100.10%, ignore the discrepancy. Discrepancies of greater than plus or minus 0.10% may indicate mathematical errors, and the entire column should be computed again.

cases (42 out of 90, or 46.6% of all cases) and suggests that the dominant visiting rate is about twice a month, or approximately 24 visits per year. The third grouping is in the 50–54 class interval with 12 cases, reflecting a visiting rate of about once a week. The cumulative percentage column indicates that the majority of the residents (58.89%) were visited 24 or fewer times a year.

2.7 CHARTS AND GRAPHS   Researchers frequently use charts and graphs to present their data in ways that are visually more dramatic than frequency distributions. These devices are particularly useful for conveying an impression of the overall shape of a distribution and for highlighting any clustering of cases in a particular range of scores. Many graphing techniques are available, but we will concentrate on just four. The first two, pie and bar charts, are appropriate for nominal, ordinal, and discrete variables. The last two, histograms and frequency polygons, are used with continuous interval-ratio variables.

Pie Charts.   To construct a **pie chart**, begin by computing the percentage of all cases that fall into each category of the variable. Then divide a circle (the pie) into segments (slices) proportional to the percentage distribution. Be sure that the chart and all segments are clearly labeled. Figure 2.1 is a pie chart that displays the distribution of "marital status" from the counseling center survey. The frequency distribution (Table 2.6) is reproduced as Table 2.16, with a column added for the percentage distribution. Since a circle's circumference is 360°, we will apportion 180° (or 50%) for the first

FIGURE 2.1   SAMPLE PIE CHART: MARITAL STATUS OF RESPONDENTS ($N = 20$)

TABLE 2.16   MARITAL STATUS OF RESPONDENTS,
COUNSELING CENTER SURVEY

Status	Frequency (f)	Percentage (%)
Single	10	50
Married	7	35
Divorced	3	15
	$N = 20$	100%

category, 126° (35%) for the second, and 54° (15%) for the last category. The pie chart visually reinforces the relative preponderance of single respondents and the relative absence of divorced students in the counseling center survey.

Bar Charts.   Like pie charts, **bar charts** are relatively straightforward. Conventionally, the categories of the variable are arrayed along the horizontal axis (or abscissa) and frequencies, or percentages if you prefer, along the vertical axis (or ordinate). For each category of the variable, construct (or draw) a rectangle of constant width with height corresponding to the number of cases in the category. The bar chart in Figure 2.2 reproduces the marital status data from Figure 2.1 and Table 2.16.

This chart would be interpreted in exactly the same way as the pie chart in Figure 2.1. Whether to use pie or bar charts to display the distribution of an ordinal or nominal variable is generally left to the discretion of the researcher. However, if a variable has more than four or five categories, the bar chart would be preferred. With too many categories, the pie chart gets very crowded and loses its visual clarity.

READING STATISTICS 2: FREQUENCY DISTRIBUTIONS AND OTHER TABLES

Tables of all kinds, including frequency distributions, are very commonly used by social science researchers. As you become more involved in the literature of the social sciences, you will often need to read, interpret, and understand tables produced by others. In this section we will discuss the interpretation of frequency distributions. In a later chapter I will suggest some ideas for reading bivariate tables.

First, there are many different formats for presenting tables. Not all frequency distributions in the literature will look exactly like the tables presented in this chapter. Second, because of space limitations, frequency distributions, if presented at all, will be presented with a minimum of detail. For example, the researcher may present the distribution of percentages alone rather than frequencies and percentages.

The key to reading any frequency distribution (or, for that matter, any other table, graph, or chart) is to let the researchers be your guides. The researchers will tell you why they think the table is important enough to be included in the report, and, for the most part, you should accept their analysis.

Begin your analysis of the table itself by first reading the title, all labels (that is, row and/or column headings), and any footnotes to the table. These will tell you exactly what information is contained in the table itself. Inspect the body of the table (frequencies, percentages, and so on) with the author's analysis in mind. See if the information in the table agrees with the author's analysis. (It almost always will, but it never hurts to double-check and exercise your critical abilities.)

Finally, most research projects analyze interrelationships among many variables. Because frequency distributions display single variables, they are unlikely to be included in such research reports (or perhaps, included only as background information). Even when these tables are not reported, you can be sure that the research began with an inspection of the frequency distribution for each variable. A careful inspection of these tables will supply the researcher with a great deal of information about the sample, and, therefore, frequency distributions are almost universally used descriptive devices.

Statistics in the Professional Literature

The following table is taken from a study of criminal victimization and combines frequencies, percentages, and rates in a format that conveys a great deal of information in a limited space. Take a moment now to read through the table, including titles, labels, and footnotes. The footnotes state the source of the data. The acronym "NCS" refers to the National Crime Survey, which is conducted every year for the Bureau of Justice Statistics (BJS) of the U.S. Department of Justice. The table concerns the crime of robbery and presents average annual numbers, percentages, and rates for a variety of subgroups. Study the table for a few minutes and try to identify any important differences among the groups.

TABLE 2.2   Robbery Rates for Various Groups

Victim Characteristics	Average Annual Number 1973–1984	Percentage	Average Annual Rate 1973–1984	1987 Rate
Sex		100%		
Male	794,200	65	9.3	6.6
Female	429,200	35	4.6	3.9
Race		100%		
White	921,800	75	5.9	4.4
Black	279,300	23	14.2	11.8
Other	22,400	2	7.2	NA
Ethnicity		100%		
Hispanic	98,200	8	10.4	9.9
Non-Hispanic	1,125,200	92	6.7	4.9
Age		100%		
12–15 years old	175,000	14	11.3	7.3
16–19	180,800	15	11.3	8.9
20–24	230,300	19	11.7	9.8
25–34	249,900	20	7.2	7.6
35–49	181,900	15	5.0	3.4
50–64	127,700	10	4.0	2.5
65 and older	77,700	6	3.3	1.8
Marital status		100%		
Married	338,700	28	3.4	2.4
Widowed	56,300	5	4.6	4.0
Divorced or separated	202,300	17	16.0	6.6
Never married	622,200	51	11.9	10.2
Family income		100%		
Less than $7,500	419,800	38	10.8	11.7
$7,500–14,999	317,300	29	6.7	6.8*
$15,000–24,999	211,100	19	5.1	3.7
$25,000 and above	154,500	14	4.7	3.5*

Notes: Rates are per 1,000 people with these characteristics per year.
*Estimated by combining the categories.
Percentages for subcategories may not total exactly 100% because of rounding. Average annual figures based on NCS findings for 1973–1984. Adapted from Harlow, 1987; BJS, 1989.

(*continued*)

READING STATISTICS 2: (*Continued*)

Here's what the author wants the reader to understand from the table:

"Starting with gender, the first pattern that stands out is that males were robbed at a rate more than twice that of females (9.3 males per 1,000 per year, compared to a rate of 4.6 for females). As for race and ethnicity, the rate for black people was more than twice as high as that for white people. Hispanics also were burdened by very high robbery rates.

"With regard to age, the survey analysis revealed that younger persons between the ages of twelve and twenty-four suffered high robbery rates that declined sharply with increasing age. Senior citizens over sixty-five were robbed the least often of any age group.

"Family income exhibited a pattern similar to that found in relation to age: As income increased, the chances of being robbed decreased. In addition to gender, race, income, and age, marital status made a big difference: People who were divorced or separated (regardless of their gender, race, income, or age) faced the highest risks of any category in the study. Those who had never been married were also robbed more often than the norm. Married persons and widowed individuals were victimized much less frequently."

Note how the information supplied by the rates clarifies the patterns in the table. As the author notes, the huge majority of robbery victims are white (75%), but the *rate* of victimization is much higher for blacks (14.2 per 1000 vs. 5.9 per 1000 for whites from 1973–1984) and Hispanics (10.4 vs. 6.7 for non-Hispanics). If the author had reported only the frequencies or percentages, the fact that the groups vary in size might not have been clear and this important difference in the probability of victimization might have been obscured.

Source: Andrew Karmen: 1990 *Crime Victims: An Introduction to Victimology.* Pacific Grove, California: Brooks/Cole. Used with permission.

Bar charts are particularly effective ways of displaying the relative frequencies for two or more categories of a variable when you want to emphasize some comparisons. Suppose, for example, that you wished to make a point about homicide victimization by sex and how it changed from 1955 to 1975. Figure 2.3 displays the comparisons for males and females over the time period in a dramatic and easily comprehended way. This bar chart shows that, while the overall homicide rate rose, the victimization rate for males rose faster than the victimization rate for females.

Histograms. **Histograms** look a lot like bar charts and, in fact, are constructed in much the same way. They are appropriate for continuous interval-ratio data; and, to reflect the numerical quality of the scale of measurement,

FIGURE 2.2   SAMPLE BAR CHART: MARITAL STATUS OF RESPONDENTS, COUNSELING CENTER SURVEY ($N = 20$)

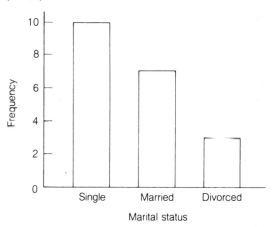

Marital status

FIGURE 2.3   SAMPLE BAR CHART: HOMICIDE VICTIMIZATION BY SEX, 1955–1975 (rates per 100,000 population, whites only)

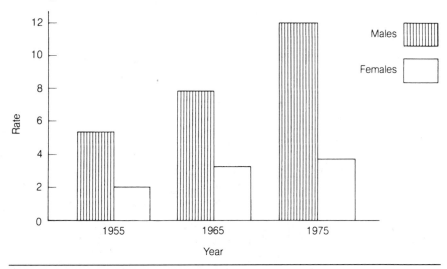

Year

Source: United States Bureau of the Census, *Statistical Abstracts of the United States: 1977* (98th edition). Washington, D.C., 1977.

the bars are contiguous to each other. To construct a histogram from a frequency distribution, follow these steps:

1. Array the class intervals or scores along the horizontal axis (abscissa) using the real class limits.

2. Array frequencies along the vertical axis (ordinate).

3. For each category in the frequency distribution, construct a bar with height corresponding to the number of cases in the category and with width corresponding to the real limits of the class intervals.
4. Label each axis of the graph.
5. Title the graph.

As an example, Figure 2.4 displays a histogram based on the visiting patterns that were presented in Table 2.15. The three clusterings of scores observed in the frequency distribution appear with clarity in the histogram, and their presence is thus reinforced visually.

Frequency Polygons.   Construction of a **frequency polygon** is similar to construction of a histogram. Instead of using bars to represent the frequencies, however, use a dot at the midpoint of each interval. The dots are then connected by straight lines and the resultant figure is closed by bringing the line down to the abscissa at the midpoints of the intervals im-

FIGURE 2.4   SAMPLE HISTOGRAM: NUMBER OF VISITS PER YEAR, RETIREMENT-COMMUNITY RESIDENTS ($N = 90$)*

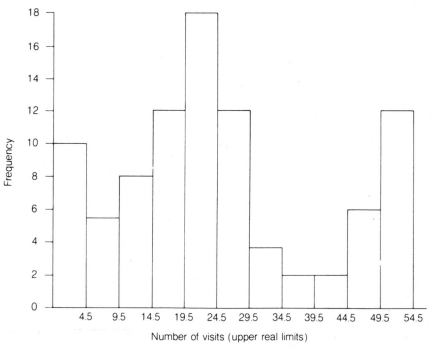

*To be perfectly accurate, the bar representing the lowest interval should begin not at the zero point but at -0.05 (the lower real limit). This increased accuracy would make the calibrations on the ordinate quite difficult to decipher (I know because I tried it), so I have simply closed the figure parallel to the vertical axis.

FIGURE 2.5   SAMPLE FREQUENCY POLYGON: NUMBER OF VISITS PER YEAR, RETIREMENT-COMMUNITY RESIDENTS ($N = 90$)

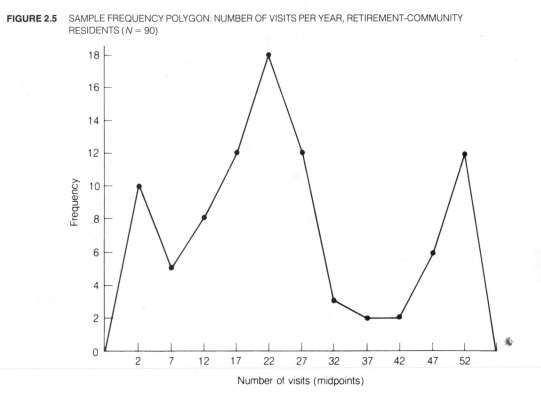

mediately beyond the lowest and highest categories. Figure 2.5 displays a frequency polygon for the visiting patterns previously displayed in the histogram. Note that the abscissa is marked off (or calibrated) in midpoints instead of real limits.

Histograms and frequency polygons represent alternative ways of displaying essentially the same message. The dominant impression conveyed by Figure 2.5 (the three clusterings of scores) is the same as that conveyed by Figure 2.4—the two graphs are functionally equivalent. Thus, the choice between the two techniques is left to the aesthetic pleasures of the researcher.

2.8 BIVARIATE TABLES

So far, this chapter has focused on ways of describing the distribution of a single variable. Social scientists are also frequently concerned with describing the relationship between two variables. A **bivariate table** is a commonly used device for displaying the joint frequency distributions of two variables. We shall see in later chapters how these tables can be used to investigate the possibility that two variables are related. For now, we will focus on how to construct bivariate tables and how completed bivariate tables display joint frequency distributions.

TABLE 2.17   SATISFACTION WITH COUNSELING CENTER BY SEX

	Sex		
Response	Males	Females	
Satisfied			13
Dissatisfied			7
	10	10	20

For purposes of illustration, suppose that a researcher wished to see if the males in the counseling center survey were more or less satisfied than the females. Although there are several ways of dealing with this issue, we will construct a bivariate table to investigate the relationship between sex and satisfaction. To construct the bivariate table, the cases will be sorted in a manner similar to that used in constructing univariate frequency distributions. The difference, of course, is that instead of sorting the cases by their scores on a single variable, the sorting will take account of the scores of each case on two variables. Our first step will be to state the categories of each of the variables. Table 2.17 displays the outlines of a bivariate table for sex and satisfaction with the categories of the latter variable collapsed into satisfied and dissatisfied students (as in Table 2.9).

Before we complete the table, let us take a moment to notice several features of bivariate table construction.

1. As should be the case with every table, chart, and graph, Table 2.17 has a descriptive title. By convention, the title includes the names of the variables contained in the table.

2. The table has two dimensions. The vertical dimension is referred to in terms of **columns** and the horizontal dimension in terms of **rows**. The names of the categories of one variable are used as column headings, and the names of the categories of the other variable as row headings. A common convention, which will be followed in this text, is to use the categories of the variable that is taken as independent (or causal) as column headings. Thus, the rows will state the categories of the dependent variable.

3. The subtotals for each column and row are reported. These subtotals are called marginals. Row and column marginals are the univariate frequency distributions for the row and column variables respectively. To illustrate, compare the marginals in Table 2.17 with the univariate frequency distributions of these two variables as displayed in Tables 2.5 and 2.9.

4. The total number of cases in the table (N) is reported at the intersection of the row and column marginals.

TABLE 2.18   SATISFACTION WITH COUNSELING CENTER BY SEX

Response	Sex		
	Males	Females	
Satisfied	6	7	13
Dissatisfied	4	3	7
	10	10	20

To complete the table, each case must be sorted and tallied according to its joint scores on both variables. Working with the data as presented in Table 2.4, we see that case A is a male who is satisfied, and he will be tallied in the upper left-hand cell of the total. Case B is a dissatisfied male and, thus, will be tallied in the lower left-hand cell. Case C, a satisfied female, will be tallied in the upper right-hand cell. Continue sorting cases in this manner until all cases have been tallied. The completed bivariate table is displayed in Table 2.18. This table suggests that there is no particular difference in the level of satisfaction between the sexes. Six out of ten of the males (or 60%) were satisfied, as were seven out of ten (70%) of the females.

Bivariate tables are used in several different contexts in statistics. So far, we have examined the construction of these tables and the way they display the joint frequency distributions of two variables. Later, we will see how bivariate tables can be used to test hypotheses (Chapter 11) and to investigate further the strength and nature of a bivariate relationship (Chapters 13 through 15).

SUMMARY

1. We considered several different ways of summarizing the distribution of a single variable and, more generally, reporting the results of our research. Our emphasis throughout is on the need to communicate our results clearly and concisely. You will often find that, as you strive to communicate statistical information to others, the meanings of the information will become clearer to you as well.

2. Percentages and proportions, ratios, and rates represent several different techniques for enhancing clarity by expressing our results in terms of relative frequency. Percentages and proportions report the relative occurrence of some category of a variable compared with the distribution as a whole. Ratios compare two categories with each other, and rates report the actual occurrences

of some phenomenon compared with the number of possible occurrences per some unit of time.

3. Frequency distributions are tables that summarize the entire distribution of some variable. It is very common to construct these tables for each variable of interest as the first step in a statistical analysis. The readability of the frequency distribution is often enhanced by adding columns for percentages, cumulative frequency, and/or cumulative percentages.

4. Pie and bar charts, histograms, and frequency polygons are graphic devices used to express the basic information contained in the frequency distribution in a compact and visually dramatic way.

5. A bivariate table is the basic device for displaying and investigating the joint frequency distribution of two variables. Bivariate tables display the scores on both variables and are used in several different contexts to probe the relationships between variables.

SUMMARY OF FORMULAS

Proportions	2.1	$p = \dfrac{f}{N}$
Percentages	2.2	$\% = \left(\dfrac{f}{N}\right) \times 100$
Ratios	2.3	$\text{Ratio} = \dfrac{f_1}{f_2}$

GLOSSARY

Bar chart. A graphic display device for nominal- and ordinal-level variables. Categories are represented by bars of equal width, the height of each corresponding to the number (or percentage) of cases in the category.

Bivariate table. A table that displays the joint frequency distributions of two variables.

Class intervals. The categories used in the frequency distributions for interval-ratio variables.

Column. The vertical dimension of a table.

Cumulative frequency. An optional column in a frequency distribution that displays the number of cases within an interval and all preceding intervals.

Cumulative percentage. An optional column in a frequency distribution that displays the percentage of cases within an interval and all preceding intervals.

Frequency distribution. A table that displays the number of cases in each category of a variable.

Frequency polygon. A graphic display device for interval-ratio variables. Class intervals are represented by dots placed over the midpoints, the height of each corresponding to the number (or percentage) of cases in the interval. All dots are connected by straight lines and the figure is closed at the midpoints of the intervals immediately above the highest interval and below the lowest interval.

Histogram. A graphic display device for interval-ratio variables. Class intervals are represented by contiguous bars of equal width (equal to the real limits), the height of each corresponding to the number (or percentage) of cases in the interval.

Midpoint. The point exactly halfway between the upper and lower limits of a class interval.

Percentage. The number of cases in a category of a variable divided by the number of cases in all categories of the variable, the entire quantity multiplied by 100.

Pie chart. A graphic display device especially for nominal and ordinal variables with only a few categories. A circle (the pie) is divided into segments proportional in size to the percentage of cases in each category of the variable.

Proportion. The number of cases in one category of a variable divided by the number of cases in all categories of the variable.

Rate. The number of actual occurrences of some phenomenon or trait divided by the number of possible occurrences per some unit of time.

Ratio. The number of cases in one category divided by the number of cases in some other category.

Real class limits. The upper and lower limits of a class interval used when the distribution is conceptualized as a continuous distribution of contiguous categories.

Row. The horizontal dimension of a table.

Stated class limits. The upper and lower boundaries of a class interval as they appear in the frequency distribution.

PROBLEMS

2.1 At St. Algebra College, the numbers of males and females in the various major fields of study are as follows:

Major	Males	Females
Humanities	117	83
Social sciences	97	132
Natural sciences	72	20
Business	156	139
Nursing	3	35
Education	30	15

Read each of the following problems carefully before constructing the fraction and solving for the answer.

a. What percentage of social science majors are male?

b. What proportion of business majors are female?

c. For the humanities, what is the ratio of males to females?

d. What percentage of the total student body are males?

e. What is the ratio of males to females for the entire sample?

f. What proportion of the nursing majors are male?

g. What percentage of the sample are social science majors?

h. What is the ratio of humanities majors to business majors?

i. What is the ratio of female business majors to female nursing majors?

j. What proportion of the males are education majors?

2.2 CJ The town of Shinbone, Kansas, has a population of 211,732 and experienced 47 bank robberies, 13 murders, and 23 auto thefts during the past year. Compute a rate for each type of crime per 100,000 population.

2.3 GER Following are reported the number of times 25 residents of a community for senior citizens left their homes for any reason during the past week.

0	2	1	7	3
7	0	2	3	17
14	15	5	0	7
5	21	4	7	6
2	0	10	5	7

a. Construct a frequency distribution to display these data.

b. What are the real limits of the class intervals?

c. Add columns to the table to display the percentage distribution, cumulative frequency, and cumulative percentages.

d. Construct a histogram and a frequency polygon to display this distribution.

e. Write a paragraph summarizing this distribution of scores.

2.4 PA/CJ As part of an evaluation of the efficiency of your local police force, you have gathered the following data on police response time to calls for assistance during two different years. Convert both frequency distributions into percentages and construct pie charts and bar charts to display the data. Write a paragraph comparing the changes in response time between the two years.

Response Time, 1977	f
More than 21 minutes	35
16–20 minutes	75
11–15 minutes	180
6–10 minutes	375
Less than 5 minutes	210
	875

Response Time, 1987	f
More than 21 minutes	45
16–20 minutes	95
11–15 minutes	155
6–10 minutes	350
Less than 5 minutes	250
	895

2.5 SOC Fifty students completed a questionnaire that measured their attitudes toward interpersonal violence. Respondents who scored high believed that in many situations a person could legitimately use physical force against another person. Respondents who scored low believed that in no situation (or very few situations) could the use of violence be justified.

52	47	17	8	92
53	23	28	9	90
17	63	17	17	23
19	66	10	20	47
20	66	5	25	17
10	82	90	40	45
8	91	82	52	20
75	32	75	60	60
80	30	70	65	52
90	29	70	66	55

a. Construct a frequency distribution to display these data.

b. What are the real limits of the class intervals?

c. Add columns to the table to display the percentage distribution, cumulative frequency, and cumulative percentage.

d. Construct a histogram and a frequency polygon to display these data.

e. Write a paragraph summarizing this distribution of scores.

2.6 [SOC/CJ] Following are reported the official homicide rates for all 50 states for 1975 and 1991.*

Homicide Rates
(per 100,000 population)

State	1975	1991
ME	2.8	1.2
NH	2.9	3.6
VT	2.1	2.1
MA	4.2	4.2
RI	3.0	3.7
CT	3.9	5.7
NY	11.0	14.2
NJ	6.8	5.2
PA	6.8	6.3
OH	8.1	7.2
IN	8.5	7.5
IL	10.6	11.3
MI	11.9	10.8
WI	3.3	4.8
MN	3.3	3.0
IA	2.5	2.0
MO	10.6	10.5
ND	0.8	1.1
SD	3.7	1.7
NE	4.3	3.3
KS	5.4	6.1
DE	7.3	5.4
MD	10.7	11.7
VA	11.5	9.3
WV	7.4	6.2
NC	12.4	11.4
SC	14.7	11.3
GA	14.4	12.8
FL	13.5	9.4
KY	10.2	6.8
TN	11.4	11.0
AL	16.0	11.5
MS	13.9	12.8
AR	10.1	11.1
LA	12.6	16.9
OK	9.4	7.2

Homicide Rates
(per 100,000 population)

State	1975	1991
TX	13.4	15.3
MT	5.2	2.6
ID	5.2	1.8
WY	10.2	3.3
CO	7.4	5.9
NM	13.3	10.5
AZ	8.6	7.8
UT	2.7	2.9
NV	13.0	11.8
WA	5.7	4.2
OR	6.2	4.6
CA	10.4	12.7
AK	12.2	7.4
HI	7.7	4.0

*Source: United States Bureau of the Census, *Statistical Abstracts of the United States: 1977* (98th edition) and *1993* (113th edition). Washington, D.C., 1977 and 1993.

a. Construct a frequency distribution for each year to display these data. Include percentage and cumulative percentage columns.

b. Construct histograms and frequency polygons for each year.

c. Compare the homicide rates for the two years. What can you tell about homicide from comparing these frequency distributions and graphs? Write a paragraph of interpretation for these data.

2.7 [SW] A psychiatric social worker rated the mental functioning of 50 patients on their release from a state mental hospital and again six months after release. He used a scale on which higher scores indicated higher levels of functioning and competence. The scores are reported below.

Case:	Release Date	Six Months
1	10	12
2	40	42
3	17	16
4	35	33
5	35	45
6	17	15
7	22	47
8	22	25

Case:	Release Date	Six Months
9	10	15
10	15	18
11	25	20
12	29	33
13	16	35
14	27	14
15	43	35
16	45	19
17	28	35
18	41	31
19	41	30
20	30	35
21	36	31
22	21	21
23	27	32
24	12	15
25	15	25
26	21	24
27	29	32
28	34	44
29	12	18
30	34	24
31	14	27
32	16	35
33	20	32
34	19	50
35	18	28
36	28	38
37	10	20
38	11	19
39	28	38
40	10	20
41	11	19
42	12	41
43	40	42
44	25	26
45	17	25
46	35	36
47	20	22
48	22	27
49	22	22
50	31	41

a. Construct a frequency distribution for each set of scores. Include percentages and cumulative percentages.

b. Construct histograms and frequency polygons to display these data.

c. Compare the two sets of scores and write a paragraph of interpretation.

2.8 Using the data from the counseling center survey (Table 2.4), construct a bivariate table to display the relationship between "marital status" and "satisfaction." Use the collapsed categories of both variables (that is, married versus not married as in Table 2.7, and satisfied versus dissatisfied, as in Table 2.9). What percentage of married students are satisfied? What percentage of nonmarried students are satisfied? Is there any pattern in this relationship?

2.9 [PS] Each of the 20 cities below has been categorized in terms of type of government and city size. Are large cities more likely to have the mayoral form of government?

City	Type of Government	City Size
A	Mayoral	Large
B	Mayoral	Small
C	City Manager	Large
D	Mayoral	Large
E	Mayoral	Small
F	Mayoral	Large
G	City Manager	Small
H	City Manager	Large
I	Mayoral	Large
J	City Manager	Small
K	City Manager	Small
L	Mayoral	Small
M	Mayoral	Large
N	City Manager	Small
O	City Manager	Large
P	Mayoral	Large
Q	City Manager	Small
R	Mayoral	Large
S	City Manager	Small
T	City Manager	Small

2.10 [CJ] The number of major felonies reported to the police in two different towns last year are listed below. Calculate the rate for each crime per 100,000 population. Relatively speaking, which town has the greater apparent crime problem? With which crimes? Write a paragraph describing the differences.

	Town A	Town B
Populations =	20,109	764,213
Homicide	13	78
Robbery	102	617
Auto Theft	125	314
Rape	23	79
Burglary	178	537

2.11 SOC The tables below report the marital status of 20 respondents in two different apartment complexes.

	Complex A	Complex B
Married	5	10
Unmarried ("living together")	8	2
Single	4	6
Separated	2	1
Widowed	0	1
Divorced	1	0
	20	20

a. What percentage of the respondents in each complex are married? _____

b. What is the ratio of single to married respondents at each complex? _____

c. What proportion of each sample are widowed? _____

d. What percentage of the single respondents live in Complex B? _____

e. What is the ratio of the "unmarried–living together" to the married at each complex? _____

2.12 SOC For the 20 College Board scores below, construct a frequency distribution with columns for percentages and cumulative percentages. Construct a histogram and frequency polygon for these data.

420	345	560	650
459	499	500	657
467	480	505	555
480	520	530	589
500	550	545	600

2.13 SW A local youth service agency has begun a sex education program for teenage girls who have been referred by the juvenile courts. The girls were given a 20-item test for general knowledge about sex, contraception, and anatomy and physiology upon admission to the program and again after completing the program. The scores of the first 15 girls to complete the program are listed below. Construct frequency distributions to display these scores.

	Pretest	Posttest
A	8	12
B	7	13
C	10	12
D	15	19
E	10	8
F	10	17
G	3	12
H	10	11
I	5	7
J	15	12
K	13	20
L	4	5
M	10	15
N	8	11
O	12	20

2.14 SW The 15 girls mentioned in Problem 2.13 were interviewed two years after the program ended and were asked if they had experienced any unplanned pregnancies since that time. Each respondent's answer is listed below along with whether the respondent had a high or low score on the posttest mentioned above. Does the girl's knowledge about sex and contraception have anything to do with whether or not she unintentionally became pregnant? Construct a bivariate table for these two sets of variables to examine the relationship.

	Posttest	Unplanned Pregnancy?
A	H	Y
B	L	N
C	H	N
D	L	N
E	H	Y
F	H	N

	Posttest	Unplanned Pregnancy?
G	H	N
H	H	N
I	L	N
J	L	Y
K	L	Y
L	L	N
M	L	Y
N	H	Y
O	H	N

2.15 SOC The scores of 25 respondents on three variables are reported below. These scores were taken from a public opinion survey called the General Social Survey, or the GSS. This data set, which is described in some detail in Appendix G, is used for the computer exercises in this text. I will use small subsamples from the GSS to provide you with the opportunity to practice your statistical skills on "real" data. For the actual questions and other details, see Appendix G. The numerical codes for these three variables are as follows:

> Sex: 1 = Male
>   2 = Female
> Support for Gun Control (item 25):
>   1 = In favor
>   2 = Opposed
> Education (item 9):
>   0 = Less than HS
>   1 = HS
>   2 = Jr. college
>   3 = Bachelor's
>   4 = Graduate

Case No.	Sex	Support for Gun Control	Level of Education
1	2	1	1
2	1	2	1
3	2	1	3
4	1	1	2
5	2	1	3
6	1	1	1
7	2	2	0
8	1	1	1
9	1	2	0
10	2	1	1
11	1	1	4

Case No.	Sex	Support for Gun Control	Level of Education
12	1	1	4
13	1	1	0
14	2	1	1
15	1	1	1
16	2	1	4
17	2	1	1
18	1	1	0
19	1	2	1
20	1	1	1
21	2	1	4
22	1	1	3
23	1	1	2
24	2	2	3
25	2	1	1

a. Construct a frequency distribution for each of the three variables. Include a column for percentages.

b. Construct pie and bar charts to display the distributions of the three variables.

2.16 SOC Construct a bivariate table to display the relationship between sex and support for gun control. What percentage of males are in favor? What percentage of females? Is there a pattern in these relationships? Repeat the exercise using education and support for gun control.

2.17 SOC The scores below are taken from the General Social Survey. The first variable is a prestige score for the respondent's occupation (item 2 in Appendix G), and the second is a prestige score for the respondent's father's occupation (item 5).

Case No.	Respondent's Occupational Prestige	Father's Occupational Prestige
1	26	29
2	39	45
3	47	30
4	36	45
5	60	40
6	12	23
7	32	70
8	15	51
9	58	50
10	43	42

Case No.	Respondent's Occupational Prestige	Father's Occupational Prestige	Case No.	Respondent's Occupational Prestige	Father's Occupational Prestige
11	78	56	36	69	17
12	29	50	37	23	16
13	17	41	38	16	28
14	32	41	39	36	41
15	40	50	40	48	47
16	50	39	41	51	33
17	47	61	42	54	41
18	55	41	43	57	34
19	50	50	44	18	50
20	50	27	45	37	50
21	26	50	46	36	50
22	25	61	47	36	45
23	62	45	48	22	29
24	51	47	49	37	37
25	34	41	50	41	41
26	50	32			
27	61	32			
28	62	51			
29	78	34			
30	32	17			
31	29	22			
32	36	27			
33	12	17			
34	28	16			
35	48	18			

a. Construct a frequency distribution for each variable. Include a column for percentages and cumulative percentages.
b. Construct histograms and frequency polygons for each variable.
c. Compare the two variables. What can you tell about the distribution of prestige for these two generations? Write a paragraph of interpretation for these variables.

SPSS/PC+ PROCEDURES FOR FREQUENCY DISTRIBUTIONS AND GRAPHS

DEMONSTRATION 2.1 Producing Frequency Distributions for the Counseling Center Data

To construct univariate frequency distributions, bar charts, and histograms, SPSS/PC+ provides a procedure called FREQUENCIES. To get a detailed look at the operation and logic of SPSS/PC+, let's use the counseling center data from Table 2.4 for our first example. In Appendix F, we saw how to create a file that included both the data and some SPSS/PC+ definitional commands. Here we will begin with this file and add some commands that will produce a frequency distribution along with some other output.

To begin, you need to retrieve the file. After you've started up the SPSS/PC+ program, press ALT-E to move to EDIT MODE. Touch the F3 key and SPSS/PC+ will ask if you want to 'edit a different file?' Press ENTER and type the file name (I suggested 'CCSURVEY' in Appendix F) in the window provided. Press ENTER again when you are done, and the file should appear in the bottom window. Remember to refer to the file with exactly the same name you used to save it in previous sessions. If nothing appears in the EDIT window and SPSS/PC+ says that you are working with a new file, you probably spelled the file name inconsistently. Within SPSS/PC+, you can use the ALT-F facility (press 'f' while holding ALT down) to list file names and remind yourself of the spelling. The MAIN MENU will disappear and all file names in the current directory will be listed in the top window. You can move around in the file list with the arrow keys. If the file is on a floppy disk, you can use the DOS 'directory' command (or 'dir') to list file names.

When you have successfully retrieved the file, it will be listed in the bottom half of the screen (the "EDIT mode" window). In this exercise, we will add a command to the file that will instruct SPSS/PC+ to create a frequency distribution for marital status (MARITAL). Touch ALT-E to return to the MAIN MENU and then

1. Select 'analyze data' from the Main Menu.
2. Select 'descriptive statistics.'
3. Select and paste the FREQUENCIES command. You are now in the FREQUENCIES submenu.
4. Select !/VARIABLES and use ALT-T to type the names. That is, when you touch 't' while holding the ALT key down, a typing window will appear in the middle of the screen. Type the variable name MARITAL in this window and then touch ENTER. The variable name will be pasted to the bottom window.
5. If you want graphic representations of the frequency distribution, select /BARCHART for nominal, ordinal, or discrete variables and /HISTOGRAM for continuous variables. For MARITAL, select /BARCHART.

When you are finished, the file should look like this:

```
DATA LIST FIXED / idno 1-2 GENDER 3 MARITAL 4 SATIS 5
AGE 6-7.
BEGIN DATA.
0111418
0212219
0321418
0421219
0512120
0611320
0722418
0821321
0911319
1023323
1121324
1212318
```

```
1321122
1422326
1511318
1612419
1722219
1813119
1923321
2011220
END DATA.
VARIABLE LABELS MARITAL 'Marital Status'
 SATIS 'Satisfaction with Counseling Center Services'.
VALUE LABELS  SEX 1 'Male' 2 'Female'/
              MARITAL 1 'Single' 2 'Married'
              3 'Divorced'/
              SATIS 1 'Very Dissatisfied'
              2 'Dissatisfied' 3 'Satisfied'
              4 'Very Satisfied'/.
MISSING VALUES SEX TO AGE (9).
FREQUENCIES /VARIABLES MARITAL /BARCHART.
```

When you run this command (switch to EDIT mode by touching ALT-E, move the cursor to the first line of the file, and then push F10), the following table and graph will be produced:

MARITAL Marital Status

Value Label	Value	Frequency	Percent	Valid Percent	Cum Percent
Single	1	10	50.0	50.0	50.0
Married	2	7	35.0	35.0	85.0
Divorced	3	3	15.0	15.0	100.0
		-------	-------	-------	
	TOTAL	20	100.0	100.0	

```
     Single ████████████████████████████████████████ 10
    Married ████████████████████████████ 7
   Divorced ████████████ 3
```

Valid Cases   20   Missing Cases   0

The 'Value' column indicates the category (for example, '1' means 'single'), the 'Frequency' column shows the count for each category (there are 10 single people in the sample), and columns for percents and cumulative percents are automatically produced. The 'Percent' column shows the percent of cases in each category, including cases that are missing scores in the total number of cases (N). Percents in the 'Valid Percent' column are computed with missing cases removed and are therefore usually more meaningful than the latter column.

**DEMONSTRATION 2.2 Producing Frequency Distributions
for the 1993 General Social Survey (GSS) Data**

The same procedures and commands described above can be used to begin to analyze the 1993 GSS data file. Let's illustrate by producing a frequency distribution for marital status in this data file. Begin by making the GSS 1993 system file active. From the MAIN MENU, move to the 'read or write data' menu and select the GET command. Select "!/FILE", and the typing window will appear. Type GSS93 (or whatever file name was used to create the system file) and then touch the F10 key. Some information describing the number of cases and variables will appear on the screen, and you can return to the MAIN MENU by pressing any key. To create a frequency distribution, follow the instructions in Demonstration 2.1. By coincidence, the variable name (MARITAL) is the same for both data files.

Unlike Demonstration 2.1, the only commands that will appear on the screen as you go through this exercise will be the FREQUENCIES command. The data and other commands are in the system file but do not appear on the screen. This feature might be a little confusing at first, but it is designed to save time and computer resources. The file is quite large—hundreds of lines long—and it would take many additional minutes (hours?) if it scrolled across the screen every time you executed a command. So the system is designed to run with the system file in the background and only the specific commands supplied in the session visible to the user.

When you are done, your command should look like this:

```
FREQUENCIES /VARIABLES MARITAL /BARCHART.
```

Execute this command (press F10) and the following table and graph will be produced:

Value Label	Value	Frequency	Percent	Valid Percent	Cum Percent
Married	1.00	406	51.5	51.5	51.5
Widowed	2.00	92	11.7	11.7	63.2
Divorced	3.00	116	14.7	14.7	77.9
Separated	4.00	22	2.8	2.8	80.7
Never Married	5.00	152	19.3	19.3	100.0
		-------	-------	-------	
	TOTAL	788	100.0	100.0	

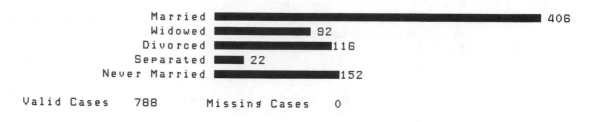

Valid Cases   788   Missing Cases   0

How would you describe this sample in terms of this variable?

DEMONSTRATION 2.3 Frequency Distributions for Interval-Ratio Variables and the RECODE Command

As we saw in Section 2.5, the categories or scores of interval-ratio level variables will normally have to be collapsed before readable frequency distributions can be produced. SPSS/PC+ provides a number of ways to manipulate the original scores on a variable. One of the most useful of these is the RECODE command. This command, in effect, changes the scores on a variable and can be used for a variety of purposes. Here, we will use RECODE to collapse the scores of an interval-ratio level variable in the GSS93 file so that we can produce a sensible and readable frequency distribution. The general format for the command is

```
RECODE VARIABLE NAME (old value = new value) (old
value = new value) . . . .
```

After SPSS/PC+ reads a RECODE command, it treats the variable as if it had only the new values or scores stipulated in the command. To return to the old scores, simply delete the command.

To illustrate the use of this procedure, let's use the AGE variable from the GSS93 file. For this sample, the youngest respondents are 18 and the oldest are 89. (I learned this by running FREQUENCIES on AGE without recoding the variable.) The range of scores is thus 71. If we wanted to use about ten intervals to display AGE, we could use an interval size of 7 or 8 and begin the first interval at 18 or at any age less than 18. Since it is more conventional, let's use an interval of 10, beginning with 10–19 as the first interval.

From the Main Menu, select 'modify data or files' and then select 'modify data values.' From the 'modify data values' menu, select RECODE. Be sure to read the brief explanation of this command in the right-hand window. From the RECODE menu, select '!variable(s)' and type the variable name (AGE) in the window provided. Now, select the parentheses and '!old values.' Use the typing window to enter the original (or 'old') values. These would be 10

THRU 19. Now, select the '!=' sign and select '!new values,' which would be 1 in this case. Repeat this process for every combination of old and new values. When you are finished, the RECODE command for AGE should look like this:

```
RECODE AGE (10 THRU 19 = 1) (20 THRU 29 = 2) (30 THRU
39 = 3) (40 THRU 49 = 4) (50 THRU 59 = 5) (60 THRU 69
= 6) (70 THRU 79 = 7) (80 THRU 89 = 8).
```

When the program reads this command, it will group the scores as instructed and treat AGE as if these were the only scores associated with this variable. As far as the program is concerned, AGE now has eight scores ranging from 1 to 8. If you need to run other tasks with the original scores, remove this RECODE command and the program will revert to the scores as actually recorded in the system file.

To clarify output, it is almost always useful to revise the VALUE LABELS for any recoded variables. For the recoding scheme above, we could add a new command that would provide appropriate labels for the collapsed version of the variable:

```
1
VALUE LABELS AGE 1 '10 TO 19' 2 '20 TO 29' 3 '30 TO
                39' 4 '40 TO 49' 5 '50 TO 59' 6 '60
                TO 69' 7 '70 TO 79' 8 '80 TO 89'/.
```

To run the task, select FREQUENCIES from the 'descriptive statistics' menu, specify AGE as the variable, and request a HISTOGRAM. The command would be

```
FREQUENCIES /VARIABLES AGE /HISTOGRAM.
```

and this command would produce the following output:

Value Label	Value	Frequency	Percent	Valid Percent	Cum Percent
10 TO 19	1.00	14	1.8	1.8	1.8
20 TO 29	2.00	137	17.4	17.4	19.2
30 TO 39	3.00	171	21.7	21.7	40.9
40 TO 49	4.00	176	22.3	22.4	63.3
50 TO 59	5.00	97	12.3	12.3	75.6
60 TO 69	6.00	89	11.3	11.3	86.9
70 TO 79	7.00	73	9.3	9.3	96.2
80 TO 89	8.00	30	3.8	3.8	100.0
	0.00	1	.1	MISSING	
		-------	-------	-------	
	TOTAL	788	100.0	100.0	

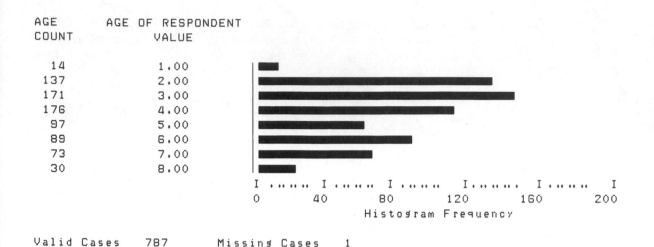

```
AGE          AGE OF RESPONDENT
COUNT              VALUE

   14             1.00
  137             2.00
  171             3.00
  176             4.00
   97             5.00
   89             6.00
   73             7.00
   30             8.00

                          I ...... I ...... I ...... I ....... I ...... I
                          0        40      80      120       160      200
                                      Histogram Frequency


Valid Cases     787      Missing Cases    1
```

If the table seems less than optimum for any reason, we could change interval sizes by simply altering the RECODE command.

Exercises

2.1 Run the FREQUENCIES procedure on the counseling center data for GENDER and AGE. Use the RECODE command to produce a second frequency distribution with AGE collapsed into just two categories. Make the categories about equal in number of cases and use the uncollapsed frequency distribution to determine reasonable cutting points for the new categories. Be sure to include a new VALUE LABELS command.

2.2 Using the GSS93 system file, run the FREQUENCIES program on 5 or 10 nominal- or ordinal-level discrete variables, including RACE, RELIG, SEX, and DEGREE. Be sure to get BARCHARTs for each variable. Write a sentence or two summarizing each frequency distribution.

2.3 Run the FREQUENCIES program for PRESTG80. First find the range of this variable by running the FREQUENCIES command without recoding. Use this information to determine reasonable class intervals by following the procedures described in Section 2.5 and then write the RECODE statement and produce the final frequency distribution. Include a HISTOGRAM and write a sentence or two of interpretation.

2.4 Run the FREQUENCIES program for several of the variables with more than 3 or 4 values. Examples include CHILDS, INCOME91, and ATTEND.

Be sure that you understand the coding scheme of these variables. Some of them seem like interval ratio variables at first glance but are not (e.g., CHILDS). Using the frequency distributions for the uncollapsed scores, develop some RECODE statements and generate some additional tables with scores collapsed. Compare with the "raw score" tables. How much detail has been lost? Does it matter?

3 Measures of Central Tendency

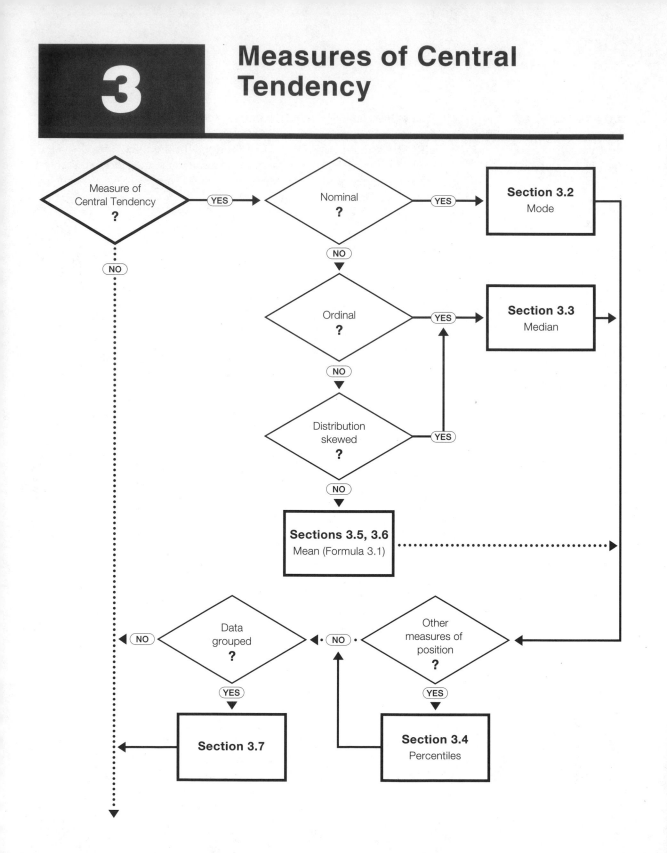

3.1 INTRODUCTION

One clear benefit of frequency distributions and the graphic devices used to display them is that they summarize the overall shape of a distribution of scores in a way that can be quickly comprehended. Often, however, you will need to report more detailed information about the distribution. Specifically, two additional kinds of statistics are almost always useful: some idea of the typical or average case in the distribution (for example, the average starting salary for social workers is $18,000 per year) and some idea of how much variety or heterogeneity there is in the distribution (starting salaries for social workers range from $15,000 per year to $20,000 per year). The first kind of statistic, called **measures of central tendency**, will be the subject of this chapter. The second kind of statistic, measures of dispersion, will be presented in Chapter 4.

The three commonly used measures of central tendency—the mode, median, and mean—are all probably familiar to you. All three summarize an entire distribution of scores by describing the most typical, or central, or representative value of that distribution. These statistics are powerful because they can reduce huge arrays of data to a single, easily understood number. Remember that this kind of data reduction or summarizing function is the central purpose of descriptive statistics.

Even though they share a common purpose, the three measures of central tendency are quite different from each other. In fact, they will be the same value only under specific and limited conditions. As we shall see, they vary in terms of level-of-measurement considerations and, perhaps more importantly, in terms of how they define *typical* (that is, they will not necessarily identify the same score or category as "the average"). Thus, your choice of an appropriate measure of central tendency will depend in part on the way you measure your variable and in part on the purpose of the research.

3.2 THE MODE

The **mode** of any distribution is the value that occurs most frequently. For example, in the set of scores 58, 82, 82, 90, 98, the mode is 82 since it occurs twice and the other scores occur only once.

The mode is, relatively speaking, a rather simple statistic, most useful when you want a "quick and easy" indicator of central tendency and when you are working with nominal-level variables. In fact, the mode is the only measure of central tendency that can be used with nominal-level variables. Such variables do not, of course, have numerical "scores" per se, and the mode of a nominally measured variable would be its largest category. For example, Table 3.1 reports the religious affiliations of a fictitious sample of 242 respondents. The mode of this distribution, the single largest category, is Protestant.

If a researcher desires to report only the most popular or common value of a distribution, or if the variable under consideration is nominal, then the mode is the appropriate measure of central tendency. However, keep in

TABLE 3.1   RELIGIOUS PREFERENCE (fictitious data)

Denomination	Frequency
Protestant	128
Catholic	57
Jew	10
None	32
Other	15
	N = 242

mind that the mode does have several limitations. First, some distributions have no mode at all (see Table 2.5) or so many modes that the statistic loses its meaning. Second, with ordinal and interval-ratio data, the modal score may not be central to the distribution as a whole. That is, *most common* does not necessarily mean "typical" in the sense of identifying the center of the distribution. For example, consider the following rather unusual (but not impossible) distribution of scores on a statistics test:

Scores (% correct)	Frequency
58	2
60	2
62	3
64	2
66	3
67	4
68	1
69	1
70	1
93	5
	N = 24

The mode of the distribution is 93. Is this score very close to the majority of the scores? If the instructor summarized this distribution by reporting only the modal score, would he or she be conveying an accurate picture of the distribution as a whole?

3.3 THE MEDIAN

Unlike the mode, the **median** (Md) always represents the exact center of a distribution of scores. The median is defined as the score of the case having half the cases above it and half below it, after all cases have been ordered from high to low. Thus, if the median family income for a community is $20,000, half the families earn more than $20,000 and half, less than $20,000.

To find the median, first locate the central or middle case. The median is the score associated with that case. When the number of cases (N) is odd, the value of the median is unambiguous because there will always be a middle case. With an even number of cases, however, there will be two middle cases; and, in this situation, the median is defined as the score exactly halfway between the scores of the two middle cases.

To illustrate, assume that seven students were asked to indicate their level of support for the intercollegiate athletic program at their universities on a scale ranging from 10 (indicating great support) to 0 (no support). After arranging their responses from high to low, you can find the median by locating the case that divides the distribution into two equal halves. With a total of seven cases, the middle case would be the fourth case, since there will be three cases above and three cases below this case. If the seven scores were 10, 10, 8, 7, 5, 4, and 2, then the median is 7, the score of the fourth case. To find the middle case when N is odd, increase the value of N by 1 and divide by 2. With an N of 7, the median is the score associated with the $(7 + 1)/2$, or fourth, case. If N had been 25, the median would be the score associated with the $(25 + 1)/2$, or 13th, case.

Now, if we added a subject whose support for athletics was measured as a 1 to the group above and thereby made N an even number (8), we would no longer have a single middle case. The ordered distribution of scores would now be 10, 10, 8, 7, 5, 4, 2, 1; any value between 7 and 5 would technically satisfy the definition of a median (that is, would split the distribution into two equal halves of four cases each). This ambiguity is resolved by defining the median as the average of the scores of the two middle cases. In the example above, the median would be defined as $(7 + 5)/2$, or 6.

To identify the two middle cases when N is an even number, divide N by 2 to find the first middle case and then increase that number by 1 to find the second middle case. In the example above with eight cases, the first middle case would be the fourth case ($N/2 = 4$) and the second middle case would be the $(N/2) + 1$, or fifth, case. If N had been 142, the first middle case would have been the 71st case and the second the 72nd case. Remember that the median is defined as the average of the scores associated with the two middle cases.*

As I have implied throughout this discussion, the median cannot be calculated for variables measured at the nominal level; such variables do not have scores that can be ordered or ranked. This measure can be calculated for either ordinal or interval-ratio data but is more commonly used with the former.

---

*If the middle cases have the same score, that score is defined as the median. In the distribution 10, 10, 8, 6, 6, 4, 2, 1 the middle cases both have scores of 6 and, thus, the median would be defined as 6.

3.4 OTHER MEASURES OF POSITION: PERCENTILES, DECILES, AND QUARTILES*

In addition to serving as a measure of central tendency, the median is also a member of a class of statistics that measure position or location. The median identifies the exact middle of a distribution, but it is sometimes useful to locate other points as well. We may want to know, for example, the scores that split the distribution into thirds or fourths or the point below which a given percentage of the cases fall. A familiar application of these measures would be scores on standardized tests, which are often reported in terms of location (for example, "a score of 476 is higher than 46% of the scores").

One commonly used statistic for reporting position is the **percentile**. A percentile identifies the point below which a specific percentage of cases fall. If a score of 476 is reported as the 46th percentile, this means that 46% of the cases had scores lower than 476. To find a percentile for ungrouped data, multiply the number of cases (N) by the proportional value of the percentile. For example, the proportional value for the 46th percentile would be 0.46. The resultant value identifies the number of the case that marks the percentile. If we had a sample of 78 cases and wanted to find the 37th percentile, we would multiply 78 by .37. The result is 28.86, and the 37th percentile would be 86/100 of the distance between the scores of the 28th and 29th cases. In most cases, we would probably avoid this extra calculation, round off 28.86 to 29, and call the score of the 29th case the 37th percentile. The slight inaccuracy would be worth the considerable savings in time and calculational effort.

Note that, if we think in terms of percentiles, then the median is simply the 50th percentile and, in our example above, we would find the median by multiplying 78 by .50, finding the 39th case, and declaring the score of that case to be the 50th percentile. Notice again that we are cutting some corners here. Technically, the median would be the score halfway between the two middle cases (the 39th and 40th cases), but it is unlikely that this inaccuracy would be very significant.

Some other commonly used measures of position are **deciles** and **quartiles**. Deciles divide the distribution of scores into tenths. So, the first decile is the point below which 10% of the cases fall and is equivalent to the 10th percentile. The fifth decile is also the same as the 50th percentile, which is the same as (you guessed it) the median. Quartiles divide the distribution into quarters, and the first quartile is the same as the 25th percentile. Any of these measures can be found by the method described above for percentiles. Remember that multiplying N by the proportional value of the percentile, decile, or quartile gives the number of the appropriate *case*, and it's the *score* of the case that actually marks the location. Also remember that this technique cuts some (probably minor) computational corners, and use it with caution.

---

*This section is optional.

3.5 THE MEAN

The **mean** ($\bar{X}$—read this as "ex-bar")*, or arithmetic average, is by far the most commonly used measure of central tendency. It reports the average score of a distribution, and its calculation is straightforward and probably familiar to most of you. To compute the mean, add the scores and then divide by the number of scores (N). To illustrate: a birth control clinic administered a 20-item test of general knowledge about contraception to 10 clients. The number of correct responses was 2, 10, 15, 11, 9, 16, 18, 10, 11, 7. To find the mean of this distribution, add the scores (total = 109) and divide by the number of scores (10). The result (10.9) is the average score on the test.

The mathematical formula for the mean is

FORMULA 3.1

$$\bar{X} = \frac{\Sigma(X_i)}{N}$$

where $\bar{X}$ = the mean
$\Sigma(X_i)$ = the summation of the scores
N = the number of scores

Since this formula introduces some new symbols, let us take a moment to consider it. First, the symbol Σ (uppercase Greek letter sigma) is a mathematical operator just like the plus sign (+) or divide sign (÷). It stands for "the summation of" and directs us to add whatever quantities are stated immediately following it. The second new symbol is $\boldsymbol{X_i}$ (X sub i), which refers to any single score—the "ith" score. If we wished to refer to a particular score in the distribution, the subscript could be replaced by the specific number of the score. Thus, X_1 would refer to the first score, X_2 to the second, X_{26} to the 26th, and so forth. The operation of adding all the scores is symbolized as $\Sigma(X_i)$. This combination of symbols directs us to sum the scores, beginning with the first score and ending with the last score in the distribution. Thus, Formula 3.1 states in symbols what has already been stated in words (to calculate the mean, add the scores and divide by the number of scores), but in a very succinct and precise way.[†]

Since the computation of a mean requires the mathematical procedures of addition and division, its use is fully justified only when working with interval-ratio data. However, researchers sometimes calculate the mean for variables measured at the ordinal level, because the mean is much more flexible than the median and is a central feature of many interesting and powerful advanced statistical techniques. Thus, if the researcher plans to do any more than merely describe his or her data, the mean will probably be the preferable measure of central tendency. When we violate level-of-mea-

---

*This is the symbol for the mean of a sample. The mean of a population is symbolized with the Greek letter mu (μ—read this symbol as "mew").
[†]See Appendix H for a further review of the summation sign.

surement considerations, of course, we are obligated to demonstrate that our conclusions still make statistical and logical sense.

3.6 SOME CHARACTERISTICS OF THE MEAN

Because the mean is the most commonly used measure of central tendency, let us consider its mathematical and statistical characteristics in some detail. First, the mean is always the center of any distribution of scores. The mean is the point around which all of the scores (X_i) cancel out. Symbolically:

$$\Sigma(X_i - \bar{X}) = 0$$

Or, if we take each score in a distribution, subtract the mean from it, and add all of the differences, the resultant sum will always be zero. To illustrate, consider the following set of test scores: 65, 73, 77, 85, and 90. The mean of these five scores is 390/5, or 78. So

X_i	$(X_i - \bar{X})$
65	$65 - 78 = -13$
73	$73 - 78 = -5$
77	$77 - 78 = -1$
85	$85 - 78 = 7$
90	$90 - 78 = 12$
$\Sigma X = 390$	$\Sigma(X_i - \bar{X}) = 0$

The total of the negative differences (-19) is exactly equal to the total of the positive differences ($+19$) and will always be equal regardless of the distribution. This algebraic relationship between the scores and the mean indicates that the mean is a good descriptive measure of the centrality of scores. You may think of the mean as a fulcrum that exactly balances all of the scores.

A second characteristic of the mean will only be mentioned at this point but is of great importance in later chapters. This characteristic is expressed in the statement

$$\Sigma(X_i - \bar{X})^2 = \text{minimum}^*$$

which says that, if the differences between the scores and the mean are squared and then added, the resultant sum will be less than the sum of the squared differences between the scores and any other point in the distribu-

---

*To illustrate this principle, let us use the distribution of five test scores mentioned above. The differences between the scores and the mean have already been found. If we square and sum these differences, we would get $(-13)^2 + (-5)^2 + (-1)^2 + (7)^2 + (12)^2$, or $(169 + 25 + 1 + 49 + 144)$, or 388. So, $\Sigma(X_i - \bar{X})^2 = 388$. If we performed the same operation with any other number, the resultant sum would be greater than 388. For example, if we used 77 instead of the mean, the sum of the squared differences would be $(65 - 77)^2 + (73 - 77)^2 + (77 - 77)^2 + (85 - 77)^2 + (90 - 77)^2$, or $(-12)^2 + (-4)^2 + (0)^2 + (8)^2 + (13)^2$, or $(144 + 16 + 0 + 64 + 169)$, or 393.

tion. That is, the mean is the point in a distribution around which the variation of the scores (as indicated by the squared differences) is minimized. In a sense, this algebraic property merely underlines the fact that the mean is closer to all of the scores than the other measures of central tendency. However, this characteristic of the mean is of central importance when we take up the topics of correlation and regression.

The final important characteristic of the mean is that it is affected by every score in the distribution. The mode (which is only the most common score) and the median (which deals only with the middle score or scores) are not so affected. This quality is both an advantage and a disadvantage. On one hand, the mean utilizes all the available information—every score in the distribution affects the mean. On the other hand, when a distribution has a few extreme cases (very high or low scores), the mean may become very misleading as a measure of centrality. To illustrate, consider the following set of five scores: 15, 20, 25, 30, 35. Both the mean and median of this distribution are 25. ($\bar{X} = 125/5 = 25$. Md = score of third case = 25.) What will happen if we change the last score from 35 to 3500? Note that nothing happens to the median. It remains at exactly 25. The median is based only on the score of the middle case and is not affected by changes in the scores of other cases in the distribution. The mean, since it takes all scores into account, is very much affected. It becomes 3590/5, or 718. Clearly, the mean is disproportionately affected by the one extreme score in the data set. In this case, which of the two measures presents the more accurate or truer measure of central tendency for this distribution?

What general point can we derive from these considerations? Relative to the median, the mean is always pulled in the direction of extreme scores. The two measures of central tendency will have the same value when and only when a distribution is symmetrical. When a distribution has some extremely high scores (a positive **skew**), the mean will always have a greater numerical value than the median. If the distribution has some very low scores (a negative skew), the mean will be lower in value than the median. Figures 3.1 to 3.3 depict three different frequency polygons that demonstrate these relationships.

These relationships between medians and means also have a practical value. For one thing, a quick comparison of the median and mean will always tell you if a distribution is skewed and the direction of the skew. Second, they provide a simple and effective way to "lie" with statistics. For example, if you want to minimize the average score of a positively skewed distribution, report the median. Income data usually have a positive skew (there are only a few very wealthy people). If you want to impress someone with the general affluence of a mixed-income community, report the mean. If you want a lower figure, report the median. For the good and honest researcher, the selection of a measure of central tendency for a badly skewed distribution will hinge on what he or she wishes to show and, in most cases, either both statistics or the median alone should be reported.

FIGURE 3.1 A POSITIVELY SKEWED DISTRIBUTION (The mean is greater in value than the median.)

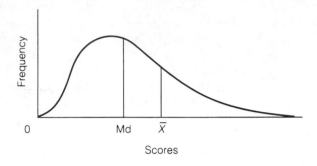

FIGURE 3.2 A NEGATIVELY SKEWED DISTRIBUTION (The mean is less than the median.)

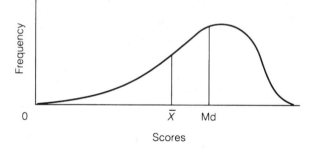

FIGURE 3.3 AN UNSKEWED, SYMMETRICAL DISTRIBUTION (The mean and median are equal.)

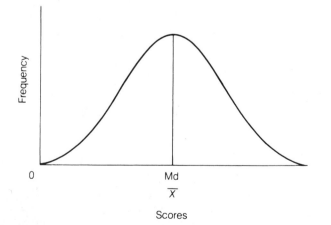

APPLICATION 3.1

Ten students have been asked how many hours they spent in the college library during the past week. What is the average "library time" for these students? The hours are reported in the following list, and we will find the mode, the median, and the mean for these data.

X_i
0
2
5
5
7
10
14
14
20
30
107

By scanning the scores, we can see that two scores, 5 and 14, occurred twice, and no other score occurred more than once. This distribution has two modes, 5 and 14.

Because the number of cases is even, the median will be the average of the two middle cases

after all cases have been ranked in order. With 10 cases, the first middle case will be the ($N/2$), or (10/2), or fifth case. The second middle case is the ($N/2$) + 1, or (10/2) + 1, or sixth case. The median will be the score halfway between the scores of the fifth and sixth cases. Counting down from the top, we find that the score of the fifth case is 7 and the score of the sixth case is 10. The median for these data is (7 + 10)/2, or (17/2), or 8.5. The median, the score that divides this distribution in half, is 8.5.

The mean is found by first adding all the scores and then dividing by the number of scores. The sum of the scores is 107, so the mean is

$$\bar{X} = \frac{\Sigma X_i}{N} = \frac{107}{10} = 10.7$$

These 10 students spent an average of 10.7 hours in the library during the week in question.

Note that the mean is a higher value than the median. This indicates a positive skew in the distribution (a few extremely high scores). By inspection we can see that the positive skew is caused by the two students who spent many more hours (20 hours and 30 hours) in the library than the other eight students.

3.7 COMPUTING MEASURES OF CENTRAL TENDENCY FOR GROUPED DATA*

In this section, we'll consider the basic techniques for computing means and medians for data presented in frequency distribution form. These techniques are useful because they can result in a significant savings in time and energy when you are working with a large data set and do not have access to a computer or calculator. Also, there are occasions when the data of interest have been organized into a frequency distribution and you do not have access to the data in ungrouped form. Under this latter condition, which occurs most often when the data are presented in the reports of other researchers, there simply is no other way to calculate means or medians.

---

*This section is optional.

TABLE 3.2   TOTAL DAYS TRUANT FOR ALL HIGH-SCHOOL STUDENTS WHO WERE TRUANT AT LEAST ONCE DURING THE PAST SCHOOL YEAR

Days	Frequency
1–5	25
6–10	30
11–15	30
16–20	20
21–25	10
26–30	5
31–35	4
	$N = 124$

Nevertheless, a small price is attached to using these techniques. Specifically, certain assumptions must be made about the way the scores are distributed within the intervals of the frequency distribution. These assumptions will not exactly reflect the way in which the scores are actually distributed and, technically, the statistics we calculate based on them should be regarded as approximations to the true mean or median.

To illustrate, consider the following situation: as a new employee of a Youth Service Bureau, you are faced with the task of preparing a report on the truancy levels in local high schools. The report is due in two weeks—leaving you no time to gather data. Fortunately, you stumble across some truancy data left behind by the person you replaced, but the data are in the form of a frequency distribution (Table 3.2), and the ungrouped data are not available. What will you do?

Computing the Mean for Grouped Data.   To compute a mean for any distribution, we must find two values: the summation of the scores and the number of scores. For the distribution in Table 3.2, we know only the latter ($N = 124$). Because the data have been grouped, we do not know the exact distribution of the original scores. We know from the table that 25 students were truant five days or fewer, but we do not know exactly how many days. For all we know, these 25 students could have been truant for only one day, or three or five days, or any combination of days between one and five.

We can bypass this dilemma by assuming that all the scores in each interval are located at the midpoint of the interval. Then, by multiplying each midpoint by the number of cases in the interval, we can obtain an approximation of the total original scores in the interval. To do this, add two columns to the original frequency distribution, one for the midpoints and one for the midpoints times the frequencies. This is done for the truancy data in Table 3.3.

The summation of the last column can be labeled $\Sigma(fm)$ and will be

TABLE 3.3   COMPUTING A MEAN FOR GROUPED DATA

Days	Frequency (f)	Midpoint (m)	Frequency × Midpoint (fm)
1–5	25	3	75
6–10	30	8	240
11–15	30	13	390
16–20	20	18	360
21–25	10	23	230
26–30	5	28	140
31–35	4	33	132
	N = 124		1567

approximately the same value as the summation of the original scores or $\Sigma(X_i)$. So, we have*

$$\bar{X} = \frac{\Sigma(X_i)}{N} \approx \frac{\Sigma(fm)}{N}$$

The final step in computing the mean for grouped data is to divide the total of the last column, $\Sigma(fm)$, by the number of cases:

$$\bar{X} = \frac{\Sigma(fm)}{N} = \frac{1567}{124} = 12.64$$

Thus, these 124 students were truant an average of 12.64 days over the year. To put it another way, the students who were truant at least once missed an average of about two and a half weeks of school.

Computing the Median for Grouped Data.   To locate the median, we must first find the middle case of the distribution. With 124 cases in the truancy sample, we will identify the middle case as the $(N/2)$, or 62nd case.[†] Our problem is first to locate this case and then find the score associated with it. To do so, we make the assumption that the cases are evenly spaced throughout each interval in the frequency distribution. We then locate the interval that contains the middle case and, based on the assumption of equal spacing, find the score associated with that case.

To locate the interval that contains the median, it is helpful to add a cumulative frequency column to the original frequency distribution as in Table 3.4. Looking at the cumulative frequency column, we see that 55

---

*The symbol $\approx$ means "approximately equal to."
[†]Technically, with an even N, the median would be the average score of the two middle cases—the 62nd and 63rd. Designating only the $(N/2)$ case as the middle case will, however, result in a significant simplification of the computations without much loss in accuracy.

TABLE 3.4 COMPUTING A MEDIAN FOR GROUPED DATA

Days	Frequency (f)	Cumulative Frequency (cf)
1–5	25	25
6–10	30	55
11–15	30	85
16–20	20	105
21–25	10	115
26–30	5	120
31–35	4	124
	N = 124	

cases are accumulated below the upper limit of the 6–10 interval and that 85 cases have been accumulated below the upper limit of the 11–15 interval. We know now that the median is some value between 10.5 (the lower real limit of this interval) and 15.5 (the upper real limit), but we do not know the exact value.

To resolve this dilemma, assume that all 30 of the cases in this interval are evenly spaced throughout the interval, with case 56 located at the lower real limit (10.5) and case 85 at the upper real limit (15.5). Case 62 (our middle case) is the seventh case of the 30 in the interval. If the scores are evenly spaced, then the 62nd case will be 7/30 of the distance from 10.5 to 15.5.

Now, 7/30 of 5 (the interval size) is 1.17, so we can approximate the median at 10.5 + 1.17, or 11.67. The formula below summarizes these steps.

FORMULA 3.2

$$Md = \text{real lower limit} + \left(\frac{N(.50) - cf\ \text{below}}{f} \right) i$$

where cf below = cumulative frequency below the interval containing the median
f = number of cases in the interval containing the median
i = interval width

Applying this formula to our example, we would have

$$Md = 10.5 + \left(\frac{124(.50) - 55}{30} \right) 5$$

$$Md = 10.5 + \left(\frac{62 - 55}{30} \right) 5$$

$$Md = 10.5 + \left(\frac{7}{30} \right) 5$$

$$Md = 10.5 + 1.17$$

$$Md = 11.67$$

Thus, half of this sample was truant fewer than 11.67 days and half more than 11.67 days.

Since the value of the median (11.67) is almost a full day lower than the value of the mean (12.6), we are justified in concluding that this distribution has a positive skew or a few very high scores. This skew is reflected in the frequency distribution itself, with most cases in the lower intervals. For purposes of description, the median may be the preferred measure of central tendency for these data, since the mean is affected by the relatively few very high scores.

Let us restate the procedures above as a series of steps. Assuming that all cases are evenly spaced through the intervals, the median for grouped data is found by these steps:

1. Find the middle case, given by $N(.50)$.
2. Find the interval that contains the middle case. A cumulative frequency column will be quite helpful for this.
3. Find the number of cases you will have to go into the interval to locate the middle case. This is given by $N(.50)$ minus the cumulative frequency below (*cf* below) the interval that contains the median.
4. Divide the number you found in step 3 by the number of cases in the interval (f).
5. Multiply the number you found in step 4 by the interval size (i).
6. Add the number you found in step 5 to the lower real limit of the interval that contains the median. The result is the median.

These same techniques can be used to locate any percentile, decile, or quartile in a distribution of scores. Simply change the proportional value in the "$N(.50)$" expression in Formula 3.2 to correspond to the location you are seeking and solve the formula. To illustrate, suppose we wanted to find the first quartile (or 25th percentile) for the data displayed in Table 3.4. Making the appropriate change, Formula 3.2 would become:

$$\text{First quartile} = \text{real lower limit} + \left(\frac{N(.25) - cf \text{ below}}{f} \right) i$$

The case marking the first quartile is given by $N(.25)$, or $(124)(.25) = 31$. By inspection we see that the 31st case is in the interval 6–10. Substituting the proper values, we would have:

$$\text{First quartile} = 5.5 + \left(\frac{31 - 25}{30} \right) 5$$

$$= 5.5 + (6/30)5$$

$$= 5.5 + (.20)5$$

$$= 5.5 + 1$$

$$\text{First quartile} = 6.5$$

Twenty-five percent of the sample had scores lower than 6.5, and 75% of the sample had scores higher than this value.

Admittedly, this procedure is rather formidable the first time it is attempted. With a little practice, however, it will become routine. Of course, if you do not have access to the ungrouped data, there simply is no other way to find the median.

3.8 CHOOSING A MEASURE OF CENTRAL TENDENCY

The selection of a measure of central tendency should, in general, be based on level-of-measurement considerations and on an evaluation of what each of the three statistics shows. Remember that the mode, median, and mean are three different statistics that will be the same value only under certain, specific conditions (that is, for symmetrical distributions with one mode). Each of the three has its own message to report and, in many circumstances, you might want to report all three. When choosing a single measure of central tendency, the following guidelines may be helpful:

Use the mode when

1. Variables are measured at the nominal level.
2. You want a quick and easy measure for ordinal and interval-ratio variables.
3. You want to report the most common score.

Use the median when

1. Variables are measured at the ordinal level.
2. Variables measured at the interval-ratio level have highly skewed distributions.
3. You want to report the central score. The median always lies at the exact center of a distribution.

Use the mean when

1. Variables are measured at the interval-ratio level (except for highly skewed distributions).
2. You want to report the typical score. The mean is "the fulcrum that exactly balances all of the scores."
3. You anticipate additional statistical analysis.

SUMMARY

1. The three measures of central tendency presented in this chapter share a common purpose. Each reports some information about the most typical or representative value in a distribution. Appropriate use of these statistics permits the researcher to report important information about an entire distribution of scores in a single, easily understood number.

2. The mode reports the most common score and is used most appropriately with nominally measured variables.

3. The median (Md) reports the score that is the exact center of the distribution. It is most appropriately used with variables measured at the ordinal level and with variables measured at the interval-ratio level when the distribution is skewed.

4. The mean ($\bar{X}$), the most frequently used of the three measures, reports the most typical score. It is used most appropriately with variables measured at the interval ratio level (except when the distribution is highly skewed).

5. The mean has a number of mathematical characteristics that are significant for statisticians. First, it is generally more stable than the mode or median. Second, it is the point in a distribution of scores around which all other scores cancel out. Third, the mean is the point of minimized variation. Last, as distinct from the mode or median, the mean is affected by every score in the distribution and is therefore pulled in the direction of extreme scores.

SUMMARY OF FORMULAS

Mean $\quad\quad\quad$ 3.1 $\bar{X} = \sum\dfrac{(X_i)}{N}$

Median
(grouped data) $\quad$ 3.2 Md = real lower limit

$$+ \left(\frac{N(.50) - cf \text{ below}}{f} \right) i$$

GLOSSARY

Deciles. The points that divide a distribution of scores into 10ths.

Mean. The arithmetic average of the scores. $\bar{X}$ represents the mean of a sample, and μ, the mean of a population.

Measures of central tendency. Statistics that summarize a distribution of scores by reporting the most typical or representative value of the distribution.

Median (Md). The point in a distribution of scores above and below which exactly half of the cases fall.

Mode. The most common value in a distribution or the largest category of a variable.

Percentile. A point below which a specific percentage of the cases fall.

Quartiles. The points that divide a distribution into quarters.

Σ (uppercase Greek letter sigma). "The summation of."

Skew. The extent to which a distribution of scores has a few scores that are extremely high (positive skew) or extremely low (negative skew).

X_i ("X sub i"). Any score in a distribution.

PROBLEMS

An asterisk indicates an optional problem.

3.1 Compute the mode, median, and mean from the ungrouped data presented in problems 2.3 and 2.5. (Remember to order the data from high to low to find the median.)

    a. Compare the mean and median for each data set. Can you tell if the distribution is skewed? In what direction?

    b. For each distribution, write a sentence or two reporting the central tendency as you would in a formal research report.

    *c. For each distribution, find the third quartile, the fourth decile, and the 23rd percentile.

    *d. For each distribution, compute the mean and the median from the frequency distribution you constructed as part of the original exercise. Compare the grouped measures to the ungrouped (true) measures. How accurate are the grouped techniques for approximating measures of central tendency?

3.2 Compute the median and mean homicide rates for each of the years presented in problem 2.6.

    a. For each year, compare the mean and the median. Are these distributions skewed? How?

    b. Compare the statistics for 1975 with 1991. Describe what happened to the homicide rates between these two years.

3.3 For problem 2.7, compute the mean and median for both sets of scores. Describe what happened to the sample between release and six months later.

3.4 SOC A variety of information has been gathered from a sample of college students. For each set of variables below, compute the indicated measure of central tendency.

a. Find the modal region of birth for each class.

Freshmen

Region of Birth	Frequency
North	7
South	10
Midwest	15
West	2
Foreign-born	1
$N =$	35

Sophomores

Region of Birth	Frequency
North	8
South	9
Midwest	12
West	5
Foreign-born	2
$N =$	36

Juniors

Region of Birth	Frequency
North	6
South	7
Midwest	9
West	6
Foreign-born	3
$N =$	31

Seniors

Region of Birth	Frequency
North	7
South	7
Midwest	8
West	6
Foreign-born	5
$N =$	33

b. About one-third of the members of each class were asked to indicate their position on the issue of legalizing marijuana by placing themselves on a scale that ranged from a score of 7 (strongly pro-legalization) to a score of 1 (strongly anti-legalization). Find the modal and median scores for each class. (*Note:* Scores must be ordered from high to low before middle cases can be located.)

Freshmen	Sophomores	Juniors	Seniors
3	7	7	7
2	4	7	6
3	3	6	7
7	4	3	5
5	7	4	1
6	5	7	5
1	1	3	6
4	2	2	7
4	4	1	7
3	2	2	
	6	5	

c. The members of the sophomore and the senior classes were asked how much money they spend for food and drink each week. Find the modal, median, and mean for each class. (Costs have been rounded to the nearest dollar.) Briefly explain and interpret each statistic.

Sophomores				Seniors			
23	29	35	82	55	52	50	80
37	52	38	42	52	47	30	39
55	82	92	51	42	50	28	63
68	75	72	82	35	38	45	55
92	90	88	75	75	39	68	54
83	88	74	75	70	68	62	31

3.5 PA The data below represent the percentage of all workers in each city who use public transportation to commute to work.

City	Workers Using Public Transportation (%)
New York	61.8
Chicago	36.2
Los Angeles	9.3
Philadelphia	37.1
Detroit	18.4
Houston	7.8
Washington, D.C.	37.8
Dallas	10.6

City	Workers Using Public Transportation (%)
San Francisco	35.7
Boston	39.3
Cleveland	22.1
Baltimore	27.0
St. Louis	21.3
Atlanta	21.3
San Diego	5.5
Milwaukee	19.2
Indianapolis	7.9
Seattle	14.9
Pittsburgh	29.5
Denver	8.2
Kansas City	10.5
Minneapolis	19.0

Source: United States Bureau of the Census, *Statistical Abstracts of the United States: 1977* (98th edition). Washington, D.C., 1977.

a. Calculate the mean and median of this distribution.

b. Compare the mean and median. Which is the higher value? Why?

c. If you removed New York from this distribution and recalculated, what would happen to the mean? to the median? Why?

d. Report the mean and median as you would in a formal research report.

3.6 PS The data below represent the percentage of eligible voters who turned out in each of 50 precincts for a recent election.

20	42	57	47	40
40	17	49	72	80
21	65	56	18	25
45	30	36	60	24
62	55	60	55	50
25	60	46	43	56
59	57	45	67	25
55	56	70	56	46
89	87	58	80	45
53	54	42	47	30

a. Calculate the mean and median of this distribution.

b. Compare the mean and median. Which is the higher value? Why?

*c. Group these data into a frequency distribution and calculate the mean and median using the grouped method. Compare these measures with those you calculated in 3.6a.

d. Report the mean and median as you would in a formal research report.

3.7 SOC The data below represent the percentage of the population living in large cities for 48 nations in 1990.

12.7	3.3	27.5	21.3	5.8
54.3	43.1	14.4	51.5	10.3
57.9	0	0	10.9	25.7
37.6	9.6	34.2	27.5	28.9
28.5	0	12.1	12.4	40.2
11.6	31.5	17.6	3.6	21.7
25.6	10.6	9.7	6.0	5.5
14.0	24.3	2.7	8.2	35.6
5.4	6.4	20.1	22.0	
16.2	21.9	34.0	9.0	

a. Calculate the mean and median of this distribution.

b. Compare the mean and median. Which is the higher value? Why?

c. Report the mean and median as you would in a formal research report.

3.8 SOC A sample of 25 freshmen at a major university completed a survey that measured their degree of racial prejudice (the higher the score, the greater the prejudice).

a. Compute the median and mean scores for these data.

10	43	30	30	45
40	12	40	42	35
45	25	10	33	50
42	32	38	11	47
22	26	37	38	10

b. These same students completed the same survey during their senior year. Compute the median and mean for this second set of scores and compare them to the earlier set. What happened?

10	40	35	27	50
35	10	50	40	30
40	10	10	37	10
40	15	30	20	43
23	25	30	40	10

3.9 SW As the head of a social services agency, you believe that your staff of 20 social workers is very much overworked compared to 10 years ago. The case loads for each worker are reported below for each of the two years in question. Has the average case load increased?

1983		1993	
52	55	42	82
50	49	75	50
57	50	69	52
49	52	65	50
45	59	58	55
65	60	64	65
60	65	69	60
55	68	60	60
42	60	50	60
50	42	60	60

3.10 An automobile manufacturer claims that a new economy-model car will get an average of 37 miles per gallon in highway driving. Being something of a skeptic as well as the proud owner of the model car in question, you decide to conduct your own road test. Below are the results of your 23 road tests. Compute the mode, mean, and median for these data. Do your results tend to confirm or refute the manufacturer's claim? Why?

42	37	40	38
39	42	37	36
37	39	38	40
35	38	37	21
38	40	40	20
27	22	22	

3.11 Professional athletes are threatening to strike because they claim that they are underpaid. The team owners have released a statement that says, in part, "the average salary for players was $1.2 million last year." The players counter by issuing their own statement that says, in part, "the average player earned only $753,000 last year." Is either side necessarily lying? If you were a sports reporter and had just read Chapter 3 of this text, what questions would you ask about these statistics?

3.12 SOC For 15 respondents, data have been gathered on four variables (table below). Find and report the appropriate measure of central tendency for each of the four variables.

Data for Problem 3.12

Respondent	Marital Status	Racial or Ethnic Group	Age	Attitude on Abortion Scale*
A	Single	White	18	10
B	Single	Spanish-speaking	20	9
C	Widowed	White	21	8
D	Married	White	30	10

Respondent	Marital Status	Racial or Ethnic Group	Age	Attitude on Abortion Scale*
E	Married	Spanish-speaking	25	7
F	Married	White	26	7
G	Divorced	Black	19	9
H	Widowed	White	29	6
I	Divorced	White	31	10
J	Married	Black	55	5
K	Widowed	Asian American	32	4
L	Married	American Indian	28	3
M	Divorced	White	23	2
N	Married	White	24	1
O	Divorced	Black	32	9

*This scale is constructed so that a high score indicates strong opposition to abortion under any circumstances.

3.13 Find the appropriate measure of central tendency for each of the four variables displayed in Table 2.4. Report each statistic as you would in a formal research report.

3.14 SOC Compute the median and mean for the data presented in problem 2.12. Is there a positive or negative skew in the distribution of these data?

3.15 SW For the test scores presented in problem 2.13, compute a median and mean for both the pretest and posttest. Interpret these statistics.

3.16 PS You have been observing the local Democratic Party in a large city and have compiled some information about a small sample of party regulars. For each variable included in the chart below, find the appropriate measure of central tendency.

Respondent	Sex	Social Class	Number of Years in the Party
A	M	High	32
B	M	Medium	17
C	M	Low	32
D	M	Low	50
E	M	Low	25
F	M	Medium	25
G	F	High	12
H	F	High	10
I	F	Medium	21
J	F	Medium	33

Respondent	Sex	Social Class	Number of Years in the Party
K	M	Low	37
L	F	Low	15
M	F	Low	31

3.17 SOC The administration is considering a total ban on student automobiles. You have conducted a poll on this issue of fellow students and the neighbors who live around the campus and have calculated scores for your respondents. On the scale you used, a high score indicates strong opposition to the proposed ban. The scores are presented below for both groups. Calculate an appropriate measure of central tendency and compare the two groups.

Students		Neighbors	
10	11	0	7
10	9	1	6
10	8	0	0
10	11	1	3
9	8	7	4
10	11	11	0
9	7	0	0
5	1	1	10
5	2	10	9
0	10	10	0

3.18 SOC You have managed to compile the information below on each of the graduates voted "most likely to succeed" by the local high school for a 10-year period. For each variable, find the appropriate measure of central tendency.

Case	Present Income	Marital Status	Owns a BMW?	Years of Schooling Completed after HS
A	24,000	Single	No	8
B	48,000	Divorced	No	4
C	54,000	Married	Yes	4
D	45,000	Married	No	4
E	30,000	Single	No	4
F	35,000	Separated	Yes	8
G	30,000	Married	No	3
H	17,000	Married	No	1
I	33,000	Married	Yes	6
J	48,000	Single	Yes	4

3.19 For problem 2.15, find the most appropriate measure of central tendency for each of the three variables.

3.20 For problem 2.17, compute the mean and median for both variables. How do these statistics help to describe and summarize these distributions? What information do the mean and the median add?

3.21 SOC Below are four variables for 30 cases from the General Social Survey. For each variable, find the appropriate measure of central tendency. See Appendix G for the meanings of the numerical codes. Summarize this statistical information as you would in a research report.

Respondent	Age (Item 7)	Income (Item 12)	Region (Item 13)	Political Party Identification (Item 16)
1	20	9	2	1
2	32	12	8	1
3	31	14	1	2
4	34	17	5	6
5	34	21	5	3
6	31	5	8	4
7	35	16	4	1
8	42	15	3	5
9	48	10	4	2
10	27	21	2	1
11	41	12	5	6
12	42	16	4	1
13	29	14	9	2
14	28	15	7	5
15	47	18	6	1
16	69	2	5	1
17	44	19	4	1
18	21	4	9	5
19	33	20	5	1
20	56	12	4	1
21	73	15	2	4
22	31	14	2	3
23	53	13	8	4
24	78	9	8	3
25	47	21	3	5
26	88	9	3	4
27	43	19	5	2
28	24	16	3	5
29	24	11	8	6
30	60	12	2	5

**SPSS/PC+ PROCEDURES FOR MEASURES
OF CENTRAL TENDENCY AND PERCENTILES**

**DEMONSTRATION 3.1 Producing Measures of Central Tendency
for the Counseling Center Survey**

The only procedure in SPSS/PC+ that will produce the mode, the median, and the mean is FREQUENCIES. In Chapter 2 we used this procedure to produce frequency distributions. Here we will use an option on this command to calculate the three measures of central tendency.

Begin by retrieving the Counseling Center Survey file. Once the MAIN MENU is available, go to EDIT mode by pressing ALT-E. Press the F3 key and answer the prompts to retrieve the counseling center file. Place the cursor at the end of the file. If you have any FREQUENCIES commands at the end of the file from the previous session, you can erase them by using the DELETE key or the BACKSPACE key. As another option, SPSS/PC+ can also delete lines with the F4 key. Place the cursor on the line you wish to delete and press the F4 key. A short menu will appear at the bottom of the screen with four choices listed horizontally. With the right-arrow key, move the cursor to DELETE and then press ENTER. The line will disappear. If you made a mistake, you can restore a deleted line by pressing the UNDELETE key. If you wish to delete many lines at once, use the F7 key to mark the area you want to erase and then use the F8 key to delete the text. Consult the SPSS/PC+ Manual for further information.

The instructions below will generate all three measures of central tendency for three of the variables in the data set. This might seem like statistical overkill since MARITAL does not justify computation of a median or mean, and SATIS is only ordinal. However, it is sometimes easier for us humans to generate "too much" output. Tailoring the commands to produce only the most appropriate measure would actually require more preparation, more keypunching, and longer commands. So, we are making things easier for ourselves without wasting very much in the way of computer time, paper, and other resources.

To produce measures of central tendency, press ALT-E to return to the MAIN MENU and

1. From the main menu, select 'analyze data' and then select 'descriptive statistics'.
2. Select and paste FREQUENCIES and !/VARIABLES. Press ALT-T and use the typing window to chose MARITAL, SATIS, and AGE as your variables.
3. We produced frequency distributions in a previous session and we don't need to duplicate this output. To suppress the printing of the frequency

distributions, select and paste /FORMAT and /NOTABLE. Return to the FREQUENCIES menu.

4. Select and paste /STATISTICS and MEAN, MEDIAN, and MODE.

At this point, your FREQUENCIES command should read

```
FREQUENCIES /VARIABLES MARITAL SATIS AGE/FORMAT
NOTABLE /STATISTICS MEAN MEDIAN MODE.
```

Touch F9 to save and then move the cursor to the beginning of the file. Press F10 to run the file and the output will be

```
MARITAL       Marital Status
Mean             1.650    Median        1.500    Mode        1.000
Valid Cases        20     Missing Cases     0

SATIS         Satisfaction with Counseling Center Serv
Mean             2.700    Median        3.000    Mode        3.000
Valid Cases        20     Missing Cases     0

AGE
Mean            20.050    Median       19.000    Mode       19.000
Valid Cases        20     Missing Cases     0
```

Looking only at the most appropriate measures for each variable, the modal category for MARITAL is reported as '1.' As you may recall, single (coded as a '1') was the most common marital status in the sample and, thus, is the mode. The median satisfaction score is reported as a 3 and the mean age is 20.05. The median age (19) is lower than the mean, so there is a positive skew in the distribution. You can confirm this by referring to Table 2.10, which shows most respondents in the 18 to 20 age range but a few cases in their mid 20s.

Note that SPSS/PC+ did not hesitate to compute means and medians for the nominal and ordinal variables in the data set. This is a good time to remind you that computers try to do exactly what you tell them to do. SPSS/PC+ does not have the ability to distinguish between the numerical codes for MARITAL and the actual numbers that represent AGE—to the program, all numbers are the same. The program has no sense of judgment or ability to screen your commands to see if they are appropriate. This blind willingness to simply follow instructions makes it easy for you, the user, to request statistics that are completely meaningless. Computers don't care about meaning; they just crunch the numbers. If you tell SPSS/PC+ to do something that is completely senseless, it will try its best to carry out your commands.

In this case, it was more convenient for us to produce statistics indiscriminately and then ignore the ones that are nonsensical. This will not always be the case and the point of all this, of course, is to caution you to use this powerful tool wisely. As the manager of your local computer center will be quick to remind you, computer resources (not to mention paper) are not unlimited.

**DEMONSTRATION 3.2 Measure of Central Tendency
for the 1993 General Social Survey**

Use the GET command and then press F10 to make the 1993 GSS system file active. (See Demonstration 2.2 for a review of retrieving system files.) Follow the instructions above for using the FREQUENCIES command to produce summary statistics without tables for SEX, ABNOMORE, and AGE. Your command should look like this:

```
FREQUENCIES \VARIABLES SEX ABNOMORE AGE \FORMAT
NOTABLE \STATISTICS MEAN MEDIAN MODE.
```

Remember that the system file will not appear on the screen so, once the system file is active, this FREQUENCIES command is all you should have in the EDIT window. Press the F10 key and the program will produce the following results:

```
SEX  RESPONDENTS SEX
Mean           1.585      Median        2.000      Mode    2.000
Valid Cases     788       Missing Cases      0

ABNOMORE  MARRIED_WANTS NO MORE CHILDREN
Mean           1.544      Median        2.000      Mode    2.000
Valid Cases     500       Missing Cases    288

AGE  AGE OF RESPONDENT
Mean          46.177      Median       43.000      Mode   28.000
Valid Cases     787       Missing Cases      1
```

   The modal category on sex is reported as a score of 2, meaning that women outnumber men in this sample. The median and the modal score on ABNOMORE is 2, indicating that the sample tends to be opposed to abortion for the reason stated in the question (see Appendix G for the exact wording). According to the output, the variable has 500 "valid cases." Remember that not every item is given to every respondent—some of the 288 missing cases may not have answered the question, but most were never presented with this item.

The median of ABNOMORE is the score halfway between the 250th case (500/2) and the 251st case. This variable has only two values (see Appendix G), so the median of 2 means that the middle case has a score of 2 and that fewer than half of the cases have a score of 1.

Although ABNOMORE is an ordinal variable, the mean might actually be more meaningful than the median in this situation. The mean score is 1.544, a little more than halfway between the two possible scores of 1 and 2. This indicates, intuitively, that the sample is roughly split between the "pro" and "anti" positions. Although perhaps not fully justified by level-of-measurement criteria, the mean seems a more meaningful insight than the median. You can check out the actual scoring pattern associated with ABNOMORE by running the FREQUENCIES task again without the \FORMAT NOTABLE option.

The mean age is 46.177 and the median age is 43.000, indicating that the variable has a positive skew (a few older people). You can verify this by running the FREQUENCIES again with an appropriate graph for AGE.

DEMONSTRATION 3.3 Finding Percentiles, Deciles, and Quartiles

We can also use FREQUENCIES to find percentiles, deciles, and quartiles for any variable. These would be especially useful for continuous variables with broad ranges of scores. One of the few variables that fits this description in the 1993 GSS data file is AGE, and we will use this variable for our illustrations.

SPSS/PC+ provides two different subcommands that can produce these statistics. With the PERCENTILES subcommand, the program will report the score below which the user-designated percentage of the cases fall. With NTILES, the program divides the distribution into the specified number of categories and displays the score associated with each category.

First, let's find the scores that mark, say, the 10th, 30th, 45th, and 67th percentiles for AGE. With the GSS93 system file active, follow the instructions in Demonstration 2.2 for the FREQUENCIES command, erase the STATISTICS subcommand, and add the PERCENTILES subcommand. Use the typing window (press ALT-T) to indicate your choice of percentiles. Your final command should look like this:

```
FREQUENCIES /VARIABLES AGE /FORMAT NOTABLE /
PERCENTILES 10 30 45 67.
```

Your output from this command will be

```
AGE
Percentile      Value     Percentile      Value     Percentile       Value
   10.00       25.000       30.00        35.000        45.00        41.000
   67.00       52.000
Valid Cases       787     Missing Cases       1
```

These results indicate that the 10th percentile (the score below which 10% of the cases fall) is associated with a score of 24, the 30th percentile lies at an age of 35, 45% of the cases are younger than 41, and 67% (about two-thirds) of the cases are younger than 52 years.

To use the NTILES subcommand, use the typing window to indicate the number of categories into which you wish to divide the distribution. For example, if you wanted the program to calculate deciles, specify 10 categories. For quartiles, specify 4 categories. The following command will generate deciles for AGE:

```
FREQUENCIES /VARIABLES AGE /FORMAT NOTABLE /
NTILES 10.
```

The output from this command is

```
AGE
Percentile  Value  Percentile  Value  Percentile   Value
   10.00   25.000    20.00    30.000    30.00     35.000
   40.00   39.000    50.00    43.000    60.00     47.800
   70.00   54.000    80.00    64.000    90.00     72.000
Valid Cases    787  Missing Cases    1
```

These results show the scores that mark the deciles. Thus, 10% of the cases fall below 25, 20% below 30, and so forth. Note that 50% of the cases (the fifth decile) are younger than 43. This value is also the value of the median as well as the second quartile and the 50th percentile.

Exercises

3.1 Using Appendix F as a guide, create an SPSS/PC+ file from the homicide data presented in problem 2.6. The file should have three variables; an identification number for the states beginning with 01 for Maine and ending with 50 for Hawaii, the 1975 rates, and the 1991 rates. Use four columns for the homicide variables, including one column for the decimal point. (There are other ways to handle decimals. See the Manual.) Make certain that you enter the rates in the proper column. For example, enter "02.8" for Maine, not "2.8." This will help keep the columns aligned properly. Save the file and then run FREQUENCIES to get measures of central tendency. Write a sentence or two describing the changes in average rate between 1975 and 1991.

3.2 Using the GSS93 system file, write a FREQUENCIES command to find all three measures of central tendency for MARITAL, INCOME 91, RACE, REGION, POLVIEWS, CAPPUN, PRESTG80, TRAUMA5, TVHOURS, and five more variables of your own choosing. Select the most appropriate measure for each of these variables and write a sentence or two reporting and summarizing the measure.

3.3 Revise the command you wrote in exercise 3.2 to also find the quartiles and the deciles for each of the variables listed.

Measures of Dispersion

4

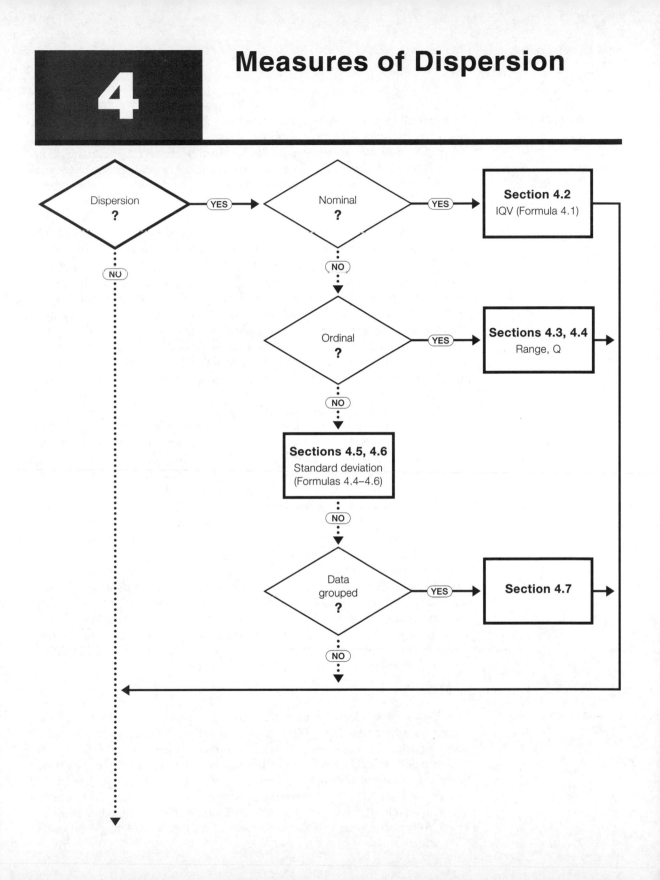

4.1 INTRODUCTION

By themselves, measures of central tendency are incomplete summarizers of data. To describe fully a distribution of scores, measures of central tendency must be paired with **measures of dispersion**. Whereas the mean, median, and mode are designed to locate the typical and/or central scores, measures of dispersion provide an indication of the amount of heterogeneity or variety within a distribution of scores.

The importance of the concept of **dispersion** might be easier to grasp if we consider a brief example. Suppose that the director of public safety for a city has been given the task of evaluating two ambulance services that have contracted with the city to provide emergency medical aid. As a part of the investigation, she has collected data on the response time of both services to calls for assistance. Data collected for the past year show that the mean response time is 7.4 minutes for Service A and 7.6 minutes for Service B. Since the average response times are essentially the same, there would seem to be no grounds for judging one service more or less efficient than the other. However, measures of dispersion might reveal substantial differences in the underlying distributions even when the measures of central tendency are equivalent. For example, consider Figure 4.1, which displays the distribution of response times for the two services.

Compare the shapes of these two figures. Note that the frequency polygon for Service B is much flatter than that for Service A. The scores for Service B are more spread out over the range of scores, while the scores for Service A are clustered or grouped around the mean. Comparatively speaking, Service B was much more variable in response time. Thus, even though both distributions have essentially the same mean, the distribution of scores for Service B is more heterogeneous. Had the dispersion of these distributions not been investigated, a possibly important difference in the performance of the two ambulance services might have gone unnoticed.

Although you can derive a general notion of what is meant by dispersion from Figure 4.1, the concept is not easily described in words alone. In this chapter we will introduce some of the more common measures of dispersion, each providing a precise, quantitative indication of heterogeneity. We will begin with the index of qualitative variation, mention two measures—the range and the interquartile range—quite briefly, and devote most of our attention to the standard deviation and the variance.

4.2 THE INDEX OF QUALITATIVE VARIATION (IQV)

The **index of qualitative variation (IQV)** is essentially the ratio of the amount of variation actually observed in a distribution of scores to the maximum variation that could exist in that distribution. The index varies from 0.00 (no variation) to 1.00 (maximum variation) and is used most commonly with variables measured at the nominal level. However, the IQV can be used with any variable when scores have been grouped into a frequency distribution.

To illustrate the logic of this statistic, assume that a researcher is interested in comparing the racial heterogeneity of the three small neighborhoods

FIGURE 4.1   RESPONSE TIME FOR TWO AMBULANCE SERVICES

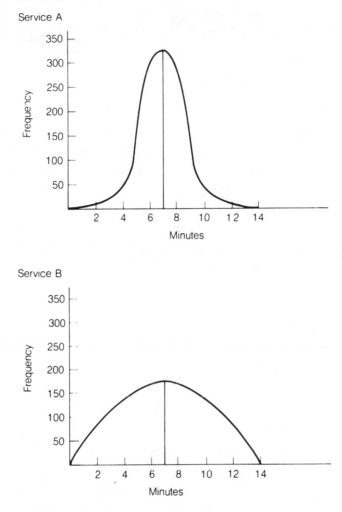

displayed in Table 4.1. By inspection, you see that neighborhood A is the least heterogeneous of the three. In fact, since all residents of A are white, there is no variation at all in this distribution. By way of contrast, neighborhood B is more heterogeneous than A, and neighborhood C is the most heterogeneous of the three. Let us see how the IQV substantiates these observations. The computational formula for the IQV is

FORMULA 4.1

$$IQV = \frac{k(N^2 - \Sigma f^2)}{N^2(k - 1)}$$

where k = the number of categories
N = the number of cases
Σf^2 = the sum of the squared frequencies

TABLE 4.1 RACIAL COMPOSITION OF THREE NEIGHBORHOODS

Neighborhood A		Neighborhood B		Neighborhood C	
Race	Frequency	Race	Frequency	Race	Frequency
White	90	White	60	White	30
Black	0	Black	20	Black	30
Other	0	Other	10	Other	30
	$N = 90$		$N = 90$		$N = 90$

To use this formula, the sum of the squared frequencies must first be computed. Do so by adding a column to the frequency distribution for the squared frequencies and then sum this column. This procedure is illustrated in Table 4.2.

The sum of the frequency column is, of course, N, and the sum of the squared frequency (Σf^2) is the total of the second column. Substituting these values into Formula 4.1 for neighborhood A, we would have an IQV of 0.00:

$$IQV = \frac{3 (8100 - 8100)}{8100(2)}$$

$$IQV = \frac{3(0)}{16,200}$$

$$IQV = \frac{0}{16,200}$$

$$IQV = 0.00$$

Since the values of k and N are the same for all three tables, the IQV for the other two neighborhoods can be found by changing the values for Σf^2. For neighborhood B,

$$IQV = \frac{3 (8100 - 4100)}{8100(2)}$$

$$IQV = \frac{12,000}{16,200}$$

$$IQV = 0.74$$

and, similarly, for neighborhood C,

$$IQV = \frac{3 (8100 - 2700)}{16,200}$$

$$IQV = \frac{16,200}{16,200}$$

$$IQV = 1.00$$

TABLE 4.2   FINDING THE SUM OF THE SQUARED FREQUENCIES

	Neighborhood A	
Race	Frequency (f)	Squared Frequency (f^2)
White	90	8100
Black	0	0
Other	0	0
	$N = 90$	8100

	Neighborhood B	
Race	Frequency (f)	Squared Frequency (f^2)
White	60	3600
Black	20	400
Other	10	100
	$N = 90$	4100

	Neighborhood C	
Race	Frequency (f)	Squared Frequency (f^2)
White	30	900
Black	30	900
Other	30	900
	$N = 90$	2700

Thus, the IQV, in a quantitative and precise way, substantiates our impressions. Neighborhood A exhibits no variation on the variable "race" (IQV = 0.00), neighborhood B has substantial variation (IQV = 0.74), and neighborhood C exhibits the maximum amount of variation possible (IQV = 1.00) for these data.

4.3 THE RANGE (R) AND INTERQUARTILE RANGE (Q)

The **range (R)** is defined as the distance between the highest and lowest scores in a distribution. It is quite easy to calculate (high score minus low score) and is perhaps most useful when used to gain a quick and general notion of variability while scanning many distributions.

Unfortunately, the range is often deceptive as a measure of dispersion. It is based on only two scores in the distribution (the highest and lowest scores). Since almost any sizable distribution will contain some atypically high and low scores, the range might be quite misleading. Also, R yields no information about the nature of the scores between these two extremes.

The **interquartile range (Q)** is a kind of range. It avoids some of the problems associated with R by taking into consideration only the middle 50% of the cases in a distribution. To find Q, arrange the scores from highest to lowest and then divide the distribution into quarters (as distinct from halves, as in locating the median). The first quartile (Q_1) is the point below which 25% of the cases fall and above which 75% of the cases fall. The second quartile (Q_2) divides the distribution into halves (thus, Q_2 is equal in value to the median). The third quartile (Q_3) is the point below which 75% of the cases fall and above which 25% of the cases fall. Thus, if line *LH* represents a distribution of scores, the quartiles are located as shown:

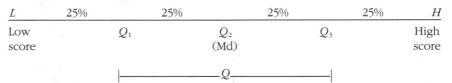

The interquartile range is defined as the distance from the third to the first quartile ($Q = Q_3 - Q_1$). Thus, Q essentially extracts the middle 50% of the cases and, like R, is based on only two scores. Q avoids the problem of being based on the most extreme scores; but, unfortunately, it has all the other disadvantages associated with R. Most importantly, Q also fails to yield any information about the nature of the scores other than the two upon which it is based.

4.4 COMPUTING THE RANGE AND INTERQUARTILE RANGE*

Table 4.3 presents per capita school expenditures for 20 states. What are the range and interquartile range of these data?

Note that the scores have already been ordered from high to low. This ordering makes the range easy to calculate and is necessary for finding the interquartile range. Of the 20 states listed in Table 4.3, Wyoming spent the most, per capita, on public education ($1440) and Arizona spent the least ($480). The range is therefore $1440 - 480$, or $960 (R = \$960)$.

To find Q, we must locate the first and third quartiles (Q_1 and Q_3). In a manner analogous to the technique for finding the median, we can define both of these points in terms of the scores associated with certain cases. Q_1 is determined by multiplying N by (.25). Since (20) (.25) is 5, Q_1 is the score associated with the fifth case, counting up from the lowest score. The fifth case is North Carolina, with a score of 565. So, $Q_1 = 565$.

The case that lies at the third quartile (Q_3) is given by multiplying N by (.75): (20) (.75) = 15th case. The 15th case is Pennsylvania, with a score of 682 ($Q_3 = 682$). Therefore:

$$Q = Q_3 - Q_1$$
$$Q = 682 - 565$$
$$Q = 117$$

---

*This section is optional.

TABLE 4.3 PER CAPITA EXPENDITURES ON PUBLIC EDUCATION, 1987

State	Expenditures per Capita
Wyoming	$1440
New Jersey	857
Texas	766
Michigan	717
Oregon	689
Pennsylvania	682
Maine	665
California	641
Virginia	634
Ohio	628
Illinois	585
Idaho	583
Nebraska	577
Louisiana	576
Florida	574
North Carolina	565
New Hampshire	529
Mississippi	496
Alabama	492
Arizona	480

Source: United States Bureau of the Census, *Statistical Abstracts of the United States: 1988* (108th edition). Washington, D.C., 1988.

In most situations, the locations of Q_1 and Q_3 will not be as obvious as they are when $N = 20$. For example, if N had been 157, then Q_1 would be (157)(1/4), or the score associated with the 39.25th case, and Q_3 would be (157)(3/4), or the score associated with the 117.75th case. Since there cannot be fractions of cases, these numbers present some problems. The easy solution to this difficulty is to take the score of the closest case to the numbers found above as marking the quartiles. Thus, Q_1 would be defined as the score of the 39th case and Q_3 as the score of the 118th case.

The more accurate solution would be to take the fractions of cases into account. For example, Q_1 could be defined as the score that is one-quarter of the distance between the scores of the 39th and 40th cases, and Q_3 could be defined as the score that is three-quarters of the distance between the scores of the 117th and 118th cases. (This procedure could be analogous to defining the median—which is also Q_2—as halfway between the two middle scores when N is even.) In most cases, the differences in the values of Q for these methods would be quite small.

4.5 THE STANDARD DEVIATION

The basic limitation of both Q and R is their failure to use all the scores in the distribution—they do not capitalize on all the available information. Also, neither Q nor R yields any information about the average or typical deviation of the scores. In that context, we might stipulate some characteristics of a good measure of dispersion. It should

1. Use all the scores in the distribution.
2. Describe the average or typical deviation of the scores.
3. Increase in value as the distribution of scores becomes more heterogeneous.

One way to develop a statistic to meet these criteria would be to find the distances between each score and some central point (that is, the mean) and sum these individual distances. The distances between the scores and the mean are called **deviations**, and their sum can be symbolized as $\Sigma(X_i - \bar{X})$. A measure of dispersion based on these deviations would meet all the criteria above. Most importantly, it would increase in value as the distribution of scores became more heterogeneous (that is, the greater the distances, the greater the sum of the deviations). This handy feature in a measure of dispersion would enable us to judge quickly the relative dispersion of different distributions by merely glancing at the summed deviations.

Unfortunately, there are two disadvantages to $\Sigma(X_i - \bar{X})$, by itself, as a measure of dispersion. First, regardless of relative variabilities, the measure would increase with sample size (that is, the larger the number of scores, the greater the value of the measure). Comparing the relative variability of distributions of different sizes would be very difficult. This problem can be solved by simply dividing the measure by N (sample size) and thus standardizing for samples of different sizes. The measure now is $\Sigma(X_i - \bar{X})/N$.

The second disadvantage is more serious. Algebraically, the deviations of scores around the mean will always total zero—thus making this eminently logical measure completely useless (see Section 3.6). To illustrate, consider a distribution of five scores: 10, 20, 30, 40, and 50. If the deviations of the scores from the mean were summed, we would have:

Scores (X_i)	Deviations $(X_i - \bar{X})$
10	$(10 - 30) = -20$
20	$(20 - 30) = -10$
30	$(30 - 30) = 0$
40	$(40 - 30) = 10$
50	$(50 - 30) = 20$
$\Sigma(X_i) = 150$	$\Sigma(X_i - \bar{X}) = 0$

$$\bar{X} = 150/5 = 30$$

Since the basic logic introduced above still has some appeal, we should search for a way to bypass this algebraic law by eliminating the minus signs of the deviations. We have two options. First, we can ignore the signs of the deviations. Second, we can square the deviations—since any negative number will become positive when multiplied by itself.

Our first solution yields the formula $\Sigma|X_i - \bar{X}|/N$, which directs us to sum the absolute values of the deviations (that is, disregard signs during the addition). This formula is, in fact, the arithmetic definition of a statistic known as the **average deviation (*AD*):**

FORMULA 4.2

$$AD = \frac{\Sigma|X_i - \bar{X}|}{N}$$

For the sample problem above, *AD* would be

$$AD = \frac{|20| + |10| + |0| + |10| + |20|}{5}$$

$$AD = \frac{60}{5} = 12$$

The average deviation indicates that, on the average, the scores lie 12 units from the mean.

The average deviation fulfills all of the criteria for a good measure. It uses all available information and is, by definition, a measure of the typical or average distance from the mean to the scores. Also, *AD* will increase in value as the distribution becomes more heterogeneous and the deviations from the mean increase. Unfortunately, however, absolute values are difficult to manipulate algebraically and, for this reason, the average deviation is not often used in the social sciences.

The second method for eliminating negative signs (squaring all deviations) yields a statistic known as the **variance**, which is symbolized as s^2 when referring to a sample and σ^2 (lowercase Greek letter sigma squared) when referring to a population. The variance is used primarily in inferential statistics, although it is a central concept in the design of some measures of association. For purposes of describing the dispersion of a distribution, a closely related statistic called the **standard deviation** (symbolized *s* for samples and σ for populations) is typically used, and this statistic will be our focus for the remainder of the chapter. The formulas for the sample variance and standard deviation are

FORMULA 4.3

$$s^2 = \frac{\Sigma(X_i - \bar{X})^2}{N}$$

FORMULA 4.4

$$s = \sqrt{\frac{\Sigma(X_i - \bar{X})^2}{N}}$$

Strictly speaking, Formulas 4.3 and 4.4 are for the variance and standard deviation of a population. Slightly different formulas, with $N - 1$ instead of N in the denominator, must be used when we are working with random samples rather than entire populations. The fact that there are two different sets of formulas for these measures of dispersion is important because many of the electronic calculators and statistical software packages to which you might have access (including the SPSS/PC+ software used in this text) use the "$N - 1$" formulas and produce results that are at least slightly different from ours.

At any rate, for the problem above, the standard deviation is

$$s = \sqrt{\frac{\Sigma(X_i - \bar{X})^2}{N}}$$

$$s = \sqrt{\frac{(-20)^2 + (-10)^2 + (0)^2 + (10)^2 + (20)^2}{5}}$$

$$s = \sqrt{\frac{400 + 100 + 100 + 400}{5}}$$

$$s = \sqrt{\frac{1,000}{5}}$$

$$s = \sqrt{200}$$

$$s = 14.14$$

To find the variance, square the standard deviation. For this problem, the variance is $s^2 = (14.14)^2 = 200$.

4.6 COMPUTING THE STANDARD DEVIATION

Formula 4.4 is the definitional formula for the standard deviation, but it is tedious to use in actual computations. For greater speed and ease of calculation, a number of computational formulas can be derived from Formula 4.4. Two of the most commonly used are Formulas 4.5 and 4.6:

FORMULA 4.5

$$s = \sqrt{\frac{\Sigma X_i^2}{N} - \bar{X}^2}$$

FORMULA 4.6

$$s = \frac{1}{N} \sqrt{N\Sigma X_i^2 - (\Sigma X_i)^2}$$

These formulas are algebraically equivalent to Formula 4.4 but are easier to use, since they eliminate computational steps. The formulas also introduce two new symbols, which we should consider before proceeding. The first new symbol is ΣX_i^2, which is contained in both formulas, and the second is $(\Sigma X_i)^2$, which is in Formula 4.6. Although these symbols might seem quite similar, they refer to two very different operations. The first—ΣX_i^2—is read as "the sum of the squared scores" and is found by first squaring and then summing the scores. On the other hand, $(\Sigma X_i)^2$ is read as "the sum of the scores, squared" and is found by first summing the scores and then squaring the sum. See Appendix H, section 3 for a more comprehensive treatment of the differences between these symbols.

To illustrate, we will work through the computation of the standard de-

TABLE 4.4   COMPUTING THE STANDARD DEVIATION BY THE DEFINITIONAL FORMULA (Formula 4.4)

Test Score (X_i)	$(X_i - \bar{X})$	$(X_i - \bar{X})^2$
57	−19	361
60	−16	256
65	−11	121
70	−6	36
78	2	4
80	4	16
82	6	36
85	9	81
90	14	196
93	17	289
$\Sigma(X_i) = 760$	$\Sigma(X_i - \bar{X}) = 0$	$\Sigma(X_i - \bar{X})^2 = 1396$

$$\bar{X} = 760/10 = 76.0$$

viation using the definitional formula and computational Formula 4.6. The data in Table 4.4 represent the scores of 10 students on a statistics test.

Using the figures calculated in Table 4.4 and Formula 4.4, we get

$$s = \sqrt{\frac{\Sigma(X_i - \bar{X})^2}{N}}$$

$$s = \sqrt{\frac{1396}{10}}$$

$$s = \sqrt{139.6}$$

$$s = 11.82$$

Formula 4.6 eliminates the need to subtract the mean from each score and, thus, is a significantly faster computational routine. Using data from Table 4.5 and Formula 4.6, we get

$$s = \frac{1}{N}\sqrt{N\Sigma X_i^2 - (\Sigma X_i)^2}$$

$$s = \frac{1}{10}\sqrt{(10)(59,156) - (760)^2}$$

$$s = \frac{1}{10}\sqrt{591,560 - 577,600}$$

$$s = \frac{1}{10}\sqrt{13,960}$$

$$s = \frac{1}{10}(118.15)$$

$$s = 11.82$$

TABLE 4.5 COMPUTING THE STANDARD DEVIATION BY THE
COMPUTATIONAL FORMULA (Formula 4.6)

Test Score (X_i)	X_i^2
57	3249
60	3600
65	4225
70	4900
78	6084
80	6400
82	6724
85	7225
90	8100
93	8649
$\Sigma X_i = 760$	$\Sigma X_i^2 = 59{,}156$

Note that the value of the standard deviation is exactly the same as that computed using the definitional formula. Formula 4.6 (like 4.5) is equivalent to but significantly faster than 4.4. Also note that the variance can be found by squaring the value of the standard deviation. In the example above, the variance would be $(11.82)^2$, or 139.71.

4.7 COMPUTING THE STANDARD DEVIATION FROM GROUPED DATA*

When data have been grouped, we face the same problem in computing the standard deviation that we faced with the mean and median: the exact value of the scores is no longer known. You can resolve this problem by using the midpoints of the intervals as approximations of the original scores.

To solve for the standard deviation by Formula 4.6, three quantities are required: N, ΣX_i, and ΣX_i^2. The number of cases is the total of the frequency (f) column. The sum of the scores (ΣX_i) is approximated by multiplying each midpoint (m) by the number of cases in the interval and then adding these products (Σfm). The sum of the squared scores (ΣX_i^2) is found by squaring the midpoints, multiplying by the number of cases in the interval, and then summing (Σfm^2). We thus have

$$\Sigma X_i \cong \Sigma fm$$

$$\Sigma X_i^2 \cong \Sigma fm^2$$

and Formula 4.6 can now be solved by making the appropriate substitutions.

To illustrate, Table 4.6 is a frequency distribution with several columns added for the purpose of organizing the required computations. The data are

---

*This section is optional.

APPLICATION 4.1

At a local preschool, 10 children were observed for 1 hour and the number of aggressive acts committed by each was recorded in the following list. What is the standard deviation of this distribution? We will use Formula 4.6 to compute the standard deviation

Number of Aggressive Acts (X_i)	X_i^2
0	0
3	9
5	25
2	4
7	49
10	100
0	0
8	64
2	4
0	0
$\Sigma X_i = 37$	$\Sigma X_i^2 = 255$

Substituting these values into Formula 4.6, we have

$$s = \frac{1}{N} \sqrt{N\Sigma X_i^2 - (\Sigma X_i)^2}$$

$$s = \frac{1}{10} \sqrt{(10)(255) - (37)^2}$$

$$s = \frac{1}{10} \sqrt{2550 - 1369}$$

$$s = \frac{1}{10} \sqrt{1181}$$

$$s = \left(\frac{1}{10}\right)(34.37)$$

$$s = 3.44$$

The standard deviation for these data is 3.44.

TABLE 4.6   COMPUTING THE STANDARD DEVIATION FROM GROUPED DATA

Correct Responses	Frequency (f)	Midpoints (m)	(fm)	m^2	fm^2
0–2	1	1	1	1	1
3–5	2	4	8	16	32
6–8	3	7	21	49	147
9–11	4	10	40	100	400
12–14	3	13	39	169	507
15–17	2	16	32	256	512
18–20	2	19	38	361	722
21–23	2	22	44	484	968
24–26	1	25	25	625	625
Total	N = 20		248		3914

READING STATISTICS 3: MEASURES OF CENTRAL TENDENCY AND DISPERSION

As was the case with frequency distributions, measures of central tendency and dispersion may not be presented in research reports in the professional literature. Given the large number of variables included in a typical research project and the space limitations in the journals and other media, there may not be room for the researcher to describe each variable fully. Furthermore, the great majority of research reports focus on relationships between variables rather than the distribution of single variables. In this sense, univariate descriptive statistics will be irrelevant to the main focus of the report.

This does not mean, of course, that univariate descriptive statistics are irrelevant to the research project; nor do I mean to imply that researchers do not calculate and interpret these statistics. Measures of central tendency and dispersion will be calculated and interpreted for virtually every variable, in virtually every research project. However, in the interest of efficiency, these statistics are less likely to be included in final research reports than the more analytical statistical techniques to be presented in the remainder of this text. Furthermore, some of these statistics (for example, the mean and standard deviation) serve a dual function. They are not only valuable descriptive statistics but also form the basis for many analytical techniques. Thus, they may be reported in the latter role if not in the former.

When included in research reports, measures of central tendency and dispersion will most often be presented in some summary form for all relevant variables—often in the form of a table. Means and standard deviations for many variables might, for example, be presented in the following table format.

Variable	$\bar{X}$	s	N
Age	33.2	1.3	1078
Number of children	2.3	.7	1078
Years married	7.8	1.5	1052
Income	20,078	982	978
.	.	.	.
.	.	.	.
.	.	.	.

These tables describe the overall characteristics of the sample succinctly and clearly. If you inspect the table carefully, you will have a good sense of the nature of the sample on the traits relevant to the project. Note that the number of cases varies from variable to variable in the preceding table. This is normal and is caused by missing data or incomplete information on some of the cases.

Statistics in the Professional Literature

The table below is from an article by J. Scott Long. He presents the results of his research into the causes of gender differences in productivity in the sciences. He begins the article by noting that "The average female scientist publishes fewer articles and receives fewer citations than the average male scientist. This finding persists across disciplines and through time." (1297) Professor Long assembled a sample of 556 males and 603 females who earned Ph.D.s in biochemistry between 1950 and 1967 and gathered information about their scholarly careers, their research productivity, and their personal lives. Inspect the table carefully and see if you can distill the essential meanings of the information presented.

Sex Differences in the Levels of Major Variables

	Sex	Mean	Std. Dev.
Age at baccalaureate	♀	22.089	2.339
	♂	22.728	2.220
Age at Ph.D.	♀	31.299	5.709
	♂	29.651	3.821
Articles by student —3-year count	♀	1.529	1.627
	♂	1.932	2.232
Baccalaureate selectivity	♀	2.958	1.819
	♂	3.066	1.694
Number of children at Ph.D.	♀	0.286	0.710
	♂	0.844	1.084
Interruption between baccalaureate and Ph.D.	♀	0.527	0.500
	♂	0.344	0.476

The table summarizes some of the important differences in means and standard deviation between the males and females. Professor Long uses means and standard deviations to summarize and describe the sample and also uses these two statistics as the basis for a test that helps to identify important or consequential differences by gender. Here's what the author has to say about the differences:

Males and females come from undergraduate institutions of nearly identical selectivity. While females are about three-fourths of a year younger than males at completion of the baccalaureate, they are more than a year and one-half older than males upon completion of the Ph.D. The older age of females upon completion of the Ph.D. is the result of females interrupting their education between the baccalaureate and the Ph.D. more frequently and longer than males. While males are more likely to be married and to have children before receiving the Ph.D., females are more likely to have had their educations interrupted, possibly as a result of family obligations.

The author goes on to conclude that the key factor that accounts for the productivity difference between the genders is that females have fewer and weaker relationships with faculty mentors because they more often interrupt their education for child bearing and child rearing. Professor Long concludes: "For females, having young children strongly decreases the odds of collaborating with the mentor, this in turn decreases predoctoral productivity." Curious and want to learn more? Complete bibliographic information is given below.

J. Scott Long: 1990. "The Origins of Sex Differences in Science." *Social Forces*, Vol. 68 (4): 1297–1315. Copyright © The University of North Carolina Press. Reprinted by permission.

APPLICATION 4.2

You have just won the state lottery and are, overnight, fabulously wealthy. Even after a brief visit with the IRS you still have enough money to live wherever you want and completely ignore such mundane considerations as work. One of your first decisions is to find the "nicest place to live" in all the world. Because you are somewhat eccentric, your only criterion for "nicest place" is climate. Specifically, you want to locate a city where the temperature is exactly 78°. After much research you find three cities where the average daily temperature is exactly 78°. Which of the three cities will you choose as your permanent residence?

The preceding information reports central tendency (the average temperature) and does not provide you with a basis for deciding among the three cities. What if you also discovered that the standard deviation and the range of the daily temperature were, respectively, .7° and 3° for City A; 10.3° and 30° in City B; and 25.8° and 103° in City C? Can you choose a permanent residence now? Which city would you choose? Why?

---

the number of correct responses on a 25-item arithmetic test by a sample of 20 students. The sum of the third column (Σfm) will be used in place of the sum of the scores (ΣX_i), and the sum of the last column (Σfm^2) will replace the sum of the squared scores (ΣX_i^2). Formula 4.6 can be restated using our approximations in place of the original symbols.

FORMULA 4.7
$$s = \frac{1}{N} \sqrt{N(\Sigma fm^2) - (\Sigma fm)^2}$$

Substituting the proper numbers, we would have

$$s = \frac{1}{20} \sqrt{(20)(3914) - (248)^2}$$

$$s = \frac{1}{20} \sqrt{78,280 - 61,504}$$

$$s = \frac{1}{20} \sqrt{16,776}$$

$$s = \frac{1}{20} (129.52)$$

$$s = 6.48$$

The mean of this distribution can also be calculated from Table 4.6 by dividing the sum of column 3 (Σfm) by N (see Section 3.7). The mean is 248/20, or 12.4. These 20 students answered an average of 12.4 items correctly, and the distribution of correct responses has a standard deviation of 6.48.

4.8 INTERPRETING THE STANDARD DEVIATION

When first encountered, the standard deviation may not have any obvious intuitive meaning. You might ask at this point, Once I've gone to the trouble of calculating the standard deviation, what do I have? This measure of dispersion can be rendered meaningful in three ways. The first and most important involves the normal curve, and we will defer this interpretation until the next chapter.

A second way of thinking about the standard deviation is as an index of variability. As we saw earlier, the standard deviation increases in value as the distribution becomes more variable. The reverse relationship also holds: the less the variability in the distribution, the lower the value of the standard deviation. The limiting case in this direction would be a distribution with no dispersion—a distribution where every case had exactly the same score. In such a case, the standard deviation would be 0. Thus, 0 is the lowest value possible for the standard deviation (although there is no upper limit).

A third way to get a feel for the meaning of the standard deviation is by comparing one distribution with another. We are given the following information for the distribution of family incomes in two cities:

City A	City B
$\bar{X} = 31{,}233$	$\bar{X} = 31{,}233$
$s = 1729$	$s = 5268$

You see immediately that City B is much more heterogeneous with respect to income than City A. That is, there are more differences among family incomes in City B than in City A. Although the standard deviation might not always have a direct or obvious interpretation when considered by itself, it is an extremely useful statistic for comparing distributions of scores as well as for measuring heterogeneity.

SUMMARY

1. Measures of dispersion provide us with techniques for summarizing information about the heterogeneity or variety in a distribution of scores. When combined with an appropriate measure of central tendency, these statistics convey a large volume of information in just a few numbers. Measures of central tendency locate the central points of the distribution, whereas measures of dispersion indicate the amount of diversity in the distribution.

2. The index of qualitative variation (IQV) can be computed for any variable that has been organized into a frequency distribution. It is the ratio of the amount of variation observed in the distribution to the maximum variation possible in the distribution. The IQV is most appropriate for variables measured at the nominal level.

3. The range (R) is the distance from the highest to the lowest score in the distribution. The interquartile range (Q) is the distance from the third to the first quartile (the "range" of the middle 50% of the scores). These two ranges can be used with variables measured at either the ordinal or interval-ratio level.

4. The standard deviation (s) is the most important measure of dispersion because of its central role in many more advanced statistical applications. The standard deviation has a minimum value of zero (indicating no variation in the distribution) and increases in value as the variability of the distribution increases. It is used most appropriately with variables measured at the interval-ratio level.

5. The variance (s^2) is a statistic closely related to the standard deviation and is used primarily in inferential statistics and in the design of some measures of association.

SUMMARY OF FORMULAS

Index of qualitative variation

4.1 $IQV = \dfrac{k(N^2 - \Sigma f^2)}{N^2 (k - 1)}$

Average deviation

4.2 $AD = \dfrac{\Sigma |X_i - \bar{X}|}{N}$

Variance (population)

4.3 $s^2 = \dfrac{\Sigma (X_i - \bar{X})^2}{N}$

Standard deviation (population)

4.4 $s = \sqrt{\dfrac{\Sigma (X_i - \bar{X})^2}{N}}$

Standard deviation (computational)

4.5 $s = \sqrt{\dfrac{\Sigma X_i}{N} - \bar{X}^2}$

Standard deviation (computational)

4.6 $s = \dfrac{1}{N} \sqrt{N\Sigma X_i^2 - (\Sigma X_i)^2}$

GLOSSARY

Average deviation. The average of the absolute deviations of the scores around the mean.

Deviations. The distance between the scores and the mean.

Dispersion. The amount of variety or heterogeneity in a distribution of scores.

Index of qualitative variation (IQV). A measure of dispersion for variables that have been organized into frequency distributions.

Interquartile range (Q). The distance from the third quartile to the first quartile.

Measures of dispersion. Statistics that indicate the amount of variety or heterogeneity in a distribution of scores.

Range (R). The highest score minus the lowest score.

Standard deviation. The square root of the squared deviations of the scores around the mean divided by N. The most important and useful descriptive measure of dispersion; s represents standard deviation of a sample; σ, the standard deviation of a population.

Variance. The squared deviations of the scores around the mean divided by N. A measure of dispersion used primarily in inferential statistics and also in correlation and regression techniques; s^2 represents the variance of a sample; σ^2, the variance of a population.

PROBLEMS

Asterisks indicate optional problems.

4.1   SOC   The marital status of residents of four apartment complexes is reported below. Compute the index of qualitative variation (IQV) for each neighborhood. Which is the most heterogeneous of the four? Which is the least?

Complex A Marital Status	Frequency	Complex B Marital Status	Frequency
Single	26	Single	10
Married	31	Married	12
Divorced	12	Divorced	8
Widowed	5	Widowed	7
	$N = 74$		$N = 37$

Complex C Marital Status	Frequency	Complex D Marital Status	Frequency
Single	20	Single	52
Married	30	Married	3
Divorced	2	Divorced	20
Widowed	1	Widowed	10
	$N = 53$		$N = 85$

4.2   Compute the range and standard deviation of the 10 scores reported below. Use both the definitional and computational formulas for computing s.

$$10, 12, 15, 20, 25, 30, 32, 35, 40, 50$$

4.3 For the 10 test scores below, compute the standard deviation using both the definitional and computational formulas.

$$77, 83, 69, 72, 85, 90, 95, 75, 55, 45$$

4.4 At St. Algebra College, the math department ran some special sections of the freshman math course using a variety of innovative teaching techniques. Students were randomly assigned to either the traditional sections or the experimental sections, and all students were given the same final exam. The results of the final are summarized below. What was the effect of the experimental course?

Traditional	Experimental
$\bar{X} = 77.8$	$\bar{X} = 76.8$
$s = 12.3$	$s = 6.2$
$N = 478$	$N = 465$

4.5 Compute the range, interquartile range, standard deviations, and variance for the ungrouped data presented in problems 2.3 and 2.5.

*4.6 Compute the standard deviation from the frequency distributions you constructed in problems 2.3 and 2.5.

4.7 Compute the standard deviation for each year's homicide rates as presented in problem 2.6. Considering also the mean and median you calculated in problem 3.2, write a few sentences of interpretation and summary for these data. What happened to homicide rates over the time period?

4.8 Compute the standard deviation for each data set presented in problem 2.7. Considering also the mean and median you calculated in problem 3.3, write a few sentences of interpretation and summary for these data.

4.9 PA Compute the range and standard deviation for the data presented in problem 3.5. If you removed New York City from this distribution and recalculated, what would happen to the value of the standard deviation? Why?

4.10 PS Compute the range and standard deviation for the data presented in problem 3.6.

4.11 SOC Compute the standard deviation for the data presented in problem 3.7.

4.12 SOC Compute the standard deviation for both sets of data presented in problem 3.8. Compare the standard deviation computed for freshmen with the standard deviation computed for seniors. What happened? Why? Does this change relate at all to what happened to the mean over the four-year period? How? What happened to the shapes of the underlying distributions?

4.13 SOC Compute the standard deviation for both data sets presented in problem 3.9. Compare the standard deviations for the two time periods along with the changes in the mean. What happened? Why?

4.14 SOC Below are listed the rates of abortion per 100,000 women for 20 states in 1973 and 1975. Describe what happened to these distributions over the two-year period. Did the average rate increase or decrease? What happened to the dispersion of this distribution?

State	1973	1975
Maine	3.5	9.5
Massachusetts	10.0	25.7
New York	53.5	40.7
Pennsylvania	12.1	18.5
Ohio	7.3	17.9
Michigan	18.7	20.3
Iowa	8.8	14.7
Nebraska	7.3	14.3
Virginia	7.8	18.0
South Carolina	3.8	10.3
Florida	15.8	30.5
Tennessee	4.2	19.2

State	1973	1975
Mississippi	0.2	0.6
Arkansas	2.9	6.3
Texas	6.8	19.1
Montana	3.1	9.9
Colorado	14.4	24.6
Arizona	6.9	15.8
California	30.8	33.6
Hawaii	26.3	31.6

Source: United States Bureau of the Census, *Statistical Abstracts of the United States: 1977* (98th edition). Washington, D.C., 1977.

4.15 SW One of your goals as the new chief administrator of a large social service bureau is to equalize work loads within the various divisions of the agency. You have gathered data on case loads per worker within each division. Which division comes closest to the ideal of an equalized work load? Which is farthest away?

A	B	C	D
50	60	60	75
51	59	61	80
55	58	58	74
60	55	59	70
68	56	59	69
59	61	60	82
60	62	61	85
57	63	60	83
50	60	59	65
55	59	58	60

4.16 CJ You're the governor of the state and must decide which of four metropolitan police departments will win the annual award for efficiency. The performance of each department is summarized in monthly arrest statistics as reported below. Which department will win the award? Why?

	Departments		
A	B	C	D
$\bar{X} = 601.30$	633.17	592.70	599.99
$s = 2.30$	27.32	40.17	60.23

4.17 Compute the Index of Qualitative Variation for the distributions presented in problems 2.10 and 2.11.

4.18 Compute the standard deviation for the pretest and posttest scores presented in problem 2.13. Considering the mean you calculated in problem 3.15, briefly describe how the sample changed from test to test. What does the standard deviation add to the information you already had?

4.19 Compute a standard deviation for the two sets of scores presented in problem 3.17. What does the standard deviation add to the information you already had?

4.20 CJ Per capita expenditures for police protection for 20 states are reported below for 1980 and 1985. Compute a mean and standard deviation for each year and describe the differences in expenditures for the five-year period.

1980		1985	
180	167	210	225
95	101	110	209
87	120	124	201
101	78	131	141
52	107	197	94
117	55	200	248
115	78	119	140
88	92	87	131
85	99	125	152
100	103	150	178

4.21 SOC Labor force participation rates for males and females in 20 states are reported below. Calculate a mean and a standard deviation for both groups and describe the differences.

Male		Female	
73.9	81.1	54.2	63.2
80.9	76.6	63.2	59.9
80.0	74.1	60.8	51.5
73.5	77.6	50.6	54.9
76.1	77.4	54.1	57.6
80.2	78.3	61.4	57.0
78.3	79.2	60.4	58.2

Male		Female	
74.9	73.1	49.1	54.4
80.7	76.0	60.8	52.8
75.0	79.0	55.0	58.9

Case	Number of Children (Item 6)	Educational Level (Item 8)	Degree of Political Liberalism (Item 19)	Frequency of Church Attendance (Item 29)
1	1	10	4	2
2	0	12	6	5
3	0	12	6	7
4	0	11	4	7
5	2	12	1	4
6	2	13	3	5
7	3	8	4	4
8	1	4	5	7
9	2	12	7	0
10	2	10	4	7
11	0	16	5	7
12	2	16	4	7
13	0	13	1	1
14	0	14	4	0
15	1	12	4	6
16	2	5	3	2
17	1	8	7	3
18	1	12	3	2
19	0	9	4	1
20	0	8	5	0
21	2	12	3	7
22	2	16	5	5
23	0	16	6	0
24	2	16	5	2
25	5	12	4	7

4.22 The occupational prestige scores of 50 respondents and their fathers were presented in problem 2.17. Find the range and standard deviation for each of these variables. Combining these measures of dispersions with the mean (problem 3.20) and the frequency distributions (problem 2.17), describe these two distributions as you would in a research report.

4.23 Compute the standard deviation for the age of the 30 respondents in problem 3.21. Taking into account the mean you computed in problem 3.21, write a sentence summarizing this variable.

4.24 SOC Below are the scores of 25 individuals on four variables taken from the 1993 General Social Survey. For each variable, compute the mean and the standard deviation. Write a sentence or two of summary and interpretation for each variable. See Appendix G for exact wording of questions and meaning of codes.

SPSS/PC+ PROCEDURES FOR MEASURES OF DISPERSION

DEMONSTRATION 4.1 Producing Measures of Dispersion for the Counseling Center Survey

The FREQUENCIES task, with which you are already familiar, can be used to produce all of the usual measures of dispersion, but let me introduce you to a new procedure called DESCRIPTIVES. This command produces univariate descriptive statistics without frequency distributions so it is appropriate for interval-ratio, continuous variables. In this demonstration we will produce most of the statistics covered in this chapter for AGE from the counseling center data set. Begin by retrieving the file (press F3 and answer the prompts). Delete any old FREQUENCIES commands from prior sessions and

1. Select analyze data from the main menu and then select descriptive statistics.
2. Select and paste DESCRIPTIVES and !/VARIABLES. Use the typing window (press ALT-T) to select AGE as your variable.
3. Select !/STATISTICS and then, from the STATISTICS menu, select 5, 6, 9, 10, and 11 for measures of dispersion. The window on the right will identify each statistic.

Your command should look like this:

```
DESCRIPTIVES /VARIABLES AGE/STATISTICS 5 6 9 10 11.
```

Remember to save the file (press F9) and then submit the file to SPSS/PC+ (place the cursor at the beginning of the file and press F10). Your output should look like this:

```
Number of Valid Observations (Listwise) =    20.00
Variable     Std Dev     Variance     Range     Minimum     Maximum     N Label
AGE             2.21         4.89      8.00          18          26     20
```

Please note that SPSS/PC+ computes both the standard deviation and variance by using $N - 1$ in the denominator rather than N as we have in this chapter. This is a correction for the fact that both statistics are biased estimators when used to infer population values. This difference in the denominator may produce very different values for s and s^2 for small samples like the counseling center survey. These differences will diminish as sample size increases. At any rate, you should certainly be aware of these differences if you use SPSS/PC+ to compute measures of dispersion for small samples. The index of qualitative variation (IQV) and the interquartile range (Q) are not available in SPSS/PC+ but the latter may be found by using the NTILES subcommand in the FREQUENCIES program.

DEMONSTRATION 4.2 Measure of Dispersion for the 1993 General Social Survey

Use the GET command to retrieve the 1993 GSS and then press F10 to make the file active. Follow the steps listed in Demonstration 4.1 to write a DESCRIPTIVES command for AGE. The command should look like this:

```
DESCRIPTIVES /VARIABLES AGE /STATISTICS 5 6 9 10 11.
```

The output (press F10) is

```
Number of Valid Observations (Listwise) =    787.00
Variable       Std Dev      Variance      Range     Minimum      Maximum       N
AGE              17.37        301.83       71.00       18.00        89.00      787
```

The standard deviation for the age of these cases is 17.37, with a range from 18 to 89. Remember that the comparable statistics for the counseling center data were a standard deviation of 2.21 and a range of 18 to 26. Which of the two samples is more diverse with respect to age?

DEMONSTRATION 4.3 Using the COMPUTE Command to Create an 'Attitude Toward Abortion' Scale

SPSS/PC+ provides a variety of ways to transform and manipulate variables. In the SPSS/PC+ exercises at the end of Chapter 2, I introduced the RECODE command as a way to change the values associated with a variable. Here, I will present the COMPUTE command as a way to create new variables and summary scales.

Let's begin by considering the two items that measure attitudes toward abortion, ABNORMORE and ABPOOR (items 39 and 40 in Appendix G). The items present two different situations under which an abortion might be desired and ask the respondent to react to each situation independently. Since the situations are distinct, even though both relate to abortion, each item should be analyzed in its own right.

However, suppose you wanted to create a summary scale that would indicate a person's *overall* feelings about abortion. One obvious way to do this would be simply to add the scores together. This would create a new variable with three possible scores: a score of 2 could be acquired only by answering YES (coded as '1') to both items. Such a score would indicate a consistent "pro-abortion" position. A score of 3 would be generated by answering YES to one item and NO to the other. This might be labeled a "moderate" position. The final possibility would be a score of 4, if the respondent answered NO to both items. This would be a consistent "anti-abortion" position.

This new variable could serve as a summary score for each respondent's overall position on abortion. Once created, the new variable could be analyzed, transformed, and manipulated exactly like a variable actually recorded in the system file.

To create a summary variable, use the COMPUTE command first to name the new variable and then to state the mathematical expression that would compute the scores. For our abortion scale, which we will call ABSCALE, the command would look like this:

```
COMPUTE ABSCALE = ABNOMORE + ABPOOR.
```

To create this command, select 'modify data or files' from the Main Menu and then select 'modify data values'. Now, select COMPUTE and a submenu will appear. Take a moment to look at the examples. Select '!target' and use the window to type the name of your new variable (that is, the variable you will compute). In this case, type ABSCALE. Next, select the '!=' symbol and then use the typing window to specify your computational instructions. In this case, type ABNOMORE + ABPOOR. When you press ENTER, the full COMPUTE command will appear in the bottom window.

This command instructs SPSS/PC+ to create a new variable called ABSCALE and to set that variable's values equal to the sum of the respondent's scores for ABNOMORE and ABPOOR. The mathematical statements to the right of the equal sign can be simple or complex and can use virtually any mathematical operation. The most common symbols in this command would be

> \+ for addition
> − for subtraction
> * for multiplication
> / for division
> ( ) to control the order of operations

As always, I urge you to consult the SPSS/PC+ Manual for further information.

Below are the frequency distributions for the two abortion items and the summary abortion scale. This output was generated by using the FREQUENCIES command and naming all three variables. Here are the command and some of the output:

```
FREQUENCIES /VARIABLES ABNOMORE ABPOOR ABSCALE.
```

ABNOMORE MARRIED−WANTS NO MORE CHILDREN

Value Label	Value	Frequency
YES	1	228
NO	2	272
		500

ABPOOR LOW INCOME−CANT AFFORD MORE CHILDREN

Value Label	Value	Frequency
YES	1	248
NO	2	255
		503

```
ABSCALE
Value Label                                    Value   Frequency
                                                2.00      204
                                                3.00       54
                                                4.00      228
                                                        -----
                                                          486
```

Take a few moments to look at these tables. First, note that only about 500 respondents answered the original two abortion items. Remember that no respondent is given the entire GSS and the vast majority of the 300 or so "missing cases" received a form of the GSS that did not include these two items. Now look at ABSCALE, the summary scale. Two hundred and four cases are "pro-abortion" (scored 1 on each item), twenty-four fewer than the number of "anti-abortion" cases (each of whom scored 2 on each item). Only 54 people were "inconsistent" in their responses, answering YES to one item and NO to the other.

Finally, note that fewer cases (only 486) are included in the summary scale than in either of the two original items. When SPSS/PC+ executes a COMPUTE statement, it automatically eliminates any cases that are missing scores on any of the constituent items. If these cases were not eliminated, a variety of errors and misclassifications could result. For example, if cases with missing scores were included, a person who scored a 2 ("anti-abortion") on ABNOMORE and then failed to respond to ABPOOR would have a total score of 2 on ABSCALE. Thus, this case would be treated as "pro-abortion" when, actually, the only information we have indicates that this respondent is "anti-abortion." To eliminate this kind of error, cases with missing scores on any of the constituent variables are deleted from calculations.

Once created, a computed variable can be used for any purpose and treated just like a variable whose scores are actually recorded in the system file. For example, if we wanted some summary statistics on all three abortion measures, we could use the DESCRIPTIVES command and treat the computed variable like the others:

```
DESCRIPTIVES /VARIABLES ABNOMORE ABPOOR ABSCALE/
STATISTICS 1 5 9.
```

The output from this command is

```
Number of Valid Observations (Listwise) =    486.00
Variable         Mean      Std Dev      Range      N Label
ABNOMORE         1.54         .50        1.00      500
ABPOOR           1.51         .50        1.00      503
ABSCALE          3.05         .94        2.00      486
```

Note that the mean of ABNOMORE and ABPOOR is about halfway between the two scores, indicating that the sample is approximately split on these issues. Likewise, the mean of ABSCALE is about halfway between the extremes. What can you say about the dispersion of these variables?

Exercises

4.1 Retrieve the homicide data file you created for Exercise 3.1 and run the DESCRIPTIVES program for all relevant statistics for both years. Write a few sentences describing the differences in the rates. Did homicide rates go up or down? Did the 50 states become more or less homogeneous on this variable?

4.2 Create an SPSS/PC+ file from the abortion data presented in Problem 4.14. The file should have three variables: an identification number for the states beginning with 01 for Maine and ending with 20 for Hawaii, the 1973 rates, and the 1975 rates. Use four columns for the abortion variables, including one column for the decimal point, as you did in Exercise 3.1. Save the file and then run DESCRIPTIVES to get all relevant univariate statistics. Write a sentence or two describing the changes in rates between 1973 and 1975.

4.3 Use the DESCRIPTIVES command to produce univariate descriptive statistics for the following variables in the 1993 GSS system file: PRESTG80, PAPRES80, and TVHOURS. Compare these statistics for PRESTG80 and PAPRES80. Is the sample higher or lower than their fathers' in occupational prestige? Are they more or less homogeneous?

4.4 Use the COMPUTE command to create summary scales for SATFAM and SATFRND (items 37 and 38 in Appendix G) and then for FEHELP and FEFAM (items 68 and 69). Use DESCRIPTIVES to get univariate descriptive statistics on each of the summary scales. *Hint:* Your first COMPUTE command will look like this:

```
COMPUTE SATSCALE = SATFAM + SATFRND).
```

5

The Normal Curve

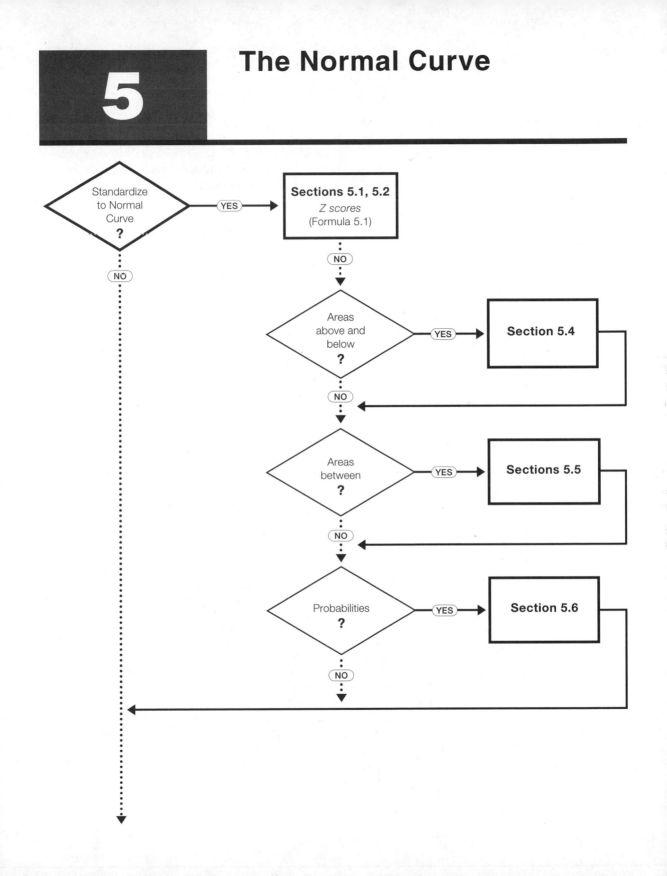

5.1 INTRODUCTION

The **normal curve** is a concept of great significance in statistics. As we shall see in Part II, it is central to the theory that underlies inferential statistics. Also, in combination with the standard deviation, the normal curve can be used to construct precise descriptive statements about empirical distributions. This chapter, then, will serve both to conclude (for the present) our treatment of statistics as descriptive devices and to lay important groundwork for Part II.

The normal curve is a theoretical model, a kind of frequency polygon that is perfectly symmetrical and smooth. It is bell shaped and unimodal, with its tails extending infinitely in both directions. Of course, no empirical distribution has a shape that perfectly matches this ideal model, but many are close enough to permit the assumption of normality. In turn, this assumption makes possible one of the most important uses of the normal curve—description of empirical distributions based on our knowledge of the theoretical normal curve.

The crucial point about the normal curve, you should remember, is that distances along the abscissa (horizontal axis) of the distribution, when measured in standard deviations from the mean, always encompass the same proportion of the total area under the curve. In other words, regardless of the precise shape of the normal curve and the nature of the underlying distribution, the distance from any given point to the mean (when measured in standard deviations) will cut off exactly the same proportion of the total area.

To illustrate, Figures 5.1 and 5.2 present two hypothetical distributions of IQ scores, one for a group of males and one for a group of females, both normally distributed (or nearly so) such that

Males	Females
$\bar{X} = 100$	$\bar{X} = 100$
$s = 20$	$s = 10$
$N = 1000$	$N = 1000$

Figures 5.1 and 5.2 are drawn with two scales on the horizontal axis (or abscissa). The first scale is stated in "IQ units" and the second in standard deviations from the mean. These scales should help you to visualize the relationships explained below between distances from the mean as measured by standard deviations from the mean and areas under the curve.

On any normal curve, the distance between ± 1 standard deviation encompasses exactly 68.26% of the total area under the curve. For the distribution of IQ scores for males, 68.26% of the total area lies between the score of 80 (-1 standard deviation) and 120 ($+1$ standard deviation). For females, the same percentage of the area lies between the scores of 90 and 110. As long as an empirical distribution is normal (or nearly so), the same percentage of the total area will always be encompassed between ± 1 standard deviation—regardless of the trait being measured and the number values of the mean and standard deviation.

FIGURE 5.1   IQ SCORES FOR A SAMPLE OF MALES

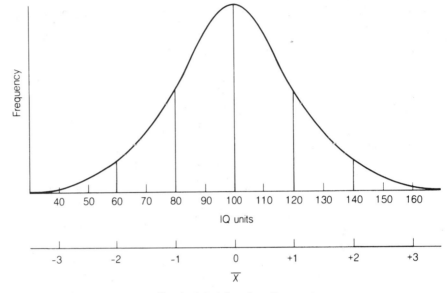

Standard deviations from the mean

FIGURE 5.2   IQ SCORES FOR A SAMPLE OF FEMALES

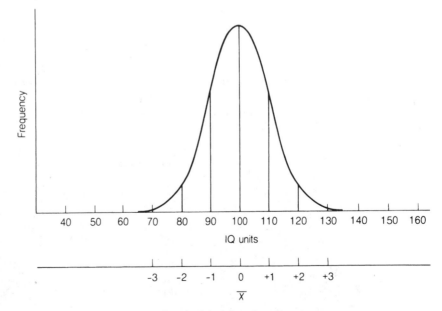

Standard deviations from the mean

FIGURE 5.3   AREAS UNDER THE THEORETICAL NORMAL CURVE

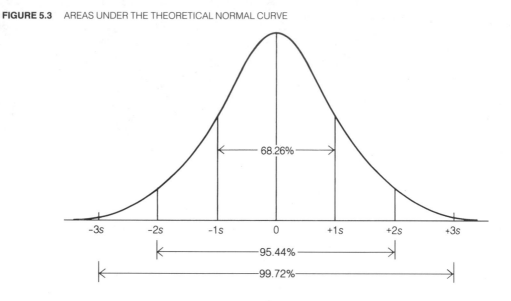

It will be useful to familiarize yourself with the following relationships between distances from the mean and areas under the curve:

between ± 1 standard deviation lies 68.26% of the area

between ± 2 standard deviations lies 95.44% of the area

between ± 3 standard deviations lies 99.72% of the area

These relationships are displayed graphically in Figure 5.3.

The regular relationship between distance from the mean and area in a theoretical normal curve allows us to describe empirical distributions that are at least approximately normal. The position of individual scores can be described with respect to the mean, the distribution as a whole, or any other score in the distribution.

The areas between scores can also be expressed, if desired, in numbers of cases rather than percentage of total area. For example, a normal distribution of 1000 cases will contain about 683 cases (68.26% of 1000 cases) between ± 1 standard deviation of the mean, about 954 between ± 2 standard deviations, and about 997 between ± 3 standard deviations. Thus, for any normal distribution, only a few cases will be farther away from the mean than ± 3 standard deviations.

5.2 COMPUTING Z SCORES   To find the percentage of the total area (or number of cases) above, below, or between scores in an empirical distribution, we must use a standardizing technique to transform the original scores into units of the standard deviation.

By this technique, the original scores will be converted into standard scores, or **Z scores**, and the empirical distribution will be standardized to the theoretical normal curve. Think of the conversion of the original scores into Z scores as a process of changing value scales. Most of you are probably already familiar with the general notion of a value scale. Distance can be measured in either feet and miles or meters and kilometers. These scales represent two different but equally valid value scales for measuring distance. Raw scores and Z scores are two equally valid but very different value scales for working with a normal curve.

The standardized theoretical normal curve always has a mean of 0 and a standard deviation of 1. When an empirical normal distribution is standardized, its mean will be converted to 0, its standard deviation to 1, and all values will be expressed in Z-score form. The formula for computing Z scores is

FORMULA 5.1

$$Z = \frac{X_i - \bar{X}}{s}$$

This formula will convert any score (X_i) from an empirical normal distribution into the equivalent Z score. To illustrate with the men's IQ data (Figure 5.1), the Z-score equivalent of a raw score of 120 would be

$$Z = \frac{120 - 100}{20} = +1.00$$

The equivalent Z score of positive 1.00 indicates that this score lies one standard deviation unit above (to the right of) the mean. A negative score would fall below (to the left of) the mean.

5.3 THE NORMAL CURVE TABLE

The theoretical normal curve has been very thoroughly analyzed and described by statisticians. The areas related to any Z score have been precisely determined and organized into a table format. This **normal curve table** or Z-score table is presented as Appendix A. The table consists of three columns, with Z scores in the left-hand column, areas between the Z score and the mean of the normal curve in the middle column, and areas beyond the Z score in the right-hand column. To find the area between the mean and any Z score, go down the column labeled "Z" until you find the score. For our example, go down the column until you find the score 1.00. The entry in the column labeled "Area Between Mean and Z" is 0.3413. The table presents all areas in the form of proportions, but we can easily translate these into percentages by multiplying them by 100 (see Chapter 2). We could say either that "a proportion of 0.3413 of the total area under the curve lies between a Z score of 1.00 and the mean," or "34.13% of the total area lies between a score of 1.00 and the mean."

The third column in the table presents "Areas Beyond Z." These would be areas above positive scores or below negative scores. This column will be used when we want to find an area above or below certain Z scores, an application that will be explained in Section 5.4.

To conserve space, the table includes only positive Z scores. Since the normal curve is perfectly symmetrical, however, the areas related to any negative score will be exactly the same as those of the corresponding positive score. For example, the area between a Z score of -1.00 and the mean will also be 34.13%, exactly the same as the area we found previously for a score of $+1.00$. As will be repeatedly demonstrated below, however, the sign of the Z score is extremely important and should be carefully noted.

For practice in using Appendix A to describe areas under an empirical normal curve, verify that the Z scores and areas given below are correct for the men's IQ distribution. For each IQ score, the equivalent Z score is computed using Formula 5.1, and then Appendix A is used to find areas between the score and the mean. ($\bar{X} = 100$, $s = 20$ throughout.)

IQ Score	Z Score	Area Between Z and the Mean
110	+0.50	19.15%
125	+1.25	39.44%
133	+1.65	45.05%
138	+1.90	47.13%

The same procedures apply when the Z-score equivalent of an actual score happens to be a minus value (that is, when the raw score lies below the mean).

IQ Score	Z Score	Area Between Z and the Mean
93	−0.35	13.68%
85	−0.75	27.34%
67	−1.65	45.05%
62	−1.90	47.13%

Remember that the areas in Appendix A will be the same for Z scores of the same numerical value regardless of sign. The area between the score of 138 ($+1.90$) and the mean is the same as the area between 62 (-1.90) and the mean.

5.4 FINDING TOTAL AREA ABOVE AND BELOW A SCORE

To this point, we have seen how the normal curve table can be used to find areas between a Z score and the mean. The information presented in the table can also be used to find other kinds of areas in empirical distributions that are at least approximately normal in shape. For example, suppose you need to determine the total area below the scores of two male subjects in the distribution described in Figure 5.1. The first subject has a score of 117 ($X_1 = 117$), which is equivalent to a Z score of $+0.85$:

$$Z_1 = \frac{X_1 - \bar{X}}{s} = \frac{117 - 100}{20} = \frac{17}{20} = +0.85$$

The plus sign of the Z score indicates that the score should be placed above (to the right of) the mean. To find the area below a positive Z score, the area between the score and the mean must be added to the area below the mean. Since the normal curve is symmetrical (unskewed), the mean will be equal to the median, and the area below the mean will therefore be 50%. Study Figure 5.4 carefully. We are interested in the lined area.

By consulting the normal curve table, we find that the area between the score and the mean is 30.23% of the total area. The area below a Z score of $+0.85$ is therefore 80.23% (50% + 30.23%). This subject scored higher than 80.23% of the persons tested.

The second subject has an IQ score of 73 ($X_2 = 73$), which is equivalent to a Z score of -1.35:

$$Z_2 = \frac{X_2 - \bar{X}}{s} = \frac{73 - 100}{20} = -\frac{27}{20} = -1.35$$

To find the area below a negative score, we use the column labeled "Area Beyond Z." The area of interest is depicted in Figure 5.5, and we must determine the size of the lined area. The area beyond a score of -1.35 is given as 0.0885, which we can express as 8.85%. The second subject ($X_2 = 73$) scored higher than 8.85% of the tested group.

In the examples above, we use the techniques for finding the area below a score. Essentially the same techniques are used to find the area above a score. If we need to determine the area above an IQ score of 108, for example, we would first convert to a Z score,

$$Z = \frac{108 - 100}{20} = \frac{8}{20} = +0.40$$

FIGURE 5.4   FINDING THE AREA BELOW A POSITIVE Z SCORE

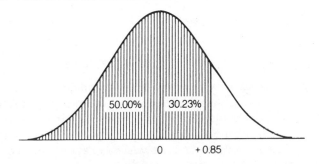

FIGURE 5.5 FINDING THE AREA BELOW A NEGATIVE Z SCORE

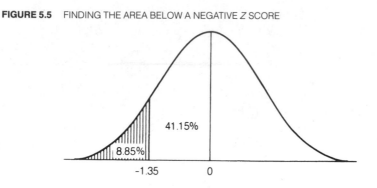

FIGURE 5.6 FINDING THE AREA ABOVE A POSITIVE Z SCORE

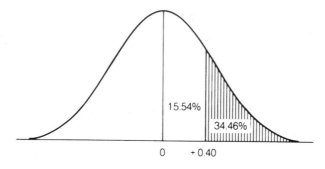

and then proceed to Appendix A. The lined area in Figure 5.6 represents the area in which we are interested. The area above a positive score is found in the "Area Beyond Z" column, and, in this case, the area is 0.3446, or 34.46%.

These techniques may be summarized as follows. To find the total area above a positive Z score or below a negative Z score, go down the "Z" column until you find the score. The area you are seeking will be in the "Area Beyond Z" column. To find the total area below a positive Z score or above a negative score, locate the score and then add the area in the "Area Between Mean and Z" to either .5000 (for proportions) or 50.00 (for percentages). These techniques might be confusing at first, and you will find it helpful to draw the curve and shade in the areas in which you are interested.

5.5 FINDING AREAS BETWEEN TWO SCORES

On occasion, you will need to determine the area between two scores rather than the total area above or below one score. In the case where the scores are on opposite sides of the mean, the area between the scores can be found by adding the areas between each score and the mean. Using the men's IQ data as an example, if we wished to know the area between the IQ scores of 93 and 112, we would convert both scores to Z scores, find the area between

APPLICATION 5.1

You have just received your score on a test of intelligence. If your score was 78 and you know that the mean score on the test was 67 with a standard deviation of 5, how does your score compare with the distribution of all test scores?

If you can assume that the test scores are normally distributed, you can compute a Z score and find the area below or above your score. The Z score equivalent of your raw score would be

$$Z = \frac{X_i - \bar{X}}{s}$$

$$Z = \frac{78 - 67}{5}$$

$$Z = \frac{11}{5}$$

$$Z = 2.20$$

Turning to Appendix A, we find that the "Area Between Mean and Z" is 0.4861, which could also be expressed as 48.61%. Since this is a positive Z score, we need to add this area to 50.00% in order to find the total area below. Your score is higher than (48.61 + 50.00), or 98.61%, of all the test scores. You did pretty well!

each score and the mean from Appendix A, and add these two areas together. The first IQ score of 93 converts to a Z score of -0.35:

$$Z_1 = \frac{X_1 - \bar{X}}{s} = \frac{93 - 100}{20} = -\frac{7}{20} = -0.35$$

The second IQ score (112) converts to $+0.60$:

$$Z_2 = \frac{112 - 100}{20} = \frac{12}{20} = +0.60$$

Both scores are placed on Figure 5.7. We are interested in the total lined area. The total area between these two scores is 13.68% + 22.57%, or 36.25%. Therefore, 36.25% of the total area (or about 363 of the 1000 cases) lies between the IQ scores of 93 and 112.

When the scores of interest are on the same side of the mean, a different procedure must be followed to determine the area between them. For example, if we were interested in the area between the scores of 113 and 121, we would begin by converting these scores into Z scores:

$$Z_1 = \frac{X_1 - \bar{X}}{s} = \frac{113 - 100}{20} = \frac{13}{20} = +0.65$$

$$Z_2 = \frac{X_2 - \bar{X}}{s} = \frac{121 - 100}{20} = \frac{21}{20} = +1.05$$

The scores are noted in Figure 5.8; we are interested in the lined area. To find the area between two scores on the same side of the mean, find the area

FIGURE 5.7 FINDING THE AREA BETWEEN TWO SCORES

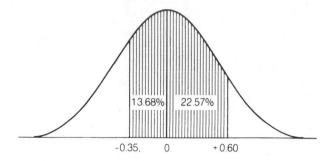

13.68%   22.57%

-0.35.   0   + 0.60

FIGURE 5.8 FINDING THE AREA BETWEEN TWO SCORES

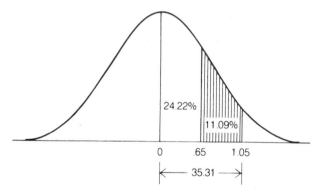

24.22%

11.09%

0   .65   1.05

⊢—— 35.31 ——⊣

between each score and the mean (given in Appendix A) and then subtract the smaller area from the larger. Between the Z score of $+0.65$ and the mean lies 24.22% of the total area. Between $+1.05$ and the mean lies 35.31% of the total area. Therefore, the area between these two scores is 35.31% − 24.22%, or 11.09% of the total area. The same technique would be followed if both scores had been below the mean.

5.6 USING THE NORMAL CURVE TO ESTIMATE PROBABILITIES

To this point, we have thought of the theoretical normal curve as a way of describing the percentage of total area above, below, and between scores in an empirical distribution. We have also seen that these areas can be converted into the number of cases above, below, and between scores. The theoretical normal curve may also be thought of as a distribution of probabilities. Specifically, we may use the known properties of the theoretical normal curve (Appendix A) to estimate the probability that a case randomly selected from an empirical normal distribution will have a score that falls in a defined range of scores. In terms of techniques, these probabilities will be found in exactly the same way as areas were found. Before we consider these mechanics, however, let us examine what is meant by the concept of probability.

Although we are rarely systematic or rigorous about it, we all attempt to deal with probabilities every day and, indeed, we base our behavior on our estimates of the likelihood that certain events will occur. We constantly ask (and answer) questions such as, What is the probability of precipitation? of drawing to an inside straight in poker? of the cheap retreads on my car going flat? of passing the final exam in Introduction to Nuclear Physics if I don't study?

To estimate the probabilities of events such as these, we must first be able to define the events that would constitute a "success." The examples above contain several different definitions of a success (that is, precipitation, drawing a certain card, flat tires, and passing grades). To determine a probability per se, a fraction must be established, with the numerator equalling the number of events that would constitute a success and the denominator equalling the total number of possible events where a success could theoretically occur.

To illustrate, assume that we wish to know the probability of selecting a specific card—say, the king of hearts—in one draw from a well-shuffled deck of cards. Our definition of a success is quite specific (drawing the king of hearts); and with the information given, we can establish a fraction. Only one card satisfies our definition of success, so the number of events that would constitute a success is 1; this value will be the numerator of the fraction. There are 52 possible events (that is, 52 cards in the deck), so the denominator will be 52. The fraction is thus 1/52, which represents the probability of selecting the king of hearts on one draw from a well-shuffled deck of cards. Our probability of success is 1 out of 52.

Now, we can leave the fraction established above as it is or we can express it in several other ways. For example, we can express it as an odds ratio by inverting the fraction, showing that the odds of selecting the king of hearts on a single draw are 52:1 (or fifty-two to one). We can express the fraction as a proportion by dividing the numerator by the denominator. For our example above, the corresponding proportion is .0192, which is the proportion of all possible events that would satisfy our definition of a success. In the social sciences, probabilities are usually expressed as proportions, and we will follow this convention throughout the remainder of this section.

As conceptualized here, probabilities have an exact meaning: over the long run, the defined successful events will bear a certain proportional relationship to the total number of events. The probability of .0192 for selecting the king of hearts in a single draw really means that, over thousands of selections of one card at a time from a full deck, the proportion of successful draws would be .0192. Or, for every 10,000 draws, 192 would be the king of hearts and the remaining 9808 selections would be other cards. Thus, when we say that the probability of drawing the king of hearts in one draw is .0192, we are essentially applying our knowledge of what would happen over thousands of draws to a single draw.

Probabilities are a type of proportion and they range from 0.00 (meaning that the event has absolutely no chance of occurrence) to 1.00 (a certainty). As the value of the probability increases, the likelihood that the defined event will occur also increases. A probability of .0192 approaches zero, and this means that the event (drawing the king of hearts) is unlikely or improbable.

Combining this way of thinking about probability with our knowledge of the theoretical normal curve allows us to estimate the likelihood of selecting a case from a defined range of scores. For example, suppose we wished to estimate the probability that a randomly chosen subject from the distribution of men's IQ scores would have an IQ score between 95 and the mean score of 100. Our definition of a success here would be the selection of any subject with a score in the specified range. Normally, we would next establish a fraction (that is, determine the number of subjects with scores in the defined range and the total number of subjects in order to establish the numerator and denominator, respectively). However, if the empirical distribution is normal in form, we can skip this step since the probabilities, in proportion form, are already stated in Appendix A. That is, the areas in Appendix A can be interpreted as probabilities.

To determine the probability that a randomly selected case will have a score between 95 and the mean, we would convert the original score to a Z score:

$$Z = \frac{95 - 100}{20} = -\frac{5}{20} = -0.25$$

The area between this score and the mean (0.0987) is the probability we are seeking. The probability that a randomly selected case will have a score between 95 and 100 is 0.0987 (or, rounded off, 0.1, or one out of 10). In the same fashion, the probability of selecting a subject from any range of scores can be estimated. Note that the techniques by which probabilities are estimated are exactly the same as those by which areas were found. The only new information introduced in this section is the idea that the areas in the normal curve table can also be thought of as probabilities.

Let us close by stressing a very important point about probabilities and the normal curve. The probability is very high that any case randomly selected from a normal distribution will have a score close in value to that of the mean. The shape of the normal curve is such that most cases are clustered around the mean and decline in frequency as we move farther away from the mean value. In fact, given the known properties of the normal curve, the probability that a randomly selected case will have a score within ± 1 standard deviation of the mean is 0.6826, whereas the probability of the case having a score beyond ± 3 standard deviations from the mean is 0.0028 $(1.000 - 0.9972)$. Thus, if we randomly select a number of cases from a normal distribution, we will often select cases that have scores close to the mean but rarely select cases that have scores far above or below the mean.

SUMMARY

1. The normal curve, in combination with the mean and standard deviation, can be used to construct precise descriptive statements about empirical distributions that are normally distributed. This chapter also lays some important groundwork for Part II.

2. To work with the theoretical normal curve, raw scores must be transformed into their equivalent Z scores. Z scores allow us to find areas under the theoretical normal curve (Appendix A).

3. We considered three uses of the theoretical normal curve: finding total areas above and below a score, finding areas between two scores, and expressing these areas as probabilities. This last use of the normal curve is especially germane because inferential statistics are centrally concerned with estimating the probabilities of defined events in a fashion very similar to the process introduced in Section 5.6.

SUMMARY OF FORMULAS

Z scores	5.1	$Z = \dfrac{X_i - \overline{X}}{s}$

GLOSSARY

Normal curve. A theoretical distribution of scores that is symmetrical, unimodal, and bell shaped. The standard normal curve always has a mean of 0 and a standard deviation of 1.

Normal curve table. Appendix A; a detailed description of the area between a Z score and the mean of any standardized normal distribution.

Z scores. Standard scores; the way scores are expressed after they have been standardized to the theoretical normal curve.

PROBLEMS

5.1 Assume that the distribution of a college entrance exam is normal with a mean of 500 and a standard deviation of 100. For each score below, find the equivalent Z score, the percentage of the area above the score, and the percentage of the area below the score.

X_i	Z Score	% Area Above	% Area Below
650			
400			
375			
586			
437			
526			
621			
498			
517			
398			

5.2 For a normal distribution where the mean is 50 and the standard deviation is 10, what is the area

a. between the scores of 40 and 47? _____

b. above a score of 47? _____

c. below a score of 53? _____

d. between the scores of 35 and 65? _____

e. above a score of 72? _____

f. below a score of 31 and above a score of 69? _____

g. between the scores of 55 and 62? _____

h. between the scores of 32 and 47? _____

5.3 At St. Algebra College, the 200 freshmen enrolled in Introductory Biology took a final exam on which their mean score was 72 and their standard deviation was 6. Below are the grades of 10 students. Convert each into a Z score and determine the number of people who scored higher or lower than each of the 10 students.

X_i	Z Score	Number of Students Above	Number of Students Below
60			
57			
55			
67			
70			
72			
78			
82			
90			
95			

5.4   CJ The local police force in Shinbone, Kansas, gives all applicants an entrance exam and accepts only those applicants who score in the top 15% on this test. If the mean score this year is 87 and the standard deviation is 8, would an individual with a score of 110 be accepted?

5.5   SW After taking the state merit examinations for the positions of social worker and employment counselor, you receive the following information on the tests and on your performance. On which of the tests did you do better?

Social Worker	Employment Counselor
$\bar{X} = 118$	$\bar{X} = 27$
$s = 17$	$s = 3$
Your score $= 127$	Your score $= 29$

5.6   PA For the past 10 years, your community has averaged 47 house fires per month, with a standard deviation of 5. Assuming the factors that cause house fires are constant, what is the probability that next month the number of house fires will be
a. 60 or more?
b. 47 or more?
c. between 42 and 52?
d. fewer than 30?
e. fewer than 55?
f. between 45 and 50?
g. 50 or more?

5.7   SOC A 25-item scale measuring political conservatism has been administered to a sample of 500 respondents. The scores of eight respondents are listed below.
a. For each score, determine the equivalent Z score and the percentage of cases above and below.

$$\bar{X} = 17$$
$$s = 3$$
$$N = 500$$

X_i	Z Score	Percentage of Cases Below	Percentage of Cases Above
19			
10			
14			
15			
18			
20			
22			
23			

b. Find areas between the following scores.

Scores	Z Scores	% Area Between
8 and 12		
9 and 13		
11 and 17		
15 and 19		
16 and 20		
17 and 23		
18 and 19		
19 and 22		

c. If you randomly selected subjects from the sample one at a time, what is the probability that their scores would be

Scores	Z Scores	Probability
Less than 17		
Less than 24		
Less than 10		
Less than 8		
Between 8 and 12		
Between 11 and 17		
Between 16 and 18		
Between 20 and 24		
More than 24		

More than 20
More than 15
More than 9

5.8 To be accepted into an honor society, students must have GPAs in the top 10% of the school. If the mean GPA is 2.78 and the standard deviation is .33, which of the following GPAs would qualify?

3.20   3.21   3.25   3.30   3.35

5.9 One thousand students have taken a language placement test, and the average score was 35 with a standard deviation of 4. What is the probability that a randomly selected student from this sample will have a score

a. between 31 and 37? _____

b. between 41 and 46? _____

c. less than 40? _____

d. less than 33? _____

e. more than 45? _____

f. more than 29? _____

5.10 In the distribution of scores mentioned in problem 5.9, which event is more likely: that a randomly selected score will be between 29 and 31 or that a randomly selected score will be between 40 and 42?

5.11 [CJ] The average burglary rate for a jurisdiction has been 311 per year with a standard deviation of 50. What is the likelihood that next year the number of burglaries will be

a. less than 250? _____

b. less than 300? _____

c. more than 350? _____

d. more than 400? _____

e. between 250 and 350? _____

f. between 300 and 350? _____

g. between 350 and 375? _____

5.12 If a distribution of test scores is normal with a mean of 78 and a standard deviation of 11, what percentage of the area lies

a. below 60? _____

b. below 70? _____

c. below 80? _____

d. below 90? _____

e. between 60 and 65? _____

f. between 65 and 79? _____

g. between 70 and 95? _____

h. between 80 and 90? _____

i. above 99? _____

j. above 89? _____

k. above 75? _____

l. above 65? _____

5.13 [SOC] A scale measuring prejudice has been administered to a large sample of respondents. The distribution of scores is approximately normal with a mean of 31 and a standard deviation of 5. What percentage of the sample had scores

a. below 20? _____

b. below 40? _____

c. between 30 and 40? _____

d. between 35 and 45? _____

e. above 25? _____

f. above 35? _____

5.14 [SOC] On the scale mentioned in problem 5.13, if a score of 40 or more is considered "highly prejudiced," what is the probability that a person selected at random will have a score in that range?

5.15 [SOC] The senior class has just finished a battery of standardized tests designed to assess their learning experience in college. The summary statistics and the scores of 15 members of the class are reported below. Assuming that the three distributions are approximately normal, each student scored higher than what percentage of the class for each test?

Math and Science	Language Skills and Writing Ability	History and Social Sciences
$\bar{X} = 58.90$	$\bar{X} = 39.06$	$\bar{X} = 74.77$
$s = 3.90$	$s = 2.76$	$s = 10.11$

		% Area			% Area			% Area
X_i	Z	Below	X_i	Z	Below	X_i	Z	Below
55	—	—	44	—	—	64	—	—
52	—	—	45	—	—	60	—	—
49	—	—	41	—	—	58	—	—
48	—	—	43	—	—	55	—	—
45	—	—	42	—	—	50	—	—
44	—	—	38	—	—	46	—	—
43	—	—	35	—	—	40	—	—
47	—	—	50	—	—	75	—	—
58	—	—	40	—	—	77	—	—
60	—	—	37	—	—	80	—	—
61	—	—	39	—	—	81	—	—
65	—	—	34	—	—	89	—	—
67	—	—	33	—	—	90	—	—
70	—	—	30	—	—	94	—	—
71	—	—	29	—	—	88	—	—

5.16 SOC For the Math and Science test scores reported in problem 5.15, what is the probability that a student randomly selected from this class will have a score

a. between 55 and 65? _____

b. between 60 and 65? _____

c. above 65? _____

d. between 60 and 50? _____

e. between 55 and 50? _____

f. below 55? _____

5.17 SOC For the History and Social Sciences test scores reported in problem 5.15, what percentage of the students had scores

a. between 75 and 85? _____

b. between 80 and 85? _____

c. above 80? _____

d. above 83? _____

e. between 80 and 70? _____

f. between 75 and 70? _____

g. below 75? _____

h. below 77? _____

i. below 80? _____

j. below 85? _____

USING SPSS/PC+ TO TRANSFORM RAW SCORES INTO Z SCORES

DEMONSTRATION 5.1 Computing Z Scores

The DESCRIPTIVES program introduced at the end of Chapter 4 can also be used to compute Z scores for any variable. These Z scores are then available for further operations and may be used in other tasks. SPSS/PC+ will create a new variable consisting of the transformed scores of the original variable. The program uses the letter Z and the first seven letters of the variable name to designate the normalized scores of a variable.

To use this feature, request OPTIONS 3 when submitting the DESCRIPTIVES command to SPSS/PC+. Almost every command in SPSS/PC+ has a variety of options available. For the most part, the options are used to handle missing values. As you execute the instructions below, take a few minutes to review the OPTIONS available in DESCRIPTIVES.

Beginning with the 1993 GSS active (use GET and name the file), you can produce Z scores for AGE by the following steps:

1. From the main menu, select 'analyze data' and then select 'descriptive statistics.'
2. Select and paste DESCRIPTIVES and !/VARIABLES. Use the typing window (ALT-T) to select AGE as the variable to be processed.
3. Select and paste /OPTIONS and then select 3 from the options menu. Note the explanations associated with the other options.

Your command should look like this:

```
DESCRIPTIVES /VARIABLES AGE /OPTIONS 3.
```

After you execute this command (press F10), the program will report some information on AGE and then compute the new variable. This variable will be named ZAGE and will be added to the active file. To confirm this, press ALT-V and SPSS/PC+ will display a list of all variables in the file. The new variable should be the last one on the list.

This transformed variable, ZAGE, can be treated just like any other variable. For example, use DESCRIPTIVES to get the mean and standard deviation for ZAGE. The command would be

```
DESCRIPTIVES /VARIABLES ZAGE /STATISTICS 1 5.
```

Since ZAGE is a normalized variable, the mean and standard deviation will be, respectively, 0.00 and 1.00.

If you would like a record of the Z scores, use the LIST procedure introduced in Appendix F (section F3). This command does not compute statistics but simply lists or displays the scores of the designated variables for as many cases as you would like. From the main menu, select 'analyze data' and then select 'reports and tables.' Now, select and paste LIST and designate AGE and ZAGE as the variables. Unless you specify otherwise, the scores of all cases will be listed. To limit the number of cases, use the /CASES subcommand, select the TO option, and type 20 to indicate the last case you want listed. To get case numbers along with scores, choose /FORMAT and NUMBERED. The command should look like this:

```
LIST /VARIABLES AGE ZAGE /CASES TO 20/FORMAT NUMBERED.
```

and the output should be

Case No.	AGE	ZAGE
1	44.00	-.12529
2	43.00	-.18285
3	78.00	1.83175
4	55.00	.50787

Case No.	AGE	ZAGE
5	31.00	-.87356
6	54.00	.45031
7	29.00	-.98868
8	23.00	-1.33404
9	61.00	.85323
10	63.00	.96835
11	33.00	-.75844
12	36.00	-.58576
13	55.00	.50787
14	44.00	-.12529
15	32.00	-.81600
16	49.00	.16251
17	36.00	-.58576
18	23.00	-1.33404
19	40.00	-.35552
20	26.00	-1.16136

Scan the list of scores and note that the scores that are close in value to the mean of AGE (46.18) are very close to the mean of ZAGE (0.00) and, the further away the score is from 46.18, the greater the numerical value of the Z score. Also note that, of course, scores below the mean (less than 46.18) have negative signs and scores above the mean (greater than 46.18) have positive signs.

Exercises

5.1 Use SPSS/PC+ to compute Z scores for PRESTG80, PAPRES80, and TVHOURS. Use LIST to display the normalized and "raw" scores for each variable for 25 cases.

5.2 Use the FREQUENCIES command to create histograms for each of the *normalized* variables you created in exercise 5.1. How close are the histograms to smooth, bell-shaped, normal curves? Run the FREQUENCIES command again and, this time, choose the /NORMAL subcommand from the histogram menu. This will superimpose a normal curve on the graph and should help you to ascertain the degree of normality in the histograms. Look up "skewness" and "kurtosis" in the SPSS/PC+ Manual. These statistics are designed to help quantify the approximation to normality of any frequency distribution.

PART I CUMULATIVE EXERCISES

1. A survey measuring attitudes toward interracial dating was administered to 1000 people. Parts of the survey and the scores of 30 respondents are reproduced below.

SURVEY

1. What is your age? ____
2. What is your sex?
    1. ___Male
    2. ___Female
3. Marriages between people of different racial groups just don't work out and should be banned by law.
    1. ___Strongly agree
    2. ___Agree
    3. ___Undecided
    4. ___Disagree
    5. ___Strongly disagree
4. How many years of schooling have you completed? ____
5. Which category below best describes the place where you grew up?
    1. ___Large city
    2. ___Medium-size city
    3. ___Suburbs of a city
    4. ___Small town
    5. ___Rural area
6. What is your marital status?
    1. ___Married
    2. ___Separated or divorced
    3. ___Widowed
    4. ___Never married

Case	Age	Sex	Attitude on interracial dating	Years of school	Area	Marital status
1	17	1	5	12	1	4
2	25	2	3	12	2	1
3	55	2	3	14	2	1
4	45	1	1	12	3	1
5	38	2	1	10	3	1
6	21	1	1	16	5	1

Case	Age	Sex	Attitude on interracial dating	Years of school	Area	Marital status
7	29	2	2	16	2	2
8	30	2	1	12	4	1
9	37	1	1	12	2	1
10	42	2	3	18	5	4
11	57	2	4	12	2	3
12	24	2	2	12	4	1
13	27	1	2	18	3	2
14	44	1	1	15	1	1
15	37	1	1	10	5	4
16	35	1	1	12	4	1
17	41	2	2	15	3	1
18	42	2	1	10	2	4
19	20	2	1	16	1	4
20	21	2	1	16	1	4
21	25	2	1	16	1	4
22	65	1	1	16	5	1
23	70	2	2	12	5	1
24	68	1	3	12	3	1
25	42	1	4	8	5	3
26	39	2	4	16	2	1
27	26	1	2	12	3	1
28	21	2	3	12	3	2
29	33	1	1	16	4	4
30	45	2	5	16	5	4

a. For each variable, construct a frequency distribution and select and calculate an appropriate measure of central tendency and a measure of dispersion. Summarize each variable in a sentence.

b. For all 1000 respondents, the mean age was 34.70 with a standard deviation of 3.4 years. Assuming the distribution of age is approximately normal, compute Z scores for each of the first ten respondents above and determine the percentage of the area below (younger than) each respondent.

2. The data set below is taken from the 1993 General Social Survey. Abbreviated versions of the questions along with the meanings of the codes are also presented. See Appendix G for the codes and the complete question wordings. The numbers in parentheses are the item numbers from Appendix G. For each variable, construct a frequency distribution and select and calculate an appropriate measure of central tendency and a measure of dispersion. Summarize each variable in a sentence.

1. How many children have you ever had? (6) (Values are actual numbers.)
2. Respondent's educational level (9):
    0. Less than HS
    1. HS
    2. Jr. college
    3. Bachelor's degree
    4. Graduate school
3. Race (11):
    1. White
    2. Black
    3. Other
4. "... methods of birth control should be available to teenagers ..." (54)
    1. Strongly agree
    2. Agree
    3. Disagree
    4. Strongly disagree
5. Number of hours of TV watched per day. (67) (Values are actual numbers of hours.)
6. What is your religious preference? (28)
    1. Protestant
    2. Catholic
    3. Jewish
    4. None
    5. Other

Case	Number of children	Educational level	Race	Birth control OK for teens	Hours of TV each day	Religious preference
1	3	1	1	3	3	1
2	2	0	1	4	1	1
3	4	2	1	2	3	1
4	0	3	1	1	2	1
5	5	1	1	3	2	1
6	1	1	1	3	3	1
7	9	0	1	1	6	1
8	6	1	2	3	4	1
9	4	3	1	1	2	4
10	2	1	3	1	1	1
11	2	0	1	2	4	1
12	4	1	2	1	5	2
13	0	1	1	3	2	2
14	2	1	1	4	2	1

Case	Number of children	Educa- tional level	Race	Birth control OK for teens	Hours of TV each day	Reli- gious pref- erence
15	3	1	2	3	4	1
16	2	0	1	2	2	1
17	2	1	1	2	2	1
18	0	3	1	3	2	1
19	3	0	1	3	5	2
20	2	1	2	1	10	1
21	2	1	1	3	4	1
22	1	0	1	3	5	1
23	0	2	1	1	2	2
24	0	1	1	2	0	4
25	2	4	1	1	1	2
26	1	0	1	1	10	2
27	4	4	1	4	3	2
28	0	2	1	2	3	2
29	2	3	1	4	1	1
30	3	0	1	4	9	1
31	4	1	1	2	3	1
32	0	4	2	3	4	1
33	2	1	2	1	1	1
34	2	1	1	3	2	1
35	1	1	1	4	4	1
36	4	0	1	2	1	1
37	3	1	2	4	0	1
38	0	1	2	2	2	5
39	3	4	1	2	3	2
40	0	1	1	3	0	1
41	4	1	1	1	2	1
42	0	4	1	2	0	4
43	2	1	1	1	6	2
44	3	0	1	2	3	1
45	6	1	1	4	7	2

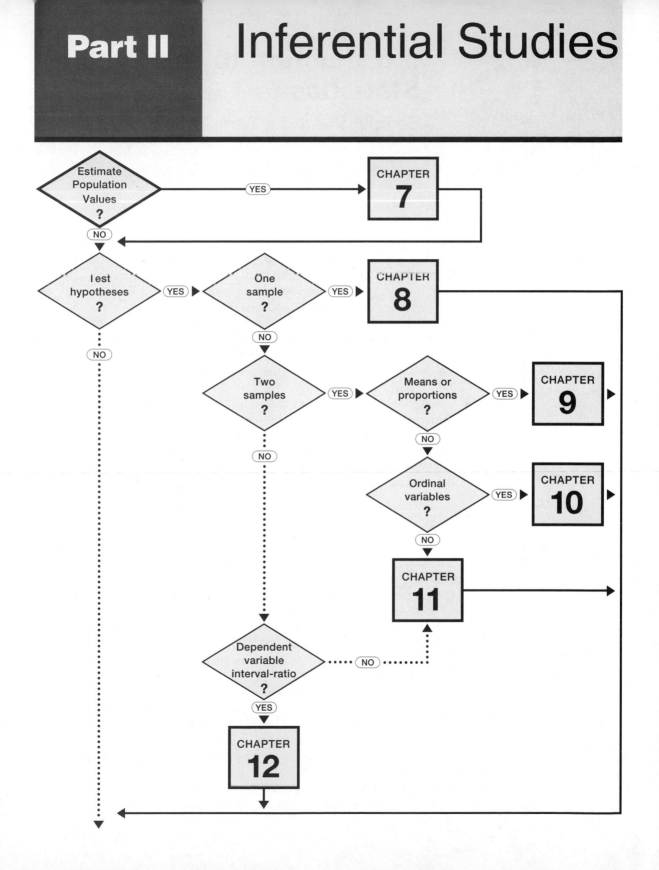

6 Introduction to Inferential Statistics
Sampling and the Sampling Distribution

Researchers in the social sciences constantly confront the problem of populations that are too large to test. The researcher almost never has enough time or money to collect data on every case of concern. Indeed, even for relatively small populations (such as a local community) the logistics of testing every case are staggering to contemplate. To deal with this problem, social scientists select samples, or subsets of cases, from the populations of interest. Our goal in inferential statistics is to generalize to the characteristics of a population (these are often called **parameters**), based on what we can learn from our samples.

Two applications of inferential statistics are covered in this text. In estimation procedures, a "guess" of the population parameter is made, based on what is known about the sample. In hypothesis testing, the validity of a hypothesis about the population is tested against sample outcomes. We will consider estimation techniques in Chapter 7 and the various forms of hypothesis testing in Chapters 8 through 12. In the present chapter, we will look briefly at sampling and then introduce the concept of the sampling distribution.

Social scientists have developed a variety of techniques for selecting samples from populations. In this chapter, we will review the basic procedures for selecting probability samples, the only type of sample that fully supports the use of inferential statistical techniques to generalize to populations. These types of samples are often described as "random" and you may be more familiar with this terminology. Because of its greater familiarity, I will often use the phrase "random sample" in the following chapters. The term "probability sample" is preferred, however, because, in everyday language, "random" is often used to mean "by coincidence" or to give a connotation of unpredictability. As you will see, probability samples are selected by techniques that are careful and methodical and leave no room for haphazardness. Interviewing the people you happen to meet in a mall one afternoon may be "random" in some sense but this technique will not result in a sample that could support inferential statistics.

Before considering probability sampling techniques, let me point out that social scientists often use nonprobability samples and that such samples are very useful for a number of purposes. Nonprobability samples are typically less costly and easier to assemble and are appropriate in many different research situations. The major limitation of nonprobability samples is that they do not permit the use of inferential statistics to generalize to populations.

With probability sampling techniques, our goal is to select cases so that the final sample is **representative** of the population from which it was drawn. A sample is representative if it reproduces the important characteristics of the population. For example, if the population consists of 60% females and 40% males, the sample should contain essentially the same proportions. Representativeness is crucial for inferential statistics. If the sample is not representative, generalizing to the population becomes, at best, extremely hazardous.

The fundamental principle of probability sampling is that a sample is very likely to be representative if it is selected by a principle called **EPSEM**, which stands for the "**E**qual **P**robability of **SE**lection **M**ethod." The EPSEM principle is that every element or case in the population must have an equal probability of being selected for the sample. So, our goal is to select a representative sample, and the technique we use to achieve that goal is to follow EPSEM, or the equal probability of selection method.

I should stress that the EPSEM selection technique and the representativeness of the sample are two different things. In other words, the fact that a sample is properly selected does not guarantee that it will be an exact representation or microcosm of the population. The probability is very high that an EPSEM sample will be representative but, just as a perfectly honest coin will sometimes show ten heads in a row when flipped, an EPSEM sample will occasionally present an inaccurate picture of the population. One great strength of inferential statistics is that they allow the researcher to estimate the probability of this type of error and interpret results accordingly.

6.3 EPSEM SAMPLING TECHNIQUES

The most basic EPSEM sampling technique produces what is called the **simple random sample**. To implement the EPSEM principle, we need a list of all elements or cases in the population and a system for selecting cases from the list that will guarantee that every case has an equal chance of being selected for the sample. The selection process could be based on a number of different kinds of operations (for example, drawing cards from a well-shuffled deck, flipping coins, throwing dice, drawing numbers from a hat, and so on). Most researchers use tables of random numbers to select cases. These tables are lists of numbers that have no pattern to them (that is, they are random), and an example of such a table is Appendix E. To use the table, first assign each case on the population list a unique identification number. Then, select cases for the sample when their identification number corresponds to the number chosen from the table. Since the numbers in the table are in random order and any number is just as likely as any other number, we will meet the EPSEM criteria by following this procedure. Stop selecting cases when you have reached your desired sample size and, if an identification number is selected more than once, ignore the repeats.*

Rigidly following the procedures above will generate random samples as long as you select from a complete list of the population. However, these procedures can be very cumbersome when there is a long list of cases. Consider the situation when your population numbers 10,000 cases. It is perfectly possible that the first case you select will come from the front of the list, the second from the back, the third from the front again, and so on—leading to a great deal of paper shuffling and a fair amount of confusion. To save time and money in such a situation, researchers often use a technique called **systematic sampling**, where only the first case is randomly selected. Thereafter, every kth case is selected. For example, if you are drawing from a list of 10,000 and desire a sample of 200, select the first case randomly and every 10,000/200th, or 50th, case thereafter. If you randomly start with case 013, then your second case will be 063, your third 113, and so on, until you reach the end of the list.

Note that systematic sampling does not strictly conform to the criterion of EPSEM. That is, once the first case has been selected, the other cases no longer have an equal probability of being chosen. In our example above, cases other than the 13th, the 63rd, the 113th, and so on will not be selected for the sample. In general, this increased probability of error is very slight as long as the list from which cases are chosen is itself random, or at least non-cyclical with respect to the traits you wish to measure. For example, if you are concerned with ethnicity and are drawing your sample from an alphabetical list, you might encounter difficulties because there is a tendency for certain ethnic names to begin with the same letter (for example, the Irish prefix *O*). Therefore, when using systematic sampling, pay careful attention to the nature of the population list as well as your sampling technique.

A third type of EPSEM sample is the **stratified sample**. This technique is very desirable because it guarantees that the sample will be representative on the selected traits. To apply this technique, you first stratify (or divide) the population list into sublists according to some relevant trait and then sample from the sublists. If you select a number of cases from each sublist proportional to the numbers for that characteristic in the population, the sample will be representative of the population.

For example, suppose that you are drawing a sample of 300 of your classmates and you wish to have proportional representation from every major

---

*Ignoring identification numbers when they are repeated is called "sampling without replacement." Technically, this practice compromises the randomness of the selection process. However, if the sample is a small fraction of the total population, we will be unlikely to select the same case twice, and ignoring repeats will not bias our conclusions.

field on campus. If only 10% of the student body is majoring in zoology, the sampling techniques discussed above could conceivably result in a sample with very few (or even no) zoologists. If, however, you first divide the population into sublists by major, you can use EPSEM to select exactly 30 zoologists from the appropriate sublist. Following the same procedure with other majors will create a sample which is, by definition, representative of the population on this characteristic. Thus, stratified samples are guaranteed to meet the all important criterion of representativeness (at least for the traits that are used to stratify the samples).

A major limitation for this technique is that the exact composition of the population is often unknown. If we have no information about the nature of the population, we will not be able to establish a scheme for stratification.

To this point, sampling techniques have been presented as straightforward processes of randomly selecting cases from a list or sublists of the population. However, sampling is rarely as uncomplicated as I have implied, and the major difficulty almost always centers on what might appear, at first glance, to be the easiest part: establishing the list of the population. For many of the populations of interest to the social sciences, there are no complete, up-to-date lists. There is no list of United States citizens, no list of the residents of any given state, and no list of residents of your local community. Devices such as telephone books or city directories might appear to contain complete lists of local residents. However, the former will omit unlisted numbers and a disproportionate number of low-income households, and the latter is very likely to be outdated.

Social scientists have devised several ways of dealing with the limitations imposed by the scarcity of lists. Probably the most significant of these is **cluster sampling**, which involves selecting geographical units rather than elements or cases from a list. Cluster sampling often proceeds in stages. For example, you might draw a cluster sample of your city or town by first numbering all of the voting precincts within the political boundaries. Next, you would use EPSEM to select a sample of precincts. The second stage of selection would involve numbering the blocks within each of the selected precincts and, following EPSEM, selecting a sample of blocks. A third stage might involve the selection of households within each selected block. When these stages are completed, you would have a sample that had a very high probability of being representative of the entire city without ever using a list of residents of the city.

Cluster sampling is less trustworthy than the other techniques summarized above. A cluster sample is a less accurate representation of the population than a simple random sample of comparable size. In part, this decreased accuracy is a result of the multiple selection stages described above. With a simple random sample, the sample is drawn in one selection from the list of the population. In a multistage cluster sample, each stage in the selection process (e.g., first the precincts, then the blocks, and then the households) has a probability of error. That is, each time we sample, we run the risk of selecting an unrepresentative sample. In simple random sampling, we run this risk once; with cluster sampling we will run the risk anew at each stage.

Although we have to treat inferences to populations based on cluster samples with some additional caution, we often have no alternative method of sampling. While it may be extremely difficult (or even impossible) to construct an accurate list of an entire city population, all you need for compiling a cluster sample is a map (or a list of voting precincts, census tracts, and so forth).

By way of summary, let me return to a major point. The purpose of inferential statistics is to acquire knowledge about populations, based on the information derived from samples drawn from that population. Each of the statistics to be presented in the following chapters requires that samples be selected according to EPSEM. While even the most painstaking and sophisticated sampling techniques will not guarantee representativeness, the probability is high that EPSEM samples will be representative of the populations from which they are selected.

6.4 THE SAMPLING DISTRIBUTION

In inferential statistics, researchers face a puzzling dilemma. On one hand, they have a great deal of information about the sample distribution. On the other hand, they know virtually nothing about the population—and let me stress, it is the population that is of interest. The sample distribution is interesting primarily insofar as it allows the researcher to generalize to the population.

Generally, the information necessary to characterize a distribution adequately would include (1) the shape of the distribution, (2) some measure of central tendency, and (3) some measure of dispersion. Clearly, all three kinds of information can be gathered (or computed) for the sample distribution. Just as clearly, none of the information can be gathered for the population. Except in rare cases (for example, IQ and height are known to be approximately normal in distribution), nothing can be known about the exact shape of the population distribution. Its mean and standard deviation are also unknown. Indeed, if any of this information were available for the population, inferential statistics would probably be unnecessary.

In statistics, this vast ocean of ignorance is bridged by a device known as the **sampling distribution**. Although we might be totally ignorant about the population distribution, the characteristics of the sampling distribution, being based on the laws of probability and not on empirical information, are very well known indeed. In fact, the sampling distribution is the central concept in inferential statistics, and a prolonged examination of its characteristics is certainly in order.

The general strategy of all applications of inferential statistics is to move from the sample to the population via the sampling distribution. Thus, three separate and distinct distributions are involved in every application of inferential statistics:

1. The sample distribution, which is empirical and known. Indeed, it is collected by researchers themselves but is important primarily insofar as it allows the researcher to learn about the population.

2. The population distribution, which, while empirical (that is, it exists in reality), is unknown. Amassing information about or making inferences to the population is the sole purpose of inferential statistics.

3. The sampling distribution, which is nonempirical (theoretical). Because of the laws of probability, a great deal is known about this distribution. Specifically, the shape, central tendency, and dispersion of the distribution can be deduced and, therefore, the distribution can be adequately characterized.

The sampling distribution can be formally defined as a theoretical, probabilistic distribution of all possible sample outcomes (with constant sample size, N) for the statistic that is to be generalized to the population. The utility of the distribution is implied in the definition. Because it encompasses all possible sample outcomes, the sampling distribution enables us to estimate the probability of any particular sample outcome for that statistic. Exactly how this is done and why it is important forms, in fact, the subject matter of the next five chapters.

The sampling distribution is theoretical, which means that it is never obtained in reality by the researcher. However, to understand better the structure and function of the distribution, let's consider how one might be constructed. Suppose that a researcher is concerned with a specific trait—for example, the average age of a particular population. Following EPSEM, the researcher draws a sample ($N = 100$) from the population, asks all respondents their age, computes the mean, and notes the resultant score on a graph. For the sake of illustration, assume that the average age of this sample is 27 ($\bar{X} = 27$). This score is noted on the graph in Figure 6.1.

Now, replace the 100 respondents in the first sample and draw another sample ($N = 100$) from the same population and again compute the average age. Assume that the mean for the second sample is 32 ($\bar{X} = 32$) and note this sample outcome on Figure 6.1. Replace the respondents from the second sample and draw still another sample ($N = 100$), calculate and note the $\bar{X}$, replace this third sample, and draw a fourth sample, continuing these operations an infinite number of times, calculating and noting the mean of each sample. Now, try to imagine what Figure 6.1 would look like after thousands of individual samples had been collected and the mean had been computed

FIGURE 6.1   CONSTRUCTING A SAMPLING DISTRIBUTION

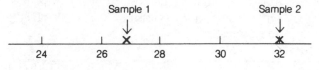

for each sample. What would be the shape, mean, and standard deviation of this distribution?

For one thing, we know that each sample will be at least slightly different from every other sample, since we will probably never sample exactly the same 100 people twice. Hence, each sample mean will be slightly different. We also know that even though the samples are random, they are not always exact representations of the population. Some of the sample means will be much lower than the population mean, others will be much higher. Common sense suggests, however, that across these thousands of individual samples, most of the sample means will cluster around the true population value.

To illustrate further, assume that we come to know that the true mean age of the population is 30. As we have seen above, most of the sample means will also be approximately 30. Thus, the sampling distribution of these sample outcomes ($\bar{X}$'s) should peak at 30. Some of the sample means will "miss the mark," but the frequency of such misses should decline as we get farther away from 30. That is, the distribution should slope to the base as we get farther away from the population value (sample means of 29 or 31 should be common; means of 20 or 40 should be rare). Since the samples are random, the means should miss an equal number of times on either side of the population value, and the distribution itself should therefore be roughly symmetrical. In other words, the sampling distribution of all possible sample means should be approximately normal and will resemble the distribution presented in Figure 6.2. These common-sense notions about the shape of the sampling distribution and other very important information about central tendency and dispersion are stated in two theorems. The first of these theorems states that

> if repeated random samples of size N are drawn from a normal population with mean μ and standard deviation σ, then the sampling distribution of sample means will be normal with a mean μ and a standard deviation of $\sigma/\sqrt{N}$.

To translate: if we begin with a trait that is normally distributed across a population (like IQ, height, or weight, for example) and take an infinite number of equally sized random samples from that population, then the sampling distribution of sample means will be normal. If it is known that the trait is distributed normally in the population, it can be assumed that the sampling distribution will be normal.

The theorem tells us more than the shape of the sampling distribution,

FIGURE 6.2   A SAMPLING DISTRIBUTION OF SAMPLE MEANS

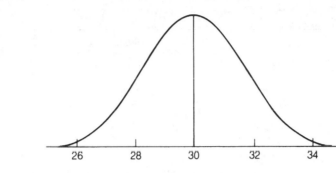

however. It also defines its mean and standard deviation. In fact, it says that the mean of the sampling distribution will be exactly the same value as the mean of the population. That is, if we know that the mean IQ of the entire population is 100, then we also know that the mean of any sampling distribution of sample mean IQ's will also be 100. Exactly why this should be so is not a matter that can be fully explicated at this level. Recall, however, that most sample means will cluster around the population value over the long run. Thus, the fact that these two values are equal should have intuitive appeal. As for dispersion, the theorem says that the standard deviation of the sampling distribution, also called the **standard error of the mean,** will be equal to the standard deviation of the population divided by the square root of N (symbolically, $\sigma/\sqrt{N}$).

If the mean and standard deviation of a normally distributed population are known, the theorem allows us to compute the mean and standard deviation of the sampling distribution.* Thus, we will know exactly as much about the sampling distribution (shape, central tendency, and dispersion) as we ever knew about any normal curve.

But what if the population distribution is not normal? This eventuality (very common, in fact) is covered by a second theorem, called the Central Limit Theorem:

> If repeated random samples of size N are drawn from any population, with mean μ and standard deviation σ, then, as N becomes large, the sampling distribution of sample means will approach normality, with mean μ and standard deviation $\sigma/\sqrt{N}$.

The importance of the Central Limit Theorem is that it removes the constraint of normality in the population. Whenever sample size is large, it can

---

*In the typical research situation, the values of the population mean and standard deviation are, of course, unknown. However, these values can be estimated from sample statistics, as we shall see in the chapters that follow.

TABLE 6.1 SYMBOLS FOR MEANS AND STANDARD DEVIATIONS OF THREE DISTRIBUTIONS

	Mean	Standard Deviation	Proportion
1. Samples	$\bar{X}$	s	P_s
2. Population	μ	σ	P_u
3. Sampling distributions:			
of means	$\mu_{\bar{X}}$	$\sigma_{\bar{X}}$	
of proportions	μ_p	σ_p	

be assumed that the sampling distribution is normal, with a mean equal to the population mean and a standard deviation equal to $\sigma/\sqrt{N}$ regardless of the shape of the population. Thus, even if we are working with a characteristic with a skewed distribution (such as income), we can still assume a normal sampling distribution.

The issue remaining, of course, is to define what is meant by a large N. A good rule of thumb is that if N is 100 or more, the Central Limit Theorem applies and you can assume that the sampling distribution is normal in shape. When N is less than 100, you must have good evidence of a normal population distribution before you can assume that the sampling distribution is normal. Thus, a normal sampling distribution can be ensured by the expedient of using fairly large samples.

6.5 SYMBOLS AND TERMINOLOGY

In the following chapters keep in mind that we will be working with three entirely different distributions. The purpose of inferential statistics is to acquire knowledge of the population distribution from the sample distribution by means of the sampling distribution. Furthermore, we will be concerned with several different kinds of sampling distributions—including the sampling distribution of sample means and the sampling distribution of sample proportions.*

To distinguish clearly among these various distributions, we will often use symbols. The symbols used for the means and standard deviations of samples and populations have already been introduced in Chapters 3 and 4. Here, I would like to introduce some of the symbols that will be used for the sampling distribution and to present the relevant symbols in a summary table that you can use for reference. Basically, as you can see from Table 6.1, the sampling distribution is denoted with Greek letter symbols that are subscripted according to the sample statistic of interest.

---

*Symbols for the sampling distributions of statistics other than means and proportions will be introduced at the appropriate time.

SUMMARY

1. Since populations are almost always too large to test, a fundamental 'strategy of social science research is to select a sample from the defined population and then use information from the sample to generalize to the population. This is done either by estimation or by hypothesis testing.

2. Several techniques are commonly used for selecting random samples. Each of these techniques involves selecting cases for the sample according to EPSEM. Even the most rigorous technique, however, cannot guarantee representativeness. One of the great strengths of inferential statistics is that the probability of this kind of error (nonrepresentativeness) can be estimated.

3. The sampling distribution, the central concept in inferential statistics, is a theoretical distribution of all possible sample outcomes. Since its overall shape, mean, and standard deviation are known (under the conditions specified in the two theorems), the sampling distribution can be adequately characterized and utilized by researchers.

4. The two theorems that were introduced in this chapter state that when the variable of interest is normally distributed in the population or when sample size is large, the sampling distribution will be normal in shape, its mean will be equal to the population mean, and its standard deviation will be equal to the population standard deviation divided by the square root of N.

5. All applications of inferential statistics involve generalizing from the sample to the population by means of the sampling distribution. Both estimation procedures and hypothesis testing incorporate the three distributions, and it is crucial that you develop a clear understanding of each distribution and its role in inferential statistics.

GLOSSARY

Central Limit Theorem. A theorem that specifies the mean, standard deviation, and shape of the sampling distribution, given that the sample is large.

Cluster sampling. A method of sampling by which geographical units are randomly selected and all cases within each selected unit are tested.

EPSEM. The Equal Probability of SElection Method for selecting samples. Every element or case in the population must have an equal probability of selection for the sample.

μ. The mean of a population.

$\mu_{\bar{x}}$. The mean of a sampling distribution of sample means.

μ_p. The mean of a sampling distribution of sample proportions.

P_s. (P-sub-s) Any sample proportion.

P_u. (P-sub-you) Any population proportion.

Representative. The quality a sample is said to have if it reproduces the major characteristics of the population from which it was drawn.

Sampling distribution. The distribution of all possible sample outcomes of a given statistic. Under specified conditions, the sampling distribution will be normal in shape with a mean equal to the population value and a standard deviation equal to the population standard deviation divided by the square root of N.

Simple random sampling. A method for choosing cases from a population by which every case and every combination of cases has an equal chance of being included.

Standard error of the mean. The standard deviation of a sampling distribution of sample means.

Stratified sampling. A method of sampling by which cases are selected from sublists of the population.

Systematic sampling. A method of sampling by which the first case from a list of the population is randomly selected. Thereafter, every kth case is selected.

PROBLEMS

6.1 Imagine that you had to gather a random sample ($N = 300$) of the student body at your school. How would you acquire a list of the population? Would the list be complete and accurate? What procedure would you follow in selecting cases (that is, simple or systematic random sampling)? Would cluster sampling be an appropriate technique (assuming that no list was available)? Describe in detail how you would construct a cluster sample.

6.2 This exercise is extremely tedious and hardly ever works out the way it ought to (mostly because not many people have the patience to draw an "infinite" number of even very small samples). However, if you want a more concrete and tangible understanding of sampling distributions and the two theorems presented in this chapter, then this exercise may have a significant payoff. Below are listed the ages of a population of college students ($N = 50$). By a random method (such as a table of random numbers), draw at least 50 samples of size 2 (that is, 50 pairs of cases), compute a mean for each sample, and plot the means on a frequency polygon. (Incidentally, this exercise will work better if you draw 100 or 200 samples and/or use larger samples than $N = 2$.)

a. The curve you've just produced is a sampling distribution. Observe its shape—after 50 samples, it should be approaching normality. What is your estimate of the population mean (μ) based on the shape of the curve?

b. Calculate the mean of the sampling distribution ($\mu_{\bar{x}}$). Be careful to do this by summing the sample means (not the scores) and dividing by the number of samples you've drawn. Now compute the population mean (μ). These two means should be very close in value because $\mu = \mu_{\bar{x}}$ by the Central Limit Theorem.

c. Calculate the standard deviation of the sampling distribution (use the means as scores) and the standard deviation of the population. Compare these two values. You should find that $\sigma_{\bar{x}} = \sigma/\sqrt{N}$.

d. If none of the above exercises turned out as they should have, it is for one or more of the following reasons:

1. You didn't take enough samples. You may need as many as 100 or 200 (or more) samples to see the curve begin to look "normal."

2. Sample size (2) is too small. An N of 5 or 10 would work much better.

3. Your sampling method is not truly random and/or the population is not arranged in random fashion.

17	20	20	19	20
18	21	19	20	19
19	22	19	23	20
20	23	18	20	20
22	19	19	20	20
23	17	18	21	20
20	18	20	19	20
22	17	21	21	21
21	20	20	20	22
18	21	20	22	21

USING SPSS/PC+ TO DRAW RANDOM SAMPLES

DEMONSTRATION 6.1 Estimating Average Age

There is a procedure called SAMPLE in SPSS/PC+ which will draw EPSEM samples of any size from a data set. We can use this procedure to illustrate some points about sampling and to convince the skeptics in the crowd that properly selected samples will produce statistics that are close approximations of the corresponding population values. The instructions below will calculate the actual average age of the entire 1993 GSS sample (this will be the population mean or μ) and the mean of three separate samples drawn from this file. The samples are roughly 10%, 25%, and 50% of the population size, and the program selects them by a process that is quite similar to a table of random numbers. Therefore, these samples may be considered "simple random samples."

As a part of this procedure we also request the "standard error of the mean" or SEMEAN (statistic 2 on the menu). This is the standard deviation of the sampling distribution ($\sigma/\sqrt{N}$) for a sample of this size. The value of this statistic will be of interest to us because we can expect our sample means to be within this distance of the population value.

I've included the sample means I found to illustrate some results. When you run this task, you will, of course, draw samples that are at least slightly different from mine, so your results will be at least slightly different.

Use the GET command on the 'read and write data' menu to make the 1993 GSS file active and

1. Run DESCRIPTIVES for AGE with STATISTICS 1 to get the population mean (46.18).
2. To draw a random sample, select 'modify data or files' from the main menu and then choose 'select or weight data'.
3. Select and paste SAMPLE and then use ALT-T to specify the proportional size of the sample. Begin with a 10% sample (type .10 in the window), then take a 25% sample (type .25) and a 50% sample (type .50). For each sample, use the DESCRIPTIVES command to find a mean and the standard error of the mean for that sample. That is, after you have selected your sample size, go back to the 'analyze data' menu and select DESCRIPTIVES and the proper subcommands.

When you're done, you should have a set of commands in the EDIT window that look like this:

```
DESCRIPTIVES /VARIABLES AGE /STATISTICS 1.
SAMPLE .10.
DESCRIPTIVES /VARIABLES AGE /STATISTICS 1 2.
SAMPLE .25.
DESCRIPTIVES /VARIABLES AGE /STATISTICS 1 2.
SAMPLE .50.
DESCRIPTIVES /VARIABLES AGE /STATISTICS 1 2.
```

Place the cursor on the first line of the file and these commands will execute. Be sure to note the means as they are displayed on the screen or have SPSS/PC+ print the results (see the Manual for instructions).

My results are presented below. Remember that the population mean is 46.18.

Sampling Fraction	Sample Size	Sample Mean	Standard Error	Sample Mean Plus and Minus Standard Error
.10	87	47.66	1.95	45.71 - 49.61
.25	219	46.89	1.14	45.75 - 48.03
.50	388	46.30	0.90	45.40 - 47.20

Note that standard error (or the standard deviation of the sampling distribution) decreases as sample size increases. This should reinforce the commonsense notion that larger samples will provide more accurate estimates of population values. All three samples produced estimates (sample means) that are quite close in value to the population value of 46.18. All three sample means are within a standard error of the population mean.

Furthermore, the smallest sample is the least accurate (47.66 is 1.48 years above the true population value of 46.18), the 25% sample is closer (almost ¾ year too high), and the largest sample is very close (only .12 year too high). This relationship between accuracy and size of sample should make sense intuitively, but remember that random samples will not always be so cooperative and it is perfectly possible that your samples will behave quite differently.

This demonstration should reinforce one of the main points of this chapter: statistics calculated on samples that have been selected according to the principle of EPSEM will (almost always) be reasonable approximations of their population counterparts.

7

Estimation Procedures

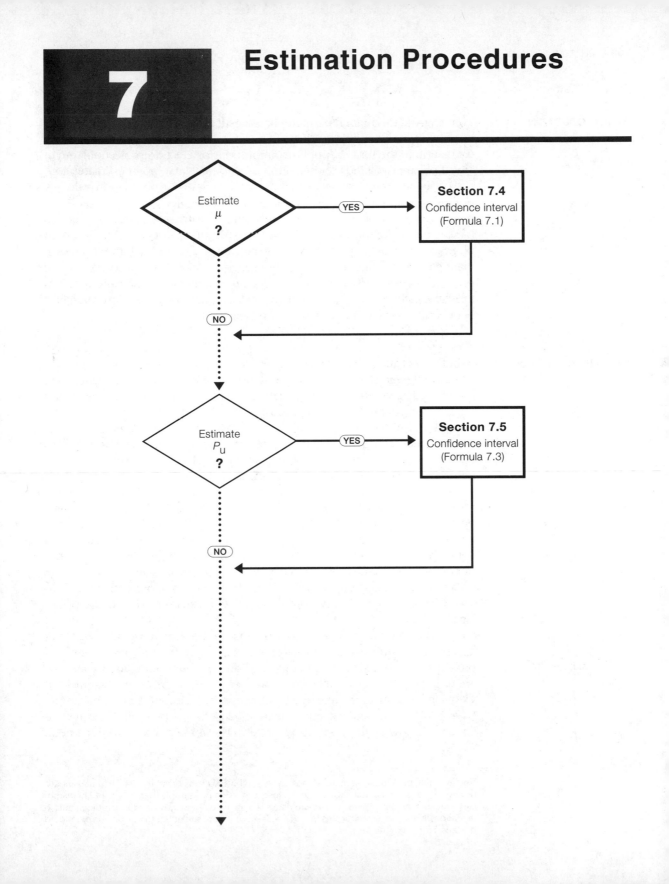

7.1 INTRODUCTION

The object of this branch of inferential statistics is to estimate population values from sample outcomes. Most of you are probably familiar with the use of estimation in the form of public-opinion polls or election projections. Actually, there are two kinds of estimation procedures. First, point estimates consist of single-sample statistics that are used to guess a population value. A polling organization that projects that 42% of the electorate will vote for a particular presidential candidate is reporting a **point estimate**. The second kind of estimation procedure involves **confidence intervals**, which consist of a range of values (an interval) instead of a single point. Rather than estimating a specific figure as in a point estimate, an interval estimate might be phrased as "from 39% to 45% of the electorate will vote for the candidate." In this latter estimate, we are essentially guessing that the population value falls between 39% and 45%, but we do not specify where.

7.2 BIAS AND EFFICIENCY

In both point and interval estimation, sample statistics serve as the estimators. Which of the many available sample statistics should be used? Estimators can be selected according to the rule that a good estimator must be unbiased and relatively efficient. An estimator is unbiased if and only if the mean of its sampling distribution is equal to the population value of interest. We know from the theorems presented in Chapter 6 that sample means conform to this criterion. The mean of the sampling distribution of sample means (which we will note symbolically as $\mu_{\bar{x}}$) is the same as the population mean (μ).

In addition, sample proportions (P_s) are also unbiased. That is, if we take repeated random samples of size N and calculate and array the resultant proportions, the sampling distribution of sample proportions will have a mean (μ_p) equal to the population proportion (P_u). Thus, if we are concerned with coin flips and sample honest coins 10 at a time ($N = 10$), the sampling distribution will have a mean equal to 0.5, which is the probability that an honest coin will be heads (or tails) when flipped. All other sample statistics are biased (that is, have sampling distributions with means not equal to the population value).*

Bias is important for a number of reasons. For one thing, knowing that an estimator is unbiased means that we can determine the probability that our sample statistic lies within a given distance of the population value we are trying to estimate. To illustrate, consider a specific problem. Assume that we wish to estimate the average income of a population. A random sample ($N = 500$) is taken and a sample mean of $20,000 is computed. Note that we have no idea what the population value (μ) is (if we did, we wouldn't need

---

*In particular, the sample standard deviation (s) is a biased estimator of the population standard deviation (σ). As you might expect, there is less dispersion in a sample than in a population and, as a consequence, s will underestimate σ. As we shall see below, however, sample standard deviation can be corrected for this bias and still serve as an estimate of the population standard deviation for large samples.

the sample), but it is μ that we are interested in. The sample mean of $20,000 is important and interesting primarily insofar as it can give us information about the population. However, because of the two theorems presented in Chapter 6, we do know that the sampling distribution of all possible sample means is normal and that its mean is equal to the population mean. We also know that all normal curves contain about 68% of the cases (the cases here are sample means) within ± 1 Z, 95% of the cases within ± 2 Z's, and 99% of the cases within ± 3 Z's of the mean. Remember that we are discussing the sampling distribution here—the distribution of all possible sample outcomes or, in this instance, sample means. Thus, the probabilities are very good (approximately 68 out of 100 chances) that our sample mean of $20,000 is within ± 1 Z, excellent (95 out of 100) that it is within ± 2 Z's, and overwhelming (99 out of 100) that it is within ± 3 Z's of the mean of the sampling distribution (which is the same value as the population mean). These relationships are graphically depicted in Figure 7.1.

If an estimator is unbiased, it is probably an accurate estimate of the population parameter (μ in this case). Note that in less than 1% of the cases, a sample mean will be more than ± 3 Z's away from the mean of the sampling distribution (very inaccurate) by random chance alone. We literally have no idea if our sample mean of $20,000 is in this small minority. We do know, however, that the odds are high that our sample mean is considerably closer than ± 3 Z's to the mean of the sampling distribution and, thus, to the population mean.

The second desirable characteristic of an estimator is **efficiency**, which is the extent to which the sampling distribution is clustered about its mean. Efficiency or clustering is essentially a matter of dispersion. The smaller the standard deviation of a sampling distribution, the greater the clustering and

FIGURE 7.1   AREAS UNDER THE SAMPLING DISTRIBUTION OF SAMPLE MEANS

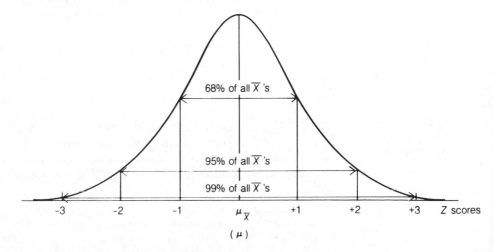

the higher the efficiency. Remember that the standard deviation of the sampling distribution of sample means, or the standard error of the mean, is equal to the population standard deviation divided by the square root of N ($\sigma_{\bar{x}} = \sigma/\sqrt{N}$). Therefore, the standard deviation of this sampling distribution is an inverse function of N. As sample size increases, $\sigma_{\bar{x}}$ will decrease. We can improve the efficiency (or decrease the standard deviation of the sampling distribution) for any estimator by increasing sample size.

An example should make this clearer. Consider two samples of different sizes:

Sample 1	Sample 2
$\bar{X}_1 = \$20{,}000$	$\bar{X}_2 = \$20{,}000$
$N_1 = \quad 100$	$N_2 = \quad 1{,}000$

Both sample means are unbiased, but which is the more efficient estimator? Consider sample 1 and assume, for the sake of illustration, that the population standard deviation (σ) is \$500.* In this case, the standard deviation of the sampling distribution of all possible sample means with an N of 100 would be $\sigma/\sqrt{N}$, or $500/\sqrt{100}$, or \$50.00.

For sample 2, the standard deviation of all possible sample means with an N of 1000 would be much smaller. Specifically, it would be equal to $500/\sqrt{1000}$ or \$15.81.

Sampling distribution 2 is much more clustered than sampling distribution 1. In fact, distribution 2 contains 68% of all possible sample means within ± 15.81 of the $\mu_{\bar{x}}$ while distribution 1 requires a much broader interval of ± 50 to do the same. The estimate based on a sample with 1000 cases is much more likely to approximate the population value than is an estimate based on a sample of 100 cases. Figures 7.2 and 7.3 illustrate these relationships graphically.

To summarize, since the standard deviation of all sampling distributions is an inverse function of N, the larger the sample, the greater the clustering and the higher the efficiency. In part, these relationships between sample size and the standard deviation of the sampling distribution do nothing more than underscore our common-sense notion that much more confidence can be placed in large samples than in small (as long as both have been randomly selected).

7.3 ESTIMATION PROCEDURES: INTRODUCTION

The procedure for constructing a point estimate is straightforward. Draw an EPSEM sample, calculate either a proportion or a mean, and estimate that the population value is the same as the sample statistic. Remember that the larger the sample, the greater the efficiency and the more likely that the estimator is

---

*In reality, of course, the value of σ would be unknown.

FIGURE 7.2   A SAMPLING DISTRIBUTION WITH N = 100 AND $\sigma_{\bar{x}}$ = $50.00

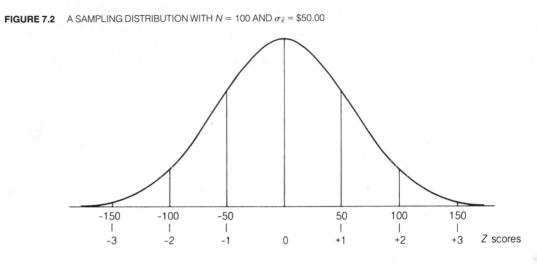

-150	-100	-50		50	100	150
-3	-2	-1	0	+1	+2	+3 Z scores

FIGURE 7.3   A SAMPLING DISTRIBUTION WITH N = 1000 AND $\sigma_{\bar{x}}$ = $15.81

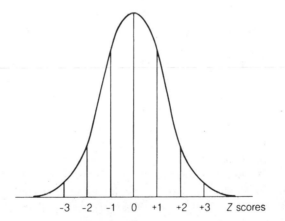

-3 -2 -1 0 +1 +2 +3 Z scores

approximately the same as the population value. Also remember that, no matter how rigid the sampling procedure or how large the sample, there is always some chance that the estimator is very inaccurate.

Compared to point estimates, interval estimates are more complicated but safer in the sense that, by guessing a range of values, we are more likely to include the population value. The first step in constructing an interval estimate is to decide on the risk you are willing to take of being wrong. An interval estimate is wrong if it does not include the population value. This probability of error is called **alpha** (symbolized α). The exact value of alpha will depend on the nature of the research situation, but a 0.05 probability is commonly used. Setting alpha equal to 0.05, also called using the 95% **con-**

FIGURE 7.4 THE SAMPLING DISTRIBUTION WITH ALPHA (α) EQUAL TO 0.05

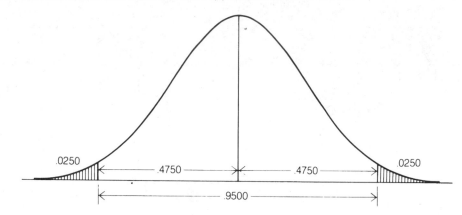

fidence level, means that over the long run the researcher is willing to be wrong only 5% of the time. Or, to put it another way, if an infinite number of intervals were constructed at this alpha level (and with all other things being equal), 95% of them would contain the population value and 5% would not. Since, in reality, only one interval is constructed, by setting the probability of error very low, we are setting the odds in our favor that the interval will include the population value. The second step is to picture the sampling distribution and divide the probability of error equally into the upper and lower tails of the distribution and then find the corresponding Z score. For example, if we decided to set alpha equal to 0.05, we would place half (0.025) of this probability in the lower tail and half in the upper tail of the distribution. The sampling distribution would thus be divided as illustrated in Figure 7.4.

We need to find the Z score that marks the beginnings of the lined areas. In Chapter 5, we learned how to calculate Z scores and find areas under the normal curve. Here, we will simply reverse that process. We need to find the Z score beyond which lies a proportion of .0250 of the total area. To do this, go down column c of Appendix A until you find this proportion (.0250). The associated Z score is 1.96. Since the curve is symmetrical and we are interested in both the upper and lower tails, we designate the Z score that corresponds to an alpha of .05 as ±1.96 (see Figure 7.5).

We now know that 95% of all possible sample outcomes fall within ±1.96 Z-score units of the population value. In reality, of course, there is only one sample outcome but, if we construct an interval estimate based on ±1.96 Z's, the probabilities are that 95% of all such intervals will trap the population value. Thus, we can be 95% confident that our interval contains the population value.

Besides the 95% level, there are two other commonly used confidence

FIGURE 7.5 FINDING THE *Z* SCORE THAT CORRESPONDS TO AN ALPHA (α) OF 0.05

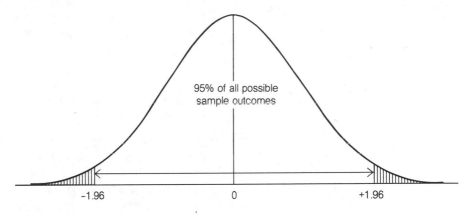

TABLE 7.1 *Z* SCORES FOR VARIOUS LEVELS OF ALPHA (α)

Confidence Level	Alpha	α/2	Z Score
90%	.10	.0500	± 1.65
95%	.05	.0250	± 1.96
99%	.01	.0050	± 2.58

Very important

levels: the 90% level (α = .10) and the 99% level (α = 0.01). To find the corresponding *Z* scores for these levels, follow the procedures outlined above for an alpha of 0.05. Table 7.1 summarizes all the information you will need.

You should turn to the *Z*-score table and confirm for yourself that these scores do indeed correspond to the alpha levels. As you do, note that, in the cases where alpha is set at 0.10 and 0.01, the precise areas we seek do not appear in the table. For example, with an alpha of 0.10, we are looking for the area .0500. Instead we find an area of .0505 (Z = ±1.64) or an area of .0495 (Z = ±1.65). The *Z* score we are seeking is somewhere between these two other scores. When this condition occurs, take the larger of the two scores as *Z*. This will make the interval as wide as possible under the circumstances and is thus the most conservative course of action. In the case of an alpha of 0.01 we encounter the same problem (the exact area .0050 is not in the table), resolve it the same way, and take the larger score as *Z*.

The third step is to actually construct the confidence interval. In the sections that follow, we illustrate how to construct an interval estimate first with sample means and then with sample proportions.

7.4 INTERVAL ESTIMATION PROCEDURES FOR SAMPLE MEANS

The formula for constructing a confidence interval based on sample means is given in Formula 7.1:

FORMULA 7.1

$$\text{c.i.} = \bar{X} \pm Z(\sigma/\sqrt{N})$$

where c.i. = confidence interval
$\bar{X}$ = the sample mean
Z = the Z score as determined by the alpha level
$\sigma/\sqrt{N}$ = the standard deviation of the sampling distribution or the standard error of the mean

As an example, suppose you desired to estimate the average IQ of a community and had randomly selected a sample of 200 residents, with a mean IQ of 105 ($\bar{X} = 105$). From extensive testing we know that the population standard deviation for IQ scores is about 15, so we can set σ equal to 15. If we are willing to run a 5% chance of being wrong and set alpha at 0.05, the corresponding Z score will be ± 1.96. These values can be directly substituted into Formula 7.1 and an interval can be constructed:

$$\text{c.i.} = \bar{X} \pm (Z)(\sigma/\sqrt{N})$$
$$\text{c.i.} = 105 \pm (1.96)(15/\sqrt{200})$$
$$\text{c.i.} = 105 \pm (1.96)(15/14.14)$$
$$\text{c.i.} = 105 \pm (1.96)(1.06)$$
$$\text{c.i.} = 105 \pm 2.08$$

That is, our estimate is that the average IQ for the population in question is somewhere between 102.92 ($105 - 2.08$) and 107.08 ($105 + 2.08$). Since 95% of all possible sample means are within ± 1.96 Z's (or 2.08 IQ units in this case) of the mean of the sampling distribution, the odds are very high that our interval will contain the population mean. In fact, even if the sample mean is as far off as ± 1.96 Z's (which is unlikely), our interval will still contain $\mu_{\bar{x}}$ and, thus, μ. Only if our sample mean is one of the few that is more than ± 1.96 Z's from the mean of the sampling distribution will we have failed to include the population mean.

Note that in the example above, the value of the population standard deviation was supplied. Needless to say, it is unusual to have such information about a population. In the great majority of cases, we will have no knowledge of σ. In such cases, however, we can estimate σ with s, the sample standard deviation. Unfortunately, s is a biased estimator of σ and the formula must be changed slightly to correct for the bias. For larger samples, the bias of s will not affect the interval very much. The revised formula for cases in which σ is unknown is

FORMULA 7.2

$$\text{c.i.} = \bar{X} \pm Z(s/\sqrt{N-1})$$

In comparing this formula with 7.1, note that there are two changes. First, σ is replaced by s and, second, the denominator of the last term is the square root of $N - 1$ rather than the square root of N. The latter change is the correction for the fact that s is biased.

Let me stress here that the substitution of s for σ is permitted only for large samples (that is, samples with 100 or more cases). For smaller samples, when the value of the population standard deviation is unknown, the standardized normal distribution summarized in Appendix A cannot be used in the estimation process. It is perfectly possible to construct meaningful interval estimates for samples smaller than 100 but, to do so, we must use a different theoretical distribution, called the Student's t distribution, to find areas under the sampling distribution. We will defer the presentation of the t distribution until Chapter 8 and confine our attention here to estimation procedures for large samples only.

Let us close this section by working through a sample problem with Formula 7.2. Average income for a random sample of a particular community is $20,000, with a standard deviation of $200. What is the 95% interval estimate of the population mean, μ?

Given that

$$\bar{X} = \$20,000$$

$$s = \$200$$

$$N = 500$$

and using an alpha of 0.05, the interval can be constructed.

$$c.i. = \bar{X} \pm (Z)\left(\frac{s}{\sqrt{N-1}}\right)$$

$$c.i. = 20,000 \pm (1.96)\left(\frac{200}{\sqrt{499}}\right)$$

$$c.i. = 20,000 \pm 17.55$$

The average income for the community as a whole is between $19,982.45 $(20,000 - 17.55)$ and $20,017.55 $(20,000 + 17.55)$. Remember that this interval has only a 5% chance of being wrong (that is, of not containing the population mean).

7.5 INTERVAL ESTIMATION PROCEDURES FOR SAMPLE PROPORTIONS (LARGE SAMPLES)

Estimation procedures for sample proportions are essentially the same as those for sample means. The major difference is that, since proportions are different statistics, we must use a different sampling distribution. In fact, again based on the Central Limit Theorem, we know that sample proportions have sampling distributions that are normal in shape with means (μ_p) equal to the population value (P_u) and standard deviations (σ_p) equal to $\sqrt{P_u(1 - P_u)/N}$. The formula for constructing confidence intervals based on sample proportions is

FORMULA 7.3

$$c.i. = P_s \pm Z\sqrt{\frac{P_u(1 - P_u)}{N}}$$

APPLICATION 7.1

A study of the leisure activities of Americans was conducted on a sample of 1000 households. The respondents identified television viewing as a major form of leisure activity. If the sample reported an average of 6.2 hours of television viewing a day, what is the estimate of the population mean? The information from the sample is

$$\bar{X} = 6.2$$

$$s = 0.7$$

$$N = 1000$$

If we set alpha at 0.05, the corresponding Z score will be ± 1.96, and the 95% confidence interval will be

$$\text{c.i.} = \bar{X} \pm (Z)\left(\frac{s}{\sqrt{N-1}}\right)$$

$$\text{c.i.} = 6.2 \pm (1.96)\left(\frac{0.7}{\sqrt{1000-1}}\right)$$

$$\text{c.i.} = 6.2 \pm (1.96)\left(\frac{0.7}{31.61}\right)$$

$$\text{c.i.} = 6.2 \pm (1.96)(.022)$$

$$\text{c.i.} = 6.2 \pm .04$$

Based on this result, we would estimate that the population spends an average of $6.2 \pm .04$ hours per day viewing television. The lower limit of our interval estimate (6.2 − .04) is 6.16, and the upper limit (6.2 + .04) is 6.24. Thus, another way to state the interval would be

$$6.16 \leq \mu \leq 6.24$$

The population mean is greater than or equal to 6.16 and less than or equal to 6.24. This estimate has a 5% chance of being wrong (that is, of not containing the population mean).

The values for P_s and N come directly from the sample, and the value of Z is determined by the confidence level using the same techniques we introduced when discussing estimation with sample means. This leaves one unknown in the formula, P_u—the value we are trying to estimate. Practically, this dilemma can be resolved by setting the value of P_u at 0.5. Since the second term in the numerator under the radical $(1 - P_u)$ is the reciprocal of P_u, the entire expression will always have a value of 0.5 × 0.5, or 0.25, which is the maximum value this expression can attain. That is, if we set P_u at any value other than 0.5, the expression $P_u(1 - P_u)$ will decrease in value. If we set P_u at 0.4, for example, the second term $(1 - P_u)$ would be 0.6, and the value of the entire expression would decrease to 0.24. Setting P_u at 0.5 ensures that the expression $P_u(1 - P_u)$ will be at its maximum possible value and, consequently, the interval will be at maximum width. This is the most conservative solution possible to the dilemma posed by having to assign a value to P_u in the estimation equation.

To illustrate these procedures, assume that you wish to estimate the proportion of students at your university who missed at least one day of classes because of illness last semester. Out of a random sample of 200 students, 60 reported that they had been sick enough to miss classes at least once during the previous semester. The sample proportion upon which we will base our estimate is thus 60/200, or 0.30. At the 95% level, the interval estimate will be

$$\text{c.i.} = P_s \pm Z \sqrt{\frac{P_u(1 - P_u)}{N}}$$

$$\text{c.i.} = 0.30 \pm 1.96 \sqrt{\frac{(0.5)(0.5)}{200}}$$

$$\text{c.i.} = 0.30 \pm 1.96 \sqrt{\frac{.25}{200}}$$

$$\text{c.i.} = 0.30 \pm (1.96)(0.035)$$

$$\text{c.i.} = 0.30 \pm 0.07$$

Based on this sample proportion of 0.30, you would estimate that the proportion of students who missed at least one day of classes because of illness was between 0.23 and 0.37. The estimate could, of course, also be phrased in percentages by reporting that between 23% and 37% of the student body was affected by illness at least once during the past semester.

7.6 CONTROLLING THE WIDTH OF INTERVAL ESTIMATES

The width of a confidence interval for either sample means or sample proportions can be partly controlled by manipulating two terms in the equation. First, the confidence level can be raised or lowered and, second, the interval can be widened or narrowed by gathering samples of different size. The researcher alone determines the risk he or she is willing to take of being wrong (that is, of not including the population value in the interval estimate). The exact confidence level (or alpha level) will depend, in part, on the purpose of the research. For example, if drugs with potentially harmful effects are being tested, the researcher would naturally demand very high levels of confidence (99.9% or 99.99%). On the other hand, if intervals are being constructed only for loose "guesstimates," then much lower confidence levels can be tolerated (such as 90%). The relationship between interval size and confidence level is that intervals widen as confidence levels increase. This relationship should make intuitive sense. Wider intervals are more likely to trap the population value; hence, more confidence can be placed in them.

To illustrate this relationship, let us return to the example where we estimated the average income for a community. In this problem, we were working with a sample of 500 residents, and the average income for this sample was $20,000, with a standard deviation of $200. We constructed the 95% confidence interval and found that it extended 17.55 around the sample mean (that is, the interval was $20,000 ± 17.55).

APPLICATION 7.2

If 45% of a random sample of 1000 Americans reports that walking is their major physical activity, what is the estimate of the population value? The sample information is

$$P_s = 0.45$$

$$N = 1000$$

Note that the percentage of "walkers" has been stated as a proportion. If we set alpha at 0.05, the corresponding Z score will be ± 1.96, and the interval estimate of the population proportion will be

$$\text{c.i.} = P_s \pm Z\sqrt{\frac{P_u(1 - P_u)}{N}}$$

$$\text{c.i.} = 0.45 \pm 1.96\sqrt{\frac{(0.5)(0.5)}{1000}}$$

$$\text{c.i.} = 0.45 \pm 1.96\sqrt{0.00025}$$

$$\text{c.i.} = 0.45 \pm (1.96)(0.016)$$

$$\text{c.i.} = 0.45 \pm 0.03$$

We can now estimate that the proportion of the population for which walking is the major form of physical exercise is between 0.42 and 0.48. That is, the lower limit of the interval estimate is (0.45 − 0.03) or 0.42, and the upper limit is (0.45 + 0.03) or 0.48. We may also express this result in percentages and say that between 42% and 48% of the population walk as their major form of physical exercise. This interval has a 5% chance of not containing the population value.

Now, if we had constructed the 90% confidence interval for these sample data (a lower confidence level), the Z score in the formula would have decreased to ± 1.65, and the interval would have been narrower:

$$\text{c.i.} = 20,000 \pm 1.65(200/\sqrt{499})$$

$$\text{c.i.} = 20,000 \pm (1.65)(8.95)$$

$$\text{c.i.} = 20,000 \pm 14.77$$

On the other hand, if we had constructed the 99% confidence interval, the Z score would have increased to ± 2.58, and the interval would have been wider:

$$\text{c.i.} = 20,000 \pm 2.58(200/\sqrt{499})$$

$$\text{c.i.} = 20,000 \pm 2.58(8.95)$$

$$\text{c.i.} = 20,000 \pm 23.09$$

TABLE 7.2 INTERVAL ESTIMATES FOR THREE CONFIDENCE LEVELS
($\bar{X}$ = $20,000, s = $200, N = 500 throughout)

Alpha	Confidence Level	Interval	Interval Width
.10	90%	$20,000 $\pm$ 14.77	$29.54
.05	95%	$20,000 $\pm$ 17.55	$35.10
.01	99%	$20,000 $\pm$ 23.09	$46.18

TABLE 7.3 INTERVAL ESTIMATES FOR FOUR DIFFERENT SAMPLES ($\bar{X}$ = $20,000, s = $200, alpha = 0.05 throughout)

Sample 1 (N = 100)	Sample 2 (N = 500)
c.i. = $20,000 $\pm$ 1.96 $(200/\sqrt{99})$	c.i. = $20,000 $\pm$ 1.96 $(200/\sqrt{499})$
c.i. = $20,000 $\pm$ 39.40	c.i. = $20,000 $\pm$ 17.55

Sample 3 (N = 1000)	Sample 4 (N = 10,000)
c.i. = $20,000 $\pm$ 1.96 $(200/\sqrt{999})$	c.i. = $20,000 $\pm$ 1.96 $(200/\sqrt{9999})$
c.i. = $20,000 $\pm$ 12.40	c.i. = $20,000 $\pm$ 3.92

Sample	N	Interval Width
1	100	$78.80
2	500	$35.10
3	1,000	$24.80
4	10,000	$ 7.84

These three intervals are grouped together in Table 7.2, and the increase in interval size can be readily observed. Although sample means have been used to illustrate the relationship between interval width and confidence level, exactly the same relationships apply to sample proportions.

Sample size bears the opposite relationship to interval width. As sample size increases, interval width decreases. Larger samples give more precise (narrower) estimates. Again, an example should make this clearer. In Table 7.3, confidence intervals for four samples of various sizes are constructed and then grouped together for purposes of comparison. The sample data are the same as in Table 7.2, and the confidence level is 95% throughout. The relationships illustrated in Table 7.3 also hold true, of course, for sample proportions.

Notice that the decrease in interval width (or, increase in precision) does not bear a constant or linear relationship with sample size. With sample 2 as compared to sample 1, the sample size was increased by a factor of 5, but the

READING STATISTICS 4: PUBLIC-OPINION POLLS

You are most likely to encounter estimates to the population in the mass media in the form of public-opinion polls, election projections, and the like. Professional polling firms typically use interval estimates. Responsible reporting by the media will usually emphasize the estimate itself (for example, "In a recently conducted survey of the American public, 45% of the respondents said that they approve of the president's performance") but also will report the width of the interval ("This estimate is accurate to within ±3%," or "Figures from this poll are subject to a sampling error of ±3%"), the alpha level (usually as the confidence level of 95%), and the size of the sample ("1458 households were surveyed").

Election projections and voter analyses, at least at the presidential level, have been common for several decades. More recently, public opinion polling has been increasingly used to gauge reactions to everything from the newest movies to the hottest gossip to the President's conduct during the latest crisis. News magazines routinely report poll results as an adjunct to news stories, and similar stories are regular features of TV news and newspapers. I would include an example or two of these applications here, but polls have become so pervasive that you can do your own example. Just pick up a news magazine or newspaper, leaf through it casually, and I bet that you'll find at least one poll. Read the story and see if you can identify the population, the confidence interval width, the sample size, and the confidence level. Bring the news item to class and dazzle your instructor.

As a citizen, as well as a social scientist, you should be extremely suspicious of estimates that do not include such vital information as sample size or interval width. You should also check to see how the sample was assembled. If the sample was not selected in some random fashion from the population as a whole (for example, a magazine polling only its subscribers), it cannot be regarded as representative of the American public.

Advertisements, commercials, and reports published by partisan groups sometimes report statistics that seem to be estimates to the population. Often, however, such estimates are based on woefully inadequate sampling sizes and biased sampling

interval is not five times as narrow. This is an important relationship because it means that N might have to be increased many times to improve the accuracy of an estimate. Since the cost of a research project is a direct function of sample size, this relationship implies a point of diminishing returns in estimation procedures. A sample of 10,000 will cost about twice as much as a sample of 5000, but estimates based on the larger sample will not be twice as precise.

procedures, and the data are collected under circumstances that evoke a desired response. "Person in the street" (or shopper in the grocery store) interviews have a certain folksy appeal to them but must not be accorded the same credibility as surveys conducted by reputable polling firms.

Public Opinion Surveys in the Professional Literature

For the social sciences, probably the single most important consequence of the growth in opinion polling is that many nationally representative data bases are now available for research purposes. These days, access to nationally representative data bases is easy and relatively cheap even for researchers working with very small budgets (the latter is a major factor for most sociologists I know).

It is common practice for researchers and public-opinion polling firms to make these data bases available (for nominal fees) to other researchers. It is now possible to conduct research on "state-of-the-art" samples without the expense and difficulty of collecting data yourself. This is an important development because our research can now be conducted by using these large, representative samples. Our theories can be tested against very high-quality data, and our conclusions will thus have a stronger empirical basis. Our research efforts will have greater credibility with our colleagues, with policy makers, and with the public at large.

One of the more important and widely used data bases of this sort is called the General Social Survey, or the GSS. In virtually every year since 1972, the National Opinion Research Council has questioned a nationally representative sample of Americans about a wide variety of issues and concerns. Since many of the questions are asked every year, the GSS offers a longitudinal record of American sentiment and opinion about a large variety of topics. Each year, new topics of current concern are added and explored, and the variety of information available thus continues to expand.

Like other nationally representative samples, the GSS sample is chosen by a complex probability design that resembles cluster sampling (see Chapter 6). Sample size varies from 1400 to 1600, and estimates based on samples this large will be accurate to within ± 3% (see Table 7.4 and Section 7.7). The computer exercises in this text are based on the 1993 GSS, and this data base is described more fully in Appendix F.

7.7 ESTIMATING SAMPLE SIZE*

The relationships displayed in Table 7.3 suggest a useful way to estimate sample size for any desired level of precision. These techniques would be most useful in the planning stage of a research project because they could yield information about approximate costs.

*This section is optional.

As an example, suppose a national polling organization wants to be able to predict the outcome of presidential races to within ±3%. How large a sample will be needed to ensure this level of precision? In setting the precision, the researcher is essentially predetermining the width of the interval. In this case, suppose we desire that the interval be no wider than ±3%. Starting with the formula for estimating confidence intervals for sample proportions,

$$c.i. = P_s \pm (Z)\sqrt{\frac{P_u(1 - P_u)}{N}}$$

we are predetermining the value of the second and third terms in the equation. Therefore, a new equation can be written:

$$0.03 = Z\sqrt{\frac{P_u(1 - P_u)}{N}}$$

This equation can be algebraically manipulated to isolate the unknown of interest (N). The formula for estimating N for sample proportions when the desired accuracy is ±3% would be

FORMULA 7.4
$$N = \frac{(Z^2)(P_u)(1 - P_u)}{(0.03)^2}$$

Setting alpha at 0.05 ($Z = \pm 1.96$) and P_u equal to 0.5, we can solve for N:

$$N = \frac{(1.96)^2(0.5)(0.5)}{0.0009}$$

$$N = 1067.11$$

Thus, to get precision (or interval width) of ±3%, a sample size of about 1000 is needed. Table 7.4 summarizes the relationship between precision and sample size for sample proportions.

Note that sample size and precision do not increase at the same rate. In

TABLE 7.4   PRECISION AND SAMPLE SIZE (alpha = 0.05, P_u = 0.5 throughout)

Precision (Interval Width)	Approximate Sample Size
± 10%	100
± 7%	200
± 5%	400
± 3%	1000
± 2%	2400
± 1%	9600

fact, sample size increases much faster. To double the precision of the estimate, the sample size must be increased by a factor of 4. To double the precision from 10% to 5%, for example, the sample size must be quadrupled. This relationship is significant for planning purposes because sample size is a rough index of cost, and what is suggested here is a point of diminishing returns. Samples that are impressively large might be a waste of resources, with no real gain in precision over a much smaller (less costly) sample. Of course, the same relationships apply to sample means.

SUMMARY

1. Population values can be inferred from sample values. With point estimates, a single sample statistic is used to estimate the corresponding population value. With confidence intervals, we estimate that the population value falls within a certain range of values.

2. Estimates based on sample statistics must be unbiased and relatively efficient. Of all the sample statistics, only means and proportions are unbiased. The means of the sampling distributions of these statistics are equal to the respective population values. Efficiency is largely a matter of sample size. The greater the sample size, the lower the value of the standard deviation of the sampling distribution, the more tightly clustered the sample outcomes will be around the mean of the sampling distribution, and the more efficient the estimate.

3. With point estimates, we estimate that the population value is the same as the sample statistic (either a mean or proportion). With interval estimates, we construct a confidence interval, a range of values into which we estimate that the population value falls. The width of the interval is a function of the risk we are willing to take of being wrong (the alpha level) and the sample size. The interval widens as our probability of being wrong decreases and as sample size decreases.

*4. These techniques also provide us with a way of estimating sample size for any desired level of precision. This application of estimation techniques is especially useful in the planning stages of a research project.

---

*This section is optional.

SUMMARY OF FORMULAS

Confidence interval for a sample mean, population standard deviation known

$$7.1 \quad \text{c.i.} = \bar{X} \pm Z\left(\frac{\sigma}{\sqrt{N}}\right)$$

Confidence interval for a sample mean, population standard deviation unknown

$$7.2 \quad \text{c.i.} = \bar{X} \pm Z\left(\frac{s}{\sqrt{N-1}}\right)$$

Confidence interval for a sample proportion, large samples

$$7.3 \quad \text{c.i.} = P_s \pm Z\sqrt{\frac{P_u(1-P_u)}{N}}$$

Estimating N for accuracy of $\pm 3\%$

$$7.4 \quad N = \frac{(Z^2)(P_u)(1-P_u)}{(0.03)^2}$$

GLOSSARY

Alpha (α). The probability of error or the probability that a confidence interval does not contain the population value. Alpha levels are usually set at 0.10, 0.05, or 0.01.

Bias. A criterion used to select sample statistics as estimators. A statistic is unbiased if the mean of its sampling distribution is equal to the population value of interest.

Confidence interval. An estimate of a population value in which a range of values is specified.

Confidence level. A frequently used alternate way of expressing alpha, the probability that an interval estimate will not contain the population value. Confidence levels of 90%, 95%, and 99% correspond to alphas of 0.10, 0.05, and 0.01, respectively.

Efficiency. The extent to which the sample outcomes are clustered around the mean of the sampling distribution.

Point estimate. An estimate of a population value where a single value is specified.

PROBLEMS

An asterisk indicates an optional problem.

7.1 For each set of sample outcomes below, construct the 95% confidence interval for estimating μ, the population mean.

a. $\bar{X} = 5.2$ b. $\bar{X} = 100.1$ c. $\bar{X} = 20$
  $s = .75$ $s = 9.8$ $s = 3$
  $N = 157$ $N = 620$ $N = 220$

d. $\bar{X} = 1023$ e. $\bar{X} = 7.3$ f. $\bar{X} = 33$
  $s = 53$ $s = 1.23$ $s = 6.2$
  $N = 329$ $N = 105$ $N = 220$

7.2 For each set of sample outcomes below, construct the 99% confidence interval for estimating P_u.

a. $P_s = .14$ b. $P_s = .37$ c. $P_s = .79$
  $N = 100$ $N = 522$ $N = 121$

d. $P_s = .43$ e. $P_s = .40$ f. $P_s = .63$
  $N = 1049$ $N = 578$ $N = 300$

7.3 For each confidence level below, determine the corresponding Z score.

Confidence Level	Alpha	Area Beyond Z	Z score
95%	.05	.0250	±1.96
94%			
92%			
97%			
98%			
99.9%			

7.4 For the sample data below, construct three different interval estimates of the population mean, one each for the 90%, 95%, and 99% level. What hap-

pens to the interval width as confidence level increases? Why?

$$\bar{X} = 100$$
$$s = 10$$
$$N = 500$$

7.5 For each of the three sample sizes below, construct the 95% confidence interval. Use a sample proportion of 0.40 throughout. What happens to interval width as sample size increases? Why?

$$P_s = 0.40$$
Sample A: $N = 100$
Sample B: $N = 1000$
Sample C: $N = 10,000$

7.6 CJ A random sample of 500 residents of Shinbone, Kansas, shows that exactly 50 of the respondents had been the victims of violent crime over the past year. Estimate the proportion of victims for the population as a whole, using the 90% confidence level.

7.7 SW You have developed a series of questions that measures "burnout" in social workers. A random sample of 100 social workers working in greater metropolitan Shinbone, Kansas, has an average score of 10.3, with a standard deviation of 2.7. What is your estimate of the average burnout score for the population as a whole? Use the 95% confidence level.

7.8 PS Two individuals are running for mayor of Shinbone. You conduct an election survey a week before the election and find that 51% of the respondents prefer candidate A. Can you predict a winner? Use the 99% level.

$$P_s = 0.51$$
$$N = 578$$

7.9 SOC The fraternities and sororities at St. Algebra College have been plagued by declining membership over the past several years and want to know if the incoming freshman class will be a fertile recruiting ground. Not having enough money to survey all 1600 freshmen, they commission you to survey the interests of a random sample of 150.

You find that 35 respondents are "extremely" interested in social clubs. At the 95% level, what is your estimate of the number of freshmen who would be extremely interested?

7.10 SOC You are the hard-hitting, painfully honest consumer-affairs reporter for a Shinbone daily newspaper. Part of your job is to investigate the claims of manufacturers, and you are particularly suspicious of a new economy car that the manufacturer claims will get 78 miles per gallon. After checking the mileage figures for a sample of 120 owners of this car, you find the average miles per gallon of your sample is 75.5, with standard deviation of 3.7. At the 99% level, do your results tend to confirm or refute the manufacturer's claims?

7.11 SOC A random sample of 178 households finds that an average of 2.3 people reside in each household. What is your estimate of the population mean? Use the 90% level.

$$\bar{X} = 2.3$$
$$s = .35$$
$$N = 178$$

7.12 SOC The same survey revealed that 25 of the households consisted of unmarried couples who were living together. What is your estimate of the population proportion? Use the 95% level.

7.13 PA A random sample of 324 residents of a community revealed that 30% were very satisfied with the quality of trash collection. At the 99% level, what is your estimate of the population value?

*7.14 GER You are the head of the research division for a local agency that serves senior citizens. The agency wants to conduct a survey to determine the number of potential clients it has for its various services. The agency head wants the estimates to be accurate to within ±3%. How large a sample will you need to draw?

7.15 SOC A random sample of 100 television programs contained an average of 2.37 acts of physical violence per program. At the 99% level, what is your estimate of the population value?

$$\bar{X} = 2.37$$
$$s = .30$$
$$N = 100$$

7.16 SOC A random sample of 1000 registered voters in a state were asked if they favored a proposal to restrict the sale of handguns. If 510 approved of the legislation, is it safe, at the 95% level, to conclude that "a majority" of the population approves?

7.17 SOC A random sample of 429 college students reported that they had spent an average of $178.23 on textbooks during the previous semester. If the sample standard deviation for these data is $15.78, construct an estimate of the population mean at the 99% level.

7.18 SOC A random sample of 1496 respondents was asked to agree or disagree with the statement "Explicit sexual books and magazines lead to rape and other sex crimes" and 823 agreed. At the 90% level, construct an estimate to the population proportion.

7.19 SW A random sample of 100 patients treated in a program for alcoholism and drug dependency over the past 10 years was selected. It was determined that 53 of the patients had been readmitted to the program at least once. At the 95% level, construct an estimate to the population proportion.

7.20 GER A sample of 103 residents of a retirement community was interviewed to assess the level of satisfaction with present services and the need for additional services. For each finding below, construct the appropriate confidence interval at the 95% level.
  a. A total of 54 residents reported that they were "very satisfied" with the quality of health care in the facility.
  b. The sample reported using the shuttle bus to town an average of 5.7 times per week with a standard deviation of .70.
  c. Only 17 of the respondents reported that they had ever attended any of the free crafts classes.
  d. The sample reported watching TV an average of 6.9 hours per day with a standard deviation

of 1.3. Also, 37 reported that they were "very satisfied" with the quality of programming available through the cable system.

e. When asked what services or facilities they would like to see added, 50 wanted a heated swimming pool, 40 wanted expanded health education programs, and 13 wanted a dating service established within the community.

7.21 SOC Let's use the General Social Survey (GSS) as the basis for some estimation problems. The GSS is administered to an EPSEM sample that should be representative of American society. The entire sample consists of about 1500 respondents, but the version used here has less than 800 respondents. Our statistics will be calculated on a "sample of a sample" and our estimates will have an additional risk of error. To judge the accuracy of our estimates, information on population parameters is given where possible in the Answers to Odd-Numbered Problems. See Appendix G for the codes and the actual wording of the questions. For each sample statistic reported below, construct the 95% confidence interval estimate of the population value.

a. The average occupational prestige score (item 2) for 745 respondents was 43.06, with a standard deviation of 13.21.

b. The respondents ($N = 784$) reported watching an average of 2.80 hours of TV per day (item 67) with a standard deviation of 1.96.

c. The average number of children (item 6) for 787 respondents was 1.86, with a standard deviation of 1.70.

d. Of the 788 respondents who answered the question, 188 identified themselves as Catholic (item 28).

e. One hundred fifty-two of the 788 respondents said that they had never married (item 3).

f. Of the 532 respondents for whom we have information, 224 said that they voted for Clinton (item 18) in the 1992 presidential election.

g. When asked if they had ever been hit by another person, 191 of 520 respondents said yes (item 63).

SPSS/PC+ PROCEDURES FOR CONFIDENCE INTERVALS

DEMONSTRATION 7.1 Generating Statistical Information for Use in Constructing Confidence Intervals

SPSS/PC+ does not provide any programs specifically for constructing confidence intervals. Some of the procedures we'll cover in future chapters do include confidence intervals as part of the output (see the SPSS/PC+ section for Chapter 12). Rather than make use of these programs, I want to use this section to confirm something you may already suspect: fancy computer programs are not always helpful or even particularly useful. The arithmetic required by estimation procedures (let's face it) is not particularly difficult, and you could probably complete the formulas faster by hand than by SPSS/PC+.

On the other hand, who wants to do all the calculations it would take to get the summary statistics on which the estimates will be based? Calculating the mean age for the GSS sample would require the addition of almost 800 numbers. The dimensions of that task should suggest the proper role of the computer. Once you know the mean age, the rest of the calculations for confidence intervals are not very formidable. So, how do you get the sample statistics?

Take a look at problem 7.21. I used a number of variables from the GSS data set for this problem and got the information I needed by using DESCRIPTIVES for the interval-ratio variables and FREQUENCIES for the nominal variables. Assuming that the 1993 GSS is the active file, the commands I used were

```
DESCRIPTIVES /VARIABLES PRESTG80 TVHOURS CHILDS /
STATISTICS 1 5.
FREQUENCIES /VARIABLES RELIG MARITAL PRES92 HIT.
```

The first command gives means and standard deviations, and the second produces frequency distributions. The latter command requires no subcommands for STATISTICS since the program automatically computes percentages that we can convert to proportions to construct our confidence intervals.

Exercises

7.1 Use the DESCRIPTIVES command to calculate means and standard deviations for PAPRES80, AGE, and INCOME. Use the output to formulate confidence interval estimates for the population values. NOTE: your estimate based on AGE will not reflect the total American population. The GSS sample is restricted to people age 18 and older, and the mean of the sample is much higher than the mean of the population as a whole.

7.2 Use the FREQUENCIES command to get sample proportions (convert the percentages in the frequency distributions) to estimate the population parameter for each of the following: proportion employed full time (WRKSTAT), proportion with college degree (DEGREE), proportion black (RACE), and proportion favoring gun control (GUNLAW).

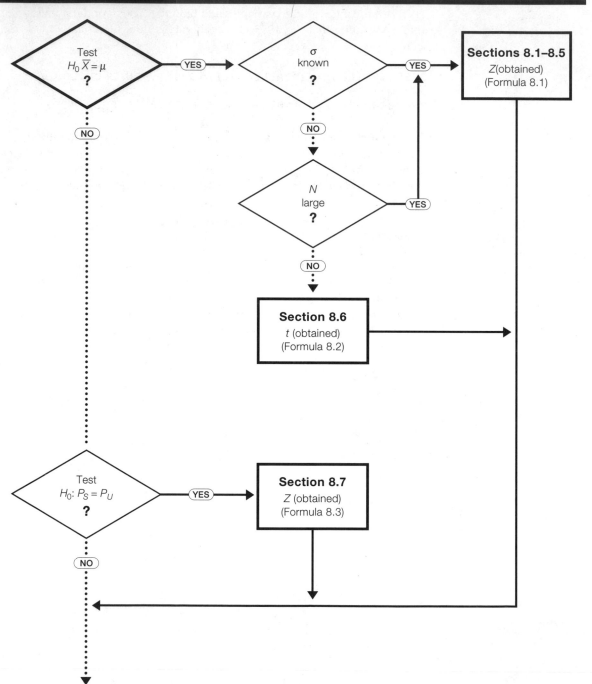

8.1 INTRODUCTION

In Chapter 7, I introduced the basic techniques for estimating population values from sample statistics. We will now investigate a rather different application of inferential statistics called **hypothesis testing** or **significance testing.** In this chapter, the techniques for hypothesis testing in the one-sample case will be introduced. The central problem in these situations is to decide (with a known probability of error) if a sample came from a population with specific characteristics. We will have some information about a population and some statistics calculated from a sample. If the differences between the population values and the sample values are large enough, we will conclude that the sample did not come from that specific population. If the differences are small, or if there are no differences, we'll conclude that the sample did come from the defined population.

These techniques might be used in such research situations as the following:

(a) A researcher has selected a sample of 789 older citizens who live in a particular city and also has some information on the rate of criminal victimization for the city as a whole. Do older citizens, as represented by the sample, have the same rate of victimization as the city as a whole?

(b) Are the GPAs of college athletes different from the GPAs of the general student body? To investigate, the academic records of a random sample of 129 student athletes from a large state university are compared with the characteristics of all students at the school.

In each of these situations, we would compare sample and population information and decide if the sample was taken from a population with the specified characteristic.

No matter what our final decision, there is a chance that we will make an error. We cannot avoid risk because, as we saw in Chapter 6, even the most painstaking sampling methods will occasionally produce a nonrepresentative sample. One of the great advantages of inferential statistics, however, is that we will be able to estimate the probability of this type of error and evaluate our decisions accordingly.

8.2 AN OVERVIEW OF HYPOTHESIS TESTING

Let's begin by considering a situation in which a single sample test of hypotheses would be appropriate. Suppose that a researcher has been hired to evaluate a treatment program for alcoholics. The program serves a very large metropolitan area, and the researcher is unable to test every client. A sample of 127 is drawn from a list of all those who have completed the program. While examining the work records of the subjects, the researcher notes that they appear to have lower absentee rates from work than the community as a whole. Specifically,

Community	Sample
$\mu = 7.2$ days per year	$\bar{X} = 6.8$ days per year
$\sigma = 1.43$	$N = 127$

Consider the information we have at this point. We have information about a population ($\mu = 7.2$) and information about a sample ($\bar{X} = 6.8$). Our question is this: Do treated alcoholics, as represented by this sample of 127, have lower absentee rates than the community as a whole? Clearly, we would be in a better position to answer this question if we had information about *everyone* who had completed the program (the population of all treated alcoholics). Then, we could compare one mean (the community) with another mean (all treated alcoholics), and we would know, without doubt, if the program was effective on this criterion. By using a sample, we create the possibility of error since samples are sometimes not representative of the populations from which they are selected.

Given the fact that we are working with a sample rather than a population, there are two ways to account for the difference in absentee rates. The first, which we shall label explanation A, is that alcoholics treated in the program really are different from the community as a whole in terms of absentee rates. Another way to say this would be "the difference between absentee rates is statistically significant." The phrase "statistically significant" means that the difference is very unlikely to have occurred by random chance alone. If explanation A is true, the sample of treated alcoholics did *not* come from a population with a mean absentee rate of 7.2 days.

In direct contradiction to explanation A is explanation B: there is really no difference between alcoholics treated in this program and the community as a whole in terms of absenteeism. The difference noted above is trivial and due to the effects of random chance. If explanation B is true, the sample *did* come from a population with a mean absentee rate of 7.2 days.

Which explanation for the observed difference is correct? As long as we are working with a sample rather than a population, we cannot know which explanation is correct on an absolute basis. However, we can set up a decision-making procedure so conservative that one of the two explanations can be chosen, with the knowledge that the probability of making an incorrect choice is very low.

This decision-making process, in broad outline, begins with the assumption that explanation B is correct. Symbolically, this assumption can be stated as

$$\mu = 7.2 \text{ days per year}$$

Remember that μ refers to the mean of the population of all alcoholics who have completed the program. This assumption, $\mu = 7.2$, can be tested statistically. If explanation B is true, then the probability of getting the observed sample outcome ($\bar{X} = 6.8$) can be found. Let us add an objective decision rule in advance. If the odds of getting the observed difference are less than .05 (5 out of 100, or 1 to 20), we will declare explanation B unlikely. If this explanation were true, a difference of this size (7.2 days vs. 6.8 days) would be a very rare event and, in hypothesis testing as well as gambling, we always bet against rare events.

The remaining problem is to determine the probability of the observed sample outcome if B is correct. We can determine this probability with precision because of what we know about the sampling distribution of all possible sample outcomes. Looking back at the information we have, we can see that we know a great deal about the sampling distribution of sample means. Based on the Central Limit Theorem and on our assumption that explanation B is true, we may assume that the sampling distribution is normal in shape, has a mean of 7.2 ($\mu_{\bar{X}} = \mu$), and a standard deviation of $1.43/\sqrt{127}(\sigma_{\bar{X}} = \sigma/\sqrt{N}$). We also know that the standard normal distribution can be interpreted as a distribution of probabilities and that the particular sample outcome noted above ($\bar{X} = 6.8$) is one of thousands of possible sample outcomes. The sampling distribution, with the sample outcome noted, is depicted in Figure 8.1.

Using our knowledge of the standardized normal distribution, we can add further useful information to this sampling distribution of sample means. Specifically, with Z scores, we can depict the decision rule stated previously—any sample outcome with probability less than 0.05 if explanation B is true will be cause to declare explanation B unlikely. The probability of 0.05 can be translated into an area and divided equally into the upper and lower tails of the sampling distribution. Using Appendix A, we find that the Z-score equivalent of this area is ± 1.96. The areas and Z scores are depicted in Figure 8.2.

FIGURE 8.1   THE SAMPLING DISTRIBUTION OF ALL POSSIBLE SAMPLE MEANS

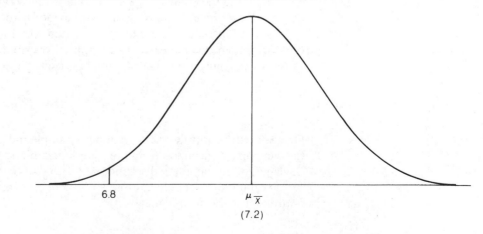

FIGURE 8.2   THE SAMPLING DISTRIBUTION OF ALL POSSIBLE SAMPLE MEANS

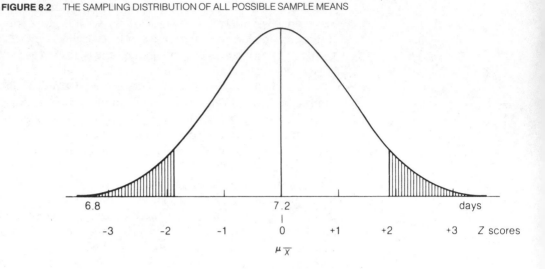

The decision rule can now be rephrased. Any sample outcome falling in the lined areas depicted in Figure 8.2 by definition has a probability of occurrence of less than 0.05. Such an outcome would be a rare event and would cause us to reject explanation B. All that remains is to translate our sample outcome into a Z score so we can see where it falls on the curve. To do this, we use the standard formula for locating any particular raw score under a normal distribution. When we use known or empirical distributions, this formula is expressed as

$$Z = \frac{X_t - \bar{X}}{s}$$

Or, to find the equivalent Z score for any raw score, subtract the mean of the distribution from the raw score and divide by the standard deviation of the distribution. Since we are now concerned with the sampling distribution of all sample means rather than an empirical distribution, the symbols in the formula will change, but the form remains exactly the same:

FORMULA 8.1
$$Z = \frac{\bar{X} - \mu_{\bar{x}}}{\sigma/\sqrt{N}}$$

Or, to find the equivalent Z score for any sample mean, subtract the mean of the sampling distribution from the sample mean and divide by the standard deviation of the sampling distribution.

Recalling the data given on this problem, we can now find the Z-score equivalent of the sample mean:

Community	Sample
$\mu = 7.2\ (=\mu_{\bar{x}})$	$\bar{X} = 6.8$
$\sigma = 1.43$	$N = 127$

$$Z = \frac{\bar{X} - \mu_{\bar{x}}}{\sigma/\sqrt{N}}$$

$$Z = \frac{6.8 - 7.2}{1.43/\sqrt{127}}$$

$$Z = -\frac{.400}{.127}$$

$$Z = -3.15$$

In Figure 8.3, this Z score of -3.15 is noted on the distribution of all possible sample means, and we see that the sample outcome does fall in the shaded area. If explanation B is true, this particular sample outcome has a probability of occurrence of less than 0.05 (how much less than 0.05 is irrelevant, since the decision rule was established in advance). The sample outcome ($\bar{X} = 6.8$ or $Z = -3.15$) would therefore be rare if explanation B were true, and the researcher is justified in rejecting explanation B. If B were true, this sample outcome would be extremely unlikely. Therefore, the researcher would bet that B is not true and that our 127 treated alcoholics come from a population that is significantly different from the community on the trait of absenteeism.

Keep in mind that in inferential statistics, our conclusions must be based on information gathered from random samples. On rare occasions, a given sample may not be representative of the population from which it was se-

FIGURE 8.3   THE SAMPLING DISTRIBUTION OF SAMPLE MEANS WITH THE SAMPLE OUTCOME ($\bar{X} = 6.8$) NOTED IN Z SCORES

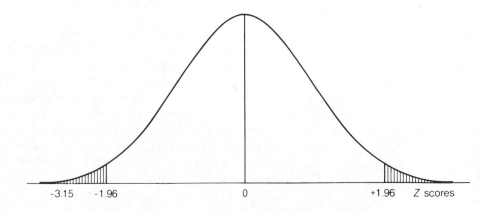

lected. Let me repeat that the decision-making process outlined above has a very high probability of resulting in correct decisions, but, as long as we must work with samples rather than populations, we face an element of risk in the process. That is, the decision with respect to explanation B might be incorrect if this sample happens to be one of the few that is unrepresentative of the population of alcoholics treated in this program. One important strength of hypothesis testing is that the probability of making an incorrect decision can be estimated. In the example at hand, explanation B was rejected and the probability of this decision being incorrect is 0.05—the decision rule established at the beginning of the process. To say that the probability of having rejected B incorrectly is 0.05 means that, if we repeated this same test an infinite number of times, we would incorrectly reject B only 5 times out of every 100.

8.3 THE FIVE-STEP MODEL FOR HYPOTHESIS TESTING

All the formal elements and concepts used in hypothesis testing were surreptitiously sneaked into the preceding discussion. This section presents their proper names and introduces a **five-step model** for organizing all hypothesis testing:

Step 1. Making assumptions

Step 2. Stating the null hypothesis

Step 3. Selecting the sampling distribution and establishing the critical region

Step 4. Computing the test statistic

Step 5. Making a decision

We will look at each step individually, using the problem from Section 8.2 as an example throughout.

Step 1. Making assumptions. Any application of a statistical technique requires that certain assumptions be made about the data. When conducting a test of hypothesis with a single sample mean, we must assume that the sample has been randomly selected from the defined population. Also, to justify computation of a mean, we must assume that the data are interval-ratio in level of measurement. Finally, we must assume that the sampling distribution of all possible sample means is normal in shape so that we may use the standardized normal distribution to find areas under the sampling distribution. We will refer to these assumptions about the sampling process, the level of measurement, and the shape of the sampling distribution as the "mathematical model." You should be very certain of the correctness of these model assumptions.

Usually, we will state these assumptions in abbreviated form. For example:

Model: Random sampling
          Level of measurement is interval ratio
          Sampling distribution is normal

Step 2. Stating the null hypothesis (H_0). The **null hypothesis** is the formal name for explanation B and is always a statement of "no difference." The exact form of the null hypothesis will vary, depending on the test being conducted. In the single-sample case, the null hypothesis states that the sample comes from a population with a certain characteristic. In our example, the null is that treated alcoholics are "no different" from the community as a whole, that the average days of absenteeism for this population is also 7.2, and that the difference between 7.2 and the sample mean of 6.8 is caused by random chance. Symbolically, the null would be stated as

$$H_0: \mu = 7.2$$

where μ refers to the mean of the population of treated alcoholics. The null hypothesis is the central element in any test of hypothesis because the entire process is aimed at rejecting or failing to reject the H_0.

Usually, the researcher believes that there is a significant difference and desires to reject the null hypothesis. At this point in the five-step model, the researcher's belief is stated in a **research hypothesis** (H_1), a statement that directly contradicts the null hypothesis. Thus, the researcher's goal in hypothesis testing is often to gather evidence for the research hypothesis by rejecting the null hypothesis. The research hypothesis can be stated in several ways. One form would simply assert that the sample was selected from a population which did not have a certain characteristic or, in terms of our example, had a mean which was not equal to a specific value:

$$(H_1: \mu \neq 7.2)$$
where ≠ means "not equal to"

Symbolically, this statement asserts that the sample does not come from a population with a mean of 7.2, or that treated alcoholics are different from the community as a whole. The research hypothesis is enclosed in parentheses to emphasize that it has no formal standing or role in the hypothesis-testing process (except, as we shall see in the next section, in choosing between one-tailed and two-tailed tests). It serves as a reminder of what the researcher believes to be the truth.

Step 3. **Selecting the sampling distribution and establishing the critical region.** The sampling distribution is, as always, the probabilistic yardstick against which a particular sample outcome is measured. By assuming that the null hypothesis is true (and *only* by this assumption), we can attach values to the mean and standard deviation of the sampling distribution and thus measure the probability of any specific sample outcome. There are several different sampling distributions, but for now we will confine our attention to the sampling distribution of sample means. We can measure areas under this distribution by use of the standard normal curve as summarized in Appendix A.

The **critical region** consists of the areas under the sampling distribution that include all unlikely sample outcomes. Prior to the test of hypothesis, we must define what is meant by unlikely. That is, we must specify in advance those sample outcomes so unlikely that they will lead us to reject the H_0. This decision rule is nothing more than establishing a critical region or a region of rejection. The word *region* is used because, essentially, we are describing those areas under the sampling distribution that contain unlikely sample outcomes. In the example above, this area corresponded to a Z score of ± 1.96, called **Z (critical)**, which was graphically displayed in Figure 8.2. The lined area was the critical region. Any sample outcome for which the Z-score equivalent fell in this area (that is, below -1.96 or above $+1.96$) would have caused us to reject the null hypothesis.

By convention, the size of the critical region is reported as alpha (α), the proportion of all of the area included in the critical region. In the example above, our **alpha level** was 0.05. Other commonly used alphas are 0.01 and 0.10.

In abbreviated form, all the decisions made in this step are noted below. The critical region is noted by the Z scores that mark its beginnings.

$$\text{Sampling distribution} = Z \text{ distribution}$$

$$\alpha = 0.05$$

$$Z \text{ (critical)} = \pm 1.96$$

Step 4. Computing the test statistic. To evaluate the probability of any given sample outcome, the sample value must be converted into a Z score. Solving the equation for Z-score equivalents is called computing the **test statistic**, and the resultant value will be referred to as **Z (obtained)** in order to differentiate the test statistic from the critical region. In our example above, we found a Z (obtained) of -3.15.

Step 5. Making a decision. As the last step in the hypothesis-testing process, the test statistic is compared with the critical region. If the test statistic falls in the critical region, our decision will be to reject the null hypothesis. If the test statistic does not fall in the critical region, we fail to reject the null hypothesis. In our example, the two values were

$$Z \text{ (critical)} = \pm 1.96$$

$$Z \text{ (obtained)} = -3.15$$

and we saw that the Z (obtained) fell in the critical region whose beginnings are marked by Z (critical). Our decision was to reject the null hypothesis and conclude that the difference observed with respect to absenteeism was unlikely to have occurred by chance alone. On the trait of absenteeism, treated alcoholics are different from the community as a whole.

This five-step model will serve us as a framework for decision making throughout the hypothesis testing chapters. The exact nature and method of expression for our decisions will, of course, change as we encounter the various tests. However, familiarity with this model will assist you in mastering this material by providing a common frame of reference for all significance testing.

8.4 ONE-TAILED AND TWO-TAILED TESTS OF HYPOTHESIS

In reviewing the five-step model for hypothesis testing, note that the researcher has little room for making choices. The model is fairly rigid, resembling a recipe. Follow the steps one by one as specified above and you cannot make a mistake. In spite of the inflexibility of the procedure, however, two crucial choices must be made. First, the researcher must decide between a one-tailed and a two-tailed test. Second, an alpha level must be selected. In this section, we will discuss the former decision and, in Section 8.5, the latter.

The choice between a one- and two-tailed test is based on the researcher's expectations about the population from which the sample was selected. These expectations are reflected in the research hypothesis (H_1), which usually states what the researcher believes to be "the truth" and is contradictory to the null hypothesis. In most situations, the researcher will wish to support the research hypothesis by rejecting the null hypothesis.

The format for the research hypothesis may take either of two forms, depending on the relationship between what the null hypothesis states and what the researcher believes to be the truth. The null hypothesis states that the population has a specific characteristic. In the example that has served us throughout this chapter, the null stated, in symbols, that "treated alcoholics have the same absentee rate (7.2 days) as the community." The researcher might believe that the population value is actually greater than or less than the value stated in the null, or he or she might be unsure about the direction of the difference.

If the researcher is unsure about the direction, the research hypothesis would state only that the population is "not equal" to the value or does not have the specific characteristic stated in the null. The research hypothesis stated in Section 8.3 ($\mu \neq 7.2$) was in this format. This is called a **two-tailed test** of significance. The researcher will be equally concerned with the possibility that the true population value is greater than the value specified in the null hypothesis and the possibility that the true population value is less than the value specified in the null hypothesis.

In other situations, the researcher might be concerned only with differences in a specific direction. If the direction of the difference can be predicted, or if the researcher is concerned only with differences in one direction, a **one-tailed test** can be used. A one-tailed test may take one of two forms, depending on the researcher's expectations about the specific direction of the difference. If the researcher believes that the true population value is higher than the value specified in the null, the research hypothesis would reflect that belief. In our example, if we had predicted that treated alcoholics had higher absentee rates than the community, our research hypothesis would have been

$$(H_1: \mu > 7.2)$$
where $>$ signifies "greater than"

For a one-tailed test in the opposite direction, the research hypothesis would be

$$(H_1: \mu < 7.2)$$
where $<$ signifies "less than"

This research hypothesis reflects the belief that treated alcoholics have lower absentee rates.

An example of a situation in which a one-tailed test is called for might be in program evaluation. Consider the evaluation of a random sample of programs designed to reduce unemployment. The evaluators would be concerned only with outcomes that show a decrease in the unemployment rate. If the rate shows no change, or if unemployment increases, the program is a failure. Both of these outcomes are equally negative with regard to the stated program goals.

In terms of the five-step model, the choice of a one-tailed or two-tailed test determines what we do with the critical region under the sampling distribution in step 3. As you recall, in a two-tailed test, we split the critical region equally into the upper and lower tails. In a one-tailed test, we place the entire critical area in one tail of the sampling distribution. If we believe that the population characteristic is "greater than" the value stated in the null (if the H_1 includes the $>$ symbol), we place the entire critical region in the upper tail. If we believe that the difference is "less than" the null value (if the H_1 includes the $<$ symbol), the entire critical region goes in the lower tail.

For example, in a two-tailed test with alpha equal to 0.05, the critical region begins at Z (critical) $= \pm 1.96$. In a one-tailed test at the same alpha level, the Z (critical) is $+1.65$ if the upper tail is specified and -1.65 if the lower tail is specified. The contrast in the placement of the critical region is graphically summarized in Figure 8.4.

You should immediately see that a one-tailed test is advantageous because it moves the critical region closer to the mean of the sampling distribution and thus improves the probability of rejecting the H_0 without affecting the alpha level. Note also that, if the wrong tail has been specified, the probability of rejection is zero. Thus, there is a risk involved in using the one-tailed

FIGURE 8.4   ESTABLISHING THE CRITICAL REGION, ONE-TAILED TESTS VERSUS TWO-TAILED TESTS (alpha = 0.05)

A.   The two-tailed test, Z (critical) = ±1.96

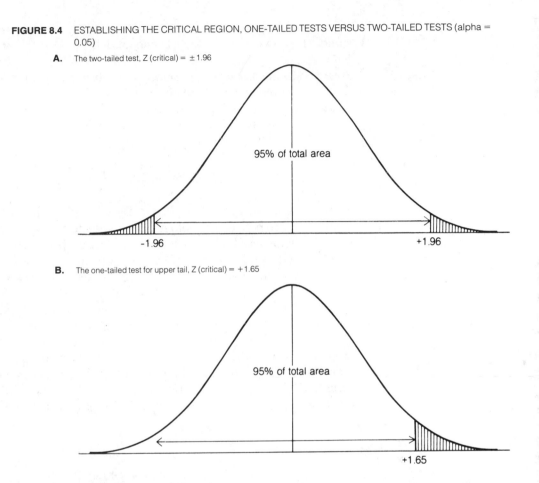

95% of total area

-1.96   +1.96

B.   The one-tailed test for upper tail, Z (critical) = +1.65

95% of total area

+1.65

test. But, assuming that the proper tail has been specified, one-tailed tests are a way of statistically both having and eating your cake and should be used whenever (1) the direction of the difference can be confidently predicted or (2) the researcher is concerned only with differences in one tail of the sampling distribution.

An example may clarify the use of the one-tailed test. After many years of work, a sociologist has noted that sociology majors seem more sophisticated, charming, and cosmopolitan than the rest of the general student body. A "Sophistication Scale" test has been administered to the entire student body and to a random sample of 100 sociology majors, and these results have been obtained:

Student body	Sociology majors
$\mu = 17.3$	$\bar{X} = 19.2$
$\sigma = 7.4$	$N = 100$

To test the H_0 of no difference between sociology majors and the general student body, we will use the five-step model.

Step 1. Making assumptions. Since we are using a mean to summarize the sample outcome, we must assume that the Sophistication Scale generates interval-ratio level data. With a sample size of 100, the Central Limit Theorem applies and we can assume that the sampling distribution is normal in shape.

> Model: Random sampling
> Level of measurement is interval ratio
> Sampling distribution is normal

Step 2. Stating the null hypothesis (H_0). The null hypothesis states that there is no difference between sociology majors and the general student body. The research hypothesis (H_1) will also be stated at this point. The researcher has predicted a direction for the difference ("Sociology majors are more sophisticated"), so a one-tailed test is justified. The two hypotheses may be stated as

$$H_0: \mu = 17.3$$

$$(H_1: \mu > 17.3)$$

Step 3. Selecting the sampling distribution and establishing the critical region. We will use the standardized normal distribution (Appendix A) to find areas under the sampling distribution. If alpha is set at 0.05, the critical region will begin at the Z score +1.65. These decisions may be summarized as

$$\text{Sampling distribution} = Z \text{ distribution}$$

$$\alpha = 0.05 \text{ (one-tailed)}$$

$$Z \text{ (critical)} = +1.65$$

Step 4. Computing the test statistic.

$$Z \text{ (obtained)} = \frac{\bar{X} - \mu}{\sigma/\sqrt{N}}$$

$$Z \text{ (obtained)} = \frac{19.2 - 17.3}{7.4/\sqrt{100}}$$

$$Z \text{ (obtained)} = +2.57$$

Step 5. Making a decision. Comparing the Z (obtained) with the Z (critical):

$$Z \text{ (critical)} = +1.65$$

$$Z \text{ (obtained)} = +2.57$$

We see that the test statistic falls in the critical region. This outcome is depicted graphically in Figure 8.5. We will reject the null hypothesis because,

FIGURE 8.5   *Z* (OBTAINED) VERSUS *Z* (CRITICAL) (alpha = 0.05, one-tailed test)

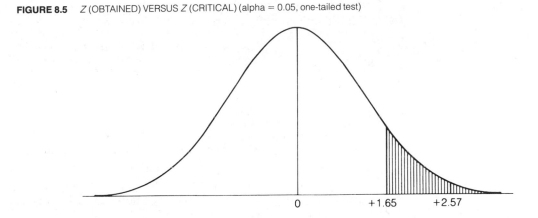

if the H_0 were true, a difference of this size would be very unlikely. There is a significant difference between sociology majors and the general student body in terms of sophistication. Since the null hypothesis has been rejected, the research hypothesis (sociology majors are more sophisticated) is supported.

8.5 SELECTING AN ALPHA LEVEL

In addition to deciding between one-tailed and two-tailed tests, the researcher must also select an alpha level. We have seen that the alpha level plays a crucial role in hypothesis testing. By assigning a value to alpha, we in effect define what we will mean by an "unlikely" sample outcome. If the probability of the observed sample outcome is lower than the alpha level (if the test statistic falls in the critical region), we reject the null hypothesis as untrue. Thus, the alpha level will have important consequences for our decision in step 5.

How can reasonable decisions be made with respect to the value of alpha? Recall that, in addition to defining what will be meant by *unlikely*, the alpha level is also the probability that, if the test statistic falls in the critical region, the decision to reject the null will be incorrect. In hypothesis testing, this kind of error is called **Type I error**, or **alpha error**. Type I error can be defined as the rejection of a null hypothesis that is, in fact, true or, more simply if somewhat redundantly, as falsely rejecting a true null. To minimize this type of error, very small alphas should be used.

To elaborate, when an alpha level is specified, the sampling distribution is divided into two sets of possible sample outcomes. The critical region includes all sample outcomes that we will define as unlikely or rare and that will cause us to reject the null hypothesis. The remainder of the area consists of all sample outcomes that we have defined as "nonrare." The lower the alpha level, the smaller the critical region and the greater the distance between the mean of the sampling distribution and the beginnings of the critical region. Compare, for the sake of illustration, the following alpha levels and Z (criticals) for two-tailed tests.

If alpha equals	The two-tailed critical region will begin at Z (critical) equal to
0.10	± 1.65
0.05	± 1.96
0.01	± 2.58

As alpha goes down, the critical region moves farther away from the mean of the sampling distribution. The lower the alpha level, the harder it will be to reject the null and, since a Type I error can be made only if our decision in step 5 is to reject, the lower the probability of Type I error. To minimize the probability of rejecting a null hypothesis that is in fact true, use very low alpha levels.

However, there is a complication. As the critical region decreases in size (as alpha levels decrease), the noncritical region must become larger. All other things being equal, the lower the alpha level, the less likely that the sample outcome will fall in the critical region. This raises the possibility of a second type of incorrect decision, called **Type II error**, or **beta error**, which can be defined as failing to reject a null that is, in fact, false. Although the probability of Type I error decreases with lower levels of alpha, the probability of Type II error increases. Thus, the two types of error are inversely related, and it is not possible to minimize both in the same test. As the probability of one type of error decreases, the other increases, and vice versa.

What all of this means, finally, is that the selection of an alpha level must be conceived of as an attempt to balance these two sources of error. Higher alpha levels will minimize the probability of Type II error, and lower alpha levels will minimize the probability of Type I error. Normally, in social science research, we will want to minimize Type I error, and lower alpha levels will be used.

As a final note, social scientists conventionally set alpha at 0.05 or, somewhat less frequently, 0.10 or 0.01. The 0.05 level in particular seems to have emerged as a generally recognized indicator of a significant result. However, the widespread use of the 0.05 level must be recognized as a convention, and there is no reason that alpha cannot be set at virtually any sensible level (such as 0.04, 0.027, 0.083). The researcher has the responsibility of selecting the alpha level that seems most reasonable in terms of the goals of the research project.

8.6 THE STUDENT'S t DISTRIBUTION

To this point, we have considered only one type of hypothesis test. Specifically, we have focused on situations involving single sample means where the value of the population standard deviation (σ) was known. Needless to say, in most research situations, the value of σ will not be known. However, a value for σ is required in order to compute the standard error of the mean

$(\sigma/\sqrt{N})$, convert our sample outcome into a Z score, and place the Z (obtained) on the sampling distribution (step 4). How can a value for the population standard deviation reasonably be obtained?

It might seem sensible to estimate σ with s, the sample standard deviation. As we noted in Chapter 7, s is a biased estimator of σ, but the degree of bias decreases as sample size increases. For large samples (that is, samples with 100 or more cases), the sample standard deviation yields an adequate estimate of σ. Thus, for large samples, we simply substitute s for σ in the formula for Z (obtained) in step 4 and continue to use the standard normal curve to find areas under the sampling distribution.*

For smaller samples, however, when σ is unknown, an alternative distribution called the **Student's t distribution** must be used to find areas under the sampling distribution and establish the critical region. The shape of the t distribution varies as a function of sample size. The relative shapes of the t and Z distributions are depicted in Figure 8.6. For small samples, the t distribution is much flatter than the Z distribution, but, as sample size increases, the t distribution comes to resemble the Z distribution more and more until the two are essentially identical when sample size is greater than 120. As N increases, the sample standard deviation (s) becomes a more and more adequate estimator of the population standard deviation (σ), and the t distribution becomes more and more like the Z distribution.

The t distribution is summarized in Appendix B. The t table differs from the Z table in several ways. First, there is a column at the left of the table labeled df for "degrees of freedom."† As mentioned above, the exact shape of the t distribution and thus the exact location of the critical region for any alpha level varies as a function of sample size. Degrees of freedom, which are equal to $N - 1$ in the case of a single-sample mean, must first be computed before the critical region for any alpha can be properly located. Second, alpha levels are arrayed across the top of Appendix B in two rows, one row for the one-tailed tests and one for two-tailed tests. To use the table, begin by locating the selected alpha level in the appropriate row.

The third difference is that the entries in the table are the actual scores, called t **(critical)**, which mark the beginnings of the critical regions and not areas under the sampling distribution. To illustrate the use of this table with single-sample means, find the critical region for alpha equal to 0.05, two-tailed test, for $N = 30$. The degrees of freedom will be $N - 1$, or 29; reading

---

*Even though its effect will be minor and will decrease with sample size, we will always correct for the bias in s by using the term $N - 1$ rather than N in the computation for the standard deviation of the sampling distribution when σ is unknown.

†Degrees of freedom refer to the number of values in a distribution that are free to vary. For a sample mean, a distribution has $N - 1$ degrees of freedom. This means that, for a specific value of $\overline{X}$ and N, $N - 1$ scores are free to vary. For example, if $\overline{X} = 3$ and $N = 5$, the distribution of five scores would have $N - 1$, or four, degrees of freedom. When the value of four of the scores is known, the value of the fifth is fixed. If four scores are 1, 2, 3, and 4, the fifth must be 5 and no other value.

FIGURE 8.6 THE *t* DISTRIBUTION AND THE *Z* DISTRIBUTION

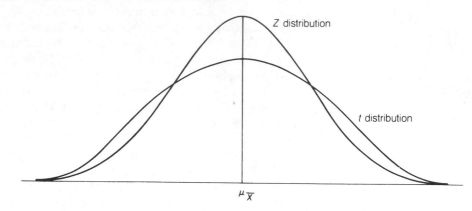

down the proper column, you should find a value of 2.045. Thus, the critical region for this test will begin at t (critical) = ± 2.045.

Take a moment to notice some additional features of the t distribution. First, note that the t (critical) we found above is larger in value than the comparable Z (critical), which—for a two-tailed test at an alpha of 0.05—would be ± 1.96. This relationship reflects the fact that the t distribution is flatter than the Z distribution (see Figure 8.6). When you use the t distribution, the critical regions will begin farther away from the mean of the sampling distribution and, therefore, the null hypothesis will be harder to reject. Furthermore, the smaller the sample size (the lower the degrees of freedom), the larger the value of t (obtained) necessary for a rejection of H_0.

Second, scan the column for an alpha of 0.05, two-tailed test. Note that, for one degree of freedom, the t (critical) is ± 12.706 and that the value of t (critical) decreases as degrees of freedom increase. For degrees of freedom greater than 120, the value of t (critical) is the same as the comparable value of Z (critical), or ± 1.96. As sample size increases, the t distribution comes to resemble the Z distribution more and more until, with sample sizes greater than 120, the two distributions are essentially identical.*

To demonstrate the uses of the t distribution in more detail, we will work through an example problem. Note that, in terms of the five-step model, the changes required by using t scores occur mostly in steps 3 and 4. In step 3,

---

*Appendix B abbreviates the t distribution by presenting a limited number of t (critical)s for degrees of freedom between 31 and 120. If the degrees of freedom for a specific problem equal 77 and alpha equals 0.05, two-tailed, we have a choice between a t (critical) of ± 2.000 (df = 60) and a t (critical) of ± 1.980 (df = 120). In situations such as these, take the larger table value as t (critical). This will make rejection of the H_0 less likely and is therefore the more conservative course of action.

the sampling distribution will be the t distribution, and degrees of freedom (df) must be computed before locating the critical region as marked by t (critical). In step 4, a slightly different formula for computing the test statistic, **t (obtained)**, will be used. As compared with the formula for Z (obtained), s will replace σ and $N - 1$ will replace N.

Specifically,

FORMULA 8.2

$$t \text{ (obtained)} = \frac{\bar{X} - \mu}{s/\sqrt{N - 1}}$$

A researcher wonders if commuter students are different from the general student body in terms of academic achievement. She has gathered a random sample of 30 commuter students and has learned from the registrar that the mean grade-point average for all students is 2.00 ($\mu = 2.00$) but the standard deviation of the population (σ) has never been computed. Sample data are reported below. Is the sample from a population which has a mean of 2.00?

Student body	Commuter students
$\mu = 2.00 \ (= \mu_{\bar{X}})$	$\bar{X} = 2.58$
$\sigma = \ ?$	$s = 1.23$
	$N = 30$

Step 1. Making assumptions.

Model: Random sampling
Level of measurement is interval ratio
Sampling distribution is normal

Step 2. Stating the null hypothesis.

$$H_0 : \mu = 2.00$$

$$(H_1 : \mu \neq 2.00)$$

You can see from the research hypothesis that the researcher has not predicted a direction for the difference. This will be a two-tailed test.

Step 3. Selecting the sampling distribution and establishing the critical region. Since σ is unknown and sample size is small, the t distribution will be used to find the critical region. Alpha will be set at 0.01.

Sampling distribution = t distribution

$$\alpha = 0.01, \text{ two-tailed test}$$

$$\text{df} = (N-1) = 29$$

$$t \text{ (critical)} = \pm 2.756$$

APPLICATION 8.1

For a random sample of 152 felony cases tried in a local court, the average prison sentence was 27.3 months. Is this significantly different from the average prison term for felons nationally? The necessary information for conducting a test of the null hypothesis is

$$\bar{X} = 27.3 \qquad \mu = 28.7$$

$$s = 3.7$$

$$N = 152$$

The null hypothesis would be

$$H_0: \mu = 28.7$$

The null states that the sample comes from a population which has a mean of 28.7.

The test statistic, Z (obtained), would be

$$Z \text{ (obtained)} = \frac{\bar{X} - \mu}{s/\sqrt{N - 1}}$$

$$Z \text{ (obtained)} = \frac{27.3 - 28.7}{3.7/\sqrt{152 - 1}}$$

$$Z \text{ (obtained)} = -\frac{1.40}{3.7/\sqrt{151}}$$

$$Z \text{ (obtained)} = -\frac{1.40}{0.30}$$

$$Z \text{ (obtained)} = -4.67$$

If alpha were set at the 0.05 level, the critical region would begin at Z (critical) ± 1.96. With an obtained Z score of -4.67, the null would be rejected. The difference between the prison sentences of felons convicted in the local court and felons convicted nationally is statistically significant. The difference is so large that we may conclude that it did not occur by random chance. The decision to reject the null hypothesis has a 0.05 probability of being wrong.

Step 4. Computing the test statistic.

$$t \text{ (obtained)} = \frac{\bar{X} - \mu}{s/\sqrt{N - 1}}$$

$$t \text{ (obtained)} = \frac{2.58 - 2.00}{1.23/\sqrt{29}}$$

$$t \text{ (obtained)} = \frac{.58}{.23}$$

$$t \text{ (obtained)} = +2.52$$

Step 5. Making a decision. The test statistic does not fall in the critical region. Therefore, the researcher fails to reject the H_0. The difference between the sample mean (2.58) and the hypothesized population mean (2.00) is no greater than what would be expected if only random chance were operating. The test statistic and critical regions are displayed in Figure 8.7.

To summarize, when testing single sample means we must make a

FIGURE 8.7   SAMPLING DISTRIBUTION SHOWING t (OBTAINED) VERSUS t (CRITICAL) ($\alpha = 0.05$, two-tailed test, df = 29)

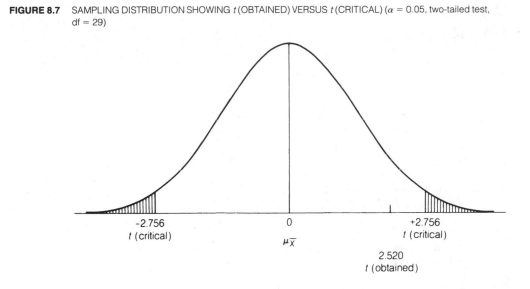

choice regarding the theoretical distribution we will use to establish the critical region. The choice is straightforward. If the population standard deviation (σ) is known or sample size is large, the Z distribution (summarized in Appendix A) will be used. If σ is unknown and the sample is small, the t distribution (summarized in Appendix B) will be used.

8.7 TESTS OF HYPOTHESES FOR SINGLE-SAMPLE PROPORTIONS (LARGE SAMPLES)

In many cases, the characteristic of interest in the sample will not be measured in a way that justifies the assumption of interval-ratio level of measurement. One alternative in this situation would be to use a sample proportion (P_s) rather than a sample mean as the test statistic. As we shall see below, the overall procedures for testing single-sample proportions are the same as those for testing means. The central question is still "Does the population from which the sample was drawn have a certain characteristic?" We still conduct the test based on the assumption that the null hypothesis is true, and we still evaluate the probability of the obtained sample outcome against a sampling distribution of all possible sample outcomes. Our decision at the end of the test is also the same. If the obtained test statistic falls in the critical region (is unlikely, given the assumption that the H_0 is true), we reject the H_0.

Having stressed the continuity in procedures and logic, I must hastily point out the important differences as well. These differences are best related in terms of the five-step model for hypothesis testing. In step 1, we assume only nominal level of measurement when working with sample proportions. In step 2, the symbols used to state the null hypothesis are different even though the null is still a statement of "no difference."

APPLICATION 8.2

In a random sample drawn from the most affluent neighborhood in a community, 76% of the respondents reported that they had voted Republican in the most recent presidential election. For the community as a whole, 66% of the electorate voted Republican. Was the affluent neighborhood significantly more likely to have voted Republican? The information necessary for a test of the null hypothesis, expressed in the form of proportions, is

Neighborhood	Community
$P_s = 0.76$	$P_u = 0.66$
$N = 103$	

Our null hypothesis would be

$$H_0: P_u = 0.66$$

or, the sample was taken from a population that voted 66% Republican.

The test statistic, Z (obtained), would be

$$Z \text{ (obtained)} = \frac{P_s - P_u}{\sqrt{\dfrac{P_u \,(1 - P_u)}{N}}}$$

$$Z \text{ (obtained)} = \frac{0.76 - 0.66}{\sqrt{\dfrac{(0.66)\,(1 - 0.66)}{103}}}$$

$$Z \text{ (obtained)} = \frac{0.100}{0.047}$$

$$Z \text{ (obtained)} = 2.13$$

Because we want to see if the affluent neighborhood was "significantly more likely to have voted Republican," a one-tailed test of hypothesis is called for. With alpha set at the 0.05 level, our critical region will begin at Z (critical) $= +1.65$. (The critical region is placed in the upper tail because we are predicting that the sample comes from a population that voted Republican at a higher rate than the community.

With an obtained Z score of 2.13, the null hypothesis would be rejected. The difference between the affluent neighborhood and the community as a whole is statistically significant and in the predicted direction. Residents of the affluent neighborhood were significantly more likely to have voted Republican in the last presidential election.

In step 3, we will use only the standardized normal curve (the Z distribution) to find areas under the sampling distribution and locate the critical region. This will be appropriate as long as sample size is large. We will not consider small-sample tests of hypothesis for proportions in this text.

In step 4, computing the test statistic, the form of the formula remains the same. That is, the test statistic, Z (obtained), equals the sample statistic minus the mean of the sampling distribution divided by the standard deviation of the sampling distribution. However, the symbols will change because we are basing the tests on sample proportions. The formula can be stated as

FORMULA 8.3
$$Z \text{ (obtained)} = \frac{P_s - P_u}{\sqrt{\dfrac{P_u (1 - P_u)}{N}}}$$

Step 5, making a decision, is exactly the same as before. If the test statistic, Z (obtained), falls in the critical region, as marked by Z (critical), reject the H_0.

An example should clarify these procedures. A random sample of 122 households in a low-income neighborhood revealed that 53, or a proportion of 0.43, of the households were headed by females. In the city as a whole, the proportion of female-headed households is .39. Are households in the lower-income neighborhood significantly different from the city as a whole in terms of this characteristic?

Step 1. Making assumptions.

Model: Random sampling
Level of measurement is nominal
Sampling distribution is normal in shape

Step 2. Stating the null hypothesis. The research question, as stated above, asks only if the sample proportion is different from the population proportion. Since no direction is predicted for the difference, a two-tailed test will be used.

$$H_0: P_u = .39$$

$$(H_1: P \neq .39)$$

Step 3. Selecting the sampling distribution and establishing the critical region.

$$\text{Sampling distribution} = Z \text{ distribution}$$

$$\alpha = 0.10, \text{ two-tailed}$$

$$Z \text{ (critical)} = \pm 1.65$$

Step 4. Computing the test statistic.

$$Z \text{ (obtained)} = \frac{P_s - P_u}{\sqrt{\dfrac{P_u (1 - P_u)}{N}}}$$

$$Z \text{ (obtained)} = \frac{0.43 - 0.39}{\sqrt{\dfrac{(0.39)(0.61)}{122}}}$$

$$Z \text{ (obtained)} = +0.91$$

FIGURE 8.8   SAMPLING DISTRIBUTION SHOWING Z(OBTAINED) VERSUS Z(CRITICAL) ($\alpha = 0.10$, two-tailed test)

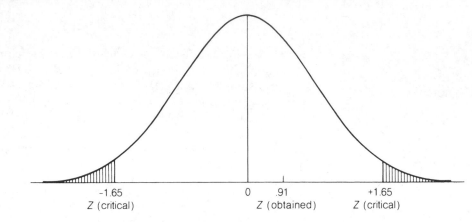

$$-1.65 \qquad\qquad 0 \quad .91 \qquad\qquad +1.65$$
$$Z \text{ (critical)} \qquad\qquad Z \text{ (obtained)} \qquad Z \text{ (critical)}$$

Step 5. **Making a decision.** The test statistic, Z (obtained), does not fall in the critical region. Therefore, we fail to reject the H_0. There is no statistically significant difference between the low-income community and the city as a whole in terms of the proportion of households headed by females. Figure 8.8 displays the sampling distribution, the critical region, and the Z (obtained).

SUMMARY

1. All the basic concepts and techniques for testing hypotheses were presented. We saw how to test the null hypothesis of "no difference" for both sample means and proportions.

2. All tests of a hypothesis involve finding the probability of the observed sample outcome, given that the null hypothesis is true. If the outcomes have low probabilities, we reject the null hypothesis. In the usual research situation, we will wish to reject the null and thereby support the research hypothesis.

3. The five-step model will be our framework for decision making throughout the hypothesis-testing chapters. We will always (1) make assumptions, (2) state the null hypothesis, (3) select a sampling distribution, specify alpha, and find the critical region, (4) compute a test statistic, and (5) make a decision. What we do during each step, however, will vary, depending on the specific test being conducted.

4. If we can predict a direction for the difference in stating the research hypothesis, a one-tailed test is called

for. If no direction can be predicted, a two-tailed test is appropriate. There are two kinds of errors in hypothesis testing. Type I, or alpha, error is rejecting a true null; Type II, or beta, error is failing to reject a false null. The probabilities of committing these two types of error are inversely related and cannot be simultaneously minimized in the same test. By selecting an alpha level, we try to balance the probability of these two kinds of error.

5. When testing sample means, the t distribution must be used to find the critical region when the population standard deviation is unknown and sample size is small.

6. Sample proportions can also be tested for significance. Tests are conducted using the five-step model. Compared to the test for the sample mean, the major differences lie in the level-of-measurement assumption (step 1), the statement of the null (step 2), and the computation of the test statistic (step 4).

7. If you are still confused about the uses of inferential statistics described in this chapter, don't be alarmed or discouraged. A sizable volume of rather complex material has been presented and only rarely will a beginning student fully comprehend the unique logic of hypothesis testing on the first exposure. After all, it is not every day that you learn how to test a statement you don't believe (the null hypothesis) against a distribution that doesn't exist (the sampling distribution)!

SUMMARY OF FORMULAS

Single-sample means, large samples   8.1 $Z = \dfrac{\bar{X} - \mu_{\bar{x}}}{\sigma/\sqrt{N}}$

Single-sample means when samples are small and population standard deviation is unknown   8.2 $t \text{ (obtained)} = \dfrac{\bar{X} - \mu}{s/\sqrt{N - 1}}$

Single sample proportions, large samples   8.3 $Z \text{ (obtained)} = \dfrac{P_s - P_u}{\sqrt{\dfrac{P_u(1 - P_u)}{N}}}$

GLOSSARY

Alpha level. The proportion of area under the sampling distribution that contains unlikely sample outcomes, given that the null hypothesis is true. Also, the probability of Type I error.

Critical region (region of rejection). The area under the sampling distribution that, in advance of the test itself, is defined as including unlikely sample outcomes, given that the null hypothesis is true.

Five-step model. A step-by-step guideline for conducting tests of hypotheses. A framework that organizes decisions and computations for all tests of significance.

Null hypothesis (H_0). A statement of "no difference." In the context of single-sample tests of significance, the population from which the sample was drawn is assumed to have a certain characteristic or value.

One-tailed test. A type of hypothesis test used when (1) the direction of the difference can be predicted or (2) concern focuses on outcomes in only one tail of the sampling distribution.

Research hypothesis (H_1). A statement that contradicts the null hypothesis. In the context of single-sample tests of significance, the population from which the sample was drawn does not have a certain characteristic or value.

t (critical). The t score that marks the beginning of the critical region of a t distribution.

t distribution. A distribution used to find the critical region for tests of sample means when σ is unknown and sample size is small.

t (obtained). The test statistic computed in step 4 of the five-step model. The sample outcome expressed as a t score.

Test statistic. The value computed in step 4 of the five-step model that converts the sample outcome into either a t score or a Z score.

Two-tailed test. A type of hypothesis test used when (1) the direction of the difference cannot be predicted or (2) concern focuses on outcomes in both tails of the sampling distribution.

Type I error (alpha error). The probability of rejecting a null hypothesis that is, in fact, true.

Type II error (beta error). The probability of failing to reject a null hypothesis that is, in fact, false.

Z (critical). The Z score that marks the beginnings of the critical region on a Z distribution.

Z (obtained). The test statistic computed in step 4 of the five-step model. The sample outcomes expressed as a Z score.

PROBLEMS

8.1 In your own words, define and explain each of the following terms:

Null hypothesis

Sampling distribution

Critical region

8.2 The questions below refer to the five-step model for hypothesis testing.

Step 1. a. Why is random sampling assumed?

b. Under what conditions can you assume that the sampling distribution is normal in shape?

Step 2. a. The null hypothesis is an assumption about reality that makes it possible to test sample outcomes for their significance. Explain.

b. Cite an example of a research situation where a one-tailed test of hypothesis would be appropriate. State the null and research hypotheses in symbols and in words.

Step 3. a. The t distribution is flatter than the Z distribution. What is the effect of this difference on the probability of rejecting the H_0?

b. For each situation, find Z (critical).

Alpha	Form	Z (critical)
.05	One-tailed	
.10	Two-tailed	
.06	Two-tailed	
.01	One-tailed	
.02	Two-tailed	

c. For each situation below, find the critical t score.

Alpha	Form	N
.10	Two-tailed	31
.02	Two-tailed	24
.01	Two-tailed	121
.01	One-tailed	31
.05	One-tailed	61

Step 4. For each situation below, compute the test statistic.

a. $\mu = 2.39$ $\bar{X} = 2.23$
   $\sigma = 0.75$ $N = 200$

b. $\mu = 17.1$ $\bar{X} = 16.8$
                $s = 0.92$
                $N = 105$

c. $\mu = 10.2$ $\bar{X} = 9.4$
                $s = 1.7$
                $N = 150$

d. $P_u = .57$ $P_s = 0.60$
                $N = 117$

e. $P_u = 0.32$ $P_s = 0.30$
                $N = 322$

Step 5. What can you conclude if a test statistic falls in the critical region?

8.3 [CJ] Nationally, the police clear by arrest 35% of the robberies reported to them. A researcher takes a random sample ($N = 207$) of all the robberies reported to a metropolitan police department in one year and finds that 83 of the cases were cleared by arrest. Is the local arrest rate significantly different from the national rate? Write a sentence or two interpreting your decision.

8.4 [SOC] The student body at St. Algebra College attends an average of 3.3 parties per month. A random sample of 117 sociology majors averages 3.8 parties per month with a standard deviation of 0.53. Do sociology majors have significantly more fun?

8.5 [SW] **a.** Nationally, social workers average 10.2 years of experience. In a random sample, 203 social workers in greater metropolitan Shinbone average only 8.7 years with a standard deviation of 0.52. Are social workers in Shinbone significantly less experienced?

   b. The same sample of social workers reports an average annual salary of $21,782 with a standard deviation of $622. Is this figure significantly higher than the national average of $20,509?

8.6 [SW] You are the head of an agency seeking funding for a program to reduce unemployment among teenage males. Nationally, the unemployment rate for this group is 18%. A random sample of 323 teenage males in your area reveals an unemployment rate of 21.7%. Can you demonstrate a need for the program?

8.7 [PA] The city manager of Shinbone has received a complaint from the local union of firefighters to the effect that they are underpaid. Not having much time, the city manager gathers the records of a random sample of 27 firefighters and finds that their average salary is $28,073 with a standard deviation of $575. If she knows that the average salary nationally is $28,202, how can she respond to the complaint?

8.8 [SOC] Nationally, the average score on the college entrance exams (verbal test) is 453 with a standard deviation of 95. A random sample of 152 freshmen entering St. Algebra College shows a mean score of 502. Is there a significant difference?

8.9 GER/CJ A survey shows that 10% of the population is victimized by property crime each year. A random sample of 527 older citizens (65 years or more of age) shows a victimization rate of 14%. Are older people more likely to be victimized?

8.10 SOC A random sample of 423 Chinese Americans has finished an average of 12.7 years of formal education with a standard deviation of 1.7. Is this significantly different from the national average of 12.2 years?

8.11 SOC A sample of 105 workers in the Overkill Division of the Machismo Toy Factory earns an average of $24,375 per year. The average salary for all workers is $24,230 with a standard deviation of $523. Are workers in the Overkill Division overpaid?

8.12 SOC/SW A researcher has compiled a file of information on a random sample of 317 families in a city that have chronic, long-term patterns of child abuse. Below are reported some of the characteristics of the sample along with values for the city as a whole. For each trait, test the null hypothesis of "no difference" and summarize your findings.
a. Mothers' educational level (proportion completing high school)

City	Sample
$P_u = 0.63$	$P_s = 0.61$

b. Family size (proportion of families with four or more children)

City	Sample
$P_u = 0.21$	$P_s = 0.26$

c. Mothers' work status (proportion of mothers with jobs outside the home)

City	Sample
$P_u = 0.51$	$P_s = 0.27$

d. Relations with kin (proportion of families that have contact with kin at least once a week)

City	Sample
$P_u = 0.82$	$P_s = 0.43$

e. Fathers' educational achievement (average years of formal schooling)

City	Sample
$\mu = 12.3$	$\bar{X} = 12.5$
	$s = 1.7$

f. Fathers' occupational stability (average years in present job)

City	Sample
$\mu = 5.2$	$\bar{X} = 3.7$
	$s = 0.5$

8.13 A school system has assigned several hundred "chronic and severe underachievers" to an alternative educational experience. To assess the program, a random sample of 35 has been selected for comparison with all students in the system.
a. In terms of GPA, did the program work?

Systemwide GPA	Program GPA
$\mu = 2.47$	$\bar{X} = 2.55$
	$s = .70$
	$N = 35$

b. In terms of absenteeism (number of days missed per year), what can be said about the success of the program?

Systemwide	Program
$\mu = 6.13$	$\bar{X} = 4.78$
	$s = 1.11$
	$N = 35$

8.14 CJ A random sample of 113 convicted rapists in a state prison system completed a program designed to change their attitudes towards women, sex, and violence before being released on parole. Fifty-eight eventually became repeat sex offenders. Is this recidivism rate significantly different from the rate for all offenders (57%) in that state? Summarize your conclusions in a sentence or two.

8.15 SOC A random sample of 26 local sociology graduates scored an average of 458 on the GRE advanced sociology test with a standard deviation of 20. Is this significantly different from the national average? ($\mu = 440$)

8.16 PA Nationally, the per capita property tax is $130. A random sample of 36 southeastern cities average $98 with a standard deviation of $5. Is the difference significant? Summarize your conclusions in a sentence or two.

8.17 PS In a recent statewide election, 55% of the voters rejected a proposal to institute a state lottery. In a random sample of 150 urban precincts, 49% of the voters rejected the proposal. Is the difference significant? Summarize your conclusions in a sentence or two.

8.18 GER Nationally, the population as a whole watches 6.2 hours of TV per day. A random sample of 1017 senior citizens report watching an average of 5.9 hours per day with a standard deviation of .7. Is the difference significant?

8.19 SOC A sociologist has developed a file of information on the students graduating from her university for the past several years. She has information on every member of the graduating class from last year and information on a random sample of 213 members of this year's class. On each of the characteristics noted below, is there a significant difference?

  a. Average number of major term papers written, all four years.

Last Year's Class	This Year's Class
$\mu = 25.2$	$\bar{X} = 24.7$
	$s = 2.7$

  b. Grade-point average

Last Year's Class	This Year's Class
$\mu = 2.67$	$\bar{X} = 2.71$
	$s = 0.50$

  c. Number of job offers received before commencement

Last Year's Class	This Year's Class
$\mu = 5.4$	$\bar{X} = 3.6$
	$s = 4.0$

  d. Proportion majoring in the social sciences

Last Year's Class	This Year's Class
$P_u = 0.17$	$P_s = 0.21$

  e. Proportion attending commencement ceremony

Last Year's Class	This Year's Class
$P_u = 0.88$	$P_s = 0.80$

Hypothesis Testing II

The Two-Sample Case

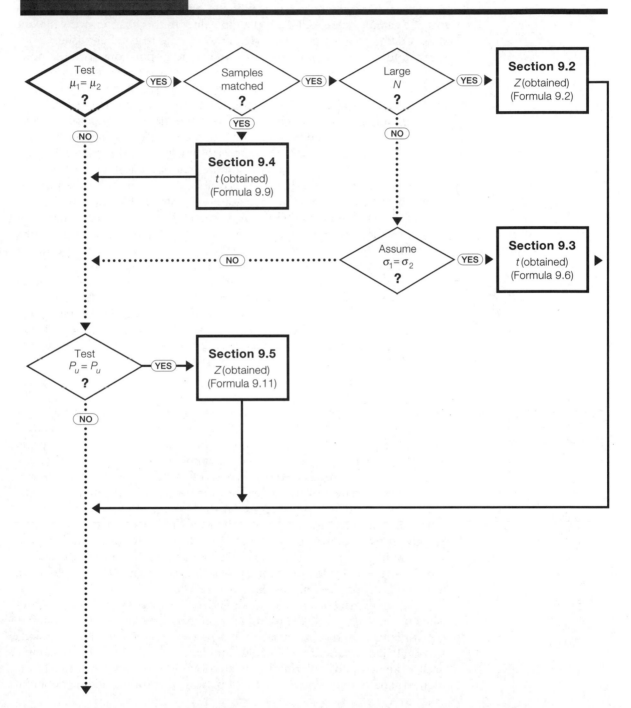

9.1 INTRODUCTION

In Chapter 8, we dealt with hypothesis testing in the one-sample case. In such a situation, our central concern is with the significance of the difference between a sample value and a population value. In this chapter, we will consider a different research situation, one where the researcher wishes to investigate the differences between two separate populations. For example, suppose a researcher wishes to see if the level of support for gun control varies between American men and women. Obviously, the researcher cannot test every male and female in the population for opinions on this issue and must instead draw random samples of men and of women. Again, we must use information gathered from random samples to infer population patterns. The central question asked in hypothesis testing in the two-sample case is "Is the difference between the samples on the trait of interest large enough so that we may conclude (with a known probability of error) that the populations represented by the samples are different on the trait in question?" Thus, if we find a large enough difference between random samples of men and women on their level of support for gun control, we will be able to argue that this sample difference does not occur by random chance alone and that it represents a real difference between men and women on this issue.

In this chapter, we will consider tests for the significance of the difference between sample means and sample proportions. In both tests, the five-step model will serve as a framework for organizing our decision making. Although the general flow of the hypothesis-testing process is very similar to the one we followed in the one-sample case, some important differences will be identified as we proceed through the material.

9.2 HYPOTHESIS TESTING WITH SAMPLE MEANS (LARGE SAMPLES)

One important difference between the one- and two-sample situations involves step 1 of the five-step model. In the one-sample case, we assumed that the sample was selected from the population of interest by the principle of EPSEM. Our assumption was "random sampling." As you recall, this means that the samples were selected so that each element or case in the population had an equal chance of being selected for the sample. In the two-sample case, we will assume **independent random sampling,** or that each of the two samples were randomly selected from their respective populations. Samples must be drawn so that the selection of a case for one sample has no effect on the selection of cases for the other sample. In our earlier example, this would mean that the selection of a specific male would have no effect on the probability of selecting any particular female.

The additional constraint of independence between samples can be satisfied by drawing two EPSEM samples from two separate lists (for example, one for females and one for males). It is usually more convenient to draw a single EPSEM sample from a single population list. As long as the sample itself is random, any subsamples created by the researcher will satisfy the requirement for independence. Thus, one way of ensuring independent random samples in our gun-control survey would be to draw a single EPSEM sample and then subdivide that sample by gender.

The second major difference in the five-step model for the two-sample case is in the form of the null hypothesis. The null is still a statement of "no difference"; but now, instead of saying that a sample is no different from a population on the trait of interest, it will say that the two populations are no different. This way of stating the null reflects the fact that our interest is centered on the populations per se and not on the samples. If the test statistic falls in the critical region, the null hypothesis of no difference between the populations can be rejected, and the argument that the populations are different on the trait of interest will be supported.

To illustrate the procedure for testing sample means, assume that a researcher has access to a nationally representative random sample and that the individuals in the sample have responded to a scale that measures attitudes toward gun control. The sample is divided by sex, and sample statistics are computed for males and females. Assuming that the scale yields interval-ratio level data, a test for the significance of the difference in sample means can be conducted.

The test statistic will be the difference in sample means. As long as sample size is large (that is, as long as the combined number of cases in the two samples exceeds 100), the sampling distribution of the differences in sample means will be normal in form, and the standard normal curve (Appendix A) can be used to find areas under the sampling distribution and establish the critical regions. The test statistic, Z (obtained), will be computed by the usual formula: sample outcome minus the mean of the sampling distribution, divided by the standard deviation of the sampling distribution. The formula is presented as Formula 9.1. Note that numerical subscripts are used to identify the two samples and the two populations they represent. The subscript $\bar{X} - \bar{X}$ is used to indicate that we are dealing with the sampling distribution of the differences in sample means.

FORMULA 9.1
$$Z \text{ (obtained)} = \frac{(\bar{X}_1 - \bar{X}_2) - (\mu_1 - \mu_2)}{\sigma_{\bar{X} - \bar{X}}}$$

where $(\bar{X}_1 - \bar{X}_2)$ = the difference in the sample means

$(\mu_1 - \mu_2)$ = the difference in the population means

$\sigma_{\bar{X} - \bar{X}}$ = the standard deviation of the sampling distribution of the differences in sample means

The second term in the numerator, $(\mu_1 - \mu_2)$, might appear to be troublesome since these two values are unknown. Fortunately, this seemingly incalculable term reduces to zero because we assume that the null hypothesis (which will be stated as $H_0: \mu_1 = \mu_2$) is true. Recall that tests of significance are always based on the assumption that the null hypothesis is true. If the means of the two populations are assumed to be equal, then the term $(\mu_1 - \mu_2)$ will be zero and can be dropped from the equation. In effect, then, the formula we will actually use to compute the test statistic in step 4 will be

FORMULA 9.2

$$Z \text{ (obtained)} = \frac{\bar{X}_1 - \bar{X}_2}{\sigma_{\bar{X}-\bar{X}}}$$

For large samples, the standard deviation of the sampling distribution of the difference in sample means is defined as

FORMULA 9.3

$$\sigma_{\bar{X}-\bar{X}} = \sqrt{\frac{\sigma_1^2}{N_1} + \frac{\sigma_2^2}{N_2}}$$

Since we will rarely be in a position to know the values of the population standard deviations (σ_1 and σ_2), we must use the sample standard deviations, suitably corrected for bias, to estimate them. Formula 9.4 displays the equation used to estimate the standard deviation of the sampling distribution in this situation. This is called a **pooled estimate** since it combines information from both samples.

FORMULA 9.4

$$\sigma_{\bar{X}-\bar{X}} = \sqrt{\frac{s_1^2}{N_1 - 1} + \frac{s_2^2}{N_2 - 1}}$$

The sample outcomes are reported below, and a test for the significance of the difference can now be conducted.

Sample 1 (men)	Sample 2 (women)
$\bar{X}_1 = 6.2$	$\bar{X}_2 = 6.5$
$s_1 = 1.3$	$s_2 = 1.4$
$N_1 = 324$	$N_2 = 317$

We see from the sample statistics that men have a lower average score on the Support for Gun Control Scale and are thus less supportive of gun control than women. By conducting the test of hypothesis, we are asking if this difference is large enough to justify the conclusion that it did not occur by random chance alone but rather reflects an actual difference between men and women on this issue.

Step 1. Making assumptions. Note that, although we now assume that the random samples are independent, the remaining model assumptions are the same as in the one-sample case.

Model: Independent random samples
Level of measurement is interval ratio
Sampling distribution is normal

Step 2. **Stating the null hypothesis.** Note that the null hypothesis states that the populations represented by the samples are not different on this variable. Since no direction for the difference has been predicted, a two-tailed test is called for, as reflected in the research hypothesis.

$$H_0: \mu_1 = \mu_2$$
$$(H_1: \mu_1 \neq \mu_2)$$

Step 3. **Selecting the sampling distribution and establishing the critical region.** For large samples, the Z distribution can be used to find areas under the sampling distribution and establish the critical region. Alpha will be set at 0.05.

$$\text{Sampling distribution} = Z \text{ distribution}$$
$$\text{Alpha} = 0.05, \text{ two-tailed}$$
$$Z \text{ (critical)} = \pm 1.96$$

Step 4. **Computing the test statistic.** Since the population standard deviations are unknown, Formula 9.4 will be used to estimate the standard deviation of the sampling distribution. Then, substituting this value into Formula 9.2, the test statistic, Z (obtained), can be computed.

$$\sigma_{\bar{X} - \bar{X}} = \sqrt{\frac{s_1^2}{N_1 - 1} + \frac{s_2^2}{N_2 - 1}}$$

$$\sigma_{\bar{X} - \bar{X}} = \sqrt{\frac{(1.3)^2}{324 - 1} + \frac{(1.4)^2}{317 - 1}}$$

$$\sigma_{\bar{X} - \bar{X}} = \sqrt{(0.0052) + (0.0062)}$$

$$\sigma_{\bar{X} - \bar{X}} = \sqrt{0.0114}$$

$$\sigma_{\bar{X} - \bar{X}} = 0.107$$

$$Z \text{ (obtained)} = \frac{\bar{X}_1 - \bar{X}_2}{\sigma_{\bar{X} - \bar{X}}}$$

$$Z \text{ (obtained)} = \frac{6.2 - 6.5}{0.107}$$

$$Z \text{ (obtained)} = -\frac{0.300}{0.107}$$

$$Z \text{ (obtained)} = -2.80$$

Step 5. **Making a decision.** Comparing the test statistic with the critical region:

$$Z \text{ (obtained)} = -2.80$$
$$Z \text{ (critical)} = \pm 1.96$$

We see that the sample outcome clearly falls in the critical region and is therefore unlikely if the null is true. The null hypothesis of no difference can be rejected, and the notion that men and women are different in terms of their support of gun control is supported. The decision to reject the null hypothesis has only a 0.05 probability (the alpha level) of being incorrect.

9.3 HYPOTHESIS TESTING WITH SAMPLE MEANS (SMALL SAMPLES)

As with single-sample means, when the population standard deviation is unknown and sample size is small, the Z distribution can no longer be used to find areas under the sampling distribution of the differences in sample means. Instead, we will use the t distribution to find the critical region and thus to identify unlikely sample outcomes. To utilize the t distribution for testing two sample means, we need to perform one additional calculation and make one additional assumption. The calculation is for degrees of freedom, a quantity required for proper use of the t distribution table (Appendix B). In the two-sample case, degrees of freedom are equal to $N_1 + N_2 - 2$.

The additional assumption is a more complex matter. With small samples, to justify the assumption of a normal sampling distribution and to form a pooled estimate of the standard deviation of the sampling distribution, we must assume that the variances of the populations of interest are equal ($\sigma_1^2 = \sigma_2^2$). The assumption of equal variance in the population can be tested by an inferential statistical technique known as the analysis of variance or ANOVA (see Chapter 12), but for our purposes here, we will simply assume equal population variances without formal testing. This assumption can be considered justified as long as sample sizes are approximately equal, so we will proceed to use the t distribution and calculate a pooled estimate of the standard deviation of the sampling distribution.

To illustrate the required procedures, assume that a researcher believes that center-city families are significantly larger than suburban families, as measured by number of children. Random samples from both areas are gathered and sample statistics computed.

Sample 1 (suburban)	Sample 2 (center-city)
$\bar{X}_1 = 2.37$	$\bar{X}_2 = 2.78$
$s_1 = 0.63$	$s_2 = 0.95$
$N_1 = 42$	$N_2 = 37$

The sample data reveal a difference in the predicted direction. The significance of this observed difference can be tested with the five-step model.

APPLICATION 9.1

An attitude scale measuring satisfaction with family life has been administered to a sample of married respondents. On this scale higher scores indicate greater satisfaction. The sample has been divided into respondents with no children and respondents with at least one child, and means and standard deviations have been computed for both groups. Is there a significant difference in satisfaction with family life between these two groups? The information necessary for conducting a test of the null hypothesis ($H_o: \mu_1 = \mu_2$) is

Sample 1 (no children)	Sample 2 (at least one child)
$\bar{X}_1 = 11.3$	$\bar{X}_2 = 10.8$
$s_1 = 0.6$	$s_2 = 0.5$
$N_1 = 78$	$N_2 = 93$

The estimate of the standard deviation of the sampling distribution would be

$$\sigma_{\bar{X}-\bar{X}} = \sqrt{\frac{s_1^2}{N_1 - 1} + \frac{s_2^2}{N_2 - 1}}$$

$$\sigma_{\bar{X}-\bar{X}} = \sqrt{\frac{(0.6)^2}{78 - 1} + \frac{(0.5)^2}{93 - 1}}$$

$$\sigma_{\bar{X}-\bar{X}} = \sqrt{0.008}$$

$$\sigma_{\bar{X}-\bar{X}} = 0.09$$

The test statistic, Z (obtained), would be

$$Z \text{ (obtained)} = \frac{\bar{X}_1 - \bar{X}_2}{\sigma_{\bar{X}-\bar{X}}}$$

$$Z \text{ (obtained)} = \frac{11.3 - 10.8}{0.09}$$

$$Z \text{ (obtained)} = \frac{0.50}{0.09}$$

$$Z \text{ (obtained)} = 5.56$$

If alpha were set at the 0.05 level, the critical region would begin at Z (critical) $= \pm 1.96$. With an obtained Z of 5.56, we would reject the null hypothesis. This test supports the conclusion that parents and childless couples are different with respect to satisfaction with family life.

Step 1. Making assumptions. Sample size is small and the population standard deviation is unknown. Hence, we must assume equal population variances in the model.

> Model: Independent random samples
> Level of measurement is interval ratio
> Equal population variances ($\sigma_1^2 = \sigma_2^2$)
> Sampling distribution is normal

Step 2. Stating the null hypothesis. Since a direction has been predicted (center-city families are larger), a one-tailed test will be used, and the research hypothesis is stated in accordance with this decision.

$$H_0: \mu_1 = \mu_2$$
$$(H_1: \mu_1 < \mu_2)$$

Step 3. Selecting the sampling distribution and establishing the critical region. With small samples the t distribution will be used to establish the critical region. Alpha will be set at 0.05, one-tailed test.

$$\text{Sampling distribution} = t \text{ distribution}$$
$$\text{Alpha} = 0.05, \text{ one-tailed}$$
$$\text{Degrees of freedom} = N_1 + N_2 - 2 = 42 + 37 - 2 = 77$$
$$t \text{ (critical)} = -1.671$$

Note that the critical region is placed in the lower tail of the sampling distribution in accordance with the direction specified in H_1.

Step 4. Computing the test statistic. With small samples, a different formula (Formula 9.5) is used for the pooled estimate of the standard deviation of the sampling distribution. This value is then substituted directly into the denominator of the formula for t (obtained) given in Formula 9.6.

FORMULA 9.5

$$\sigma_{\bar{x}-\bar{x}} = \sqrt{\frac{N_1 s_1^2 + N_2 s_2^2}{N_1 + N_2 - 2}} \sqrt{\frac{N_1 + N_2}{N_1 N_2}}$$

$$\sigma_{\bar{x}-\bar{x}} = \sqrt{\frac{(42)(.63)^2 + (37)(.95)^2}{42 + 37 - 2}} \sqrt{\frac{42 + 37}{(42)(37)}}$$

$$\sigma_{\bar{x}-\bar{x}} = \sqrt{\frac{50.06}{77}} \sqrt{\frac{79}{1554}}$$

$$\sigma_{\bar{x}-\bar{x}} = (.81)(.23)$$

$$\sigma_{\bar{x}-\bar{x}} = .19$$

FORMULA 9.6

$$t \text{ (obtained)} = \frac{\bar{X}_1 - \bar{X}_2}{\sigma_{\bar{x}-\bar{x}}}$$

$$t \text{ (obtained)} = \frac{2.37 - 2.78}{.19}$$

$$t \text{ (obtained)} = -\frac{.41}{.19}$$

$$t \text{ (obtained)} = -2.16$$

Step 5. Making a decision. Comparing the test statistic with the critical region,

$$t \text{ (obtained)} = -2.16$$
$$t \text{ (critical)} = -1.671$$

you can see that the test statistic falls in the critical region. If the null ($\mu_1 = \mu_2$) were true, this would be a very unlikely outcome, and the null can be rejected. There is a statistically significant difference (a difference so large

FIGURE 9.1 THE SAMPLING DISTRIBUTION WITH CRITICAL REGION AND TEST STATISTIC DISPLAYED

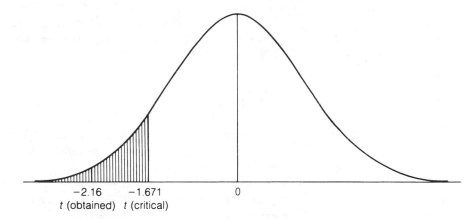

$$-2.16 \qquad -1.671 \qquad\qquad 0$$
$$\textit{t (obtained)} \quad \textit{t (critical)}$$

that it is unlikely to be due to random chance) in the sizes of center-city and suburban families. Furthermore, center-city families are significantly larger in size. The test statistic and sampling distribution are depicted in Figure 9.1.

9.4 HYPOTHESIS TESTING WITH SAMPLE MEANS (MATCHED SAMPLES)*

Some research designs require samples that are intentionally selected in violation of the assumption of independence. For example, some designs use matched samples in which subjects are selected so as to resemble each other as closely as possible. The matching might be based on any number of traits including age, gender, race, or social class. The strongest form of matching would be a "before/after" design where exactly the same subjects are tested before undergoing some experimental treatment and again after completing the treatment.

These matched-sample designs are frequently found in social psychological experiments, where they are used to eliminate the confounding effects of variables other than those that are relevant to the research project. If the two samples are matched on all dimensions and one sample receives an experimental treatment and the other does not, the researcher can more confidently attribute any change in the experimental group to the treatment itself. The greater the extent to which the samples are matched, the greater the certainty that changes were not caused by some factor extraneous to the experiment.

---

*This section is optional.

In matched designs, the techniques described in the previous two sections can no longer be used because the researcher has intentionally violated the assumption of independent random samples. Instead, an alternative procedure that focuses on the differences between the two sets of scores is used. The null hypothesis is still a statement to the effect that there is "no difference." Specifically, the null states that the average difference between the two sets of scores will be zero. That is, if we found the differences between each pair of scores, added up these differences and divided by the number of pairs of scores, the result would be zero. This would be the case if the sets of scores were exactly the same. The greater the overall differences between the two sets of scores, the greater the average difference, and the more likely we will be to reject the null hypothesis.

To illustrate, suppose that fifteen people were tested for their level of prejudice both before and after they viewed a movie that presented minority groups sympathetically. Scores from the scale used to measure prejudice increase as prejudice increases, and the results are presented in Table 9.1. Additional columns for computations have been added.

The difference, or D, column in the table displays the results of subtracting the second score (the posttest) from the first (the pretest). The values in the D^2 column are found by squaring the values in the D column. We will need these quantities to complete our calculations.

The mean difference and the standard deviation of the differences are computed by using the formulas on the following page:

TABLE 9.1 PRETEST AND POSTTEST SCORES ON PREJUDICE, WITH COMPUTATIONAL COLUMNS ADDED

Case	Pretest	Posttest	Difference (D)	D^2
1	10	10	0	0
2	11	10	1	1
3	11	11	0	0
4	15	14	1	1
5	16	10	6	36
6	17	10	7	49
7	20	15	5	25
8	20	16	4	16
9	20	19	1	1
10	25	20	5	25
11	26	25	1	1
12	27	23	4	16
13	28	20	8	64
14	30	25	5	25
15	35	30	5	25
	311	258	53	285

FORMULA 9.7
$$\bar{X}_D = \Sigma D/N$$

where $\bar{X}_D$ is the mean of the differences
ΣD is the summation of the D column

FORMULA 9.8
$$s_D = \sqrt{\frac{\Sigma D^2 - (\Sigma D)^2/N}{N}}$$

where s_D is the standard deviation of the differences
ΣD is the summation of the Difference (D) column
ΣD^2 is the summation of the D^2 column

The test statistic computed in step 4 of the five-step model will be a t score:

FORMULA 9.9
$$t \text{ (obtained)} = \frac{\bar{X}_D}{s_D/\sqrt{N-1}}$$

We can now apply the five-step model to this situation.

Step 1. Making assumptions.

> Model: Random sampling
> Level of measurement is interval ratio
> The population of all differences is
> normally distributed

Note that we are still treating the 15 cases as a random sample. The assumption of normal population distributions is necessary because of the small sample size.

Step 2. Stating the null hypothesis.

$$H_0: \mu_D = 0$$

$$(H_1: \mu_D \neq 0)$$

Step 3. Selecting the sampling distribution and establishing the critical region.

> Sampling distribution = t distribution
>
> Alpha = .05, two-tailed
>
> Degrees of freedom = $N - 1 = 15 - 1 = 14$
>
> t (critical) = ± 2.145

Note that the degrees of freedom are based on an N of 15, *not* 30, as would be the case if we had independent random samples.

Step 4. Computing the test statistic.

$$\bar{X}_D = \Sigma D/N$$

$$\bar{X}_D = 53/15 = 3.53$$

$$s_D = \sqrt{\frac{\Sigma D^2 - (\Sigma D)^2/N}{N}}$$

$$s_D = \sqrt{\frac{(285) - (53)^2/15}{15}}$$

$$s_D = \sqrt{\frac{285 - 187.27}{15}}$$

$$s_D = \sqrt{97.73/15}$$

$$s_D = \sqrt{6.52}$$

$$s_D = 2.55$$

$$t \text{ (obtained)} = \frac{\bar{X}_D}{s_D/\sqrt{N - 1}}$$

$$t \text{ (obtained)} = \frac{3.53}{2.55/\sqrt{14}}$$

$$t \text{ (obtained)} = 3.53/.68$$

$$t \text{ (obtained)} = 5.19$$

Step 5. Making a decision. The test statistic is in the critical region, so we can reject the null hypothesis. The difference in prejudice scores from the pretest to the posttest is statistically significant. The evidence suggests that the experience of viewing the movie did have an impact on prejudice.

9.5 HYPOTHESIS TESTING WITH SAMPLE PROPORTIONS (LARGE SAMPLES)

Testing for the significance of the difference between two sample proportions is analogous to testing sample means. The null hypothesis states that, on the trait being tested, no difference exists between the populations from which the samples are drawn. The sample proportions form the basis of the test statistic computed in step 4, which is then compared with the critical region. When sample sizes are large (combined N's of more than 100), the Z distribution may be used to find the critical region. We will not consider tests of significance for proportions based on small samples or matched samples in this text.

The formula for finding Z (obtained) in step 4 is

FORMULA 9.10

$$Z \text{ (obtained)} = \frac{(P_{s1} - P_{s2}) - (P_{u1} - P_{u2})}{\sigma_{p-p}}$$

where $(P_{s1} - P_{s2})$ = the difference between the sample proportions
$(P_{u1} - P_{u2})$ = the difference between the population proportions
σ_{p-p} = the standard deviation of the sampling distribution of the differences in sample proportions

As was the case with sample means, the second term in the numerator is assumed to be 0 by the null. Therefore, the formula reduces to

FORMULA 9.11

$$Z \text{ (obtained)} = \frac{P_{s1} - P_{s2}}{\sigma_{p-p}}$$

The value of the standard deviation of the sampling distribution of the differences in sample proportions is found by solving Formula 9.12:

FORMULA 9.12

$$\sigma_{p-p} = \sqrt{P_u(1 - P_u)} \sqrt{\frac{N_1 + N_2}{N_1 N_2}}$$

Solving this formula requires that a value for P_u be estimated. This is done by pooling information from both samples as specified in Formula 9.13:

FORMULA 9.13

$$P_u = \frac{N_1 P_{s1} + N_2 P_{s2}}{N_1 + N_2}$$

To find the test statistic, Z (obtained), in step 4, we must begin with Formula 9.13 and work back to Formula 9.10. Let us illustrate these procedures with a sample problem.

Random samples of black and white senior citizens have been gathered and each respondent has been classified as "highly participatory" or "not highly participatory" in terms of the number of memberships he or she holds in voluntary associations. Is there a statistically significant difference in the participation patterns of black and white elderly? Sample data for the proportions of each group classified as highly participatory are reported below.

Sample 1 (black senior citizens)	Sample 2 (white senior citizens)
$P_{s1} = 0.34$	$P_{s2} = 0.25$
$N_1 = 83$	$N_2 = 103$

Step 1. **Making assumptions.**

> Model: Independent random samples
> Level of measurement is nominal
> Sampling distribution is normal

Step 2. Stating the null hypothesis. Since no direction has been predicted, this will be a two-tailed test.

$$H_0: P_{u1} = P_{u2}$$

$$(H_1: P_{u1} \neq P_{u2})$$

Step 3. Selecting the sampling distribution and establishing the critical region. Since sample size is large, the Z distribution will be used to establish the critical region. Setting alpha at 0.05, we have

Sampling distribution = Z distribution
Alpha = 0.05, two-tailed
Z (critical) = ± 1.96

Step 4. Computing the test statistic. Remember to begin with the formula for estimating P_u (Formula 9.13), substitute the resultant value in the formula for σ_{p-p} (Formula 9.12), and then solve for Z (obtained) (Formula 9.11).

$$P_u = \frac{N_1 P_{s1} + N_2 P_{s2}}{N_1 + N_2}$$

$$P_u = \frac{(83)(.34) + (103)(.25)}{83 + 103}$$

$$P_u = .29$$

$$\sigma_{p-p} = \sqrt{P_u(1 - P_u)} \sqrt{\frac{N_1 + N_2}{N_1 N_2}}$$

$$\sigma_{p-p} = \sqrt{(.29)(.71)} \sqrt{\frac{83 + 103}{(83)(103)}}$$

$$\sigma_{p-p} = (.45)(.15)$$

$$\sigma_{p-p} = .07$$

$$Z \text{ (obtained)} = \frac{P_{s1} - P_{s2}}{\sigma_{p-p}}$$

$$Z \text{ (obtained)} = \frac{.34 - .25}{.07}$$

$$Z \text{ (obtained)} = 1.29$$

Step 5. Making a decision. Since the test statistic, Z (obtained) 1.29, does not fall in the critical region as marked by the Z (critical) of ± 1.96, we fail to reject the null hypothesis. The difference between the sample proportions is no greater than what would be expected if the null hypothesis were true and only random chance were operating. Black and white senior citizens are not significantly different in terms of participation patterns as measured in this test.

APPLICATION 9.2

Do attitudes toward sex vary by gender? The respondents in a national survey have been asked if they think that premarital sex is "always wrong" or only "sometimes wrong." The proportion of each sex that feels that premarital sex is always wrong is

Sample 1 Females	Sample 2 Males
$P_{s1} = 0.35$	$P_{s2} = 0.32$
$N_1 = 450$	$N_2 = 417$

This is all the information we will need to conduct a test of the null hypothesis ($H_0: P_{u1} \ P_{u2}$). The estimate of the population proportion is

$$P_u = \frac{N_1 P_{s1} + N_2 P_{s2}}{N_1 + N_2}$$

$$P_u = \frac{(450)(0.35) + (417)(0.32)}{450 + 417}$$

$$P_u = \frac{290.94}{867}$$

$$P_u = 0.34$$

The standard deviation of the sampling distribution would be

$$\sigma_{p-p} = \sqrt{P_u(1 - P_u)} \sqrt{\frac{N_1 + N_2}{N_1 + N_2}}$$

$$\sigma_{p-p} = \sqrt{(0.34)(0.66)} \sqrt{\frac{450 + 417}{(450)(417)}}$$

$$\sigma_{p-p} = \sqrt{0.2244} \sqrt{\frac{867}{187650}}$$

$$\sigma_{p-p} = \sqrt{0.2244} \sqrt{0.0046}$$

$$\sigma_{p-p} = (0.47)(0.068)$$

$$\sigma_{p-p} = 0.032$$

The test statistic, Z (obtained), is

$$Z \text{ (obtained)} = \frac{P_{s1} - P_{s2}}{\sigma_{p-p}}$$

$$Z \text{ (obtained)} = \frac{0.35 - 0.32}{0.032}$$

$$Z \text{ (obtained)} = \frac{0.030}{0.032}$$

$$Z \text{ (obtained)} = 0.94$$

If alpha is set at the 0.05 level, the critical region would begin at Z (critical) ± 1.96. With an obtained Z score of 0.94, we would fail to reject the null hypothesis. There is no statistically significant difference between males and females on attitudes toward premarital sex.

9.6 THE LIMITATIONS OF HYPOTHESIS TESTING

Given that we are usually interested in rejecting the null hypothesis, we should take a moment to consider systematically the factors that affect our decision in step 5. For all tests of hypothesis, the probability of rejecting the null is a function of four independent factors:

1. The size of the observed difference(s)
2. The alpha level
3. The use of one- or two-tailed tests
4. The size of the sample

READING STATISTICS 5: HYPOTHESIS TESTING

Professional researchers use a vocabulary that is much terser than ours when presenting the results of tests of significance. Because of space limitations and the volume of information that must be conveyed, results are not reported in very much detail. Furthermore, because professional researchers can assume a certain level of statistical literacy in their audiences, they can avoid stating many of the details we have been so careful about. So, in the literature it is likely that the null hypothesis, the critical region, the choice of a sampling distribution, and the underlying assumptions will not be mentioned at all.

Instead, researchers will report only the sample values (for example, means or proportion), the value of the test statistic (for example, a Z or t score), the alpha level, the degrees of freedom (if applicable), and sample size. The results of the example problem in Section 9.3 might be reported in the professional literature as "the difference between the sample means of 2.37 (suburban families) and 2.78 (center-city families) was tested and found to be significant ($t = -2.28$, df $= 77$, $p < 0.05$)." Note that the alpha level is reported as "$p < 0.05$." This is shorthand for "the probability of a difference of this magnitude occurring by chance alone, if the null hypothesis of no difference is true, is less than 0.05" and is a good illustration of how researchers can convey a great deal of information in just a few symbols. In a similar fashion, our somewhat long-winded phrase "the test statistic falls in the critical region and, therefore, the null hypothesis is rejected" is rendered tersely and simply: "the difference . . . was . . . found to be significant."

When researchers need to report the results of many tests of significance, they will often use a summary table in which they report only the sample information and whether the difference is significant at a certain alpha level. If you read the researcher's description and analysis of such tables, you should have little difficulty interpreting and understanding them. As a final note, these comments about how significance tests are reported in the literature apply to all of the tests of hypothesis covered in Part II of this text.

Statistics in the Professional Literature

Professor Jennifer Glass used data from the 1977 Quality of Employment Survey to analyze the effect of sex segregation in occupations on a variety of working conditions. The Quality of Employment Survey was administered to a nationally representative sample of about 1500 Americans in 1972–1973 by the Institute for Social Research. See Reading Statistics 4 in Chapter 7 for a discussion of this type of data base. Professor Glass was particularly concerned with the contention that ". . . sex segregation arises from women's preference for jobs that mesh easily with family caregiving . . .".(779) The table below reports the results of a series of t tests for sex differences on a variety of job characteristics as described or rated by the respondents.

TABLE 2   Means by Sex for Variables Used in the Analysis

	Women $N = 493$	Men $N = 793$
Social Rewards*	3.10 (.65)[a]	3.01 (.66)
Interpersonal Ease***	3.08 (1.01)	2.78 (.99)
Ease of Work	2.08 (.59)	2.10 (.60)
Flexibility**	2.52 (.65)	2.59 (.61)
Total Time[b] for Breaks***	63.53 (52.35)	83.47 (89.74)
Challenge and Interest***	3.16 (.87)	3.35 (.88)
Fairness of Promotion Policies*	2.32 (1.11)	2.46 (1.02)
Chances for Promotion**	2.08 (1.67)	2.39 (1.80)
Wages (in dollars)***	4.66 (2.57)	7.77 (5.07)
Benefits***	2.73 (1.29)	3.10 (1.10)
Age	36.29 (13.09)	37.54 (12.59)
Education (in years)	12.50 (2.77)	12.53 (2.92)
Years at Present Job***	5.68 (6.18)	7.67 (7.41)
Hours Worked per Week***	38.14 (8.67)	44.63 (11.73)
Years of Job Experience***	14.04 (10.53)	20.02 (12.66)

*Difference significant at $p < .05$ **$p < .01$ ***$p < .001$
[a]Standard deviations in parentheses   [b]Time in minutes per day

This table reports, in a very compact way, full information for 15 different tests of significance. Asterisks are used to identify level of statistical significance and the table report means, standard deviations, and number of cases. Here's what Professor Glass has to say about these results:

"Table 2 presents a preliminary breakdown of scores on each job description variable by sex. This table shows that women rate their jobs as slightly more socially rewarding and significantly more conflict-free than men rate their jobs. Women do not indicate that their jobs are any easier or more difficult than men's jobs.

"For the rest of the job description variables, men's scores are significantly higher than women's scores. Men report that they are given more opportunity to use their skills in challenging work, that their jobs are more flexible, that promotions are handled more fairly, and that their chances for promotion are better. Men also have significantly higher wages and more benefits." Professor Glass then goes on to report the results of other statistical tests. If you would like to learn more, bibliographical information is given below.

Jennifer Glass: 1990. "The Impact of Occupational Segregation on Working Conditions." *Social Forces*, Vol. 68(3): 779–796. Copyright © The University of North Carolina Press. Reprinted by permission.

Only the first of these four is not under the direct control of the researcher. The size of the difference (either between the sample outcome and the population value or between two sample outcomes) is a function, in part, of the testing procedures (that is, how variables are measured), but for the most part it should reflect the underlying realities we are trying to probe.

The relationship between alpha level and the probability of rejection is straightforward. The higher the alpha level, the larger the critical region, the higher the percentage of all possible sample outcomes that fall in the critical region, and the greater the probability of rejection. Thus, it is easier to reject the H_0 at the 0.05 level than at the 0.01 level, and easier still at the 0.10 level. The danger here, of course, is that higher alpha levels will lead to more frequent Type I errors, and we might find ourselves declaring rather small differences to be statistically significant. In similar fashion, use of the one-tailed test increases the probability of rejection (assuming that the proper direction of the difference has been predicted).

The final factor that affects the probability of rejecting the null hypothesis is sample size. The larger the sample, with all other factors constant, the higher the probability of rejecting H_0. This may appear to be a surprising relationship, but the reasons for it can be appreciated with a brief consideration of the formulas used to compute test statistics in step 4. In all these formulas, sample size (N) is in the "denominator of the denominator." Algebraically, this is equivalent to being in the numerator of the formula and means that the value of the test statistic is directly proportional to N and that the two will increase together. To illustrate, consider Table 9.2, which shows the value of the test statistic for single sample means from samples of various sizes.

The value of the test statistic, Z (obtained), increases as N increases even though none of the other terms in the formula were changed. This pattern of higher probabilities for rejecting H_0 with larger samples holds for all tests of significance.

Let me make two points by way of conclusion. On one hand, this relationship between sample size and the probability of rejecting the null should not alarm us unduly. Larger samples are, after all, better approximations of the populations they represent. Thus, decisions based on larger samples,

TABLE 9.2 TEST STATISTICS FOR SINGLE SAMPLE MEANS COMPUTED FROM SAMPLES OF VARIOUS SIZES ($\overline{X}$ 80, $\mu = 79$, s 5 for all problems)

Sample Size	Test Statistic, Z (obtained)
100	1.99
200	2.82
500	4.47

whether to reject or fail to reject the null, can be regarded as more trustworthy than decisions based on small samples.

On the other hand, this relationship clearly underlines what is perhaps the most significant limitation of hypothesis testing. Simply because a difference is statistically significant does not guarantee that it is important in any other sense. Particularly with very large samples (say, N's in excess of 1000), relatively small differences may be statistically significant. Even with small samples, of course, differences that are otherwise trivial or uninteresting may be statistically significant. The point I want to make is that there can be a difference between statistical significance on one hand and theoretical or practical importance on the other. In the context of quantitative research, statistical significance is a necessary but not sufficient condition for theoretical importance. Once the statistical significance of a research result has been demonstrated, the researcher still faces the task of evaluating the results in terms of the theory that guides the inquiry.

SUMMARY

1. A common research situation is to test for the significance of the difference between two populations. Sample statistics are calculated for random samples of each population, and then we test for the significance of the difference between the samples as a way of inferring differences between the specified populations.

2. When sample information is summarized in the form of sample means, and N is large, the Z distribution is used to find the critical region. When N is small, the t distribution is used to establish the critical region. In the latter circumstance, we must also assume equal population variances before forming a pooled estimate of the standard deviation of the sampling distribution.

3. When samples have been matched or paired, an alternative significance test must be used. The t test for matched samples, which is based on the differences between the sets of scores, should be used whenever the research design requires the researcher to violate the assumption of independence.

4. Differences in sample proportions may also be tested for significance. For large samples, the Z distribution is used to find the critical region.

5. In all tests of hypothesis, a number of factors affect the probability of rejecting the null: the size of the difference, the alpha level, the use of one- versus two-tailed tests, and sample size. Statistical significance is not the same thing as theoretical or practical impor-

tance. Even after a difference is found to be statistically significant, the researcher must still demonstrate the relevance or importance of his or her findings.

SUMMARY OF FORMULAS

Test statistic for two sample means, large samples

9.1

$$Z \text{ (obtained)} = \frac{(\bar{X}_1 - \bar{X}_2) - (\mu_1 - \mu_2)}{\sigma_{\bar{x} - \bar{x}}}$$

Test statistic for two sample means, large samples (simplified formula)

9.2 $\quad Z \text{ (obtained)} = \dfrac{\bar{X}_1 - \bar{X}_2}{\sigma_{\bar{x} - \bar{x}}}$

Standard deviation of the sampling distribution of the difference in sample means, large samples

9.3 $\quad \sigma_{\bar{x} - \bar{x}} = \sqrt{\dfrac{\sigma_1^2}{N_1} + \dfrac{\sigma_2^2}{N_2}}$

Pooled estimate of the standard deviation of the sampling distribution of the difference in

sample means, large
samples   9.4

$$\sigma_{\bar{x}-\bar{x}} = \sqrt{\frac{s_1^2}{N_1 - 1} + \frac{s_2^2}{N_2 - 1}}$$

Pooled estimate
of the standard
deviation of the
sampling distribution
of the difference in
sample means, small
samples   9.5

$$\sigma_{\bar{x}-\bar{x}} = \sqrt{\frac{N_1 s_1^2 + N_2 s_2^2}{N_1 + N_2 - 2}} \sqrt{\frac{N_1 + N_2}{N_1 N_2}}$$

Test statistic for two
sample means, small
samples   9.6 $t \text{ (obtained)} = \dfrac{\bar{X}_1 - \bar{X}_2}{\sigma_{\bar{x}-\bar{x}}}$

Mean of the
differences, matched
samples*   9.7 $\bar{X}_D = \dfrac{\Sigma D}{N}$

Standard deviation
of the differences,
matched samples*   9.8 $s_D = \sqrt{\dfrac{\Sigma D^2 - (\Sigma D)^2 / N}{N}}$

Test statistic,
matched samples*   9.9 $t \text{ (obtained)} = \dfrac{\bar{X}_D}{s_D / \sqrt{N - 1}}$

Test statistic for two
sample proportions,
large samples   9.10

$$Z \text{ (obtained)} = \frac{(P_{s1} - P_{s2}) - (P_{u1} - P_{u2})}{\sigma_{p-p}}$$

Test statistic for two
sample proportions,
large samples.
Simplified formula   9.11 $Z \text{ (obtained)} = \dfrac{P_{s1} - P_{s2}}{\sigma_{p-p}}$

Standard deviation
of the sampling

---

*This formula is optional.

distribution of the
difference in sample
proportions, large
samples   9.12

$$\sigma_{p-p} = \sqrt{P_u(1 - P_u)} \sqrt{\frac{N_1 + N_2}{N_1 N_2}}$$

Pooled estimate
of population
proportion, large
samples   9.13 $P_u = \dfrac{N_1 P_{s1} + N_2 P_{s2}}{N_1 + N_2}$

GLOSSARY

Independent random samples. Random samples
gathered in such a way that the selection of a particular case for one sample has no effect on the
probability that any other particular case will be selected for the other samples.

σ_{p-p}. Symbol for the standard deviation of the sampling distribution of the differences in sample
proportions.

$\sigma_{\bar{x}-\bar{x}}$. Symbol for the standard deviation of the sampling distribution of the differences in sample
means.

PROBLEMS

9.1 For each problem below, test for the significance
of the difference in sample statistics using the five-
step model.

a.

Sample 1	Sample 2
$\bar{X}_1 = 72.5$	$\bar{X}_2 = 76.0$
$s_1 = 14.3$	$s_2 = 10.2$
$N_1 = 36$	$N_2 = 57$

b.

Sample 1	Sample 2
$\bar{X}_1 = 107$	$\bar{X}_2 = 103$
$s_1 = 14$	$s_2 = 17$
$N_1 = 175$	$N_2 = 200$

c.

Sample 1	Sample 2
$P_{s1} = 0.17$	$P_{s2} = 0.20$
$N_1 = 101$	$N_2 = 114$

d.

Sample 1	Sample 2
$P_{s1} = 0.62$	$P_{s2} = 0.60$
$N_1 = 532$	$N_2 = 478$

9.2 SOC Gessner and Healey administered questionnaires to samples of undergraduates. Among other things, the questionnaires contained a scale that measured attitudes toward interpersonal violence (higher scores indicate greater approval of interpersonal violence). Test the results as reported below for sexual, racial, and social-class differences.

a.

Males	Females
$\bar{X}_1 = 2.99$	$\bar{X}_2 = 2.29$
$s_1 = 0.88$	$s_2 = 0.81$
$N_1 = 122$	$N_2 = 251$

b.

Blacks	Whites
$\bar{X}_1 = 2.76$	$\bar{X}_2 = 2.49$
$s_1 = 0.68$	$s_2 = 0.91$
$N_1 = 43$	$N_2 = 304$

c.

White Collar	Blue Collar
$\bar{X}_1 = 2.46$	$\bar{X}_2 = 2.67$
$s_1 = 0.91$	$s_2 = 0.87$
$N_1 = 249$	$N_2 = 97$

d. Summarize your results in terms of the significance and the direction of the differences. Which of these three factors seems to make the biggest difference in attitudes toward interpersonal violence?

9.3 GER Among senior citizens, do relations with relatives vary by social class? Below are reported the average number of visits with close kin per week for samples of white- and blue-collar elderly. Is the difference significant? Write a sentence or two explaining the results of your test.

White Collar	Blue Collar
$\bar{X}_1 = 1.42$	$\bar{X}_2 = 1.58$
$s_1 = 0.10$	$s_2 = 0.78$
$N_1 = 432$	$N_2 = 375$

9.4 SOC Do athletes in different sports vary in terms of intelligence? Below are reported College Board scores of random samples of college basketball and football players. Is there a significant difference? Write a sentence or two explaining the difference.

Basketball Players	Football Players
$\bar{X}_1 = 460$	$\bar{X}_2 = 442$
$s_1 = 92$	$s_2 = 57$
$N_1 = 102$	$N_2 = 117$

What about male and female college athletes?

Male	Female
$\bar{X}_1 = 452$	$\bar{X}_2 = 480$
$s_1 = 88$	$s_2 = 75$
$N_1 = 107$	$N_2 = 105$

9.5 CJ About half of the police officers in Shinbone, Kansas, have completed a special course in investigative procedures. Has the course increased their efficiency in clearing crimes by arrest? The proportions of cases cleared by arrest for samples of trained and untrained officers are reported below.

Trained	Untrained
$P_{s1} = 0.47$	$P_{s2} = 0.43$
$N_1 = 157$	$N_2 = 113$

9.6 SW A private agency in Shinbone has instituted an experimental program for divorce counseling; the key feature of the program is its counselors, who are married couples working in teams. About half of all clients have been randomly assigned to this special program and half to the regular program; the proportion of cases that eventually ended in divorce is recorded for both. The results for random samples of couples from both programs are reported below. In terms of preventing divorce, did the new program work? Write a sentence or two of explanation for your conclusion.

Special Program	Regular Program
$P_{s1} = 0.53$	$P_{s2} = 0.59$
$N_1 = 78$	$N_2 = 82$

9.7 [PA] A number of years ago, the fire department in Shinbone, Kansas, began recruiting minority group members through an affirmative action program. In terms of efficiency ratings as compiled by their superiors, how do the affirmative action employees rate? The ratings of random samples of both groups were collected and the results are reported below (higher ratings indicate greater efficiency).

Affirmative Action	Regular
$\bar{X}_1 = 15.2$	$\bar{X}_2 = 15.5$
$s_1 = 3.9$	$s_2 = 2.0$
$N_1 = 97$	$N_2 = 100$

Write a sentence or two of interpretation.

9.8 [SW] As the director of the local Boys Club, you have claimed for years that membership in your club reduces juvenile delinquency. Now, a cynical member of your funding agency has demanded proof of your claim. Fortunately, your local sociology department is on your side and springs to your aid with student assistants, computers, and hand calculators at the ready. Random samples of members and nonmembers are gathered and interviewed with respect to their involvement in delinquent activities. Each respondent is asked to enumerate the number of delinquent acts he has engaged in over the past year. The results are in and reported below (the average number of admitted acts of delinquency). What can you tell the funding agency?

Members	Nonmembers
$\bar{X}_1 = 10.3$	$\bar{X}_2 = 12.3$
$s_1 = 2.7$	$s_2 = 4.2$
$N_1 = 50$	$N_2 = 55$

9.9 [SOC] At St. Algebra College, the sociology and psychology departments have been feuding for years about the respective quality of their programs. In an attempt to resolve the dispute, you have gathered data about the graduate school experience of random samples of both groups of majors. The results are presented below: the proportion of majors who applied to graduate schools, the proportion of majors accepted in their preferred programs, and the proportion of these who completed their programs. As measured by these data, is there a significant difference in program quality?

a. Proportion of majors who applied to graduate school

Sociology	Psychology
$P_{s1} = 0.53$	$P_{s2} = 0.40$
$N_1 = 150$	$N_2 = 175$

b. Proportion accepted by program of first choice

Sociology	Psychology
$P_{s1} = 0.75$	$P_{s2} = 0.85$
$N_1 = 80$	$N_2 = 70$

c. Proportion completing the programs

Sociology	Psychology
$P_{s1} = 0.75$	$P_{s2} = 0.69$
$N_1 = 60$	$N_2 = 60$

9.10 [SW] A program designed to reduce complaints of sexual harassment in a randomly selected sample of 15 organizations has been implemented. The researchers have information on the average number of complaints per year before and after implementation. What can be said about the effectiveness of the program?

Organization	Average Number of Complaints Before Implementation	Average Number of Complaints After Implementation
A	33	34
B	25	27
C	18	7
D	45	20
E	19	5
F	10	2
G	10	0
H	15	2
I	15	12
J	14	10
K	8	2
L	2	8
M	2	0
N	0	0
O	0	1

9.11 SOC A randomly selected sample of entering college freshmen have participated in a special program designed to enhance their academic abilities, and their GPA's at the end of one year have been recorded. A group of 20 students from the same class who did not participate in the program have been selected as a control group, and they have been matched with the experimental group by gender, age, high-school class rank, College Board scores, and declared college major. The results (GPA's) are presented below. Can the program claim that it was successful?

Project Students	Control Group
3.45	3.41
3.10	3.01
3.04	2.41
3.01	2.00
3.00	2.56
3.00	3.76
2.89	2.80
2.76	2.90
2.75	2.01
2.67	1.90
2.50	2.65
2.50	2.56
2.40	2.40
2.33	1.67
2.30	1.50
2.00	1.00
2.00	2.89
2.00	1.79
1.80	1.00
1.50	1.23

9.12 SW As one part of an attempt to deal with the problem of teen pregnancies as well as other sexual issues, the local juvenile court system has begun to refer all female clients to a sex education program. The premise of the program is that, with full and accurate information, people will make better, more responsible decisions about sex. The scores on a test of general knowledge about sex given before and after the course are presented below for a random sample of participants. Are graduates of the course in fact significantly more knowledgeable?

Student	Pretest	Posttest
A	8	10
B	10	12
C	15	16
D	6	18
E	9	12
F	15	14
G	7	14
H	5	7
I	18	18
J	17	18
K	13	15

9.13 SOC Are middle-class families more likely than working-class families to maintain contact with kin? Below are the average number of visits per year with close kin for random samples of middle- and working-class families. Is the difference significant?

Middle Class	Working Class
$\bar{X}_1 = 7.3$	$\bar{X}_2 = 8.2$
$s_1 = 0.3$	$s_2 = .5$
$N_1 = 89$	$N_2 = 55$

9.14 SOC Are professional basketball teams significantly more oriented toward offense these days? Do they score more points, on the average, than teams in past years? Games are randomly selected from the most recent season and from 10 years ago. Is the difference in average points per game significant?

This Season	10 Years Ago
$\bar{X}_1 = 110$	$\bar{X}_2 = 102$
$s_1 = 10$	$s_2 = 9$
$N_1 = 35$	$N_2 = 37$

9.15 CJ The local police chief started a "crimeline" program some years ago and wonders if it's really working. The program publicizes unsolved violent crimes in the local media and offers cash rewards for information leading to arrests. Are "featured" crimes more likely to be cleared by arrest than other violent crimes? Results from random samples of both types of crimes are reported below.

Crimeline Crimes Cleared by Arrest	Noncrimeline Crimes Cleared by Arrest
$P_{s1} = .35$	$P_{s2} = .25$
$N_1 = 178$	$N_2 = 212$

9.16 SOC Are college students who live in dormitories significantly more involved in campus life than students who commute to campus? The data below report the average number of hours per week students devote to extracurricular activities. Is the difference between these randomly selected samples of commuter and residential students significant?

Residential	Commuter
$\bar{X}_1 = 12.4$	$\bar{X}_2 = 10.2$
$s_1 = 2.0$	$s_2 = 1.9$
$N_1 = 158$	$N_2 = 173$

9.17 SOC Some results from the 1993 General Social Survey are reported below in terms of differences by sex. The actual questions are listed in Appendix G. Which of these differences, if any, are significant? Write a sentence or two of interpretation for each test.

a. Proportion favoring gun control (item 25)

Males	Females
$P_{s1} = .72$	$P_{s2} = .90$
$N_1 = 214$	$N_2 = 306$

b. Proportion agreeing that premarital sex is "always wrong" (item 56)

Males	Females
$P_{s1} = .25$	$P_{s2} = .28$
$N_1 = 230$	$N_2 = 292$

c. Proportion voting for President Clinton in 1993 (item 18)

Males	Females
$P_{s1} = .36$	$P_{s2} = .47$
$N_1 = 231$	$N_2 = 301$

d. Average occupational prestige (item 2)

Males	Females
$\bar{X}_1 = 43.93$	$\bar{X}_2 = 42.41$
$s_1 = 13.40$	$s_2 = 13.05$
$N_1 = 318$	$N_2 = 427$

e. Average rate of church attendance (item 29)

Males	Females
$\bar{X}_1 = 3.53$	$\bar{X}_2 = 4.18$
$s_1 = 2.71$	$s_2 = 2.76$
$N_1 = 322$	$N_2 = 452$

f. Number of children (item 6)

Males	Females
$\bar{X}_1 = 1.76$	$\bar{X}_2 = 1.93$
$s_1 = 1.70$	$s_2 = 1.69$
$N_1 = 326$	$N_2 = 461$

SPSS/PC+ PROCEDURES FOR TESTING THE SIGNIFICANCE OF THE DIFFERENCE BETWEEN TWO MEANS

DEMONSTRATION 9.1 Do Men or Women Watch More TV?

SPSS/PC+ includes a procedure called T-TEST to test for the significance of the difference between two sample means. There are separate tests, one for independent samples and one for matched or paired samples. In this demonstration, I'll explain the use of the independent samples T-TEST, the test we covered in Sections 9.2 and 9.3. To illustrate the process, I'll use the 1993 GSS system file to test for differences between men and women in average hours of reported TV watching. If there is a statistically significant difference between the sample mean for men and the sample mean for women, we can reject the null hypothesis and conclude that the populations (men and women) are different on this variable.

Make the 1993 GSS system file active by using the GET command and

1. From the main menu, select 'analyze data' and then select 'comparing group means' and 'T-TEST.'

2. We want to divide the sample by gender and compare mean hours of TV watching. In the terms of the menu before you, the "groups" are based on sex and the "variable" will be TVHOURS. So, under 'independent test', select /GROUPS and then press ALT-T and specify SEX in the typing window. For /VARIABLES, select TVHOURS.

The command should look like this:

```
T-TEST /GROUPS SEX /VARIABLES TVHOURS.
```

and the output (press F10) will look like this:

```
Independent samples of  SEX       RESPONDENTS SEX
Group 1: SEX EQ        1.00       Group 2: SEX EQ        2.00

t-test for: TVHOURS    HOURS PER DAY WATCHING TV

                  Number                 Standard     Standard
                  of Cases      Mean     Deviation    Error
       Group 1      327        2.6789      1.826        .101
       Group 2      457        2.8906      2.047        .096

                  Pooled Variance Estimate | Separate Variance Estimate
   F     2-Tail |   t    Degrees of 2-Tail |   t    Degrees of 2-Tail
 Value   Prob.  | Value  Freedom    Prob.  | Value  Freedom    Prob.
 1.26    .028   | -1.49    782      .136   | -1.52   745.08     .129
```

The 327 men (group 1) watched an average of 2.6789 hours of TV, whereas the 457 women (group 2) averaged 2.8906 hours. The results of the test for significance are reported in the next several lines of output. SPSS/ PC+ does a separate test for each assumption about the population variance (see Sections 9.2 and 9.3), but we will look only at the Pooled Variance Estimate. This is basically the same model used in this chapter.

SPSS/PC+ reports the t value (-1.49), the Degrees of Freedom (782), and the 2-Tail Prob. (.136). This last piece of information is an alpha level (or a "p" level— see Reading Statistics 5) except that it is the *exact* probability of getting the observed difference in sample means if only chance is operating. Thus, there is no need to look up the test statistic in a t table. This value is greater than .05, our usual indicator of significance. We will fail to reject the null hypothesis and conclude that the difference is not statistically significant. On the average, men and women do not have significantly different TV-viewing habits.

DEMONSTRATION 9.2 Using the COMPUTE Command to Test for Gender Differences in Attitudes about Abortion

I introduced the COMPUTE command in Demonstration 4.3. To refresh your memory, I used COMPUTE to create a summary scale for attitudes toward abortion by adding the scores on the two constituent items (items 52 and 53 in Appendix G). The command looked like this:

```
COMPUTE ABSCALE = ABNOMORE + ABPOOR.
```

Remember that, once created, a computed variable is added to the active file and can be used like any of the variables actually recorded in the file. Here, we will test ABSCALE for the significance of the difference by gender. Our question is, Do men and women have different attitudes towards abortion? If the difference in the sample means is large enough, we can reject the null hypothesis and conclude that the populations are different. Before we conduct the test, I should point out that the abortion scale used in this test is only ordinal in level of measurement. Scales like this are often treated as interval-ratio variables, but we should still be cautious in interpreting our results.

Follow the instructions in Demonstration 9.1 for writing the T-TEST command, and choose ABSCALE rather than TVHOURS as the dependent variable. Your command should look like this:

```
T-TEST /GROUPS SEX /VARIABLES ABSCALE.
```

Your output will be:

```
Independent  samples  of  SEX        RESPONDENTS SEX
Group 1: SEX EQ        1.00      Group 2: SEX EQ        2.00
```

```
t-test for: ABSCALE
```

		Number of Cases	Mean	Standard Deviation	Standard Error
Group 1		196	3.0408	.938	.067
Group 2		290	3.0552	.947	.056

F Value	2-Tail Prob.	Pooled Variance Estimate			Separate Variance Estimate		
		t Value	Degrees of Freedom	2-Tail Prob.	t Value	Degrees of Freedom	2-Tail Prob.
1.02	.898	-.18	484	.853	.16	421.18	.869

The sample means are nearly identical in value (3.0408 versus 3.0552). Intuitively we know that this slight difference cannot be statistically significant; sure enough, the 2-Tail Prob., or alpha (or *p*), value is .869.

As a final point, let me direct your attention to the "Number of Cases" column. This test was based on only 196 men and 290 women, for a total of 486 people. The original sample included 800 people. What happened to all those cases?

Recall from Appendix F that many of the items on the GSS are given to only about two-thirds of the sample. Furthermore, as I mentioned at the end of Chapter 4, the COMPUTE command is designed to drop a case from the computations if it is missing a score on any of the constituent variables. So, most of the "missing cases" were not asked these questions, and the rest were respondents who did not answer one or both of the constituent abortion items. This phenomenon of diminishing sample size is a common problem in survey research and, at some point, the shrinking sample may jeopardize the integrity of the statistical inquiry.

DEMONSTRATION 9.3 Using COMPUTE and RECODE to Test for Differences in Attitudes on Abortion by Church Attendance

Now things are getting very tricky. In this demonstration, we will combine a computed and a recoded variable to conduct an additional test of significance. I introduced the RECODE command in Demonstration 2.3. As you may recall, this procedure allows us to change the scores on a variable. In the previous example, I used the RECODE command to collapse the scores on AGE in order to make that variable more suitable for display in a frequency distribution. Here, we will dichotomize a variable (or recode it so that it has only two scores) so that we can use it in a T-TEST.

In Demonstration 9.2, we tested for the significance of the difference between average ABSCALE scores for men and women. We found that the difference was quite trivial and not at all significant. It is common for a researcher, after getting a result like this, to search for some other variable that might have a significant impact on ABSCALE. The ATTEND variable, a measure of how often the respondents attend church, might be a likely independent variable, but it has eight possible scores, not the two required by the logic of T-TEST. What can we do?

One possibility would be to collapse the variable into a dichotomy, thus satisfying the requirements for the *t* test. At what point should we divide the scores to create this dichotomy? If you look at the coding scheme for AT-TEND (see Appendix G, item 29), you will see that there is no obvious cutting point. The scores graduate from "never" to "several times a week," and there is no point where we could unequivocally divide the sample into "low" and "high." One solution to this problem would be to divide the variable at the median. Although arbitrary, this choice does divide the sample into equal halves, an advantage for many statistical procedures.

Find the median on ATTEND by running the FREQUENCIES command. Locate the category with the middle case by using the cumulative percent column. This will be once a month, or a score of 4. Dichotomize the scores at that point with the RECODE command and prepare a new VALUE LABELS command for the recoded variable. Conduct the T-TEST with recoded AT-TEND as the independent variable. Assuming that you still have the ABSCALE variable available (if not, see Demonstration 9.2), the commands, in order, would look like this:

```
FREQUENCIES /VARIABLES ATTEND.
RECODE ATTEND (0 THRU 4 = 1)(5 THRU 8 = 2).
VALUE LABELS  ATTEND 1 'Low' 2 'High'.
T-TEST /GROUPS ATTEND /VARIABLES ABSCALE.
```

Your output from these commands would be:

```
Independent samples of  ATTEND      HOW OFTEN R ATTENDS RELIGIOUS SERVICES
Group 1: ATTEND EQ         1.00       Group 2: ATTEND EQ        2.00
```

t-test for: ABSCALE

	Number of Cases	Mean	Standard Deviation	Standard Error
Group 1	262	2.7366	.894	.055
Group 2	214	3.4112	.866	.059

		Pooled Variance Estimate			Separate Variance Estimate		
F Value	2-Tail Prob.	t Value	Degrees of Freedom	2-Tail Prob.	t Value	Degrees of Freedom	2-Tail Prob.
1.07	.626	−8.30	474	.000	−8.33	460.52	.000

People who are "low" on ATTEND are in Group 1, and those who are "high" are in Group 2. The former have an average score on ABSCALE of 2.7366, and the latter score an average of 3.4112. This difference is statistically significant (the 2-Tail Prob., or alpha, is very low, less than .000) and the "lows" on ATTEND are significantly more pro-abortion.

Exercises

9.1 Are men significantly different from women on occupational prestige? Using Demonstration 9.1 as a guide, substitute PRESTG80 for TVHOURS on the /VARIABLES list. Write a sentence or two summarizing the results of this test. (See Reading Statistics 5 for some ideas on writing up results.)

9.2 Use the COMPUTE command to create an Anomia scale with items 47 and 48 (ANOMIA5 and ANOMIA6) in the 1993 GSS file. Test the scale for the significance of the difference by sex. Write a sentence or two summarizing the results of this test.

9.3 Use the RECODE command to collapse INCOME into a dichotomy. Find the median on this variable by running the FREQUENCIES first and use this information to write the RECODE statement. Test for the significance of the difference in your ANOMIA scale, computed in Exercise 9.2, using the recoded INCOME as the /GROUPS variable. Are low- or high-income groups more anomic? Write up your results.

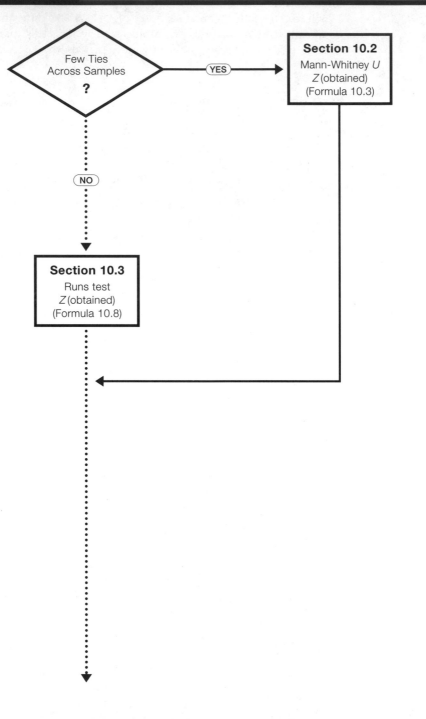

10.1 INTRODUCTION

In this chapter, two **nonparametric tests** of hypothesis, the Mann-Whitney U and the runs test, will be presented. Both tests are appropriate in the two-sample case when the variable of interest is measured at the ordinal level and, under conditions to be specified, may be considered as alternatives to the differences-in-means test. Before introducing the tests themselves, let us consider the situations where they can or should be used.

The **Mann-Whitney** U and the **runs test** represent a large class of tests of significance called "nonparametric" or "distribution free" tests. These tests differ from those introduced in Chapters 8 and 9 in that they require no particular assumption about the shape of the population distribution. Recall that, to test proportions and means for significance, we needed to assume that the sampling distribution was normal in shape. This assumption is satisfied only if sample size is large or if the additional assumption of a normally distributed population can be made (see the theorems presented in Chapter 6). Thus, means and proportions based on small samples can be tested for their significance only when we can make assumptions about the particular shape of the population distribution. Needless to say, researchers often find themselves in situations where they are uncomfortable with such precise assumptions about an unknown distribution.

Nonparametric tests, on the other hand, do not require the assumption of a normally shaped population distribution. In fact, with nonparametric tests, we do not need to assume any particular shape for the population distribution. The tests to be presented in this chapter therefore have wide applicability in situations where the researcher is unsure of the form of the population distribution. Since the requirement of a normal population can be relaxed with large samples, these tests are particularly useful when working with small samples.

Besides the assumptions of a normal sampling distribution, the test for means also requires the assumption of interval-ratio level of measurement. The Mann-Whitney U and the runs test may also be used when the researcher is uncertain of the appropriateness of this assumption.

To summarize, you should seriously consider ordinal-level, nonparametric tests of significance when either the assumption of interval-ratio measurement or the assumption of normal sampling distribution is in doubt for a test of sample means. Both of these assumptions are included in the model assumptions (step 1) for tests of means. The model can be thought of as the mathematical foundation for the rest of the test, and the researcher should be very certain of the appropriateness of these assumptions before placing confidence in the test results. If the model assumptions can be satisfied, tests of the differences in means are preferable to the nonparametric alternatives because the assumption of interval-ratio level permits more sophisticated mathematical procedures to be performed on the data; thus, a greater volume of more precise information can be amassed with respect to the variable of interest. If the model assumptions for means cannot be satisfied, the Mann-

Whitney U and the runs test are very useful alternative tests for investigating the significance of the difference between two samples.

10.2 THE MANN-WHITNEY U

This test of significance has many similarities with the test of significance for the difference in sample means. In both cases we compare random samples as a way of making inferences about the possible differences between two populations. In both cases, the test statistics are computed from the samples and compared with the sampling distribution of all possible sample outcomes. Instead of computing means as the sample statistic, however, the Mann-Whitney test is based on the ranking of the sample scores. This is an appropriate basis since ranking is the most sophisticated mathematical operation that can be performed on ordinal-level data.

The computation of U is straightforward. First, the scores from both samples are pooled and ranked from highest to lowest. Second, the ranks for the two samples are totaled and compared. Intuitively, you can see that, if the two samples represent populations not significantly different from each other, then the total ranks should be similar in value. On the other hand, if the two samples represent populations that do differ on the variable of interest, these totals will be very different.

An example should clarify the underlying logic as well as the computational routines for this test. Assume that you are concerned with racial differences in the level of satisfaction with the social life available on your campus. You have administered survey instruments to randomly selected samples of black and white students and have devised a scale on which a high score indicates great satisfaction. The scores, by race, are presented in Table 10.1.

You undoubtedly have a strong temptation to compute a mean for each

TABLE 10.1   SCORES ON SOCIAL SATISFACTION SCALE FOR BLACK AND WHITE STUDENTS

Sample 1 Black Students		Sample 2 White Students	
Case	Score	Case	Score
1	42	13	45
2	35	14	40
3	30	15	32
4	25	16	30
5	19	17	28
6	17	18	27
7	15	19	26
8	14	20	24
9	9	21	20
10	5	22	10
11	4	23	8
12	2	24	7

TABLE 10.2   RANKED SCORES ON SOCIAL SATISFACTION SCALE FOR BLACK AND WHITE STUDENTS

Sample 1 Black Students			Sample 2 White Students		
Case	Score	Rank	Case	Score	Rank
1	42	2	13	45	1
2	35	4	14	40	3
3	30	6.5	15	32	5
4	25	11	16	30	6.5
5	19	14	17	28	8
6	17	15	18	27	9
7	15	16	19	26	10
8	14	17	20	24	12
9	9	19	21	20	13
10	5	22	22	10	18
11	4	23	23	8	20
12	2	24	24	7	21
		$\Sigma R_1 = 173.5$			$\Sigma R_2 = 126.5$

of these samples and test the difference for its significance. If you keep in mind that these scores are only ordinal in level of measurement, you should be able to restrain yourself long enough to compute U.

First, pool the scores and rank them from high to low. If you encounter any tied scores, assign all of them the average of the ranks they would have used up if they had not been tied. For example, cases 3 and 16 are tied with scores of 30. If they had not been tied, they would have used up ranks 6 and 7. Therefore, assign both cases the rank 6.5 [(6 + 7)/2]. Ranked scores are presented in Table 10.2. To compute U, use Formula 10.1:

FORMULA 10.1

$$U = N_1 N_2 + \frac{N_1(N_1 + 1)}{2} - \Sigma R_1$$

where ΣR_1 = the sum of the ranks for sample 1

For our sample problem above:

$$U = (12)(12) + \frac{(12)(13)}{2} - 173.5$$

$$U = 144 + 156/2 - 173.5$$

$$U = 48.5$$

Note that we could have computed the Mann-Whitney U by using data from sample 2. This alternative solution, which we will label U' (U prime), would have resulted in a larger value for U. The smaller of the two values, U or U', is always taken as the value of U. Once U has been calculated, U' can be quickly determined by means of Formula 10.2:

FORMULA 10.2
$$U' = N_1 N_2 - U$$

Thus, for the sample problem, U' would be $(12)(12) - 48.5$, or 95.5. Remember that the lower of these values is always taken as U.

Once the value of U has been determined, we must still conduct the test of significance. In step 1 of the five-step model, we assume ordinal level of measurement and make no assumption about the shape of the population distribution. The null hypothesis in step 2 is, as always, a statement of "no difference": the two populations represented by these samples are identical. Note that if this assumption is true, then the differences in total ranks should be rather small.

The alternative or research hypothesis is usually a statement to the effect that the two populations are different. This form for H_1 would direct the use of a two-tailed test. It is perfectly possible to use one-tailed tests with Mann-Whitney U when a direction for the difference can be predicted. But, to conserve space and time, we will consider only the two-tailed case.

In step 3, we will take advantage of the fact that, when both sample N's are greater than 10, the sampling distribution of U approximates normality. This will allow us to use the Z-score tables to find the critical region as marked by Z (critical).

To compute the test statistic (step 4), the necessary formulas are

FORMULA 10.3
$$Z \text{ (obtained)} = \frac{U - \mu_u}{\sigma_u}$$

where U = the sample statistic

μ_u = the mean of the sampling distribution of sample U's

σ_u = the standard deviation of the samping distribution of sample U's

FORMULA 10.4
$$\mu_u = \frac{N_1 N_2}{2}$$

FORMULA 10.5
$$\sigma_u = \sqrt{\frac{N_1 N_2 (N_1 + N_2 + 1)}{12}}$$

For the sample problem, μ_u will be $(12)(12)/2$, or 72. The standard deviation of the sampling distribution is

$$\sigma_u = \sqrt{\frac{(12)(12)(12 + 12 + 1)}{12}}$$

$$\sigma_u = \sqrt{300}$$

$$\sigma_u = 17.32$$

We now have all the information we need to conduct a test of significance for U.

APPLICATION 10.1

Do men or women endorse stricter norms of honesty? Samples of 12 men and women have read brief descriptions of various acts that might be considered dishonest (for example, not telling a cashier that he or she has given you too much change) and have been asked to classify each on a scale from "very dishonest" to "not dishonest at all." Summary scores on the scale range from 50 (indicating very strict norms of honesty) to 0 and are reported in the following tables for each subject. Is the difference between men and women statistically significant?

Because the honesty scores are only ordinal, a Mann-Whitney U test of significance will be conducted. The scores have already been ranked and the sum of ranks calculated.

Men			Women		
Case	Score	Rank	Case	Score	Rank
1	47	2	13	48	1
2	44	4	14	45	3
3	40	7	15	43	5
4	35	10	16	42	6
5	32	12	17	39	8
6	31	13	18	36	9
7	30	14	19	33	11
8	29	15	20	28	16
9	25	17	21	23	19
10	24	18	22	21	20
11	20	21	23	15	22
12	12	24	24	14	23
		$\Sigma R_1 = 157$			$\Sigma R_2 = 143$

The value of the Mann-Whitney U is

$$U = N_1 N_2 + \frac{N_1(N_1 + 1)}{2} - \Sigma R_1$$

$$U = (12)(12) + \frac{12(13)}{2} - 157$$

$$U = 144 + 78 - 157$$

$$U = 65$$

The mean of the sampling distribution is

$$\mu_u = \frac{N_1 N_2}{2}$$

$$\mu_u = \frac{(12)(12)}{2}$$

$$\mu_u = 72$$

The standard deviation of the sampling distribution is

$$\sigma_u = \sqrt{\frac{N_1 N_2 (N_1 + N_2 + 1)}{12}}$$

$$\sigma_u = \sqrt{\frac{(12)(12)(12 + 12 + 1)}{12}}$$

$$\sigma_u = \sqrt{\frac{(144)(25)}{12}}$$

$$\sigma_u = \sqrt{300}$$

$$\sigma_u = 17.32$$

The value of the test statistic is

$$Z \text{ (obtained)} = \frac{U - \mu_u}{\sigma_u}$$

$$Z \text{ (obtained)} = \frac{65 - 72}{17.32}$$

$$Z \text{ (obtained)} = \frac{-7}{17.32}$$

$$Z \text{ (obtained)} = -0.40$$

If alpha is set at the 0.05 level, the critical region would begin at Z (critical) $= \pm 1.96$. With an obtained Z of -0.40, we would fail to reject the null hypothesis (H_o: the populations are identical). The scores of men and women on the honesty scale are not significantly different.

Step 1. Making assumptions.

Model: Independent random samples
Level of measurement is ordinal

Step 2. Stating the null hypothesis. H_0: The populations from which the samples are drawn are identical on the variable of interest.

(H_1: The populations from which the samples are drawn are different on the variable of interest.)

Step 3. Selecting the sampling distribution and establishing the critical region.

Sampling distribution = Z distribution
Alpha = 0.05
Z (critical) = ±1.96

Step 4. Calculating the test statistic. With U equal to 48.5, μ_u equal to 72, and a σ_u of 17.32,

$$Z \text{ (obtained)} = \frac{U - \mu_u}{\sigma_u}$$

$$Z \text{ (obtained)} = \frac{48.5 - 72}{17.32}$$

$$Z \text{ (obtained)} = -1.36$$

Step 5. Making a decision. The test statistic, a Z (obtained) of −1.36, does not fall in the critical region as marked by the Z (critical) of ±1.96. Therefore, we fail to reject the null of no difference. Black students are not significantly different from white students in terms of their level of satisfaction with the social life available on campus. Note that if we had used the U' value of 95.5 instead of U in computing the test statistic, the value of Z (obtained) would have been +1.36, and our decision to fail to reject the null would have been exactly the same.

10.3 THE RUNS TEST

Far from being a test of physical fitness, the runs test is very similar in logic and form to the Mann-Whitney U test. The null hypothesis is, again, the assumption that there is no significant difference between the populations from which the samples come. To conduct the test, the scores from both samples are pooled and then ranked from high to low as if they were a single sample. If the null is true, then the scores should be very intermixed and there should be many runs. A **run** is defined as any sequence of one or more scores from the same sample. If the null is false and the populations are different on the trait being measured, then there should be very few runs.

To illustrate with the data on social satisfaction presented above, if we pooled the two samples and designated a white student with a W and a black student with a B, we would get the following sequences:

$$\underset{1\ 2\ 3\ 4\ 5\ 6}{\underline{W\ B\ W\ B\ W\ B}}\ \underset{7}{\underline{W\ W\ W\ W}}\ \underset{8}{\underline{B}}\ \underset{9}{\underline{W\ W}}\ \underset{10}{\underline{B\ B\ B\ B}}\ \underset{11}{\underline{W}}\ \underset{12}{\underline{B}}\ \underset{13}{\underline{W\ W}}\ \underset{14}{\underline{B\ B\ B}}$$

where the W at the far left represents the individual from either sample with the highest score (case 13 with a score of 45), the next letter represents the individual with the second highest score (case 1 with a score of 42), and so on. The underlinings represent runs, and by counting we find that there are 14 runs in these data.

Although no prediction of the direction of the difference need be made with the runs test, the smaller the number of runs, the greater the significance of the difference between the samples. The lowest number of runs possible is, of course, two. If there had been racial differences in social satisfaction in the data above, with all white students expressing greater satisfaction than any black student, the data would have looked like this:

$$\underset{1}{\underline{W\ W\ W\ W\ W\ W\ W\ W\ W}}\ \ \ \ \underset{2}{\underline{B\ B\ B\ B\ B\ B\ B\ B\ B\ B}}$$

The fewer the differences between the two populations, the more intermixing between the two samples. This will result in a higher number of runs and, consequently, a lower probability of being able to reject the null hypothesis.

For situations in which at least one sample is larger than 20, the sampling distribution of all possible sample runs approximates normality. The mean (μ_R) and the standard deviation (σ_R) of the sampling distribution can be found by Formulas 10.6 and 10.7, respectively:

FORMULA 10.6

$$\mu_R = \frac{2\ N_1 N_2}{N_1 + N_2} + 1$$

FORMULA 10.7

$$\sigma_R = \sqrt{\frac{2\ N_1\ N_2\ (2\ N_1\ N_2 - N_1 - N_2)}{(N_1 + N_2)^2\ (N_1 + N_2 - 1)}}$$

The test statistic, Z (obtained), is computed with Formula 10.8:

FORMULA 10.8

$$Z\ (\text{obtained}) = \frac{R - \mu_R}{\sigma_R}$$

where R = the number of runs

To illustrate with an example, assume that randomly selected samples of white males and females have been tested for their racial prejudice. Is there a significant difference in levels of prejudice between the sexes? The scores are reported in Table 10.3. First, they must be pooled and then ranked; then we can count the number of runs. Tied scores can create rather serious prob-

TABLE 10.3   RACIAL PREJUDICE SCORES FOR SAMPLES OF MALES
AND FEMALES

Sample 1 Males		Sample 2 Females	
Case	Score	Case	Score
1	29	23	32
2	29	24	31
3	25	25	30
4	25	26	30
5	25	27	27
6	24	28	26
7	23	29	26
8	23	30	22
9	21	31	22
10	18	32	20
11	18	33	20
12	17	34	19
13	15	35	19
14	13	36	16
15	12	37	14
16	8	38	14
17	8	39	10
18	7	40	9
19	6	41	4
20	5	42	4
21	3	43	2
22	3	44	2

lems with this particular test because, depending on how the tied cases are arranged, the number of runs can be very much affected. If you encounter tied scores, probably the best (safest) thing to do is to compute all possible numbers of runs by rearranging the tied cases. If all possible solutions lead to the same decision (reject H_0 or fail to reject H_0), then we are on safe ground in making that decision. If the different arrangements lead to different decisions, then we clearly must opt for another test of significance. In our example, there are no tied cases across the samples (ties within samples won't make any difference in the number of runs).

Designating males with M and females with F, we can array the scores as follows:

$\underline{\text{F F F F}}$	$\underline{\text{M M}}$	$\underline{\text{F F F}}$	$\underline{\text{M M M M M M}}$	$\underline{\text{F F}}$	$\underline{\text{M}}$	$\underline{\text{F F F F}}$	$\underline{\text{M M M}}$	$\underline{\text{F}}$
1	2	3	4	5	6	7	8	9

$\underline{\text{M}}$	$\underline{\text{F F}}$	$\underline{\text{M M}}$	$\underline{\text{F F}}$	$\underline{\text{M M M M M}}$	$\underline{\text{F F}}$	$\underline{\text{M M}}$	$\underline{\text{F F}}$
10	11	12	13	14	15	16	17

There are 17 runs in these data, and we can now proceed with the formal test of significance.

Step 1. Making assumptions.

Model: Independent random sampling
Level of measurement is ordinal

Step 2. Stating the null hypothesis. H_0: The two populations are identical on racial prejudice.

(H_1: The two populations are different on racial prejudice.)

Step 3. Selecting the sampling distribution and establishing the critical region.

Sampling distribution = Z distribution
Alpha = 0.05
Z (critical) = ± 1.96

Step 4. Calculating the test statistic. Before solving for Z (obtained), both μ_R (the mean of the sampling distribution) and σ_R (the standard deviation of the sampling distribution) must be calculated. The necessary formulas were presented above as Formulas 10.6 and 10.7.

$$\mu_R - \frac{2N_1 N_2}{N_1 + N_2} + 1$$

$$\mu_R = \frac{(2)(22)(22)}{22 + 22} + 1$$

$$\mu_R = 23$$

$$\sigma_R = \sqrt{\frac{2N_1 N_2 (2N_1 N_2 - N_1 - N_2)}{(N_1 + N_2)^2 (N_1 + N_2 - 1)}}$$

$$\sigma_R = \sqrt{\frac{(2)(22)(22) \, [(2)(22)(22) - 22 - 22]}{(22 + 22)^2 \, (22 + 22 - 1)}}$$

$$\sigma_R = 3.28$$

$$Z \text{ (obtained)} = \frac{R - \mu_R}{\sigma_R}$$

$$Z \text{ (obtained)} = \frac{17 - 23}{3.28}$$

$$Z \text{ (obtained)} = -1.83$$

Step 5. Making a decision. The test statistic, a Z (obtained) of -1.83, does not fall in the critical region as marked by the Z (critical) of ± 1.96. Therefore, we must fail to reject the null hypothesis. Males and females do not differ significantly on the trait of racial prejudice.

10.4 CHOOSING AN ORDINAL TEST OF SIGNIFICANCE

Several other ordinal tests of significance have been invented by statisticians. Most are based on logic similar to that presented in Sections 10.2 and 10.3, although some are appropriate only in specific circumstances. For example, a test statistic called Wilcoxon's T is designed for the situation where the two samples are not independent (that is, when the samples have been matched on some criterion as in a "before-after" test of the same subjects). There is even a way of conducting a kind of analysis of variance on ordinal data for situations where the researcher has more than two samples.

The two tests presented above are among the more popular general tests of significance for ordinal data, and choosing between them may be something of a problem. In most cases, which test is selected makes little difference, because the decision made in step 5 will be the same. One criterion you can use in making the choice is the number of ties across the samples. As was pointed out above, a large number of such ties makes the runs test rather troublesome and, in such a case, you should opt for the Mann-Whitney U. Also, remember that, when the more restrictive model assumptions can be satisfied, the test for the significance of the difference in sample means is preferred over either of the tests presented here.

SUMMARY

1. When testing for the significance of the difference between two sample means, researchers sometimes must make assumptions (step 1) about which they are uncertain. If sample size is small and/or the appropriateness of assuming interval-ratio level of measurement is unclear, serious consideration should be given to using nonparametric tests. These tests require only ordinal level of measurement and make no particular assumptions about the shape of the population distribution. Two of these tests are the Mann-Whitney U and the runs test.

2. In the Mann-Whitney U test, the scores from both samples are pooled and then ranked from high to low. The ranks are summed for each sample, and the test is based on a comparison of the summed ranks. If the null is true (if the samples come from the same population), the summed ranks should be approximately equal. The greater the difference between the summed ranks, the greater the probability of rejecting the null.

3. The runs test also involves pooling and ranking the scores from both samples. For this test, we count the number of runs (sequences of one or more scores from

the same sample) in the pooled scores. If the null hypothesis is true, there should be many runs (the scores should be very intermixed). The fewer the runs, the greater the probability of rejecting the null.

4. If there are many ties across samples, the Mann-Whitney U is preferred to the runs test. If ties across samples are few in number, both tests will result in the same decision (reject or fail to reject) in the huge majority of cases.

SUMMARY OF FORMULAS

Mann-Whitney U	10.1	$U = N_1 N_2 + \dfrac{N_1(N_1 + 1)}{2} - \Sigma R_1$
U'	10.2	$U' = N_1 N_2 - U$
Z (obtained) for Mann-Whitney U	10.3	$Z \text{ (obtained)} = \dfrac{U - \mu_u}{\sigma_u}$

Mean of the sampling

distribution for
sample U's   10.4 $\mu_u = \dfrac{N_1 N_2}{2}$

Standard
deviation of
the sampling
distribution for
sample U's   10.5 $\sigma_u = \sqrt{\dfrac{N_1 N_2 (N_1 + N_2 + 1)}{12}}$

Mean of the
sampling
distribution for
sample runs   10.6 $\mu_R = \dfrac{2 N_1 N_2}{N_1 + N_2} + 1$

Standard
deviation of
the sampling
distribution for
sample runs   10.7 $\sigma_R = \sqrt{\dfrac{2 N_1 N_2 (2 N_1 N_2 - N_1 - N_2)}{(N_1 + N_2)^2 (N_1 + N_2 - 1)}}$

Z (obtained)
for runs test   10.8 $Z \text{ (obtained)} = \dfrac{R - \mu_R}{\sigma_R}$

GLOSSARY

Mann-Whitney U. A nonparametric test of significance for the two-sample case when the variable
of interest has been measured at the ordinal level.

μ_R. The symbol for the mean of a sampling distribution
of all possible sample runs.

μ_u. The symbol for the mean of the sampling distribution of all possible sample U's.

Nonparametric test. A type of significance test where
no assumptions about the precise shape of the
population need be made.

Run. A sequence of one or more scores from the same
sample. The basis of the test statistic used in the
runs test.

Runs test. A nonparametric test of significance for the
two-sample case when the variable of interest has
been measured at the ordinal level.

σ_R. The symbol for the standard deviation of the sampling distribution of all possible sample runs.

σ_u. The symbol for the standard deviation of the sampling distribution of all possible sample U's.

PROBLEMS

For problems 10.1 to 10.4, conduct both the Mann-
Whitney U and the runs test. Use the normal approximations for the sampling distributions throughout.

10.1   SOC Random samples of freshmen and seniors
were tested for their level of racial prejudice, and
the scores are reported below (higher scores indicate greater prejudice). Is there a significant
difference between the two samples? Write a
sentence or two interpreting your results.

Freshmen		Seniors	
Case	Score	Case	Score
1	59	21	44
2	58	22	39
3	57	23	39
4	56	24	38
5	54	25	38
6	52	26	37
7	52	27	37
8	45	28	36
9	40	29	35
10	30	30	34
11	30	31	31
12	29	32	10
13	25	33	9
14	25	34	8
15	24	35	8
16	23	36	7
17	22	37	5
18	20	38	1
19	15	39	1
20	12	40	0

10.2   SW/CJ Your local juvenile probation department sponsors summer camping trips for some
of its clients and wants to know if these wilder-
ness experiences have any effect on the attitudes
of the campers. You have gathered random
samples of participants and nonparticipants and
have administered a scale that measures respect
for authority (higher scores indicate greater respect). What can you tell the agency?

Campers		Noncampers	
Case	Score	Case	Score
1	35	21	27
2	35	22	26
3	34	23	23
4	30	24	22
5	30	25	22
6	29	26	17
7	28	27	17
8	25	28	12
9	24	29	11
10	23	30	10
11	23	31	10
12	23	32	7
13	19	33	7
14	18	34	3
15	18	35	3
16	18	36	3
17	16	37	2
18	15	38	2
19	15	39	1
20	14	40	1

10.3 GER A nursing home has an unusually large number of patients who are uncommunicative and socially isolated. In an effort to alleviate this condition, the staff began a limited program of art and music therapy for some of the patients. After six months of operation, the staff rated the patients in terms of responsiveness to social interaction. Random samples of both therapy and nontherapy patients were assembled. Is there a significant difference between the two groups? (Higher scores indicate greater responsiveness.)

Therapy		Nontherapy	
Case	Score	Case	Score
1	102	22	100
2	98	23	99
3	95	24	96
4	85	25	70
5	77	26	65
6	70	27	65
7	69	28	50
8	68	29	30
9	68	30	30
10	66	31	29

Therapy		Nontherapy	
Case	Score	Case	Score
11	51	32	28
12	49	33	21
13	45	34	15
14	42	35	10
15	40	36	9
16	39	37	3
17	35	38	1
18	35		
19	34		
20	33		
21	32		

10.4 SOC As is the case with most institutions of higher education, St. Algebra College provides sociology majors with numerous ways of avoiding courses in mathematics. This arrangement is generally satisfactory for all concerned until, of course, the majors encounter their required sociology statistics course (sound familiar?). The instructor of this course has been keeping careful track of the mathematical background of the students and is now prepared to do a little hypothesis testing. The question is, Do students with college mathematics courses perform at higher levels in the sociology statistics course than students who have taken no college mathematics courses? Random samples are collected and final course averages are used as a measure of student performance. Being extremely conservative, the instructor decides that grades must be considered ordinal-level data. What can be concluded?

Students With at Least One College Math Course		Students With No College Math Courses	
Case	Final Average in Statistics	Case	Final Average in Statistics
1	95.6	26	95.4
2	94.3	27	93.7
3	92.1	28	93.2
4	90.1	29	90.9
5	89.9	30	87.6
6	88.0	31	83.2

Students With at Least One College Math Course		Students With No College Math Courses	
Case	Final Average in Statistics	Case	Final Average in Statistics
7	85.3	32	80.9
8	82.1	33	74.2
9	81.1	34	74.1
10	79.2	35	69.5
11	78.0	36	68.4
12	76.3	37	66.3
13	76.2	38	66.0
14	75.7	39	62.0
15	75.1	40	61.0
16	74.9	41	60.0
17	73.5	42	59.2
18	72.0	43	58.7
19	71.3		
20	70.5		
21	69.3		
22	65.0		
23	64.0		
24	60.0		
25	53.4		

10.5 SOC A random sample of students in a large introductory sociology class has rated the effectiveness of the instructor on a scale that ranges from 10 (excellent) to 1 (poor). Is there a difference in the ratings for sociology majors as opposed to nonmajors? Use the Mann-Whitney U test for significance and explain your conclusions. Could the runs test be used for these data? Why or why not?

Majors		Nonmajors	
Case	Rating	Case	Rating
1	10	21	10
2	9	22	9
3	10	23	7
4	10	24	10
5	7	25	8
6	10	26	6
7	6	27	10
8	10	28	9
9	8	29	6

Majors		Nonmajors	
Case	Rating	Case	Rating
10	6	30	4
11	10	31	7
12	9	32	6
13	5	33	10
14	3	34	2
15	9	35	9
16	5	36	3
17	4	37	7
18	9	38	5
19	3	39	3
20	8	40	6

10.6 SW A family therapist used a new approach with half of her case load for a year and then had an associate rate each family for level of conflict (higher scores indicate greater conflict). Is there a significant difference between the samples?

New Therapy		Old Therapy	
Case	Score	Case	Score
1	53	16	50
2	53	17	49
3	40	18	49
4	35	19	45
5	33	20	44
6	32	21	44
7	30	22	41
8	20	23	39
9	17	24	39
10	16	25	32
11	10	26	27
12	9	27	22
13	4	28	21
14	4	29	15
15	3	30	11

10.7 SOC College fraternities and sororities are sometimes accused of being anti-intellectual in that they encourage and reward behaviors and lifestyles incompatible with scholarly pursuits. To test the accuracy of this charge, samples of "Greeks" and independents have been given a survey that measures support for scholarship and the scores are reported below. Is there a significant difference between the samples?

Greeks		Independents	
Case	Score	Case	Score
1	74	16	85
2	53	17	84
3	53	18	84
4	52	19	75
5	52	20	75
6	52	21	74
7	51	22	50
8	50	23	49
9	47	24	48
10	47	25	48
11	40	26	29
12	35	27	22
13	34	28	21
14	33	29	15
15	32	30	11

10.8 SOC A scale measuring support for legalized abortion has been administered to samples of men and women. Is there a significant difference?

Men		Women	
Case	Score	Case	Score
1	13	12	13
2	12	13	13
3	12	14	11
4	11	15	11
5	10	16	9
6	10	17	8
7	10	18	8
8	9	19	6
9	9	20	4
10	8	21	3
11	5	22	1

10.9 PS Samples of Republican and Democratic legislators have been rated on their support for a new missile system being requested by the Pentagon. Is there a significant difference between the parties?

Democrats		Republicans	
Case	Score	Case	Score
1	25	13	35
2	24	14	34
3	23	15	32

Democrats		Republicans	
Case	Score	Case	Score
4	22	16	31
5	14	17	30
6	10	18	30
7	10	19	29
8	10	20	26
9	9	21	21
10	4	22	15
11	2	23	9
12	2	24	8

10.10 SW For a number of months, some surgical patients in a large hospital have been visited shortly after release by a social worker and a public health nurse working as a team. This new program is designed to ease the recovery period and ensure that all patients have access to all necessary services. Samples of participants and nonparticipants have been asked to rate the difficulty of their recovery period. As assessed by these scores, did the program work?

Participants		Nonparticipants	
Case	Score	Case	Score
1	23	12	25
2	22	13	25
3	19	14	25
4	16	15	24
5	9	16	24
6	5	17	20
7	5	18	20
8	5	19	17
9	4	20	14
10	1	21	10
11	1	22	7

10.11 SOC Twenty males and 20 females have been randomly selected from the General Social Survey data file. Their scores on each of five variables are reported below. Conduct the Mann-Whitney U test for each variable. There are too many tied scores to justify use of the runs test. See Appendix G for the exact wording of each question. Be sure to rank order the scores before conducting the tests.
a. Occupational prestige (item 2).

Males		Females	
Case	Score	Case	Score
1	39	21	26
2	32	22	47
3	36	23	60
4	52	24	12
5	32	25	58
6	33	26	50
7	15	27	29
8	43	28	63
9	78	29	32
10	42	30	34
11	17	31	55
12	19	32	60
13	50	33	26
14	32	34	60
15	40	35	39
16	51	36	25
17	47	37	62
18	55	38	51
19	49	39	34
20	37	40	62

b. Father's occupational prestige (item 5).

Males		Females	
Case	Score	Case	Score
1	45	21	29
2	28	22	30
3	47	23	40
4	34	24	23
5	50	25	50
6	70	26	35
7	17	27	41
8	45	28	25
9	51	29	45
10	31	30	21
11	41	31	43
12	52	32	47
13	54	33	52
14	58	34	62
15	51	35	41
16	39	36	61
17	61	37	45
18	56	38	47
19	27	39	41
20	29	40	51

c. Number of children (item 6).

Males		Females	
Case	Score	Case	Score
1	1	21	2
2	2	22	0
3	2	23	2
4	0	24	3
5	3	25	0
6	1	26	0
7	0	27	2
8	2	28	2
9	0	29	3
10	0	30	1
11	0	31	0
12	0	32	0
13	2	33	1
14	0	34	3
15	7	35	1
16	4	36	4
17	1	37	0
18	2	38	2
19	3	39	2
20	2	40	3

d. Income (item 12).

Males		Females	
Case	Score	Case	Score
1	12	21	6
2	9	22	12
3	12	23	11
4	12	24	11
5	12	25	12
6	10	26	9
7	8	27	8
8	9	28	12
9	10	29	7
10	11	30	5
11	12	31	12
12	7	32	4
13	12	33	12
14	9	34	12
15	8	35	10
16	10	36	10
17	10	37	9
18	8	38	8
19	7	39	7
20	6	40	12

e. Church attendance (item 29).

Males		Females			Males		Females	
Case	Score	Case	Score		Case	Score	Case	Score
1	0	21	3		10	2	30	4
2	1	22	3		11	7	31	6
3	7	23	3		12	0	32	5
4	2	24	8		13	5	33	7
5	3	25	3		14	0	34	6
6	3	26	0		15	6	35	2
7	0	27	4		16	2	36	4
8	7	28	5		17	7	37	7
9	7	29	8		18	0	38	4
					19	8	39	4
					20	4	40	2

SPSS/PC+ PROCEDURES FOR NONPARAMETRIC TESTS OF SIGNIFICANCE

DEMONSTRATION 10.1 Does Income Vary by Educational Level?

As you can see by inspecting the menus, SPSS/PC+ includes a number of nonparametric tests for a variety of situations. In this section, we will use only the Mann-Whitney test and the runs test, but you should take some time to familiarize yourself with the wide variety of alternative tests available. As an example, I will use the INCOME91 variable from the 1993 GSS. Inspect the scores for this variable (see Appendix G, item 12) and you'll see that the categories are unequal in size. INCOME91 can be considered ordinal in level of measurement, and the tests introduced in this chapter are appropriate.

For an independent variable, let's use DEGREE. Conventional wisdom is that higher levels of education lead to more significant jobs and, therefore, to higher income levels. We can consider this idea a hypothesis and test it with the procedures presented in this chapter. Since DEGREE has five categories, we will need to use the RECODE command to collapse it into the two-category format required by the logic of these tests. It seems reasonable to divide the sample into those with a high school diploma or less (categories 0 and 1) and those with at least some college (categories 2 thru 4). The appropriate RECODE and VALUE LABELS commands, assuming that the 1993 GSS system file is active, would be

```
RECODE DEGREE (0,1 = 1)(2 THRU 4 = 2).
VALUE LABELS DEGREE 1 'HS or Less' 2 'At Least Some
College'.
```

To do the Mann-Whitney U test first:

1. Select 'analyze data' from the main menu and then select 'other'. Select and paste 'NPAR TESTS.'
2. Scroll down to the section entitled '2 independent samples' and select and paste /MANN-WHITNEY.
3. The '!test variable' is the dependent variable—the variable that is the focus of our attention. In the example at hand, INCOME would be the test variable. Specify this variable by pressing ALT-T and using the typing window.
4. Select and paste the keyword 'BY' and then specify the groups between which the test will be conducted (levels of education or the recoded DEGREE variable). Highlight '!grouping variable', press ALT-T, and type DEGREE in the typing window. Next, select the parentheses and use the typing window to specify the values that will identify the two groups. In our case, the values are simply 1 and 2. All cases with the first value go in group one and all the cases with the second value are in group two.

Your command should look like this:

```
NPAR TEST /MANN-WHITNEY INCOME91 BY DEGREE (1,2).
```

and the output will be

```
- - - - - Mann-Whitney U - Wilcoxon Rank Sum W Test

  INCOME91      TOTAL FAMILY INCOME
by DEGREE       RS HIGHEST DEGREE

   Mean Rank  Cases
      307.57    507   DEGREE = 1 less than hs
      480.92    208   DEGREE = 2 at least some college
                ---
                715   Total
                                Corrected for Ties
         U           W          Z        2-tailed P
     27161.0     100031.0   -10.2200       .0000
```

The mean rank is the sum of ranks divided by N. People with lower levels of education have the lower mean rank (307.57 versus 480.92), which indicates that, on the average, they have lower incomes than the people with higher levels of educational attainment. The output also shows the number of cases included, the U statistic (27161.0), the obtained Z score (-10.2200), and the exact probability that the Z score occurred by random chance alone. This 2-tailed P is reported as .0000, which means that it is less than .0001. The difference is very significant, so large that the probability of it having occurred by random chance alone is less than a 1 out of 10,000.

To conduct the runs test, select /WALD-WOLFOWITZ from the NPAR TESTS menu. To specify the test variable and other specifics, repeat the steps summarized above for the Mann-Whitney test. Your command should look like this:

```
NPAR TESTS /WALD-WOLFOWITZ INCOME91 BY SEX (1,2).
```

and your output will be

```
- - - - Wald-Wolfowitz Runs Test

   INCOME91     TOTAL FAMILY INCOME
  by DEGREE     RS HIGHEST DEGREE

        Cases
          507  DEGREE = 1 less than hs
          208  DEGREE = 2 at least some colleg
          ---
          715  Total

                        Runs           Z      1-tailed P
  Minimum Possible:       19     -25.1329         .0000
  Maximum Possible:      382       7.8052         .0000

WARNING -- There are 18 Inter-group Ties involving 672
cases.
```

With only 21 possible scores for INCOME91, there will be many ties across the samples; for this specific problem, the results of a runs test may be hard to interpret. If there are ties, the output will include the minimum and maximum number of runs possible in the data and the Z scores for each value. If both these Z scores are significant (this is the case for our example), we are safe in concluding that the scores of the two samples are not randomly distributed.

The results of these two tests are consistent. Not surprisingly, people with lower levels of educational attainment report significantly lower incomes. The common wisdom is affirmed by these tests of significance.

DEMONSTRATION 10.2 Can Money Buy Happiness?

As long as we are testing conventional wisdom, let's see if there is any empirical support for the widely held view that money can't buy happiness. There is a variable called HAPPY available in the GSS data set that records the respondent's level of happiness (item 34 in Appendix G). We can use the INCOME91 variable from Demonstration 10.1 as our indicator of "money." (Note that we used INCOME91 as a dependent variable in the previous demonstration, and we are now changing its role to causal, or independent.) If the adage is true, there should be no significant relationship between these two variables.

We need to collapse INCOME91 into just two categories in order to conduct the test. As was the case in Demonstration 9.3, there is no obvious place to split this variable, so we need to run a FREQUENCIES for the variable, find the median, and split the variable at that point. To save you the trouble, I'll just let you know that the median for INCOME91 is in the 30,000 to 34,999 income category. Our RECODE and VALUE LABELS statement will look like this:

```
RECODE INCOME (1 thru 16 = 1)(17 thru 21 = 2).
VALUE LABELS INCOME91 1 'lower' 2 'higher'.
```

Follow the instructions in Demonstration 10.1 to generate the two tests with HAPPY as the dependent, or test, variable and INCOME91 as independent. The results of the tests are

```
- - - - Mann-Whitney U - Wilcoxon Rank Sum W Test

     HAPPY      GENERAL HAPPINESS
by INCOME91     TOTAL FAMILY INCOME

    Mean Rank    Cases
       378.21      413  INCOME91 = 1 lower
       329.08      301  INCOME91 = 2 higher
                   ---
                   714  Total

                                    Corrected for Ties
       U            W           Z        2-tailed P
    53603.0      99054.0     -3.5707        .0004

- - - - Wald-Wolfowitz Runs Test

     HAPPY      GENERAL HAPPINESS
by INCOME91     TOTAL FAMILY INCOME

        Cases
          413  INCOME91 = 1 lower
          301  INCOME91 = 2 higher
          ---
          714  Total

                          Runs           Z    1-tailed P
Minimum Possible:           4       -26.5101      .0000
Maximum Possible:         603        19.4888      .0000

WARNING -- There are 3 Intergroup Ties involving 714 cases.
```

The Mann-Whitney U test shows that the higher-income category has a lower mean rank on HAPPY and the Z score and 2-tailed P, or exact alpha level, show that the difference is significant. Because HAPPY is coded so that lower scores indicate higher levels of reported happiness, these results indicate that higher-income respondents are significantly happier than lower-income respondents.

The Runs test indicates a significant relationship (because both the P values are less than .0001), but, with only three possible scores for HAPPY, there are many ties on this variable, and we need to be cautious in our interpretation. Overall, we could conclude that the evidence does not support the old adage that money can't buy happiness.

Exercises

10.1 As long as we have INCOME91 recoded and in a suitable form for use as an independent variable for these tests, take a look at the relationship between income and political ideology. Run both the Mann-Whitney U and the Runs test with POLVIEWS (item 19 in Appendix G) as the test variable and INCOME91 as the independent. It is commonly argued that a conservative political ideology is associated with higher levels of material wealth. Is there any support for this view in these data? Write a few sentences of interpretation.

10.2 Is church attendance (ATTEND is item 29 in Appendix G) related to gender? Use both ordinal tests of significance and write a few sentences summarizing your conclusions.

10.3 Remember ABSCALE? If not, see Demonstration 4.3. Use the COMPUTE command to recreate this scale, if necessary. Use the scale as the dependent variable and run separate tests with recoded DEGREE, recoded INCOME91, and SEX as independent variables. Write a brief report summarizing your results and indicate which of the three independent variables had the strongest effect.

Hypothesis Testing IV

11

Chi Square

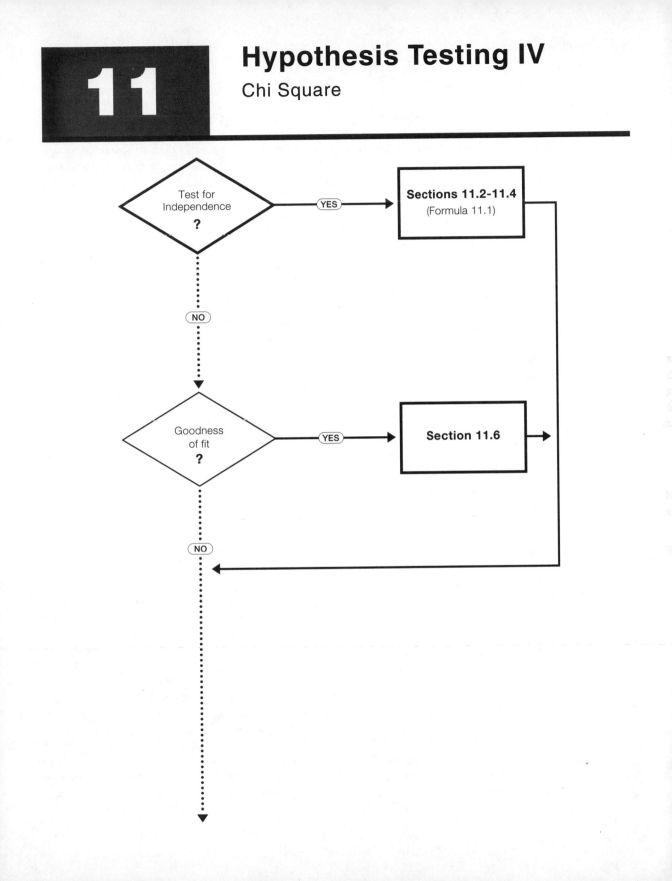

11.1 INTRODUCTION

The **chi square test** is probably the most frequently used test of hypothesis in the social sciences. In part, this popularity is due to the relative ease with which the model assumptions in step 1 can be satisfied. Chi square is a nonparametric test of hypothesis and requires no assumption about the exact shape of the population distribution. Since this test is appropriate for nominally measured variables, we make only the minimum possible assumption with respect to level of measurement.

The popularity of chi square is also related to its usefulness in a wider variety of research situations than the other tests of significance we have covered to this point. Although the chi square test can be used in the familiar two-sample case, it is also applicable to research situations where the variables of interest have more than two categories and in situations where we have more than two samples. For example, in Chapter 9, we compared a sample of black senior citizens with a sample of white senior citizens in terms of participation patterns (see Section 9.5). What if the researcher desired to expand the test to include samples of Spanish-speaking and Asian American senior citizens in the comparison? Two-sample tests are not readily adaptable to research designs with many samples, but chi square handles these situations easily and in a single set of calculations.

11.2 THE LOGIC OF CHI SQUARE

The chi square test has several different uses. We will look first at (and, in fact, spend most of this chapter on) an application called the chi square test for independence, which can be used when two variables have been organized into a bivariate table. We have encountered the term *independence* in connection with the requirements for random sampling in the two-sample case. In that context, we noted that independent random samples are gathered such that the selection of a particular case for one sample has no effect on the probability that any particular case will be selected for the other sample (see Section 9.1).

In the context of chi square, the meaning of **independence** is closely related to its earlier meaning. Two variables are independent if, for all cases, the classification of a case into a particular category of one variable has no effect on the probability that the case will fall into any particular category of the second variable. For example, assume that the variables of interest are rates of participation and racial-group membership for a sample of senior citizens. These two variables will be independent of each other if the classification of the cases on one variable (as black or white) has no effect on the classification of the cases on the other variable (as high or low on participation).

The notion of independence between variables can be more concretely illustrated by the use of bivariate tables. Recall (Section 2.8) that bivariate tables display the joint classification of the cases on two variables. The categories of one variable are used as column headings and the categories of the

TABLE 11.1   RATES OF PARTICIPATION IN VOLUNTARY ASSOCIATIONS
BY RACIAL GROUP FOR 100 SENIOR CITIZENS

Participation Rates	Racial Group		
	Black	White	
High			50
Low			50
	50	50	100

TABLE 11.2   THE CELL FREQUENCIES THAT WOULD BE EXPECTED IF RATES
OF PARTICIPATION AND RACIAL GROUP WERE INDEPENDENT

Participation Rates	Racial Group		
	Black	White	
High	25	25	50
Low	25	25	50
	50	50	100

other as row headings. The cell frequencies indicate the number of cases classified into each of the joint categories of the two variables, and the **marginals** are the univariate frequency distributions for each variable. Table 11.1 shows the outlines of a bivariate table for the variables of participation and racial group for a fictitious sample of 100 senior citizens. Note that the sample includes exactly 50 members of each racial group and that the sample is evenly divided on rates of participation. Note also that there are exactly four cross-classification categories or **cells** for these two variables.

Now, if these two variables are independent, then the cell frequencies will be determined solely by random chance. Thus, just as an honest coin will show heads about 50% of the time when flipped, about half of the black senior citizens will rank high on participation and half will rank low. The same pattern would hold for the 50 white senior citizens and, therefore, each of the four cells should have about 25 cases in it, as illustrated in Table 11.2. This pattern of cell frequencies indicates that the racial classification of the subjects has no effect on the probability that they would be either high or low in participation. The probability would be 0.5 for both blacks and whites, and the variables would therefore be independent.

In the case of chi square, the null hypothesis is that the variables are independent. Under the assumption that the null hypothesis is true, the cell frequencies we would expect to find if only random chance were operating are computed. These frequencies, called expected frequencies (symbolized

f_e), are then compared, cell by cell, with the frequencies actually observed in the table ("observed frequencies," symbolized f_o). If the null is true and the variables are independent, then there should be little difference between the expected and observed frequencies. If the null is false, however, there should be large differences between the two. The greater the differences between expected (f_e) and observed (f_o) frequencies, the less likely that the variables are, in fact, independent and the more likely that we will be able to reject the null hypothesis.

11.3 THE COMPUTATION OF CHI SQUARE

As is the case with all tests of hypothesis, the test with chi square consists of computing a test statistic, χ^2 **(obtained)**, from the sample data and placing that value on the sampling distribution of all possible sample outcomes. Specifically, the χ^2 (obtained) will be compared with the value of χ^2 **(critical)** that will be determined by consulting a chi square table (Appendix C) for a particular alpha level and degrees of freedom. Prior to conducting the formal test of hypothesis, let us take a moment to consider the calculation of chi square:

FORMULA 11.1

$$\chi^2 \text{ (obtained)} = \sum \frac{(f_o - f_e)^2}{f_e}$$

where f_o = the cell frequencies observed in the bivariate table
f_e = the cell frequencies that would be expected
if the variables were independent

Solving this formula requires that an **expected frequency** be computed for each cell in the table. In Table 11.1, because the marginals are the same value for all rows and columns, the expected frequencies are obvious by intuition and will be the same value (f_e = 25) for all four cells. In the more usual case, the expected frequencies will not be obvious, marginals will be unequal, and we must use Formula 11.2 to find the expected frequency for each cell:

FORMULA 11.2

$$f_e = \frac{(\text{row marginal})(\text{column marginal})}{N}$$

That is, the expected frequency for any cell is equal to the total of all cases in the row where the cell is located (the row marginal) times the total of all cases in the column (the column marginal), the quantity divided by the total number of cases in the table (N).

Once the expected frequencies have been calculated, Formula 11.1 directs us to subtract the expected frequency from the **observed frequency** for each cell, square this difference, divide by the expected frequency for that cell, and then sum the resultant values for all cells.

TABLE 11.3   EMPLOYMENT OF 100 SOCIAL WORK MAJORS BY ACCREDITATION STATUS
OF UNDERGRADUATE PROGRAM

| | Accreditation Status | | |
Employment Status	Accredited	Not Accredited	
Working as social worker	30	10	40
Not working as social worker	25	35	60
	55	45	100

TABLE 11.4   EXPECTED FREQUENCIES FOR TABLE 11.3

| | Accreditation Status | | |
Employment Status	Accredited	Not Accredited	
Working as social worker	22	18	40
Not working as social worker	33	27	60
	55	45	100

Let us go through these procedures step by step using the data presented in Table 11.3 as an example. In a random sample, 100 social work majors have been classified in terms of whether their undergraduate programs have been accredited by the Council on Social Work Education and whether they were hired in social work positions within three months of graduation.

Beginning with the upper left-hand cell (graduates of accredited programs who are working as social workers), the expected frequency for this cell, using Formula 11.2, is (40)(55)/100, or 22. For the other cell in this row (graduates of nonaccredited programs who are working as social workers), the expected frequency is (40)(45)/100, or 18. For the two cells in the bottom row, the expected frequencies are (60)(55)/100, or 33, and (60)(45)/100, or 27, respectively. The expected frequencies for all four cells are displayed in Table 11.4.

Note that the row and column marginals as well as the total number of cases are all exactly the same as they were in Table 11.3. The row and column marginals for the expected frequencies must always equal those of the original observed frequencies. This relationship provides a convenient way of checking your arithmetic to this point.

The value for chi square for these data can now be found by making the appropriate substitutions into Formula 11.1.

$$\chi^2 \text{ (obtained)} = \sum \frac{(f_o - f_e)^2}{f_e}$$

$$\chi^2 \text{ (obtained)} = \frac{(30 - 22)^2}{22} + \frac{(10 - 18)^2}{18} + \frac{(25 - 33)^2}{33} + \frac{(35 - 27)^2}{27}$$

$$\chi^2 \text{ (obtained)} = \frac{64}{22} + \frac{64}{18} + \frac{64}{33} + \frac{64}{27}$$

$$\chi^2 \text{ (obtained)} = 2.91 + 3.56 + 1.94 + 2.37$$

$$\chi^2 \text{ (obtained)} = 10.78$$

This sample value for chi square must still be tested for its significance.

11.4 THE CHI SQUARE TEST FOR INDEPENDENCE

As always, the five-step model for significance testing will provide the framework for organizing our decision making. The data presented in Table 11.3 will serve as our example.

Step 1. Making assumptions.

> Model: Independent random samples
> Level of measurement is nominal

Step 2. Stating the null hypothesis. As stated previously, the null hypothesis in the case of chi square states that the two variables are independent. If the null is true, the differences between the observed and expected frequencies should be small. As usual, the research hypothesis directly contradicts the null. Thus, if we reject H_0, the research hypothesis will be supported.

> H_0: The two variables are independent.
> (H_1: The two variables are dependent.)

Step 3. Choosing a sampling distribution and establishing the critical region. The sampling distribution of sample chi squares, unlike the Z and t distributions, is positively skewed, with higher values of sample chi squares in the upper tail of the distribution (to the right). Thus, with the chi square test, the critical region is established in the upper tail of the sampling distribution.

Values for χ^2 (critical) are given in Appendix C. This table is similar to the t table, with alpha levels arrayed across the top and degrees of freedom down the side. A major difference from the table of critical t scores is that, for chi square, degrees of freedom (df) are found by the following formula:

FORMULA 11.3
$$df = (r - 1)(c - 1)$$

where df = degrees of freedom
$(r - 1)$ = number of rows minus 1
$(c - 1)$ = number of columns minus 1

FIGURE 11.1   THE CHI SQUARE DISTRIBUTION WITH $\alpha = 0.05$, df = 1

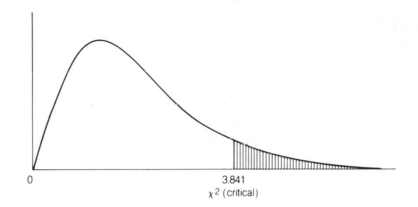

A table with two rows and two columns (a 2 × 2 table) has one degree of freedom regardless of the number of cases in the sample.* A table with two rows and three columns would have $(2 - 1)(3 - 1)$, or two, degrees of freedom. Since our sample problem involves a 2 × 2 table (df = 1), if we set alpha at 0.05, the critical chi square score would be 3.841. Summarizing these decisions, we have

$$\text{Sampling distribution} = \chi^2 \text{ distribution}$$
$$\text{Alpha} = 0.05$$
$$\text{Degrees of freedom} = 1$$
$$\chi^2 \text{ (critical)} = 3.841$$

The sampling distribution of χ^2 is depicted in Figure 11.1.

Step 4. Computing the test statistic. The mechanics of these computations were introduced in Section 11.3. As you recall, we had

$$\chi^2 \text{ (obtained)} = \sum \frac{(f_o - f_e)^2}{f_e}$$

$$\chi^2 \text{ (obtained)} = 10.78$$

---

*Degrees of freedom are the number of values in a distribution that are free to vary for any particular statistic. A 2 × 2 table has one degree of freedom because, for a given set of marginals, once one cell frequency is determined, all other cell frequencies are fixed (that is, they are no longer free to vary).

In Table 11.3, for example, if any one cell frequency is known, all others are determined. If the upper left-hand cell is known to be 30, the remaining cell in that row must be 10, since there are 40 cases total in the row and $40 - 30 = 10$. Once the frequencies of the cells in the top row are established, cell frequencies for the bottom row are determined by subtraction from the column marginals. Incidentally, this relationship can be used to good advantage when computing expected frequencies. For example, in a 2 × 2 table, only one cell expected frequency needs to be computed. The f_e's for all other cells can then be found by subtraction.

Step 5. Making a decision. Comparing the test statistic with the critical region,

$$\chi^2 \text{ (obtained)} = 10.78$$
$$\chi^2 \text{ (critical)} = 3.841$$

we see that the test statistic falls in the critical region and, therefore, we reject the null hypothesis of independence. The pattern of cell frequencies observed in Table 11.3 is unlikely to have occurred by chance alone. The variables are dependent. Specifically, based on these sample data, the probability of securing employment in the field of social work seems to be dependent on the accreditation status of the program.

11.5 THE CHI SQUARE TEST: AN EXAMPLE

To this point, we have confined our attention to 2 × 2 tables. For purposes of illustration, we will work through the computational routines and decision-making process for a larger table. As you will see, larger tables require more computations (because they have more cells); but, in all other essentials, they are dealt with in the same way as the 2 × 2 table.

A researcher is concerned with the possible effects of marital status on the academic progress of college students. Do married students, with their extra burden of family responsibilities, suffer academically as compared to unmarried students? A random sample of 453 students is gathered and each student is classified as either married or unmarried, and, using grade-point average (GPA) as a measure, as a good, average, or poor student. Results are presented in Table 11.5.

For the top left-hand cell (married students with good GPA's) the expected frequency would be (160)(175)/453, or 61.8. For the other cell in this row, expected frequency is (160)(278)/453, or 98.2. In similar fashion, all expected frequencies are computed (being very careful to use the correct row and column marginals) and displayed in Table 11.6.

With all expected frequencies computed, the formula for χ^2 (obtained)

TABLE 11.5   GRADE-POINT AVERAGE (GPA) BY MARITAL STATUS FOR 453 COLLEGE STUDENTS

GPA	Marital Status		
	Married	Not Married	
Good	70	90	160
Average	60	110	170
Poor	45	78	123
	175	278	453

TABLE 11.6   EXPECTED FREQUENCIES FOR TABLE 11.5

	Marital Status		
GPA	Married	Not Married	
Good	61.8	98.2	160
Average	65.7	104.3	170
Poor	47.5	75.5	123
	175	278	453

can be solved. In working through this formula, double-check to be certain that you are using the proper f_o's and f_e's for each cell.

$$\chi^2 \text{ (obtained)} = \sum \frac{(f_o - f_e)^2}{f_e}$$

$$\chi^2 \text{ (obtained)} = \frac{(70 - 61.8)^2}{61.8} + \frac{(90 - 98.2)^2}{98.2} + \frac{(60 - 65.7)^2}{65.7}$$

$$+ \frac{(110 - 104.3)^2}{104.3} + \frac{(45 - 47.5)^2}{47.5} + \frac{(78 - 75.5)^2}{75.5}$$

$$\chi^2 \text{ (obtained)} = \frac{(8.2)^2}{61.8} + \frac{(-8.2)^2}{98.2} + \frac{(-5.7)^2}{65.7} + \frac{(5.7)}{104.3} + \frac{(-2.5)^2}{47.5} + \frac{(2.5)^2}{75.5}$$

$$\chi^2 \text{ (obtained)} = 1.09 + 0.69 + 0.49 + 0.31 + 0.13 + 0.08$$

$$\chi^2 \text{ (obtained)} = 2.79$$

This value for the test statistic can now be tested for its significance.

Step 1. **Making assumptions.**

Model: Independent random samples
Level of measurement is nominal

Step 2. **Stating the null hypothesis.**

H_0: The variables are independent.
(H_1: The variables are dependent.)

Step 3. **Selecting the sampling distribution and establishing the critical region.**

Sampling distribution = χ^2 distribution
Alpha = 0.05
Degrees of freedom = $(r - 1)(c - 1) = (3 - 1)(2 - 1) = 2$
χ^2 (critical) = 5.991

APPLICATION 11.1

Do members of different groups have different levels of "narrow-mindedness"? A random sample of 47 individuals have been classified as white or black and have been rated as high or low on a scale that measures intolerance of viewpoints or belief systems different from their own. The results are

	Group		
Intolerance	White	Black	
High	15	5	20
Low	10	17	27
	25	22	47

The frequencies we would expect to find if the null hypothesis (H_0: the variables are independent) were true are

	Group		
Intolerance	White	Black	
High	10.64	9.36	20.00
Low	14.36	12.64	27.00
	25.00	22.00	47.00

Expected frequencies are found on a cell-by-cell basis by the formula

$$f_e = \frac{(\text{row marginal})\,(\text{column marginal})}{N}$$

for "White-High": $f_e = \dfrac{(20)\,(25)}{47} = 10.64$

for "Black-High": $f_e = \dfrac{(20)(22)}{47} = 9.36$

for "White-Low": $f_e = \dfrac{(27)(25)}{47} = 14.36$

for "Black-Low": $f_e = \dfrac{(27)(22)}{47} = 12.64$

The value of χ^2 (obtained) would be

$$\chi^2 \text{ (obtained)} = \sum \frac{(f_o - f_e)^2}{f_e}$$

$$\chi^2 \text{ (obtained)} = \frac{(15 - 10.64)^2}{10.64} + \frac{(5 - 9.36)^2}{9.36}$$
$$+ \frac{(10 - 14.36)^2}{14.36} + \frac{(17 - 12.64)^2}{12.64}$$

$$\chi^2 \text{ (obtained)} = \frac{19.01}{10.64} + \frac{19.01}{9.36} + \frac{19.01}{14.36}$$
$$+ \frac{19.01}{12.64}$$

$$\chi^2 \text{ (obtained)} = 1.79 + 2.03 + 1.32 + 1.50$$

$$\chi^2 \text{ (obtained)} = 6.64$$

If alpha is set at the 0.05 level, the critical region, with 1 degree of freedom, would begin at χ^2 (critical) = 3.841. With an obtained χ^2 of 6.64, we would reject the null hypothesis of independence. For this sample there is a statistically significant relationship between group membership and intolerance.

Step 4. Computing the test statistic.

$$\chi^2 \text{ (obtained)} = \sum \frac{(f_o - f_e)^2}{f_e}$$

$$\chi^2 \text{ (obtained)} = 2.79$$

Step 5. **Making a decision**. The test statistic, χ^2 (obtained) = 2.79, does not fall in the critical region, which, for alpha = 0.05, df = 2, begins at a χ^2 (critical) of 5.991. Therefore, we fail to reject the null. The observed frequencies are not significantly different from the frequencies we would expect to find if the variables were independent and only random chance were operating. Based on these sample results, we can conclude that the academic performance of college students is not dependent on their marital status.

11.6 AN ADDITIONAL APPLICATION OF THE CHI SQUARE TEST: THE GOODNESS-OF-FIT TEST*

To this point we have dealt with the chi square test for independence for situations involving two variables, each of which has two or more categories. Another situation in which the chi square statistic is useful, called the **goodness-of-fit test**, is one in which the distribution of scores on a single variable must be tested for significance. The logic underlying this second research situation is quite similar to that of the test for independence. The chi square statistic will be computed by comparing the actual distribution of a variable against a set of expected frequencies. The greater the difference between the observed distribution of scores and the expected distribution, the more likely that the observed pattern did not occur by random chance alone. If the observed and expected frequencies are similar, it is said that there is a "good fit" (hence the name of this application), and we would conclude that the two distributions were not significantly different.

A major difference in this new application lies in the way in which the expected frequencies are ascertained. Instead of computing these scores, we will use the null hypothesis to figure out what the expected frequencies should be. An example may make this process clear. Suppose you were gambling on coin tosses and suspected that a particular coin was biased towards heads. Over a series of tosses, what percentage of heads would you expect to observe from an unbiased coin? In an actual test of significance for this problem, the null hypothesis would be that the coin was unbiased and that half of all tosses should be heads and half tails. Notice what we have just done: we figured out the expected frequencies (half of the flips should be heads) from the null hypothesis (the coin is unbiased), rather than by computing them.

Let's consider another example of a situation in which the goodness-of-fit test is appropriate. Is there a seasonal rhythm to the crime rate? If we gathered crime statistics on a monthly basis for a given jurisdiction, what would we expect to find? Given the way in which the problem has been stated, there is only one variable (crime rate), and the focus is on the distribution of that variable over a given set of categories (by month). If the crime rate does not vary by month, we would expect that about 1/12 of all crimes committed in a year would be committed in each and every month. Our null hypothesis

*This section is optional.

TABLE 11.7 NUMBER OF CRIMES PER MONTH

Month	Number of Crimes
January	190
February	152
March	121
April	110
May	147
June	199
July	250
August	247
September	201
October	150
November	193
December	212
	2172

would be that the crime rate does not vary across time, and our expected frequencies would be calculated by dividing the actual (observed) number of crimes equally by the number of months. (To conserve time and space, we will ignore the slight complexities that would be introduced by taking account of the varying number of days per month.)

In a particular jurisdiction, a total of 2172 crimes were committed last year. The actual distribution of crimes by month is displayed in Table 11.7. The expected distribution of crimes per month would be found by dividing the total number of crimes by 12: 2172/12 = 181. Note that the expected frequency will be the same for every month. With these values determined, we can calculate chi square by using Formula 11.1.

$$\chi^2 \text{ (obtained)} = \frac{(f_o - f_e)^2}{f_e}$$

$$\chi^2 \text{ (obtained)} = \frac{(190 - 181)^2}{181} + \frac{(152 - 181)^2}{181} + \frac{(121 - 181)^2}{181}$$

$$+ \frac{(110 - 181)^2}{181} + \frac{(147 - 181)^2}{181} + \frac{(199 - 181)^2}{181}$$

$$+ \frac{(250 - 181)^2}{181} + \frac{(247 - 181)^2}{181} + \frac{(201 - 181)^2}{181}$$

$$+ \frac{(150 - 181)^2}{181} + \frac{(193 - 181)^2}{181} + \frac{(212 - 181)^2}{181}$$

$$\chi^2 \text{ (obtained)} = 81/181 + 841/181 + 3600/181 + 5041/181$$
$$+ 1156/181 + 324/181 + 4761/181 + 4356/181$$
$$+ 400/181 + 961/181 + 144/181 + 961/181$$

$$\chi^2 \text{ (obtained)} = .45 + 4.65 + 19.89 + 27.85 + 6.39 + 1.79$$
$$+ 26.30 + 24.07 + 2.21 + 5.31 + .80 + 5.31$$

$$\chi^2 \text{ (obtained)} = 125.02$$

The χ^2 of 125.02 can now be tested for significance in the usual manner.

Step 1. **Making assumptions.**

Model: Random sampling
Level of measurement is nominal

In this application of significance testing, we are assuming that the observed frequencies represent a random sample of the theoretical frequency distribution.

Step 2. **Stating the null hypothesis.**

H_0: There is no difference in the crime rate by month.
(H_1: There is a difference in crime rate by month.)

Step 3. **Selecting the sampling distribution and establishing the critical region.** In the goodness-of-fit test, degrees of freedom are equal to the number of categories minus 1, or df $= k - 1$. In the problem under consideration, there are 12 months or categories and, therefore, 11 degrees of freedom.

Sampling distribution $= \chi^2$ distribution
Alpha $= .05$
Degrees of freedom $= k - 1 = 12 - 1 = 11$
χ^2 (critical) $= 19.675$

Step 4. **Computing the test statistic.**

χ^2 (obtained) $= 125.02$

Step 5. **Making a decision.** The test statistic (125.02) clearly falls into the critical region (which begins at 19.675), so we may reject the null. These data suggest that, at least for this jurisdiction, crime rate does vary by month in a nonrandom fashion.

11.7 THE LIMITATIONS OF THE CHI SQUARE TEST

Two potential difficulties occur with the chi square test. The first occurs with small samples and the second, interestingly enough, occurs with large samples. When sample size is small, you can no longer assume that the sampling distribution of all possible sample test statistics is accurately described by the chi square distribution. In the case of the chi square test, a small sample is defined as one where a high percentage of the cells have expected frequencies (f_e) of 5 or less. Various rules of thumb have been developed to help the researcher decide what constitutes a "high percentage of cells." Probably the safest course is to take corrective action whenever any of the cells have expected frequencies of 5 or less.

In the case of 2 $\times$ 2 tables, the value of χ^2 (obtained) can be adjusted by applying Yate's correction for continuity, the formula for which is

FORMULA 11.4

$$\chi_c^2 = \sum \frac{(|f_o - f_e| - .5)^2}{f_e}$$

where χ_c^2 = corrected chi square

$|f_o - f_e|$ = the absolute value of the difference between the observed and expected frequency for each cell

The correction factor is applied by reducing the value of the term $(f_o - f_e)$ by .5 before squaring the difference and dividing by the expected frequency for the cell.

For tables larger than 2×2, there is no correction formula for computing χ^2 (obtained) for small samples. It may be possible to combine some of the categories of the variables and thereby increase cell sizes. Obviously, however, this course of action should be taken only when it is sensible to do so. In other words, distinctions that have clear theoretical justifications should not be erased merely to conform to the requirements of a statistical test. In any case where you feel that categories cannot be combined to build up cell frequencies and the percentage of cells with expected frequencies of 5 or less is small, it is probably justifiable to continue with the uncorrected chi square test as long as the results are regarded with a suitable amount of caution.

The second potential problem with chi square occurs with large samples. I pointed out at the end of Chapter 9 that all tests of hypothesis are sensitive to sample size. That is, the probability of rejecting the null hypothesis increases with sample size regardless of the size of the difference and the selected alpha level. It turns out that chi square, too, is sensitive to sample size and that larger samples may lead to the decision to reject the null when the actual relationship is trivial. In fact, chi square is more responsive to changes in sample size than the other test statistics we have encountered, since the value of χ^2 (obtained) will increase at the same rate as sample size. That is, if sample size is doubled, the value of χ^2 (obtained) will be doubled.

In a sense, this general relationship between sample size and the value of the test statistic is more a theoretical, mathematical problem than a realistic or practical one. My major purpose in stressing this relationship is really to emphasize the distinction between statistical significance and theoretical importance. On one hand, tests of significance play a crucial role in research. As long as we are working with random samples, we must know if our research results could have been produced by mere random chance.

On the other hand, tests of hypothesis, like any statistical techniques, are limited in the range of questions they can answer. Specifically, these tests will tell us whether our results are statistically significant or not. They will not necessarily tell us if the results are important in any other sense. To deal more directly with questions of importance, we must apply an additional set of statistical techniques called measures of association. These techniques will be the subject of Part III.

SUMMARY

1. The chi square test for independence is appropriate for situations in which the variables of interest have been organized into table format. The null hypothesis is that the variables are independent or that the classification of a case into a particular category on one variable has no effect on the probability that the case will be classified into any particular category of the second variable.

2. Chi square is probably the most commonly used test of hypothesis in the social sciences. Because the test is nonparametric and requires only nominally measured variables, the model assumptions are easily satisfied. Because chi square is computed from bivariate tables, in which the number of rows and columns can be easily expanded, the chi square test can be used in many situations in which other tests are inapplicable.

3. In the chi square test, we first find the frequencies that would appear in the cells if the variables were independent (f_e) and then compare those frequencies, cell by cell, with the frequencies actually observed in the cells (f_o). If the null is true, expected and observed frequencies should be quite close in value. The greater the difference between the observed and expected frequencies, the greater the possibility of rejecting the null.

4. In the chi square goodness-of-fit test, the observed frequencies are compared against a set of expected frequencies as derived from the null hypothesis. This application is used to test for the randomness of the distribution of a single variable across a series of categories.

5. As sample size (N) decreases, the chi square test becomes less trustworthy and corrective action may be required. As sample size increases, the value of the obtained chi square also increases, and with very large samples, we may declare relatively trivial relationships to be statistically significant. As is the case with all tests of hypothesis, statistical significance is not the same thing as "importance" in any other sense. As a general rule, statistical significance is a necessary but not sufficient condition for theoretical or practical importance.

SUMMARY OF FORMULAS

Chi square
obtained   11.1 $\chi^2 \text{ (obtained)} = \sum \dfrac{(f_o - f_e)^2}{f_e}$

Expected
frequencies   11.2 $f_e = \dfrac{(\text{row marginal})(\text{column marginal})}{N}$

Degrees of
freedom,
bivariate
tables   11.3 $\text{df} = (r - 1)(c - 1)$

Yate's correction for
continuity   11.4 $\chi_c^2 = \sum \dfrac{(|f_o - f_e| - .5)^2}{f_e}$

GLOSSARY

Cells. The cross-classification categories of the variables in a bivariate table.

χ^2 **(critical).** The score on the sampling distribution of all possible sample chi squares that marks the beginning of the critical region.

χ^2 **(obtained).** The test statistic as computed from sample results.

Chi square test. A nonparametric test of hypothesis for variables that have been organized into a bivariate table.

Expected frequency (f_e). The cell frequencies that would be expected in a bivariate table if the variables were independent.

Goodness-of-fit test. A chi square test to see if a variable is randomly distributed across a series of categories.

Independence. The null hypothesis in the chi square test. Two variables are independent if, for all cases, the classification of a case on one variable has no effect on the probability that the case will be classified in any particular category of the second variable.

Marginals. The row and column subtotals in a bivariate table.

Observed frequency (f_o). The cell frequencies actually observed in a bivariate table.

PROBLEMS

Asterisk indicates an optional problem.

11.1 For each table below calculate the obtained chi square:

a.

20	25	45
25	20	45
45	45	90

b.

10	15	25
20	30	50
30	45	75

c.

25	15	40
30	30	60
55	45	100

d.

20	45	65
15	20	35
35	65	100

For each of the following problems, go through the five-step model carefully and in detail.

11.2 SOC Is there a relationship between length of marriage and satisfaction with marriage? The necessary information has been collected from a random sample of 100 respondents drawn from a local community. Write a sentence or two explaining your decision.

	Length of Marriage (years)			
Satisfaction	More than 10	5–10	0–5	
Low	20	20	10	50
High	10	20	20	50
	30	40	30	100

11.3 CJ The new police chief in Shinbone is concerned about the image of the department and wants to know which sectors of the community have the least favorable attitudes toward the police. The results of a police opinion survey conducted on a random sample of 520 citizens are reported below. Are attitudes related to social class, ethnicity, or contact with members of the force? Summarize your findings. Does the police force have an "image problem"? Where?

a.

	Social class			
Attitude	Lower	Middle	Upper	
Positive	40	120	140	300
Negative	110	80	30	220
	150	200	170	520

b.

	Ethnicity			
Attitude	Black	White	Other	
Positive	70	190	40	300
Negative	60	150	10	220
	130	340	50	520

c. Have you had any contacts with the police during the past six months?

Attitude	Yes, many	Yes, few	No	
Positive	15	100	185	300
Negative	55	100	65	220
	70	200	250	520

11.4 PA Is there a relationship between salary levels and unionization for public employees? The data below represent this relationship for fire departments in a random sample of 100 cities of roughly the same size. Salary data have been dichotomized at the median. Summarize your findings.

Salary	Union	Nonunion	
High	21	29	50
Low	14	36	50
	35	65	100

11.5 SOC At a large urban college, about half of the students live off campus in various arrangements, and the other half live in dormitories on campus. Is academic performance dependent on living arrangements? The results based on a random sample of 300 students are presented below.

	Living Arrangement			
	Off Campus:			
GPA:	With Roommates	With Parents	On Campus	
High	32	10	38	80
Moderate	36	40	54	130
Low	22	20	48	90
	90	70	140	300

11.6 GER Are retired citizens who maintain close relations with their kin happier than those who do not? The table below relates reported "happiness" to kin relations for a random sample of 123 retired citizens. Is there a significant relationship?

Happiness	Relations with Kin		
	Close	Not Close	
"Very happy"	26	44	70
"Not very happy"	32	21	53
	58	65	123

Class	Frequency
Freshman	200
Sophomore	150
Junior	120
Senior	110
	580

11.7 [SOC] Does support for the legalization of marijuana vary by region of the country? The table displays the relationship between the two variables for a random sample of 1020 adult citizens. Is the relationship significant?

	Region				
	North	Midwest	South	West	
Legalize marijuana?					
Yes	60	65	42	78	245
No	245	200	180	150	775
	305	265	222	228	1020

11.8 [PS] Is there a relationship between political ideology and class standing? Are upperclass students significantly different from underclass students on this variable? The table below reports the relationship between these two variables for a random sample of 267 college students.

Political ideology	Class Standing		
	Underclass	Upperclass	
Liberal	43	40	83
Moderate	50	50	100
Conservative	40	44	84
	133	134	267

*11.9 [SOC] The director of athletics at the local high school wonders if the sports program is getting a proportional amount of support from each of the four classes. If there are roughly equal numbers of students in each of the classes, what does the following breakdown of attendance figures from a random sample of students in attendance at a recent basketball game suggest?

11.10 [SOC] A researcher is concerned with the relationship between attitudes toward violence and violent behavior. If attitudes "cause" behavior (a very debatable proposition), then people who have positive attitudes toward violence should have high rates of violent behavior. A pretest was conducted on 70 respondents and, among other things, the respondents were asked, "Have you been involved in a violent incident of any kind over the past six months?" The researcher established the following relationship:

Involvement	Attitudes Toward Violence		
	Favorable	Unfavorable	
Yes	16	19	35
No	14	21	35
	30	40	70

The chi square calculated on these data is .23, which is not significant at the .05 level (confirm this conclusion with your own calculations).

Undeterred by this result, the researcher proceeded with the project and gathered a random sample of 7000. In terms of percentage distributions, the results for the full sample were exactly the same as for the pretest:

Involvement	Attitude		
	Favorable	Unfavorable	
Yes	1600	1900	3500
No	1400	2100	3500
	3000	4000	7000

However, the chi square obtained is a very healthy 23.4 (confirm with your own calculations). Why is the full-sample chi square significant when the pretest was not? What happened? Do you think that the second result is important?

*11.11 SOC A small western town has roughly equal numbers of Hispanic, Asian, and Anglo American residents. Are the three groups equally represented at town meetings? The attendance figures for a random sample drawn from those attending a meeting were

Group	Frequencies
Hispanic	74
Asian	55
Anglo	53

Is there a statistically significant pattern here?

11.12 SW/GER A program of pet therapy has been running at a local nursing home. Are the participants in the program more alert and responsive than nonparticipants? The results, drawn from a random sample of residents, are reported below.

Alertness	Participants	Non-participants	
High	23	15	38
Low	11	18	29
	34	33	67

11.13 CJ A local judge has been allowing some individuals convicted of "driving under the influence" to work in a hospital emergency room as an alternative to fines, suspensions, and other penalties. A random sample of offenders has been drawn. Do participants in this program have lower rates of recidivism for this offense?

Recidivist?	Participants	Non-participants	
Yes	60	123	183
No	55	108	163
	115	231	346

11.14 SOC The state Department of Education has rated a sample of the local school systems for compliance with state mandated guidelines for quality. Is the quality of a school system signifi-

cantly related to the affluence of the community as measured by per capita income?

Quality	Per Capita Income		
	Low	High	
Low	16	8	24
High	9	17	26
	25	25	50

11.15 SOC An urban sociologist has built up a data base describing a sample of the neighborhoods in her city and has developed a scale by which each area can be rated for the "quality of life" (this includes measures of pollution, noise, open space, services available, and so on). She has also asked samples of residents of these areas about their level of satisfaction with their neighborhoods. Is there significant agreement between the sociologist's objective ratings of quality and the respondent's self reports of satisfaction?

Reported Satisfaction	Quality of Life			
	Low	Moderate	High	
High	8	17	32	57
Moderate	12	25	21	58
Low	21	15	6	42
	41	57	59	157

11.16 SOC Some results from the General Social Survey are presented below. The survey is administered every year to a nationally representative sample. See Appendix G for the complete questions and other details about the survey. For each table, conduct the chi square test of significance. Write a sentence or two of interpretation for each test.

a. Should US be active in world affairs? (item 27)

	Age		
	Less than 35	35 and Older	
Yes	143	211	354
No	73	71	144
	216	282	498

b. How much confidence in education? (item 40)

	Age		
	Less than 35	35 and Older	
A great deal	50	72	122
Only some	130	161	291
Hardly any	41	52	93
	221	285	506

c. How much confidence in executive branch? (item 41)

	Age		
	Less than 35	35 and Older	
A great deal	35	37	72
Only some	112	152	264
Hardly any	68	91	159
	215	280	495

d. Afraid to walk at night? (item 66)

	Age		
	Less than 35	35 and Older	
Yes	95	140	235
No	138	160	298
	233	300	533

e. Punching OK? (item 64)

	Age		
	Less than 35	35 and Older	
Yes	152	157	309
No	67	102	169
	219	259	478

SPSS/PC+ PROCEDURES FOR THE CHI SQUARE TEST FOR INDEPENDENCE

DEMONSTRATION 11.1 Are Attitudes about Marijuana Dependent on Parenthood?

The CROSSTABS procedure in SPSS/PC+ produces bivariate tables and a wide variety of statistics. This procedure is very commonly used in social science research at all levels, and you will see many references to CROSSTABS in chapters to come. I will introduce the command here, and we will return to it often in later sessions.

Does being a parent have any relationship with people's attitudes about illegal drugs? Are parents more opposed than nonparents to the temptations represented by such drugs as marijuana? We can answer this question, at least for the 1993 GSS data set, by first recoding the variable CHILDS (item 6 in Appendix G) into two categories: parents (people with at least one child) and nonparents (people with no children). The recoded CHILDS variables can then be crosstabulated against the GRASS variable and a chi square test conducted.

Let's first review the RECODE command. This command allows us to change the scores associated with a variable. In Demonstration 2.3, we used this command to collapse a continuous interval-ratio level variable (AGE) so that we could use the variable in a frequency distribution. Here, we need to collapse CHILDS into just a few categories so we can use it in a bivariate table. As recorded in the SPSS/PC+ system file, CHILDS has 9 categories. The format of the question we have raised (our "research hypothesis") dictates just two categories, parents and nonparents. Make sure the 1993 GSS file is active (use the GET command) and

1. From the main menu, select 'modify data or files' and then 'modify data values'. Select and paste RECODE and then highlight '!variables' and use ALT-T and the typing window to type the variable name (CHILDS). Next, select the '( )' and type '0 = 0' in the typing window and touch return. Repeat these steps to complete the RECODE instruction. That is, next select '( )' and type '1 THRU HIGHEST' in the typing window or use the menu options. When completed, your command should look like this:

```
RECODE CHILDS (0 = 0)( 1 THRU HIGHEST = 1).
```

2. Add a new VALUE LABELS command since the recoding will change the meaning of the scores on CHILDS. The new command should read

```
VALUE LABELS CHILDS 0 'NOT A PARENT' 1 'PARENT'.
```

3. To write the CROSSTABS command, start with the main menu and select 'analyze data' and then 'descriptive statistics'. Choose and paste CROSS-TABS and !/TABLES. Name the dependent variable (GRASS in our case) as the row variable, select and paste the keyword BY, and then name the independent or causal variable (CHILDS) as the column variable. To get chi square, include the subcommand STATISTICS 1.

The command should look like this:

```
CROSSTABS /TABLES GRASS BY CHILDS /STATISTICS 1.
```

and the results are

```
Cross-tabulation:       GRASS      SHOULD MARIJUANA BE MADE LEGAL
                     By CHILDS         NUMBER OF CHILDREN

             Count   :           :           :
CHILDS=>             :           :           : Row
             Col Pct :    .00 :    1.00 : Total
GRASS        ------- : -------- : -------- :
               1.00 :     36 :     85 :    121
    LEGAL          :   25.2 :   24.0 :   24.3
                   : -------- : -------- :
               2.00 :    107 :    269 :    376
NOT LEGAL          :   74.8 :   76.0 :   75.7
                   : -------- : -------- :
             Column     143       354       497
             Total     28.8      71.2     100.0
```

```
Chi-Square   D.F.   Significance   Min E.F.   Cells with E.F.< 5
----------   ----   ------------   --------   ------------------
  .02502      1        .8743        34.815        None
  .07486      1        .7844       ( Before Yates Correction )
Number of Missing Observations = 291
```

The table displays the number of cases in each cell and the row and column totals. If you would also like to see the expected frequencies, include OPTIONS 14 in the command. Below the table, the program reports, in the first line, the obtained chi square adjusted by Yate's correction for continuity (this adjustment is automatically made for all 2 by 2 tables), the degrees of freedom, the exact significance of the chi square, and the smallest value for expected frequency found in the table (Min E.F.). Also, in the last column, any cells with expected frequencies less than 5 are identified (Cells with E.F.< 5). In the second line, this information is repeated for the unadjusted chi square.

We can see from the results that neither the adjusted nor unadjusted chi squares are significant at less than .05 (the reported significance is .8743 for the adjusted chi square and .7844 for the unadjusted). There is no statistically significant relationship between parenthood and opposition to the legalization of marijuana. Opinions regarding legalization of marijuana are not dependent on parenthood.

DEMONSTRATION 11.2 What Are the Sources of Happiness? Is Ignorance Bliss?

In Demonstration 10.2, we tested the old adage that money can't buy happiness. Using ordinal tests of significance, we found that the adage was false and that people with higher levels of income also tended to report higher levels of happiness. What effect would education have on happiness? Would more-educated people be more or less happy? Is there any truth to the saying that "ignorance is bliss"?

To test these relationships, we will use DEGREE as our measure of education. As originally coded, the variable has five categories, but we will simplify the analysis and use only two categories, distinguishing between people with a high school education or less and people with at least some education beyond high school. The necessary RECODE command would be

```
RECODE DEGREE (0, 1 = 1)(2 THRU 4 = 2).
VALUE LABELS DEGREE 1 'High school or less' 2 'At
least some college'.
```

Follow the instructions in Demonstration 11.2 for the CROSSTABS command. Specify HAPPY as the row variable and recoded DEGREE as the column variable. Your command should look like this:

```
CROSSTABS /TABLES HAPPY BY DEGREE /STATISTICS 1.
```

and your results should be

```
                      By DEGREE        RS HIGHEST DEGREE

             Count ¦ high sch ¦ at least ¦
   DEGREE=>        ¦ ool or 1 ¦ some co  ¦  Row
                   ¦        1 ¦       2  ¦ Total
   HAPPY      ---- ¦ -------- ¦ -------- ¦
               1 ¦      168   ¦    75    ¦   243
      VERY HAPPY  ¦          ¦          ¦  31.0
                  ¦ -------- ¦ -------- ¦
               2 ¦      323   ¦   130    ¦   453
     PRETTY HAPPY ¦          ¦          ¦  57.8
                  ¦ -------- ¦ -------- ¦
               3 ¦       73   ¦    15    ¦    88
    NOT TOO HAPPY ¦          ¦          ¦  11.2
                  ¦ -------- ¦ -------- ¦
          Column       564        220       784
          Total       71.9       28.1     100.0

Chi-Square  D.F.   Significance  Min E.F.  Cells with E.F.< 5
----------  ----   ------------  --------  ------------------

  6.32646     2        .0423      24.694      None
Number of Missing Observations = 4
```

The table displays the observed frequencies for the cells, along with row and column marginals and total number of cases. Below the table are the value of chi square (6.32646), the degrees of freedom (2—note that this is a 2×3 table), the exact probability that this pattern of cell frequencies occurred by chance (.0423), and other information. The reported significance is less than .05, the standard value for alpha, so we would reject the null hypothesis of independence. There is a statistically significant relationship between level of education and degree of happiness. Further inspection of the table shows that the better-educated respondents report higher levels of happiness. For example, 75 of the 220 (or 34%) better-educated respondents report that they are "very happy," whereas 168 of the 564 (or 30%) less well-educated respondents report similar high levels of happiness. The idea that "ignorance is bliss" is not supported by these results.

DEMONSTRATION 11.3 Selecting Subgroups: Using the SELECT IF Command to Test for Differences in Racial Prejudice by Educational Level

Not all hypotheses should be tested on all cases. Sometimes it is necessary to limit the sample and test our ideas only on cases with certain characteristics. For example, the 1993 GSS has many items that measure racial prejudice in general and anti-Black prejudice specifically. Our shortened version of the survey has one item (RACSEG, item 31) that measures anti-Black prejudice and one item (RACLIVE, item 32) that measures the degree of neighborhood integration. Usually, the interviewers would screen out nonwhite respondents for questions such as these but, to demonstrate some of the case selection capabilities of SPSS/PC+, let's conduct a test with only white respondents.

The SELECT IF command provides one useful way of refining a sample or limiting respondents to those who meet certain criteria. Selection is based on whether cases meet what is called a "logical argument." That is, the user states the selection criteria and the program selects only those cases that possess the required characteristics. The general format for SELECT IF is

```
SELECT IF (selection criteria stated in logical
format).
```

You may use a variety of logical expressions to state your criteria. To conserve space, we will confine our attention to the 'EQ' or 'equal to' criterion and refer you to the manual for other possibilities. To include only white respondents, we would tell SPSS/PC+ to select only the cases that have a score of 1 on RACE. To construct this command, follow these instructions:

1. From the Main Menu, select 'modify data or files' and then select 'select or weight data' and SELECT IF. The command will appear in the bottom window.

2. Select the parentheses !( ) and use the typing window to specify your selection criteria, RACE EQ 1. When you press ENTER, the criteria will appear in the bottom window. The completed command will look like this:

```
SELECT IF (RACE EQ 1).
```

As an illustration, let's test the idea that prejudice declines with education. We'll use RACSEG to measure prejudice and DEGREE to measure education, but we need to RECODE both variables. To simplify the output, recode both variables into dichotomies and then run the CROSSTABS command. Here's what your commands should look like when you're done:

```
SELECT IF (RACE EQ 1).
RECODE DEGREE (0,1=1)(2 thru 4=2).
VALUE LABELS DEGREE 1 'HS or Less' 2 'At Least Some
College'.
RECODE RACSEG (1,2=1)(3,4=2).
VALUE LABELS RACSEG 1 'Agree' 2 'Disagree'.
CROSSTABS /TABLES RACSEG BY DEGREE/STATISTICS 1.
```

Be sure that you move the cursor to the first line of this block of commands (the SELECT IF command) and, when you press the F10 key, you will produce the following output:

```
Crosstabulation:    RACSEG         WHITES HAVE RIGHT TO
                 By DEGREE          SEG. NEIGHBORHOOD
                                    RS HIGHEST DEGREE


              Count : hs or le : at Least :
 DEGREE=>           : ss        : some co : Row
                    :        1  :      2  : Total
 RACSEG        ----- : -------- : -------- :
                 1 :      64   :     10   :     74
    agree           :          :          :    16.8
                    : -------- : -------- :
                 2 :     236   :    131   :    367
  disagree          :          :          :    83.2
                    : -------- : -------- :
              Column      300        141        441
              Total       68.0       32.0      100.0

Chi-Square   D.F.  Significance   Min E.F.   Cells with E.F.< 5
----------   ----  ------------   --------   ------------------
  12.92946     1       .0003       23.660        None
  13.93062     1       .0002      ( Before  Yates Correction )
Number of Missing Observations = 220
```

There is a very significant relationship between these two variables. The 'Significance' is less than .001. This means that the probability of the observed pattern of frequencies occurring by random chance alone is less than 1 out of 1,000. The relationship between education and prejudice is statistically significant. Level of prejudice seems to depend on level of education.

Exercises

11.1 Since we already have DEGREE recoded, let's see if education has any significant relationship with GRASS, GUNLAW, CAPPUN, or PORN-

MORL. SPSS/PC+ allows the user to request more than one table in a single command. You can request multiple dependent (row) variables by listing them in the appropriate place on the command. For example, the command below would generate all four tables necessary for this exercise.

```
CROSSTABS /TABLES GRASS, GUNLAW, CAPPUN, PORNMORL BY
DEGREE/STATISTICS 1.
```

Write a paragraph summarizing your results. Which relationships are significant? At what levels?

11.2 The user can also specify multiple independent or column variables. Write a command to run CROSSTABS with BUSING as the row variable. Use the RECODE command to dichotomize DEGREE, INCOME, and CHILDS and use the recoded versions as independent (column) variables by listing them after the "BY" specification. To make the output more meaningful, write a SELECT IF command that limits the cases to whites only. Make sure that you place the SELECT IF command before the CROSSTABS command. Write up the results as you would for a professional journal (see Reading Statistics 5 in Chapter 9).

11.3 Write the SPSS/PC+ commands that would crosstabulate recoded DEGREE against FEHELP and FEFAM for the entire sample and then use the SELECT IF command to limit cases to, first, men only and, then, women only. Use the COMPUTE command to create a summary scale that adds the scores on FEHELP and FEFAM, and then RECODE the scale scores into a dichotomy. Repeat the above analysis with the recoded DEGREE as the column variable and the recoded scale score as the row variable.

Hypothesis Training V
The Analysis of Variance

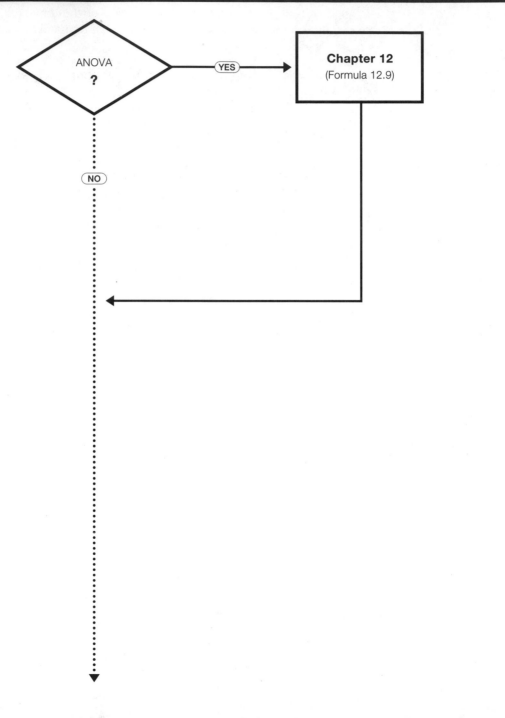

12.1 INTRODUCTION

In this chapter, we will examine a test of significance called the **analysis of variance** (often abbreviated as **ANOVA**). Like chi square, ANOVA is very flexible and widely used in social science research. It can be used in a number of situations where previously discussed tests are less than optimum or entirely inappropriate. For example, ANOVA, like chi square, easily handles situations in which we have more than two samples or categories to compare. Unlike chi square, however, ANOVA is designed to be used with interval-ratio level variables and is a powerful tool for analyzing the most sophisticated and precise measurements you are likely to encounter.

It is perhaps easiest to think of ANOVA as an extension of the t test for the significance of the difference between two sample means (see Chapter 9). As you recall, the t test is appropriate for situations in which we have divided the subjects into two groups (for example, by gender) and are interested in the significance of the difference between the two groups on some variable (such as support for gun control). The t test is limited only to this two-sample case, but ANOVA allows us to investigate the significance of differences of this sort in a much broader range of situations.

To illustrate, suppose we were interested in examining the social basis of support for capital punishment. One possibility that might occur to us is to see if a person's opinion on this issue is affected by religion. The death penalty is a moral issue, and it is certainly reasonable to expect that a person's general morality would be reflected and/or shaped by religion. Suppose that we decide to conduct an investigation by devising a scale that measures support for capital punishment at the interval-ratio level and administering the scale to a randomly selected sample that includes Protestants, Catholics, Jews, and people with no religious affiliation ("Nones"). Take a couple of seconds to think through the logic of our situation here. We will wind up with four categories of subjects, and we will want to see if a particular attitude or opinion varies significantly by the category (religion) into which a person is classified. We will also want to answer other questions such as, Which religion shows the most support for capital punishment? The least? Are Protestants significantly more supportive than Catholics? Than Jews? How do people with no religious affiliation compare to people in the other categories? ANOVA will not answer all of these questions directly but, as you will see, it will provide a very useful statistical context in which the questions can be addressed.

In considering this situation, some test options may have occurred to you. You might consider testing for the significance of the differences by collapsing the scores on the "Support of Capital Punishment Scale" into a few categories (for example, pro–capital punishment, neutral, and anti-capital punishment) and conducting a chi square test. Although certainly a possibility (especially if we are uncertain about the adequacy of our level-of-measurement assumption), this approach has the severe disadvantage of losing or obscuring potentially valuable information and detail as the data are collapsed. A second option might be to run a series of t tests between the various categories. You might begin by testing the mean score for the Protestants against the mean for the Catholics, then against the Jews, and so on, continuing until all possible combinations of religions had been compared. There are two disadvantages to this approach. The first, and more obvious, would simply be the amount of work involved. For the four religious affiliations, we would have to run six different t tests (six different sets of calculations, six trips through the five-step model, and so on) to exhaust all the possibilities.

A second disadvantage, which is less obvious but more important, has to do with Type I error (the probability of rejecting a null hypothesis that is, in fact, true). When conducting multiple tests of significance, the chance of committing at least one alpha error over the series of tests is greater than the selected alpha level for each individual test. Even if each single test is conducted at the .05 level, the combined probability of alpha error over the series of tests will be greater than .05. Thus, not only would multiple t tests require a great deal of computational work, but we would also increase our chances of making a rather serious error. In contrast to these two options, ANOVA handles situations such as this without sacrificing any detail in the data. Moreover, ANOVA has a known alpha level and requires only a single set of calculations.

12.2 THE LOGIC OF THE ANALYSIS OF VARIANCE

For ANOVA, the null hypothesis is that the populations from which the samples are drawn are equal on the characteristic of interest. As applied to our problem, the null could be phrased as, "People of the various religious denominations do not vary in their support for the death penalty." The null will be stated symbolically as $\mu_1 = \mu_2 = \mu_3 = \cdots = \mu_N$, and you should see immediately that this is an extended version of the null for the two-sample t test. As usual, the researcher will normally be interested in rejecting the null and, in the case at hand, gathering evidence to show that support for capital punishment is related to religion.

If the null hypothesis of "no difference" is true, then any means calculated from randomly selected samples should be roughly equal in value. The average score for the Protestant sample should be about the same as the average score for the Catholics and the Jews, and so forth. Note that the averages are unlikely to be exactly the same value even if the null really is true, since we will always encounter some error or chance fluctuations in the measurement process. The question we are asking by applying the ANOVA test is not "Are there differences among the religions?" but "Are the differences among the religions large enough to justify a decision to reject the null hypothesis?"

Now, consider what kinds of outcomes we might encounter if we actually administered a "Support of Capital Punishment Scale" to a sample of respondents and organized the scores by religion. Of the infinite variety of possibilities, let's focus on two extreme outcomes as exemplified by

TABLE 12.1   SUPPORT FOR CAPITAL PUNISHMENT BY RELIGION (fictitious data)

	Protestant	Catholic	Jew	None
Mean = 10.3		11.0	10.1	9.9
Standard Deviation = 2.4		1.9	2.2	1.7

TABLE 12.2   SUPPORT FOR CAPITAL PUNISHMENT BY RELIGION (fictitious data)

	Protestant	Catholic	Jew	None
Mean = 14.7		11.3	5.7	8.3
Standard Deviation = .5		1.3	.6	.8

Tables 12.1 and 12.2. In the first set of hypothetical results (Table 12.1), we see that the means and standard deviations of the groups are quite similar. The average scores are about the same for every religious group, and all four groups contain about the same dispersion. These results would be quite consistent with the null of no difference and would suggest that, on this issue, neither the typical score nor the range of scores changes in any important way by religion.

Now consider another set of fictitious results as displayed in Table 12.2. Here we see substantial differences in average score from category to category, with Jews showing the lowest support and Protestants showing the strongest. Take a minute to observe the standard deviations for each category. The values for this statistic are low and similar from religion to religion, indicating that there is not much variation within the categories. These results show marked differences *between* religions combined with homogeneity *within* religions. In other words, each of the four religions is associated with a distinctly different level of support for the death penalty. There are marked differences from religion to religion combined with strong similarities within each religion. These results would certainly support the argument that religion is an important determinant of a person's support for the death penalty and would contradict the null.

In principle, ANOVA proceeds by making the kinds of comparisons outlined above. The test compares the amount of variation *between* categories (for example, from Protestants to Catholics to Jews to "Nones") with the amount of variation *within* categories (among Protestants, among Catholics, and so forth). The greater the differences between categories, relative to the differences within categories, the more likely that the null of "no difference" is false and can be rejected. If support for capital punishment truly varies by religion, then the sample mean for each of the religions should be quite different from the others, and dispersion within the categories should be relatively low.

12.3 THE COMPUTATION OF ANOVA

Even though we have been thinking of ANOVA as a test for the significance of the difference between sample means, the computational routine actually involves developing two separate estimates of the population variance, σ^2 (hence the name analysis of variance). Recall from Chapter 4 that the variance of any distribution is the standard deviation squared and that both statistics are measures of dispersion or variability. One estimate of the population variance will be based on the pattern of variation *within* each of the samples separately. The other estimate of the population variance is based on the pattern of variation *between* samples.

Before constructing these estimates, we need to introduce some new concepts and statistics to help us deal with the notion of analyzing the variance of a distribution of scores. The first new concept is the total variation of the scores, which is measured by a quantity called the **total sum of squares**, or **SST**:

FORMULA 12.1
$$SST = \Sigma(X_i - \bar{X})^2$$

To find this quantity, we would take each score, subtract the mean, square the difference, and then add to get the total of the squared differences. If this formula seems vaguely familiar, it's because the same expression appears in the numerator of the formula for the sample variance and the standard deviation (see Section 4.5). This redundancy is not surprising, considering that the SST is also a way of measuring dispersion.

To construct the two separate estimates of the population variance, the total variation, as measured by SST, is divided into two components. One of these reflects the pattern of variation within the categories and is called the **sum of squares within (SSW)**. The other component is based on the variation between categories and is called the **sum of squares between (SSB)**. The relationship of these three sums of squares is

FORMULA 12.2
$$SST = SSB + SSW$$

This mathematical relationship reflects the fact that SSW and SSB are components of SST.

The sum of squares within is defined as

FORMULA 12.3
$$SSW = \Sigma(X_i - \bar{X}_k)^2$$

where SSW = the sum of the squares within the categories
$\bar{X}_k$ = the mean of a category

This formula directs us to take each score, subtract the mean *of the category* from the score, square the result, and then sum the squared differences. We will do this for each category separately and then summate the squared differences for all categories to find SSW.

The between sum of squares (SSB) reflects the variation between the samples, with the category means serving as summary statistics for each category. Each category mean will be treated as a "case" for purposes of this estimate, and the formula for the sum of squares between (SSB) is

FORMULA 12.4

$$SSB = \Sigma N_k (\bar{X}_k - \bar{X})^2$$

where SSB = the sum of squares between the categories
N_k = the number of cases in a category
$\bar{X}_k$ = the mean of a category

To find SSB, subtract the overall mean of all scores from each category mean, square the difference, multiply by the number of cases in the category, and add the results across all the categories.

Let's pause for a second to let me remind you of what we are after here. If the null hypothesis is true, then there should not be much variation from category to category, relative to the variation within categories, and the two estimates to the population variance based on SSW and SSB should be roughly equal. Large differences between the two estimates would indicate important differences between categories relative to the differences within. If the differences between the two estimates are large enough, we will be able to reject the null hypothesis.

The next step in the computational routine is to construct the estimates of the population variances. To do this we will divide each sum of squares by its respective degrees of freedom. To find the degrees of freedom associated with SSW, subtract the number of categories (k) from the number of cases (N). The degrees of freedom associated with SSB are the number of categories minus one. In summary,

FORMULA 12.5

$$dfw = N - k$$

where dfw = degrees of freedom associated with SSW
N = number of cases
k = number of categories

FORMULA 12.6

$$dfb = k - 1$$

where dfb = degrees of freedom associated with SSB
k = number of categories

The estimates, called the **mean square** estimates, are calculated by dividing each sum of square by its respective degrees of freedom:

FORMULA 12.7

$$\text{Mean square within} = \frac{SSW}{dfw}$$

FORMULA 12.8

$$\text{Mean square between} = \frac{SSB}{dfb}$$

The actual test statistic, which we will calculate in step 4 of the five-step model, is called the **F ratio**, and its value is determined by the following formula:

FORMULA 12.9

$$F = \frac{\text{mean square between}}{\text{mean square within}}$$

As you can see, the value of the F ratio will be a function of the amount of variation between categories to the amount of variation within the categories. The greater the variation between the categories relative to the variation within, the higher the value of the F ratio and the more likely we will be able to reject the null.

12.4 A COMPUTATIONAL SHORTCUT

The computational routine for ANOVA, as summarized in the previous section, requires a number of separate steps and different formulas and will almost certainly seem complicated and time consuming the first time you see it. As is almost always the case, if you proceed systematically from formula to formula (in the correct order, of course), then the computations won't seem nearly so formidable. Unfortunately, they will still be lengthy and time consuming. So, at the risk of stretching your patience, let me introduce a way to save some time and computational effort. This will require the introduction of even more formulas, but the eventual savings in time will be worth the effort.

The formulas presented to this point have been definitional formulas and, as we have seen in previous chapters, these are often not very convenient for computation. The inconvenience is particularly a problem for SST and SSW and, fortunately, a computational formula is available for SST:

FORMULA 12.10

$$\text{SST} = \Sigma X^2 - N\bar{X}^2$$

To solve this formula, first square the value of the overall mean, multiply the result by the total number of cases in the sample, and subtract that sum from the sum of the squared scores. In finding the latter quantity, ΣX^2, be sure that you use all of the scores, square each one, and then sum the squared scores.

Once we have the values of SST and SSB, we can find SSW by manipulating Formula 12.2 and doing some simple subtraction:

FORMULA 12.11

$$\text{SSW} = \text{SST} - \text{SSB}$$

The computational routine for ANOVA can be summarized as follows:

1. Find SST by Formula 12.10.
2. Find SSB by Formula 12.4.
3. Find SSW by subtraction (Formula 12.11).
4. Calculate the degrees of freedom (Formulas 12.5 and 12.6).
5. Construct the two mean square estimates to the populations variance by dividing SSB and SSW by their respective degrees of freedom (Formulas 12.7 and 12.8).
6. Find the obtained F ratio by dividing the between estimate by the within estimate (Formula 12.9).

These computations and the actual test of significance will be illustrated in the next sections.

TABLE 12.3   SUPPORT FOR CAPITAL PUNISHMENT BY RELIGION FOR 16 SUBJECTS (fictitious data)

	Protestant		Catholic		Jew		None	
	X	X^2	X	X^2	X	X^2	X	X^2
	8	64	12	144	12	144	15	225
	12	144	20	400	13	169	16	256
	13	169	25	625	18	324	23	529
	17	289	27	729	21	441	28	784
	50	666	84	1898	64	1078	82	1794
$\bar{X}_k =$	12.5		21.0		16.0		20.5	

$$\bar{X} = 280/16 = 17.5$$

12.5 A COMPUTATIONAL EXAMPLE

Assume that we have administered our "Support for Capital Punishment Scale" to a sample of 16 individuals who are equally divided into the four religions. (Obviously, this sample is much too small for any serious research and is intended solely for purposes of illustration.) All scores are reported in Table 12.3 along with the squared scores, the category means, and the overall mean.

To organize our computations, we'll follow the routine summarized at the end of Section 12.4. We begin by finding SST by Formula 12.10:

$$\text{SST} = \Sigma X^2 - N\bar{X}^2$$
$$\text{SST} = (666 + 1898 + 1078 + 1794) - (16)(17.5)^2$$
$$\text{SST} = 5436 - (16)(306.25)$$
$$\text{SST} = 5436 - 4900$$
$$\text{SST} = 536$$

The sum of squares between is found by Formula 12.4:

$$\text{SSB} = \Sigma N_k(\bar{X}_k - \bar{X})^2$$
$$\text{SSB} = (4)(12.5 - 17.5)^2 + (4)(21.0 - 17.5)^2 + (4)(16.0 - 17.5)^2 + (4)(20.5 - 17.5)^2$$
$$\text{SSB} = 4(-5)^2 + 4(3.5)^2 + 4(-1.5)^2 + 4(3)^2$$
$$\text{SSB} = 100 + 49 + 9 + 36$$
$$\text{SSB} = 194$$

Now SSW can be found by subtraction (Formula 12.11):

$$\text{SSW} = \text{SST} - \text{SSB}$$
$$\text{SSW} = 536 - 194$$
$$\text{SSW} = 342$$

To find the degrees of freedom for the two sums of squares, we use Formulas 12.5 and 12.6:

$$\text{dfw} = N - k = 16 - 4 = 12$$

$$\text{dfb} = k - 1 = 4 - 1 = 3$$

Finally, we are ready to construct the mean square estimates to the population variance. For the estimate based on SSW, use Formula 12.7:

$$\text{Mean square within} = \text{SSW/dfw} = 342/12 = 28.5$$

For the between estimate, we use Formula 12.8:

$$\text{Mean square between} = \text{SSB/dfb} = 194/3 = 64.67$$

The test statistic, or F ratio, is found by dividing the between estimate by the within estimate (see Formula 12.9):

$$F \text{ (obtained)} = \text{mean square between/mean square within}$$
$$F \text{ (obtained)} = 64.67/28.5$$
$$F \text{ (obtained)} = 2.27$$

This statistic must still be evaluated for its significance.

12.6 A TEST OF SIGNIFICANCE FOR ANOVA

In the previous section, we calculated a value for the test statistic F from the data on capital punishment and religion. In this section, we will see how to test that value for its significance and also take a look at some of the assumptions underlying the ANOVA test. As usual, we will follow the five-step model as a convenient way of organizing the decision-making process.

Step 1. Making assumptions.

Model: Independent random samples
Level of measurement is interval-ratio
Populations are normally distributed
Population variances are equal

The model assumptions are quite stringent and underscore the fact that this technique should be used only with variables that have been carefully and precisely measured. As long as sample sizes are equal (or nearly so), ANOVA can tolerate some violation of the model assumptions but, in situations where you are uncertain or have samples of very different size, it is probably advisable to use an alternative test. Chi square, as noted previously, might serve as an alternative if you are willing to collapse the data into a few categories.

Step 2. Stating the null hypothesis. For ANOVA, the null always states that the means of the populations from which the samples were drawn are equal. For our example problem, we are concerned with four different populations, so our null hypothesis would be

$$H_o: \mu_1 = \mu_2 = \mu_3 = \mu_4$$

where μ_1 represented the mean for the Protestants, μ_2, the mean for Catholics, and so forth.

The alternative hypothesis for this test states simply that at least one of the population means is different. The wording here is important. If we reject the null, ANOVA does not identify which of the means are significantly different.

(H_1: At least one of the population means is different.)

Step 3. **Selecting the sampling distribution and establishing the critical region.** The sampling distribution for ANOVA is the F distribution, which is summarized in Appendix D. Note that there are separate tables for alphas of .05 and .01, respectively. As with the t and chi square tables, the value of the critical F score will vary by degrees of freedom. For ANOVA, there are two separate degrees of freedom, one for each estimate of the population variance. The numbers across the top of the table are the degrees of freedom associated with the between estimate (dfb), and the numbers down the side of the table are those associated with the within estimate (dfw). In our example, the former is $(k - 1)$, or 3, and the latter is $(N - k)$, or 12 (see Formulas 12.5 and 12.6). So, if we set alpha at .05, our critical F score will be 3.49. Summarizing these considerations:

$$\text{Sampling distribution} = F \text{ distribution}$$

$$\text{Alpha} = .05$$

$$\text{Degrees of freedom (within)} = (N - k) = 12$$

$$\text{Degrees of freedom (between)} = (k - 1) \ 3$$

$$F \text{ (critical)} = 3.49$$

Taking a moment to inspect the two F tables, you will notice that all the values are greater than 1.00. This is because ANOVA is a one-tailed test and we are concerned only with outcomes in which there is more variance between categories than within categories. F values of less than 1.00 would indicate that the between estimate was lower in value than the within estimate and, since we would always fail to reject the null in such cases, we simply ignore this class of outcomes.

Step 4. **Computing the test statistic.** This was done in the previous section, where we found an obtained F ratio of 2.27.

Step 5. **Making a decision.** Compare the test statistic with the critical value:

$$F \text{ (obtained)} = 2.27$$

$$F \text{ (critical)} = 3.49$$

Since the test statistic does not fall in the critical region, our decision would be to fail to reject the null. Support for capital punishment does not differ significantly by religion, and the variation we observed in the sample means is unimportant.

APPLICATION 12.1

An experiment in teaching introductory biology was recently conducted at a large university. One section was taught by the traditional lecture-lab method, a second was taught by an all-lab/demonstration approach with no lectures, and a third was taught entirely by a series of videotaped lectures and demonstrations that the students were free to view at any time and as often as they wanted. Students were randomly assigned to each of the three sections and, at the end of the semester, random samples of final exam scores were collected from each section. Is there a significant difference in student performance by teaching method?

Final exam scores by teaching method

Lecture		Demonstration		Videotape	
X	X^2	X	X^2	X	X^2
55	3,025	56	3,136	50	2,500
57	3,249	60	3,600	52	2,704
60	3,600	62	3,844	60	3,600
63	3,969	67	4,489	61	3,721
72	5,184	70	4,900	63	3,969
73	5,329	71	5,041	69	4,761
79	6,241	82	6,724	71	5,041
85	7,225	88	7,744	80	6,400
92	8,464	95	9,025	82	6,724

$\Sigma X = 636 \quad\quad 651 \quad\quad 588$
$\Sigma X^2 = \quad 46,286 \quad\quad 48,503 \quad\quad 39,420$
$\bar{X}_k = \quad 70.67 \quad\quad 72.33 \quad\quad 65.33$
$\bar{X} = 1,875/27 = 69.44$

We can see by inspection that the "Videotape" group had the lowest average score and that the "Demonstration" group had the highest average score. The ANOVA test will tell us if these differences are large enough to justify the conclusion that they did not occur by chance alone. Following the computational routine established at the end of Section 12.4:

$SST = \Sigma X^2 - N\bar{X}^2$
$SST = (46286 + 48503 + 39420) - (27)$
$\quad\quad \times (69.44)^2$
$SST = 134209 - 130191.67$
$SST = 4017.33$
$SSB = \Sigma N_k (\bar{X}_k - \bar{X})^2$
$SSB = 9(70.67 - 69.44)^2 + 9(72.33 - 69.44)^2$
$\quad\quad + 9(65.33 - 69.44)^2$
$SSB = 13.62 + 75.17 + 152.03$
$SSB = 240.82$
$SSW = SST - SSB$
$SSW = 4017.33 - 240.82$
$SSW = 3776.51$
$dfw = N - k = 27 - 3 = 24$
$dfb = k - 1 = 3 - 1 = 2$

Mean square within $= SSW/dfw = 3776.51/24$
$\quad\quad = 157.36$
Mean square between $= SSB/dfb = 240.82/2$
$\quad\quad = 120.41$
$F =$ mean square between/mean square within
$\quad\quad = 120.41/157.36$
$F = 0.77$

If we set alpha equal to .05, the critical region for 2 and 24 degrees of freedom would begin at 3.40, and our obtained F score is less than 1.00. We would clearly fail to reject the null hypothesis ("the population means are equal") and would conclude that the observed differences among the category means were the results of random chance. Student performance in this course does not vary significantly by teaching method.

12.7 AN ADDITIONAL EXAMPLE FOR COMPUTING AND TESTING THE ANALYSIS OF VARIANCE

In this section, I want to work through an additional example of the computation and interpretation of the ANOVA test. We will first review matters of computation, find the obtained F ratio, and then test the statistic for its significance. In the computational section, we will follow the step-by-step guidelines presented at the end of Section 12.4.

A researcher has been asked to evaluate the efficiency with which each of three social service agencies is administering a particular program. One area of concern is the speed with which the agencies are processing paperwork and determining the eligibility of potential clients. The researcher has gathered information on the number of days required for processing a random sample of 10 cases in each agency. Is there a significant difference? The data are reported in Table 12.4, where I have included some additional information we will need to complete our calculations.

To find SST by the computational formula (12.10):

$$\text{SST} = \Sigma X^2 - N\bar{X}^2$$
$$\text{SST} = (524 + 1816 + 2462) - 30(11.67)^2$$
$$\text{SST} = (4802) - 30(136.19)$$
$$\text{SST} = 4802 - 4085.70$$
$$\text{SST} = 716.30$$

To find SSB by Formula 12.4:

$$\text{SSB} = \Sigma N_k(\bar{X}_k - \bar{X})^2$$
$$\text{SSB} = 10(7 - 11.67)^2 + 10(13 - 11.67)^2 + 10(15 - 11.67)^2$$
$$\text{SSB} = 10(-4.67)^2 + 10(1.33)^2 + 10(3.33)^2$$
$$\text{SSB} = 10(21.81) + 10(1.77) + 10(11.09)$$
$$\text{SSB} = 218.10 + 17.70 + 110.90$$
$$\text{SSB} = 346.7$$

TABLE 12.4   NUMBER OF DAYS REQUIRED TO PROCESS CASES FOR THREE AGENCIES (fictitious data)

Client	Agency A X	Agency A X^2	Agency B X	Agency B X^2	Agency C X	Agency C X^2
1	5	25	12	144	9	81
2	7	49	10	100	8	64
3	8	64	19	361	12	144
4	10	100	20	400	15	225
5	4	16	12	144	20	400
6	9	81	11	121	21	441
7	6	36	13	169	20	400
8	9	81	14	196	19	361
9	6	36	10	100	15	225
10	6	36	9	81	11	121
$\Sigma X =$	70		130		150	
$\Sigma X^2 =$		524		1816		2462
$\bar{X}_k =$	7.0		13.0		15.0	

$$\bar{X} = 350/30 = 11.67$$

Now we can find SSW by Formula 12.11:

$$SSW = SST - SSB$$
$$SSW = 716.30 - 346.70$$
$$SSW = 369.60$$

The degrees of freedom are found by Formulas 12.5 and 12.6:

$$dfw = N - k = 30 - 3 = 27$$
$$dfb = k - 1 = 3 - 1 = 2$$

The estimates to the population variance are found by Formulas 12.7 and 12.8:

$$\text{Mean square within} = SSW/dfw = 369.60/27 = 13.69$$

$$\text{Mean square between} = SSB/dfb = 346.7/2 = 173.35$$

The F ratio (Formula 12.9) is

$$F = \text{mean square between/mean square within} = 173.35/13.39$$

$$F = 12.95$$

And we can now test this value for its significance.

Step 1. Making assumptions.

> Model: Independent random samples
> Level of measurement is interval-ratio
> Population distributions are normal
> Population variances are equal

The researcher will always be in a position to judge the adequacy of the first two assumptions in the model. The second two assumptions are more problematical, but remember that ANOVA will tolerate some deviation from its assumptions as long as sample sizes are roughly equal.

Step 2. Stating the null hypothesis.

$$H_0: \mu_1 = \mu_2 = \mu_3$$

(H_1: At least one population mean is different.)

Step 3. Selecting the sampling distribution and establishing the critical region.

$$\text{Sampling distribution} = F \text{ distribution}$$

$$\text{Alpha} = .05$$

$$\text{Degrees of freedom (within)} = (N - k) = (30 - 3) = 27$$

$$\text{Degrees of freedom (between)} = (k - 1) = (3 - 1) = 2$$

$$F \text{ (critical)} = 3.35$$

Step 4. Computing the test statistic. We found an obtained F ratio of 12.95.

Step 5. Making a decision.

$$F \text{ (critical)} = 3.35$$

$$F \text{ (obtained)} = 12.95$$

The test statistic is in the critical region, and we would reject the null of no difference. The differences among the three agencies are very unlikely to have occurred by chance alone. The agencies are significantly different in the speed with which they process paperwork and determine eligibility.

12.8 THE LIMITATIONS OF THE TEST

The test presented in this chapter is appropriate whenever you want to test the significance of a difference across three or more categories of a single variable. This application is called **one-way analysis of variance**, since we are observing the effect of a single variable (for example, religion) on another (for example, support for capital punishment). This is the simplest application of ANOVA, and you should be aware that the technique has numerous more advanced and complex forms. For example, you may encounter research projects in which the effects of two separate variables (for example, religion and gender) on some third variable were observed.

The major limitations of ANOVA are that it requires interval-ratio measurement and roughly equal numbers of cases in each of the categories. The former condition may be difficult to meet with complete confidence for many variables of interest to the social sciences. The latter condition may create problems when the research hypothesis calls for comparisons between groups that are, by their nature, unequal in numbers (for example, white versus Black Americans) and may call for some unusual sampling schemes in the data-gathering phase of a research project. Neither of these limitations should be particularly crippling, however, since ANOVA can tolerate some deviation from its model assumptions, but you should be aware of these limitations in planning your own research as well as in judging the adequacy of research conducted by others.

A final limitation of this test relates to the research hypothesis. As you recall, when we reject the null hypothesis of no difference, we generate support for the alternative hypothesis which, in the case of ANOVA, simply asserts that at least one of the population means is different from the others. The next question that will occur to you is, of course, "Which one?" You will sometimes be able to make this determination by inspection of the sample means. In our problem involving social service agencies, for example, it is pretty clear from Table 12.4 that Agency A is the source of most of the differences in the data. There are tests by which you can ascertain exactly which differences are statistically significant, but these tests are beyond the scope of this text. You should be aware that the informal "eyeball" method can be misleading so that, if you reject the null, you should exercise caution in making conclusions about which means are significantly different.

SUMMARY

1. One-way analysis of variance is a powerful test of significance that is commonly used when comparisons across more than two categories or samples are of interest. It is perhaps easiest to conceptualize ANOVA as an extension of the test for the difference in sample means.

2. ANOVA compares the amount of variation within the categories to the amount of variation between categories. If the null of no difference is false, there should be relatively great variation between categories and relatively little variation within categories. The greater the differences from category to category relative to the differences within the categories, the more likely we will be able to reject the null.

3. The computational routine for even simple applications of ANOVA can quickly become quite complex. The basic process is to construct separate estimates to the population variance based on the variation within the categories and the variation between the categories. The test statistic is the F ratio, which is based on a comparison of these two estimates. The basic computational routine is summarized at the end of Section 12.4, and this is probably an appropriate time to mention the widespread availability of statistical packages such as SPSS^x, the purpose of which is to perform complex calculations such as these accurately and quickly. If you haven't yet learned how to use such programs, ANOVA may provide you with the necessary incentive.

4. The ANOVA test can be organized into the familiar five-step model for testing the significance of sample outcomes. Although the model assumptions (step 1) require high-quality data, the test can tolerate some deviation as long as sample sizes are roughly equal. The null takes the familiar form of stating that there is no difference of any importance between the population values, while the alternative hypothesis asserts that at least one population mean is different. The sampling distribution is the F distribution, and the test is always one-tailed. The decision to reject or to fail to reject the null is based on a comparison of the obtained F ratio with the critical F ratio as determined for a given alpha level and degrees of freedom. The decision to reject the null indicates only that one or more of the population means is different from the others. We can often determine which sample mean(s) account for the difference by inspecting the sample data, but this informal method should be used with caution.

SUMMARY OF FORMULAS

Total sum of squares	12.1	$SST = \Sigma(X_i - \bar{X})^2$
The two components of the total sum of squares	12.2	$SST = SSB + SSW$
Sum of squares within	12.3	$SSW = \Sigma(X_i - \bar{X}_k)^2$
Sum of squares between	12.4	$SSB = \Sigma N_k(\bar{X}_k - \bar{X})^2$
Degrees of freedom for SSW	12.5	$dfw = N - k$
Degrees of freedom for SSB	12.6	$dfb = k - 1$
Mean square within	12.7	$\text{Mean square within} = \dfrac{SSW}{dfw}$
Mean square between	12.8	$\text{Mean square between} = \dfrac{SSB}{dfb}$
F ratio	12.9	$F = \dfrac{\text{mean square between}}{\text{mean square within}}$
Computational formula for SST	12.10	$SST = \Sigma X^2 - N\bar{X}^2$
Finding SSW by subtraction	12.11	$SSW = SST - SSB$

GLOSSARY

Analysis of variance. A test of significance appropriate for situations in which we are concerned with the differences among more than two sample means.

ANOVA. See **analysis of variance**.

F ratio. The test statistic computed in step 4 of the ANOVA test.

Mean square. An estimate of the variance calculated by dividing the sum of squares within (SSW) or the sum of squares between (SSB) by the proper degrees of freedom.

One-way **analysis of variance.** Applications of ANOVA in which the effect of a single variable on another is observed.

Sum of squares between (SSB). The sum of the squared deviations of the sample means from the overall mean, weighted by sample size.

Sum of squares total (SST). The sum of the squared deviations of the scores from the overall mean.

Sum of squares within (SSW). The sum of the squared deviations of scores from the category means.

PROBLEMS

12.1 Conduct the ANOVA test for the scores presented below.

	Category	
A	B	C
5	10	12
7	12	16
8	14	18
9	15	20

12.2 [SOC] Random samples of cities have been drawn from each of four geographical regions. The homicide rates for each city are reported below. Is there a significant difference?

North	South	Midwest	West
5	13	18	34
5	18	25	30
10	20	30	8
6	25	20	10
3	12	10	25
4	10	11	26
8	8	10	28
7	18	12	30
12	20	20	28
15	21	21	7

12.3 [PS] Does the rate of voter turnout vary significantly by the type of election? A random sample of voting precincts displays the following pattern of voter turnout by election type. Assess the results for significance.

Local only	State	National
33	35	42
78	56	40
32	35	52
28	40	66
10	45	78
12	42	62
61	65	57
28	62	75
29	25	72
45	47	51
44	52	69
41	55	59

12.4 [SOC] A random sample of 50 colleges has been taken and, for each college, average growth in student population over the past five years has been calculated. The colleges are divided into four types by size and funding source. Is there a significant difference in growth rate by type of college?

Small Private	Large Private	Small Public	Large Public
1	2	4	2
− 2	0	3	3
− 10	5	7	2
0	− 1	− 2	0
− 1	0	5	4
− 3	− 10	0	5
− 2	7	8	6
5	4	4	− 1
4	2	3	1
1	0	5	2
2	3	− 1	4
	2	2	2
		8	− 3
		4	

12.5 [SOC] In a local community, a random sample of 25 couples has been assessed on a scale that measures the extent to which power and decision making are shared (lower scores) or monopolized by one party (higher scores). The couples were also classified by type of relationship: traditional (only the husband works outside the home), dual-career (both parties work), and cohabitational (parties living together but not legally married, regardless of work patterns). Does power structure vary significantly by type of relationship?

Traditional	Dual-career	Cohabitational
7	8	2
8	5	1
2	4	3
5	4	4
7	5	1
6	5	2
5	5	1
	8	1
	2	2

12.6 [CJ] Two separate crime-reduction programs have been implemented in the city of Shinbone. One involves a neighborhood watch program with citizens actively involved in crime prevention. The second involves officers patrolling the neighborhoods on foot rather than in patrol cars. In terms of the percentage reduction in crimes reported to the police over a one-year period, were the programs successful? The results are for random samples drawn from all neighborhoods.

Neighborhood Watch	Foot Patrol	No Program
−10	−21	+ 3
− 2	−15	−10
+ 1	− 8	+14
+ 2	−10	+ 8
+ 7	− 5	+ 5
+10	− 1	− 2
+10	+ 1	−10
+12	+ 8	− 2

12.7 [SOC] Are sexually active teenagers any better informed about AIDS and other potential health problems related to sex? A 15-item test of general knowledge about sex and health was administered to random samples of teens who are sexually inactive, teens who are sexually active but only with a single partner ("going steady"), and teens who are sexually active with more than one partner. Is there any significant difference in the test scores?

Inactive	Active— One Partner	Active— More Than One Partner
10	11	12
12	11	12
8	6	10
10	5	4
8	15	3
5	10	15
3	11	14
8	9	7
10	14	9
11	11	15

12.8 [GER] Do older citizens lose interest in politics and current affairs? A brief quiz on recent headline stories was administered to random samples of respondents from each of four different age groups. Is there a significant difference? The data below represent numbers of correct responses.

High School (15–18)	Young Adults (21–30)	Middle-aged (40–55)	Retired (65+)
0	0	2	5
1	0	3	6
1	2	3	6
2	2	4	6
2	4	4	7
2	4	5	7
3	4	6	8
5	6	7	10
5	7	7	10
7	7	8	10
7	7	8	10
9	10	10	10

12.9 [SW] A random sample of counties has been rated as predominantly urban, suburban, or rural. Does infant mortality rate (number of infant deaths per 1000 live births) in these counties vary significantly by this variable?

Rural	Suburban	Urban
15.1	11.0	12.5
14.7	10.9	12.4
14.2	10.1	12.1
13.5	10.0	11.9
12.5	9.9	9.7

Rural	Suburban	Urban
11.2	9.8	9.2
10.1	8.5	8.2
9.9	7.1	6.5
8.5	7.0	6.2
7.0	6.9	6.0

12.10 SOC A sociologist believes that interest in the American sport institution is almost entirely a phenomenon of white-collar males. A random sample has been given a list of 20 sports and has been asked to check off those they follow on a regular basis. The results for each sex/class group are reported below. Are these results consistent with the hypothesis?

Blue-Collar Males	Blue-Collar Females	White-Collar Males	White-Collar Females
0	0	0	0
1	0	1	0
1	0	3	0
1	0	3	0
1	0	3	0
2	1	3	0
3	3	4	1
3	3	5	1
3	3	6	3
3	3	6	3
5	3	7	4
5	3	10	5
8	4	11	6
10	5	12	9
10	8	15	10

12.11 SOC A small random sample of respondents has been selected from the General Social Survey data base. Using item 14 (see Appendix G), each respondent has been classified as either a city dweller, a suburbanite, or a rural dweller. Are there statistically significant differences by place of residence for any of the variables listed below?

a. Occupational prestige (item 2)

Urban	Suburban	Rural
32	40	30
45	48	40
42	50	40
47	55	45
48	55	45
50	60	50
51	65	52
55	70	55
60	75	55
65	75	60

b. Number of children (item 6)

Urban	Suburban	Rural
1	0	1
1	1	4
0	0	2
2	0	3
1	2	3
0	2	2
2	3	5
2	2	0
1	2	4
0	1	6

c. Family income (item 12)

Urban	Suburban	Rural
5	6	5
7	8	5
8	11	11
11	12	10
8	12	9
9	11	6
8	11	10
3	9	7
9	10	9
10	12	8

d. Church attendance (item 29)

Urban	Suburban	Rural
0	0	1
7	0	5
0	2	4
4	5	4
5	8	0
8	5	4
7	8	8
5	7	8
7	2	8
4	6	5

e. Hours of TV watching per day (item 67)

Urban	Suburban	Rural
5	5	3
3	7	7
12	10	5
2	2	0

Urban	Suburban	Rural
0	3	1
2	0	8
3	1	5
4	3	10
5	4	3
9	1	1

SPSS/PC+ PROCEDURES FOR THE ANALYSIS OF VARIANCE

DEMONSTRATION 12.1 Does Conservatism Increase with Age?

SPSS/PC+ provides several different ways of conducting the analysis of variance test. The procedure summarized below is the most accessible of these, but it still incorporates options and capabilities that we have not covered in this chapter. If you wish to explore these possibilities, please consult the SPSS/PC+ manual.

In designing examples for these sessions, my choices are constrained by the scarcity of interval-ratio variables in the 1993 GSS data set. Some variables which 'should be' interval-ratio are actually measured at the ordinal level (e.g., INCOME and CHILDS) while others are unsuitable dependent variables on logical grounds (e.g., AGE). To have something to discuss, I have shaded the truth with respect to level-of-measurement criteria in several of the examples that follow.

Let's begin by exploring the idea that political ideology is linked to age in American society. People are often said to become more conservative about a wide range of issues as they age, and this might result in greater political conservatism in older age groups. For a measure of ideology, we will use POLVIEWS, item 19 in Appendix G. As you inspect this variable, note that higher scores indicate higher levels of conservatism. We need to collapse AGE into a few categories to fit the ANOVA design. To find reasonable cutting points for AGE, I ran the FREQUENCIES command to find the ages that divided the sample into three groups of roughly equal size. Note that I have used the keywords 'Lo' and 'Hi' in the RECODE command instead of specifying the exact upper and lower limits of this variable. See the manual for further information.

```
RECODE AGE (Lo Thru 36=1)(37 Thru 51=2)(52 Thru Hi=3).
VALUE LABELS AGE 1 '36 and younger' 2 '37 to 51' 3 '52
and older'/.
```

The ANOVA procedure we will use is called ONEWAY. The following instructions describe how to use this command.

1. From the main menu, select 'analyze data' and then select 'comparing group means.' Finally, select and paste 'ONEWAY.'
2. From the ONEWAY menu, select and paste '!/VARIABLES.' Move the cursor over '!dependent(s)' and then press ALT-T and type POLVIEWS in the typing window. Next, select the keyword BY, move the cursor over '!factor,' press ALT-T, and type AGE, the independent variable, in the typing window. You must specify the low and high values for the independent variable. Do this by selecting and pasting () from the menu and typing the values 1,3 in the typing window. You may name as many as 100 dependent variables in ONEWAY but only one independent variable.
3. Select and paste /STATISTICS 1 for descriptive statistics. Your command should look like this:

```
ONEWAY /VARIABLES POLVIEWS BY AGE (1,3) /STATISTICS 1.
```

and the output should be:

```
- - - - - - - - - - O N E W A Y - - - - - - - - - - - - -
   Variable  POLVIEWS   THINK OF SELF AS LIBERAL OR CONSERVATIVE
 By Variable  AGE        AGE OF RESPONDENT

                             Analysis of Variance
                             Sum of        Mean           F       F
        Source        D.F.   Squares       Squares      Ratio   Prob.
Between Groups          2     48.2517       24.1258     11.1554  .0000
Within Groups         780   1686.9080        2.1627
Total                 782   1735.1596
            - - - - - - - - - - O N E W A Y - - - - - - - - - -
                             Standard   Standard
Group      Count      Mean   Deviation    Error    95 Pct Conf Int for Mean
Grp 1       265     4.1132    1.3880      .0853     3.9453  To   4.2811
Grp 2       255     4.0000    1.4954      .0936     3.8156  To   4.1844
Grp 3       263     4.5741    1.5262      .0941     4.3888  To   4.7595
Total       783     4.2312    1.4896      .0532     4.1267  To   4.3357

Group      Minimum      Maximum
Grp 1      1.0000       8.0000
Grp 2      1.0000       8.0000
Grp 3      1.0000       8.0000
Total      1.0000       8.0000
```

The report includes the various degrees of freedom, all of the sums of squares, the Mean Squares, the F ratio (11.1554), and, at the far right, the exact probability of getting these results if the null is true. This is reported as an "F Prob." of .0000, which is much lower than our usual alpha level of .05. The differences in POLVIEWS for the various AGE groups are statistically significant.

The second half of the report displays some summary statistics, with the three age groups identified as Grp 1 (this is the youngest group), Grp 2, and Grp 3 (the oldest group). An inspection of the means shows that the average score for the entire sample was 4.2312. The oldest age group is the most conservative (remember that higher scores indicate greater conservatism) and the two other age groups have similar means and standard deviations.

DEMONSTRATION 12.2 Does Political Ideology Vary by Educational Level?

Let's continue our analysis of POLVIEWS and see if there are any significant differences in political ideology by social class. Our version of the 1993 GSS has several variables which might be used as indicators of class, including DEGREE, INCOME, CLASS, and PRESTG80. We should probably investigate all these but, to conserve space, we confine our attention to DEGREE. First, we need to recode the variable into fewer categories. The recoding scheme below divides the variable into three major categories.

```
RECODE DEGREE (0=0)(1=1)(2 Thru 4=2).
VALUE LABELS DEGREE 1 'Less than HS' 2 'HS' 3 'At
least some college/.
```

Follow the previous instructions for ONEWAY and specify POLVIEWS again as the dependent variable and DEGREE as the factor, or independent variable. The command should look like this:

```
ONEWAY /VARIABLES POLVIEWS BY DEGREE (1,3)
/STATISTICS 1.
```

and the results should be:

```
ONEWAY /VARIABLES POLviews BY DEGREE (0,2) /STATISTICS 1.
            - - - - - - - - - - O N E W A Y - - - - - - - - - - -
      Variable  POLVIEWS   THINK OF SELF AS LIBERAL OR CONSERVATIVE
   By Variable  DEGREE     RS HIGHEST DEGREE
```

```
                            Analysis of Variance
                            Sum of          Mean             F        F
            Source          D.F.    Squares         Squares          Ratio    Prob.

Between Groups                2     10.9589         5.4794           2.5140   .0816

Within Groups               778   1695.7197        2.1796

Total                       780   1706.6786

          - - - - - - - - - - O N E W A Y - - - - - - - - - -
                                      Standard     Standard
Group           Count       Mean     Deviation      Error     95 Pct Conf Int for Mean

Grp 0            156       4.4231      1.5577        .1247      4.1767   To    4.6694
Grp 1            404       4.2228      1.4351        .0714      4.0824   To    4.3631
Grp 2            221       4.0769      1.4919        .1004      3.8791   To    4.2747

Total            781       4.2215      1.4792        .0529      4.1176   To    4.3254

Group         Minimum     Maximum

Grp 0          1.0000      8.0000
Grp 1          1.0000      8.0000
Grp 2          1.0000      8.0000

Total          1.0000      8.0000
```

The F Ratio (2.5140) and F Prob. (.0816) reported at the far right at the top of the output show that the differences between groups are significant at the .10 level but not at the .05 level. Inspect the group means in the bottom half of the output and you will see that the least educated (Grp 0) are the most conservative and the most educated (Grp 2) are the most liberal, with high school–educated respondents intermediate.

DEMONSTRATION 12.3 Another Test For the Sources of Personal Happiness.

In previous exercises, we investigated the relationship between money (IN-COME91), education (DEGREE), and personal happiness (HAPPY). In Demonstrations 10.2 and 11.2, we found that people with higher incomes and higher levels of education also reported higher levels of happiness. In this section, we investigate these relationships once more using ANOVA as our statistical tool, slightly different codings for DEGREE and INCOME91, and a different measure of happiness. We can be even more certain of our previous findings if we confirm them under these new conditions.

Happiness seems like a very personal and subjective characteristic and would appear to be much harder to measure than more concrete and obvious traits such as height or age. It is not uncommon to measure such elusive feelings in a variety of different ways, and our version of the 1993 GSS includes two other variables (SATFAM and SATFRND, items 37 and 38) which seem, taken at face value, to be indicators of happiness. Each has seven response categories, and if we add the two variables together, we will create a scale with 14 possible scores. Because of the coding scheme, higher scores indicate *lower* levels of satisfaction (happiness). This computed measure of happiness is not the continuous, interval-ratio level dependent variable most appropriate for the ANOVA test, but we should still be able to learn something about the relationship. The COMPUTE statement is

```
COMPUTE SATSCALE = SATFAM + SATFRND.
```

We use INCOME91 and DEGREE, each recoded into three categories, as the factor, or independent, variable. The recoding schemes I used were

```
RECODE INCOME91 (0 Thru 12 = 1)(13 thru 17 = 2)(18
thru 21 = 3).
VALUE LABELS INCOME91 1 '20000 and less' 2 '20000 to
40000' 3 '40000 and more'.

RECODE DEGREE (0 = 0)(1 = 1)(2 thru 4 = 3).
VALUE LABELS DEGREE 1 'Less than HS' 2 'HS' 3 'At
least some college'.
```

Follow the instructions in DEMONSTRATION 12.1 to create the following commands:

```
ONEWAY /VARIABLES SATSCALE BY INCOME91 (1,3) /STATISTICS 1.
ONEWAY /VARIABLES SATSCALE BY DEGREE (1,3) /STATISTICS 1.
```

Your output should look like this:

```
- - - - - - - - - - O N E W A Y - - - - - - - - - -
```

Variable   SATSCALE

By Variable   INCOME91   TOTAL FAMILY INCOME

Analysis of Variance

Source	D.F.	Sum of Squares	Mean Squares	F Ratio	F Prob.
Between Groups	2	146.0228	73.0114	14.4544	.0000
Within Groups	464	2343.7374	5.0512		
Total	466	2489.7602			

- - - - - - - - - - O N E W A Y - - - - - - - - - - - -

Group	Count	Mean	Standard Deviation	Standard Error	95 Pct Conf Int for Mea		
Grp 1	156	5.2115	2.7001	.2162	4.7845	To	5.638
Grp 2	149	4.2752	2.0920	.1714	3.9365	To	4.613
Grp 3	162	3.8889	1.8750	.1473	3.5980	To	4.179
Total	467	4.4540	2.3115	.1070	4.2438	To	4.664

Group	Minimum	Maximum
Grp 1	2.0000	13.0000
Grp 2	2.0000	12.0000
Grp 3	2.0000	11.0000
Total	2.0000	13.0000

- - - - - - - - - - O N E W A Y - - - - - - - - - - -

Variable   SATSCALE

By Variable   DEGREE   RS HIGHEST DEGREE

Analysis of Variance

Source	D.F.	Sum of Squares	Mean Squares	F Ratio	F Prob.
Between Groups	2	21.2415	10.6207	1.9750	.1398
Within Groups	513	2758.7333	5.3776		
Total	515	2779.9748			

- - - - - - - - - - O N E W A Y - - - - - - - - - - -

Group	Count	Mean	Standard Deviation	Standard Error	95 Pct Conf Int for Mean		
Grp 0	104	4.7212	2.4828	.2435	4.2383	To	5.2040
Grp 1	269	4.5130	2.3589	.1438	4.2298	To	4.7962
Grp 2	143	4.1538	2.1107	.1765	3.8049	To	4.5028
Total	516	4.4554	2.3234	.1023	4.2545	To	4.6564

Group	Minimum	Maximum
Grp 0	2.0000	12.0000
Grp 1	2.0000	14.0000
Grp 2	2.0000	12.0000
Total	2.0000	14.0000

Starting with the test for a relationship between income and satisfaction, we see that the lowest-income group had the lowest average level of happiness (5.2115) and that happiness increased as income increased. (Remember that SATFAM and SATFRND are coded so that lower scores indicate greater satisfaction.) The F Ratio and the F Prob. indicate that the differences in means is statistically significant at less than .0001. Our previous finding is confirmed.

The test for education and satisfaction shows that satisfaction increases as education increases. The least-educated group has the highest average score (4.7212)—or the lowest satisfaction—and the best-educated group had the lowest average score (4.1538). Although the pattern of these differences is similar to our previous findings, they are not significant (F Prob. = .1398) and do not support our earlier conclusion that happiness or satisfaction is higher for the better educated.

Exercises

12.1 As a follow-up on Demonstration 12.2, test INCOME91, PRESTG80, and PAPRES80 as independent variables against POLVIEWS. Do these measures of social class display the same type of relationship with POLVIEWS as DEGREE? First, recode each of these independents into three categories. Run FREQUENCIES for each and find cutting points which divide the variable into three groups of roughly equal size.

12.2 Use the COMPUTE facility to create three scales: one for ANOMIA5 and ANOMIA6; one for the six "confidence" items, CONCLERG to CONTV (you might think of this as an overall measure of confidence in societal institutions); and one for ABNOMORE and ABPOOR. Use ONEWAY to test each of these scales against the measures of social class you developed in Exercise 12.1. Remember that you can specify multiple dependent variables but only one independent variable in a single command. Summarize your results in a paragraph.

PART II CUMULATIVE EXERCISES

In this review, I've stated a number of research questions that can be answered by the techniques presented in Chapters 7–9, 11, and 12. (I omitted Chapter 10 since it is optional.) For each question, select the most appropriate test, compute the necessary statistics, and state your conclusions.

In order to complete some problems, you must first calculate the sample statistics (for example, means or proportions) that are used in the test of hypothesis. Use alpha = .05 throughout. The questions are presented in random order. There is at least one problem for each chapter, but I have not included a research situation for every statistical procedure covered in the chapters.

In selecting tests and procedures, you need to consider the question, the number of samples or categories being compared, and the level of measurement of the variables. The flowcharts at the beginning of each chapter may be helpful.

The situations refer to the data base below, which is based on the General Social Survey (GSS). The actual questions asked and the complete response codes for the GSS are presented in Appendix G. Abbreviated codes are listed below. Some variables have been recoded for this exercise. These problems are based on a small sample, and you may have to violate some assumptions about sample size in order to complete this exercise.

a. Is there a statistically significant difference in average hours of TV watching by income level? By race?
b. Is there a statistically significant relationship between age and happiness?
c. Estimate the average number of hours spent watching TV for the entire population.
d. If Americans currently average 2.3 children, is this sample representative of the population?
e. Are the educational levels of Catholics and Protestants significantly different?
f. Does average hours of TV watching vary by level of happiness?
g. Based on the sample data, estimate the proportion of Black Americans in the population.

Survey Items (numbers in parentheses refer to Appendix G)

1. How many children have you ever had? (6) (Values are actual numbers.)
2. Respondent's educational level (9)
    0. Less than HS
    1. HS
    2. At least some college

3. Race (11)

      1. White

      2. Black

4. Respondent's age (recode of 7)

      1. Younger than 35

      2. 35 and older

5. Number of hours of TV watched per day. (67) (Values are actual number of hours.)

6. What is your religious preference? (28)

      1. Protestant

      2. Catholic

7. Respondent's family income (recode of 12)

      1. 24999 or less

      2. 25000 or more

8. Respondent's overall level of happiness (34)

      1. Very happy

      2. Pretty happy

      3. Not too happy

Case	No. of Kids	Educa-tional Level	Race	Age	TV Hours	Religious Pref-erence	In-come	Happi-ness
1	3	1	1	1	3	1	1	2
2	2	0	1	1	1	1	2	3
3	4	2	1	2	3	1	1	1
4	0	2	1	1	2	1	1	1
5	5	1	1	1	2	1	2	2
6	1	1	1	2	3	1	2	1
7	9	0	1	1	6	1	1	1
8	6	1	1	2	4	1	1	2
9	4	2	1	1	2	2	2	2
10	2	1	1	1	1	1	2	3
11	2	0	1	2	4	1	1	3
12	4	1	2	1	5	2	1	2
13	0	1	1	2	2	2	1	2
14	2	1	1	2	2	1	1	2
15	3	1	2	2	4	1	1	1
16	2	0	1	2	2	1	2	1
17	2	1	1	2	2	1	2	3
18	0	2	1	2	2	1	2	1
19	3	0	1	2	5	2	1	1
20	2	1	2	1	10	1	1	3
21	2	1	1	2	4	1	2	1
22	1	0	1	2	5	1	2	1
23	0	2	1	1	2	2	1	1

Case	No. of Kids	Educa-tional Level	Race	Age	TV Hours	Reli-gious Pref-erence	In-come	Happi-ness
24	0	1	1	2	0	2	2	1
25	2	2	1	1	1	2	2	1
26	1	0	1	1	10	2	1	1
27	4	2	1	2	3	2	2	2
28	0	2	1	2	3	2	2	2
29	2	2	1	2	1	1	2	1
30	3	0	1	2	9	1	1	1
31	4	1	1	2	3	1	2	1
32	0	2	2	2	4	1	1	2
33	2	1	2	1	1	1	2	2
34	2	1	1	2	2	1	2	1
35	1	1	1	1	4	1	1	1
36	4	0	1	1	1	1	1	1
37	3	1	2	1	0	1	2	1
38	0	1	2	2	2	2	2	3
39	3	2	1	2	3	2	2	3
40	0	1	1	1	0	1	2	2
41	4	1	1	1	2	1	1	3
42	0	2	1	1	0	2	1	2
43	2	1	1	1	6	2	1	1
44	3	0	1	1	3	1	2	2
45	6	1	1	1	7	2	1	3

Part III Measures of Association

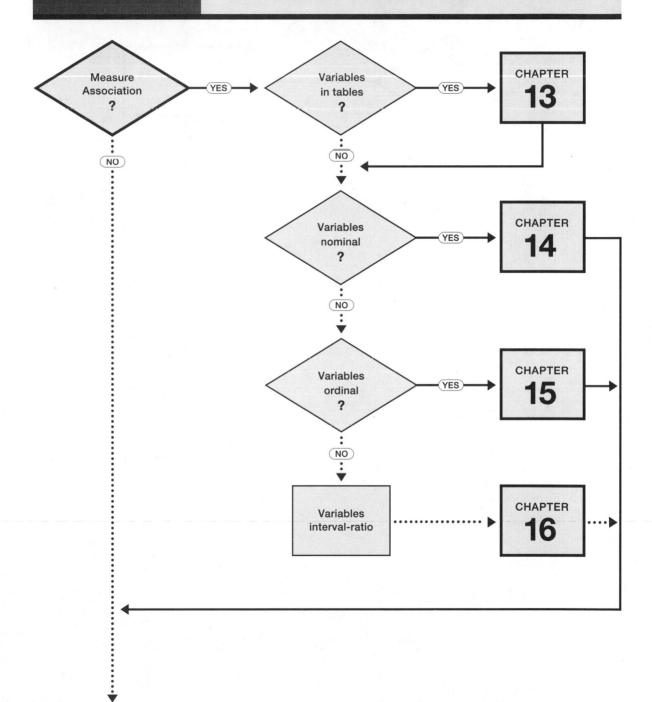

13

Bivariate Association
Introduction and Basic Concepts

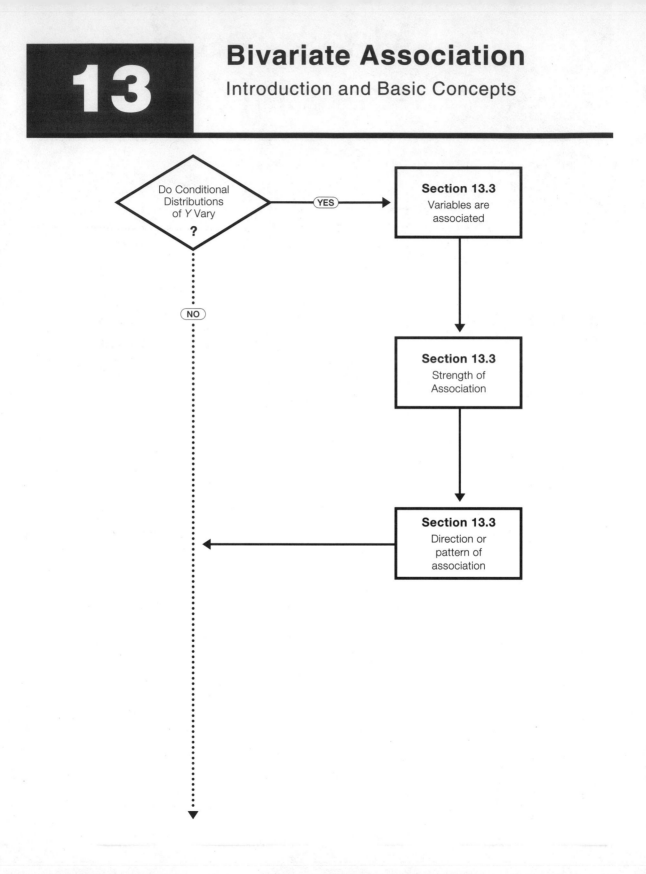

13.1 STATISTICAL SIGNIFICANCE AND THEORETICAL IMPORTANCE

As we have seen over the past several chapters, tests of statistical significance are extremely important in social science research. As long as social scientists continue to work with random samples rather than populations, these tests are indispensable for dealing with the possibility that our research results are the products of mere random chance. However, tests of significance are often merely the first step in the analysis of research results. We have seen that these tests do have limitations. Statistical significance is not necessarily the same thing as relevance or importance, and all tests of significance are affected by sample size. Tests of significance performed on large samples may result in decisions to reject the null hypothesis when, in fact, the observed differences are slight.

Beginning with this chapter, I will introduce **measures of association**, a new type of statistic that will provide us with additional information about our research results. Whereas tests of significance are designed to detect non-random relationships, measures of association are designed to quantify the strength and direction of a relationship. The two statistical techniques perform complementary functions and yield different types of information about our research results.

Measures of association are useful because they can assist the researcher in accomplishing important scientific goals. One very important scientific goal is to increase our understanding of the causal relationships among variables. Measures of association are very useful for tracing causal relationships. I should emphasize that measures of association, by themselves, cannot prove that two variables are causally related. That is, even if there is a strong statistical association between two variables, we cannot automatically conclude that one variable causes the other. We will explore causation in more detail in Part IV, but for now you should keep in mind that causation and association are two different things. Although we must be aware of this distinction, we can still use a statistical association between variables as evidence for a causal relationship.

Measures of association are also useful for predicting scores from one variable to another. Prediction is also an important goal of science and, if two variables are associated, we will be able to predict the score of a case on one variable from the score of that case on the other variable. Note that prediction and causation can be two separate goals. If variables are associated, we can predict from one to the other even if the variables are not causally related. Thus, measures of association perform a number of functions. They provide the researcher with important information about the relationships between variables, and they also help us pursue the goals of prediction and understanding causation.

In this chapter, I will introduce the concept of **association** and illustrate what an association between two variables "looks like" when the variables are presented in table format. I will also show how information about the strength and nature of an association can be gathered by calculating percentages for the table. In the chapters that follow, we will concentrate on the logic, calculation, and interpretation of the various measures of association. Finally, in Part IV, we will extend some of these ideas to the multivariate (more than two variables) case.

13.2 ASSOCIATION BETWEEN VARIABLES AND THE BIVARIATE TABLE

Most generally, two variables are said to be associated if the distribution of the categories or scores of one of them changes under the various categories or scores of the other. For example, suppose that an industrial sociologist was concerned with the relationship between job satisfaction and productivity for assembly-line workers. If these two variables are associated, then the distribution of scores on productivity will change under the different conditions of satisfaction. Workers who are classified as highly satisfied will have different patterns of productivity than workers who are low on satisfaction. In other words, if the variables are associated, then levels of productivity will vary by levels of satisfaction.

We can see these notions more concretely by setting up bivariate tables. As you recall, these tables are devices for displaying the joint frequency distribution of two variables. By convention, the **independent**, or X, variable (that is, the variable taken as causal) is arrayed in the columns and the **dependent**, or Y, variable in the rows.* That is, each column of the table (the vertical dimension) represents a score or category of the independent variable (X), and each row (the horizontal dimension) represents a score or category of the dependent variable (Y). Table 13.1 illustrates, with fictitious data, the basic features of a bivariate table.

This table displays the relationships between productivity and job satisfaction for a sample of 173 factory workers. To investigate the association between variables displayed in table format, we will focus our attention on the columns. Each column shows the pattern of scores on the dependent variable for each score on the independent variable. For example, the first column indicates that, of the 60 workers who were low on job satisfaction, 30 were low on productivity, 20 were moderately productive, and 10 were high on productivity. The second column shows that, of the 61 moderately satisfied workers, 21 were low, 25 were moderate, and 15 were high on productivity. What is the pattern for the 52 highly satisfied workers?

Inspecting the table from column to column allows us to observe the effects of the independent variable on the dependent variable (provided, of course, that the table is constructed with the independent variable in the columns). These "within-column" frequency distributions are called the **conditional distribution of Y**, since they display the distribution of scores on the dependent variable for each condition (or score) of the independent variable.

---

*In the material that follows, we will often, for the sake of brevity, refer to the independent variable as X and the dependent variable as Y.

TABLE 13.1 PRODUCTIVITY BY JOB SATISFACTION

Productivity (Y)	Job Satisfaction (X)			
	Low	Moderate	High	
Low	30	21	7	58
Moderate	20	25	18	63
High	10	15	27	58
	60	61	52	173

Table 13.1 indicates that productivity and satisfaction are associated: the distribution of scores on Y (productivity) changes across the various conditions of X (satisfaction). For example, half of the workers who were low on satisfaction were also low on productivity. On the other hand, over half of the workers who were high on satisfaction were high on productivity. Since the conditional distributions of Y change, these two variables are associated in some way.

There is, as you may have realized, another way to detect the existence of an association between two variables that have been organized into table format. In Chapter 11, we introduced the chi square statistic. Even though this statistic is designed as a test of significance, it can also be used as an indicator of association. Just as any change in the conditional distributions of the dependent variable indicates the existence of an association, so too does any nonzero value for the obtained chi square. For example, the obtained chi square for Table 13.1 would be 24.2, and this nonzero value affirms our previous conclusion, based on the observation of the conditional distributions of Y, that an association of some sort exists between job satisfaction and productivity.

Often, the researcher will have already conducted a chi square test for independence before considering matters of association. In such cases, an extended inspection of the conditional distributions of Y will not be necessary to ascertain whether or not the two variables are associated. If the obtained chi square is zero, the two variables are independent and not mutually associated. Any value other than zero indicates some association between the variables. Remember, however, that independence and association are two different things. It is perfectly possible for two variables to be associated (as indicated by a nonzero chi square) but still independent (if we fail to reject the null hypothesis).

In this section, we have defined, in a general way, the concept of association between two variables. We have also shown two different ways to detect the presence of an association. In the next section, we will extend the analysis beyond questions of the mere presence or absence of an association and, in a systematic way, show how additional very useful information about the relationship between two variables can be developed.

13.3 THREE CHARACTERISTICS OF BIVARIATE ASSOCIATIONS

Bivariate associations possess three different characteristics, and each must be analyzed for a full investigation of the relationship. Investigating these characteristics may be thought of as a process of finding answers to three questions:

1. Does an association exist?
2. If an association does exist, how strong is it?
3. What is the pattern and/or the direction of the association?

We will consider each of these questions in the rest of this section.

Does an Association Exist?   We have already discussed the general definition of association, and we have seen how a bivariate association would manifest itself in changing conditional distributions of Y in a table or in a nonzero value for chi square. In Table 13.1, since the conditional distributions of productivity (Y) are different across the various categories of satisfaction (X) and chi square is a nonzero value, we conclude that these two variables are associated to some extent.

In Table 13.1, comparisons from column to column are relatively easy to make because the column totals are roughly equal. This will not always be the case, and it is usually helpful to control for varying column totals by figuring percentages for the table. Calculating percentages also helps to make the pattern of association more visible.

To figure percentages for a bivariate table to display the effect of X on Y, the general rule is to calculate percentages in the direction of the independent variable and then compare in the opposite direction. When the independent variable has been arrayed in the columns, compute the percentages within each column separately. Table 13.2 presents percentages calculated in the direction of the independent variable based on the data in Table 13.1.

Note that in Table 13.2, the row marginals have been omitted and the column marginals are reported in parentheses. Besides controlling for any differences in column totals, the table in percentage form is usually easier to read, and so changes in the conditional distributions of Y are easier to detect.

We can readily see that the largest cell changes position from column to column. For workers who were low on satisfaction, the single largest cell is in the top row (low on productivity). For the middle column (moderate on satisfaction), the largest cell is in the middle row (moderate on productivity), and, for the right-hand column (high on satisfaction), it is in the bottom row (high on productivity). Even a cursory glance at the extent to which the conditional distributions of Y change in Table 13.2 reinforces our conclusion that an association does exist between these two variables.

If two variables are not associated, then the conditional distributions of Y will not change across the columns. The distribution of Y would be the same for each condition of X. Table 13.3 illustrates a "perfect nonassociation" between the variables hair color and productivity for assembly-line workers.

TABLE 13.2   PRODUCTIVITY BY JOB SATISFACTION (in percentages)

Productivity (Y)	Job Satisfaction (X)		
	Low	Moderate	High
Low	50.00%	34.43%	13.46%
Moderate	33.33	40.98	34.62
High	16.67	24.59	51.92
	100%	100%	100%
	(60)	(61)	(52)

TABLE 13.3   PRODUCTIVITY BY HAIR COLOR (an illustration of no association)

Productivity (Y)	Hair Color (X)		
	Brown	Blonde	Red
Low	33.33%	33.33%	33.33%
Moderate	33.33	33.33	33.33
High	33.33	33.33	33.33
	100%	100%	100%

Table 13.3 is only one of many possible patterns that would indicate "no association"; but the important point is that, comparing across the table from column to column, the conditional distributions of Y are the same. Levels of productivity do not change at all for the various hair colors and, therefore, no association exists between these variables. Also, the obtained chi square computed from this table would have a value of zero, again indicating no association between these two variables.

How Strong Is the Association?   Once we establish the existence of the association, we need to develop some idea of how strong the association is. This is essentially a matter of determining the amount of change in the conditional distributions of Y. At one extreme, of course, there is the case of "no association" where the conditional distributions of Y do not vary at all (see Table 13.3). At the other extreme is the case of "perfect association"—the strongest possible relationship between X and Y. In general, a perfect association exists between two variables if each value of Y is associated with one and only one value of X.* In a bivariate table, all cases in each column would

*Each measure of association that will be introduced in the following chapters incorporates its own definition of a "perfect association," and these definitions vary somewhat depending on the specific logic and mathematics of the statistic. That is, for different measures computed from the same table, some measures will possibly indicate perfect relationships when others will not. We will note these variations in the mathematical definitions of a perfect association at the appropriate time.

TABLE 13.4 PRODUCTIVITY BY HAIR COLOR (an illustration of perfect association)

Productivity (Y)	Hair Color (X)		
	Brown	Blonde	Red
Low	0%	0%	100%
Moderate	0	100	0
High	100	0	0
	100%	100%	100%

be located in a single cell. There would be no variation in Y for a given value of X, as illustrated in Table 13.4.

A perfect relationship would be taken as very strong evidence of a causal relationship between the variables, at least for the particular sample at hand. In fact, the results presented in Table 13.4 would indicate that, for this sample, hair color is the sole cause of productivity. Also, in the case of a perfect relationship, predictions from one variable to the other could be made without error. If we knew that a particular worker had red hair, for example, we could precisely predict that that worker would be highly productive.

Of course, the huge majority of relationships will fall somewhere between these two extremes of no association and perfect association. We need to develop some way of describing these intermediate relationships consistently and meaningfully. For example, Tables 13.1 and 13.2 show that there is an association between productivity and job satisfaction. How could this relationship be described in terms of strength? How close is the relationship to perfect? How far away from no association?

Although you might be able to describe this association impressionistically, the task would certainly be easier if you had a set of objective indices to consistently describe the strength of the association (for example, as strong, moderate, or weak). As we shall see in Chapters 14–16, these objective indices of the strength of a relationship are supplied by the various measures of association. Virtually all of these statistics are designed so that they have a lower limit of 0 and an upper limit of 1 (± 1 for ordinal and interval-ratio measures of association). A measure that equals 0 indicates no association between the variables (the conditional distributions of Y do not vary), and a measure of 1 (± 1 in the case of ordinal and interval-ratio measures) indicates a perfect relationship. The exact meaning of values between 0 and 1 varies from measure to measure, but, for all measures, the closer the value is to 1.00, the stronger the relationship (the greater the change in the conditional distributions of Y).

Since you may be tempted to regard the results of the chi square test as evidence of the strength of the relationship, I will take this opportunity to

APPLICATION 13.1

Why are many Americans attracted to movies that emphasize graphic displays of violence? One idea is that "slash" movie fans feel threatened by violence in their daily lives and use these movies as a means of coping with their fears. In the safety of the theater, violence can be vicariously experienced, and feelings and fears can be expressed privately. Also, such movies almost always, as a necessary plot element, provide a role model of one character who does deal with violence successfully (usually, of course, with more violence).

Is fear of violence associated with frequent attendance at high-violence movies? The following table reports the joint frequency distributions of "fear" and "attendance" in percentages for a sample of 600.

Attendance	Fear		
	Low	Moderate	High
Rare	50%	20%	30%
Occasional	30	60	30
Frequent	20	20	40
	100%	100%	100%
	(200)	(200)	(200)

The conditional distributions of attendance (Y) do change across the values of fear (X), so these variables are associated. The clustering of cases in the diagonal from upper left to lower right suggests a substantial relationship in the predicted direction. People who are low on fear are infrequent attenders, people who are high on fear are frequent attenders. These results do suggest an important relationship between fear and attendance.

Notice that the results, as presented here, pose an interesting causal problem. The table supports the idea that fearful and threatened people attend violent movies as a coping mechanism (X causes Y). However, the table is also consistent with the reverse causal argument: attendance at violent movies increases fears for one's personal safety (Y causes X). The results support both causal arguments, and I remind you that association is not the same thing as causation.

remind you once again that significance and association are two separate matters. Chi square, by itself, is not a measure of association. While nonzero values for chi square indicate the existence of an association, the actual numerical value for obtained chi square bears no necessary relationship to the strength of the association. That is, the obtained chi square might be quite large in value while the actual association is weak. In Chapter 14, I will introduce some ways to transform chi square into other statistics that do measure the strength of the association between two variables.

What Is the Pattern and/or the Direction of the Association? Investigating the pattern of the association requires that we ascertain which values or categories of one variable are associated with which values or cate-

READING STATISTICS 6: BIVARIATE TABLES

The conventions for constructing and interpreting bivariate tables presented in this text are commonly but not universally followed in the professional literature. Tables will usually be constructed with the independent variable in the columns, the dependent variable in the rows, and percentages calculated in the columns. However, you should be careful to check the format of every table you attempt to read to see if these conventions have been observed. If the table is presented with the independent variable in the rows, for example, you will have to reorient your analysis (or redraw the table) to account for this. Above all, you should convince yourself that the percentages have been calculated in the correct direction. It is possible for even skilled professionals to calculate percentages incorrectly and potentially misinterpret the table as a consequence.

Once you have assured yourself that the table is properly presented, you can apply the analytical techniques developed in this chapter. By comparing the conditional distributions of the dependent variable, you can ascertain for yourself if the variables are associated and check the strength and pattern of the association. You may then compare your conclusions with those of the researchers.

As an aid in the interpretation of bivariate tables, researchers will usually compute and report other statistics in addition to percentages. We'll talk about these statistics in Reading Statistics 7 in Chapter 15.

Statistics in the Professional Literature:

In 1989, a group of sociologists and lawyers began a study of community standards regarding pornography and obscenity. The results of the study were used in a court case involving charges of pornography in Charlotte, North Carolina. One of the legal criteria currently used to define pornography across the nation is that a judge or jury must determine that the ". . . average person, applying contemporary community standards, would find that the work, taken as a whole, appeals to . . . prurient interests." This standard is one of three established by the U.S. Supreme Court in *Miller vs. California* (1973).

The researchers gathered an EPSEM sample of 129 residents of the Charlotte area, and each respondent viewed either one of the sexually explicit films that were the subjects of the litigation or a nonexplicit control film. Each subject filled out a survey measuring their views of community standards before and after viewing the films. The table below presents some of the results of this study.

Note that the tables are constructed with the independent variable in the columns and the dependent variable in the rows. Percentages are calculated within the columns, and the cell frequencies are reported as well. In the top and (especially) in the bottom tables the conditional distributions change. There is no change in the middle table. For the top and bottom tables, opinions are affected by viewing the movies. The opinion measured in the middle table was not affected by the film viewing.

What do these results suggest about community standards? What do they suggest about people's attitudes towards sexually explicit material after actually viewing the films? Here's part of how the authors interpreted these results:

Comparison of "Before" and "After" Film Viewing (X-rated film viewing subjects only) on Beliefs About Community Tolerance, Viewing Materials, and Appeal to Shameful and Morbid Interest in Sex.[1]

	Before	After
"Do you think it *is* or *is not* tolerated in your community for the average adult to obtain and see adult movies, video cassettes, and magazines showing nudity and sex if they want to?"		
Tolerated	59.1%	52.7%
N	65	58
Not Tolerated	40.9%	47.3%
N	45	52
"Do you believe you *should* be able to see any such showing of actual sex acts in adult movies, video cassettes, or magazines if you want to?"		
Definitely Yes	74.5%	75.2%
N	82	86
Definitely No	25.5%	24.8%
N	28	24
"Some adult movies, video cassettes, and magazines show actual sex acts in great detail and with close-ups of sexual organs. Would viewing this type of material appeal to an unhealthy, shameful, or morbid interest in sex?"		
Definitely Yes	43.6%	16.4%
N	48	18
Definitely No	56.4%	83.6%
N	62	92

[1] Adapted from Table 4, page 96.

. . . a majority . . . believe that . . . (adult movies, video cassettes, and magazines are) . . . tolerated in their community . . . Seventy-five percent believe they personally should be able to obtain and see any such showing of (these materials), and 16% believe viewing the type of material charged in the case appealed to unhealthy, shameful, or morbid interest in sex. (P)re- and post-film . . . responses on the belief that these materials are tolerated showed no significant change (the test of significance used was chi square). Likewise, there was no statistically significant change in subjects' opinions about whether they should be allowed to see such materials. There was, however, a statistically significant change in . . . perception that the materials appealed to unhealthy, shameful, or morbid interest in sex.

In the latter case, the percentage drops from 43.6% to only 16.4%.

Source: Daniel Linz, et al.: 1991. "Estimating Community Standards: The Use of Social Science Evidence in an Obscenity Prosecution." *Public Opinion Quarterly*, Vol. 55: 80–112. Reprinted by permission.

TABLE 13.5   LIBRARY USE BY EDUCATION (an illustration of a positive relationship)

Library Use	Education		
	Low	Moderate	High
Low	60%	20%	10%
Moderate	30	60	30
High	10	20	60
	100%	100%	100%

gories of the other. We have already remarked on the pattern of the relationship between productivity and satisfaction. Table 13.2 indicates that low scores on satisfaction are associated with low scores on productivity, moderate satisfaction with moderate productivity, and high satisfaction with high productivity.

When both variables are at least ordinal in level of measurement, the pattern of association may also have a direction to it.* The direction of the association can be either positive or negative.

An association is positive if the variables vary in the same direction. That is, in a **positive association**, high scores on one variable are associated with high scores on the other variable and low scores on one variable are associated with low scores on the other. Still another way to express this relationship would be to say that, in a positive association, as one variable increases in value, the other also increases; and, as one variable decreases in value, the other also decreases. Table 13.5 displays, with fictitious data, a positive relationship between education and use of public libraries. As education increases, library use also increases. The association between job satisfaction and productivity, as displayed in Tables 13.1 and 13.2, is also a positive association.

In a **negative association**, the variables vary in opposite directions. High scores on one variable are associated with low scores on the other, and increases in one variable are accompanied by decreases in the other. Table 13.6 displays a negative relationship, again with fictitious data, between education and television viewership. The amount of television viewing decreases as education increases.

Measures of association for ordinal and interval-ratio variables are designed so that they will take on positive values for positive associations and negative values for negative associations. Thus, a measure of association preceded by a plus sign indicates a positive relationship between the two vari-

---

*Variables measured at the nominal level have no numerical order to them (by definition). Therefore, associations including nominal-level variables, while they may have a pattern, cannot be said to have a direction.

TABLE 13.6   AMOUNT OF TELEVISION VIEWING BY EDUCATION (an illustration of a negative relationship)

Television Viewing	Education		
	Low	Moderate	High
Low	10%	20%	60%
Moderate	30	60	30
High	60	20	10
	100%	100%	100%

ables, with the value $+1.00$ indicating a perfect positive relationship. A negative sign indicates a negative relationship, with -1.00 indicating a perfect negative relationship.

SUMMARY

1. The analysis of associations between variables is another statistical technique. Whereas tests of hypotheses are designed to detect nonrandom relationships, measures of association are designed to quantify the importance or strength of a relationship.

2. There are three characteristics of an association: the existence of an association, the strength of the association, and the direction or pattern of the association. These three characteristics can be investigated by calculating percentages for a bivariate table in the direction of the independent variable and then comparing in the opposite direction.

3. Tables 13.1 and 13.2 can be analyzed in terms of these three characteristics. Clearly, a relationship does exist between job satisfaction and productivity, since the conditional distributions of the dependent variable (productivity) are different for the three different conditions of the independent variable (job satisfaction). Even though we cannot quantify the strength of the relationship yet, we can see that the association is substantial in that the change in Y (productivity) across the three categories of X (satisfaction) is marked. Furthermore, the relationship is positive in direction. Productivity increases as job satisfaction rises, and workers who report high job satisfaction tend to be also high on productivity. Workers with little job satisfaction tend to be low on productivity.

4. Given the nature and strength of the relationship, it could be predicted with fair accuracy that highly sat-

isfied workers would tend to be highly productive ("happy workers are busy workers"), and these results might be taken as evidence of a causal relationship between these two variables. However, these results could not be taken as proof of a causal relationship. These results show only that the variables are associated, and association is not the same thing as causation. In fact, although we have presumed that job satisfaction is the independent variable in the relationship, we could possibly have argued the reverse causal sequence ("busy workers are happy workers"). The results presented in Tables 13.1 and 13.2 are consistent with both causal arguments. Thus, the analysis of the association between variables may be seen as a very important process by which evidence for (or against) a causal relationship is systematically gathered. Ultimate proof for causal relationships depends more on logical, theoretical, and methodological grounds (actually proving causation is a rather difficult task). As we shall see in Part IV, some of the multivariate techniques are quite useful for probing possible causal relationships, and we will return to some of these concerns at that point.

GLOSSARY

Association. The relationship between two (or more) variables. Two variables are said to be associated if the distribution of one variable changes for the various categories or scores of the other variable.

Conditional distribution of Y. The distribution of scores on the dependent variable for a specific score or category of the independent variable when the variables have been organized into table format.

Dependent variable. In a bivariate relationship, the variable that is taken as the effect.

Independent variable. In a bivariate relationship, the variable that is taken as the cause.

Measures of association. Statistics that quantify the strength of the association between variables.

Negative association. A bivariate relationship where the variables vary in opposite directions. As one variable increases, the other decreases, and high scores on one variable are associated with low scores on the other.

Positive association. A bivariate relationship where the variables vary in the same direction. As one variable increases, the other also increases, and high scores on one variable are associated with high scores on the other.

X. Symbol used for any independent variable.

Y. Symbol used for any dependent variable.

PROBLEMS

13.1 In any social science journal, find an article that includes a bivariate table. Inspect the table and the related text carefully and answer the following questions:

a. Identify the variables in the table. What values (categories) does each possess? What is the level of measurement for each variable?

b. Is the table in percentage form? In what direction? Are comparisons made between columns or rows?

c. Is one of the variables identified by the author as independent? Are the percentages in the direction of the independent variable?

d. How is the relationship characterized by the author in terms of the strength of the association? In terms of the direction (if any) of the association?

e. Find the measure of association (if any) calculated for the table. What is the numerical value of the measure? What is the sign (if any) of the measure?

13.2 In Chapter 11, two bivariate tables (Tables 11.3 and 11.5) were used to illustrate the logic and computation of chi square. Turn to these tables

and calculate percentages in the direction of the independent variable. Is there an association between the variables? What are your impressions of the strength of the association? What is the pattern of the association?

13.3 For each problem in Chapter 11, calculate percentages in the direction of the independent variable. Is there an association between the variables? Characterize each association in terms of the strength and direction of the relationship.

13.4 $\boxed{\text{SOC}}$ The table below shows the relationship between place of birth and racial prejudice for a sample of white Americans. Calculate percentages for the table in the direction of the independent variable. Characterize the relationship in terms of strength. Which region seems to be associated with high levels of prejudice? Which with low levels of prejudice?

	Place of Birth			
Prejudice	North	South	Midwest	West
Low	10	5	30	25
Moderate	20	10	10	15
High	10	20	10	5
	40	35	50	45

13.5 $\boxed{\text{SOC}}$ A researcher hypothesizes that physical attractiveness is related to academic achievement (that is, the more attractive the student, the higher the grade). One hundred students are rated on physical attractiveness and asked for their grade-point averages. The bivariate table below summarizes the data. Calculate percentages for the table in the proper direction and compare the conditional distributions. Is there an association between these variables? What is your impression of the strength of the association? What is the direction of the association? Is the hypothesis supported? Why or why not?

	Attractiveness		
GPA	Low	Moderate	High
Low	7	8	15
Moderate	10	10	16
High	8	12	14
	25	30	45

13.6 [PA] Various supervisors in the city government of Shinbone, Kansas, have been rated on the extent to which they practice authoritarian styles of leadership and decision making. The efficiency of each department has also been rated, and the results are summarized below. Figure percentages for the table so that it shows the effect of leadership style on efficiency. Is there an association between these two variables? What is the strength and direction of the relationship?

Efficiency	Authoritarianism	
	Low	High
Low	10	12
High	17	5
	27	17

13.7 [SOC] The latest fad to sweep college campuses is streaking to panty raids while swallowing live goldfish. A researcher is interested in how closely the spread of this bizarre form of behavior is linked to the amount of coverage and publicity provided by local campus newspapers. For a sample of 25 universities, the researcher has rated the amount of press coverage (as extensive, moderate, or no coverage) and how much the student body was involved in this new fad. The data for each campus are reported below. Organize the data into a properly labeled table in percentage form. Does the table indicate an association between press coverage and fad behavior?

Campus	Amount of Press Coverage	Extent of Student Involvement
1	Extensive	Extensive
2	Extensive	Some
3	Moderate	Some
4	Moderate	Some
5	Moderate	Extensive
6	Extensive	Some
7	Extensive	Extensive
8	Moderate	Some
9	None	Some
10	Moderate	None
11	None	None
12	Extensive	Extensive
13	None	Some
14	Extensive	Extensive
15	Moderate	None
16	Moderate	Some
17	Moderate	Extensive
18	Moderate	None
19	None	None
20	Extensive	Some
21	None	Extensive
22	Moderate	None
23	None	None
24	Extensive	Extensive
25	Moderate	Extensive

13.8 [SOC] Twenty-seven men and women were asked if they favored (F) or opposed (O) the death penalty. Is there any association between gender and support for capital punishment?

Subject	Gender	Support for Capital Punishment
1	M	F
2	M	F
3	M	F
4	M	F
5	M	O
6	M	O
7	M	O
8	M	F
9	M	O
10	M	F
11	M	O
12	M	O
13	M	F
14	F	F
15	F	O
16	F	F
17	F	F
18	F	O
19	F	F
20	F	F
21	F	O
22	F	O
23	F	O
24	F	F

Subject	Gender	Support for Capital Punishment
25	F	O
26	F	F
27	F	O

13.9 PS How consistent are people in their voting habits? Do people vote for the same party from election to election? Below are the results of a poll in which people were asked if they had voted Democrat or Republican in each of the last two presidential elections. Assess the strength of this relationship.

1992 Election	1988 Election	
	Democrat	Republican
Democrat	117	23
Republican	17	178
	134	201

13.10 SOC Is interest in sports associated with social class? Write a few sentences describing the relationship in terms of pattern and strength of the association.

Interest	Working Class	Middle Class	Upper Class
Low	21	14	21
Moderate	24	40	23
High	12	45	7
	57	99	51

13.11 CJ About half the neighborhoods in a large city have instituted programs to increase citizen involvement in crime prevention. Do these areas experience less crime? Write a few sentences describing the relationship in terms of pattern and strength of the association.

Crime Rate	Program	
	No	Yes
Low	29	15
Moderate	33	27
High	52	45
	114	87

13.12 GER A needs assessment survey has been distributed in a large retirement community. Residents were asked to check off the services or programs they thought should be added. Is there any association between gender and the perception of a need for more social occasions? Write a few sentences describing the relationship in terms of pattern and strength of the association.

More Parties?	Male	Female
Yes	321	426
No	175	251
	496	677

13.13 PS Are people who think of themselves as political liberals really more liberal on social issues? The table below summarizes the relationship between political ideology and the respondent's position on the legalization of marijuana. Are the variables associated? Write a few sentences describing the relationship in terms of pattern and strength of the association.

Should Marijuana Be Legal?	Liberals	Moderates	Conservatives
Yes	132	78	52
No	101	87	109
	233	165	161

13.14 SW As the state director of mental health programs, you note that some local mental health facilities have very high rates of staff turnover. You believe that part of this problem is a result of the fact that some of the local directors have very little training in administration and poorly developed leadership skills. Before implementing a program to address this problem, you collect some data to make sure that your beliefs are supported by the facts. Is there a relationship between staff turnover and the administrative experience of the directors? Describe the relationship in terms of pattern and strength of the association.

Turnover	Director Experienced? No	Yes
Low	4	9
Moderate	9	8
High	15	5
	28	22

13.15 SOC Let's see if political ideology is associated with a variety of different issues and concerns. The tables below are taken from the General Social Survey for 1993. Political ideology (item 14) has been collapsed into just three categories and is used as the "column," or independent, variable. For each table, compute percentages and describe the pattern and strength of the relationship. Overall, is political ideology an important factor in shaping people's views and opinions?

a. Attitudes about welfare, by political ideology

Spending on Welfare (item 22)	Liberal	Moderate	Conservative
Too little	22	21	16
About right	26	35	27
Too much	46	77	80
	94	133	123

b. Attitudes about the role of the United States in world affairs, by political ideology.

Should U.S. be Active? (item 27)	Liberal	Moderate	Conservative
Active	94	131	125
Stay out	42	58	38
	136	189	163

c. Attitudes about spending for national health care

Spending on Health (item 23)	Liberal	Moderate	Conservative
Too little	80	105	75
About right	15	18	27
Too much	3	10	19
	98	133	121

d. Attitudes towards pornography, by political ideology

Pornography causes breakdown. (item 59)	Liberal	Moderate	Conservative
Yes	68	121	126
No	62	41	52
	130	162	178

e. Attitudes toward women's roles, by political ideology

Women Should Take Care of Home and Family. (item 69)	Liberal	Moderate	Conservative
Strongly agree	6	9	15
Agree	37	51	60
Disagree	58	101	71
Strongly disagree	36	30	22
	137	191	168

SPSS/PC+ PROCEDURES FOR ANALYZING BIVARIATE ASSOCIATION

DEMONSTRATION 13.1 Does Support for the Death Penalty Vary by Political Ideology?

In Chapter 11, we used the CROSSTABS procedure to conduct the chi square test. Here we will use the same procedure to investigate the strength and direction of the association between two variables. The example below concerns the relationship between support for the death penalty (CAPPUN, or item 24, in Appendix G) and political ideology (POLVIEWS, or item 19). Given the way in which this issue is usually debated, we would expect liberals to oppose the death penalty, conservatives to favor it, and moderates to be intermediate in support.

To reduce the number of columns to a manageable number, I recoded POLVIEWS into three categories:

```
RECODE POLVIEWS (1 thru 3 = 1)(4 = 2)(5 thru 7 = 3).
VALUE LABELS 1 'Liberal' 2 'Moderate' 3
'Conservative'/.
```

The CROSSTABS procedure is accessed by following the steps explained in Demonstration 11.1. Specify CAPPUN as the row variable, or dependent variable, and POLVIEWS as the column, or independent, variable. The only new part of the command we need to add is 'OPTIONS 4'. This OPTION will produce column percentages so that we can observe the effect of POLVIEWS on CAPPUN. (Take a minute to check out the list of OPTIONS. Among other things, they control the way in which the percentages are calculated and missing values are handled.) For the sake of curiosity, I also requested a chi square on the table (/STATISTICS 1) even though I'm not really interested in a formal test of hypothesis at this point. The CROSSTABS commands is

```
CROSSTABS /TABLES CAPPUN BY POLVIEWS /OPTIONS 4
/STATISTICS 1.
```

In preparing these commands, be sure to name your dependent variable first. This will place the dependent variable in the rows and the independent variable in the columns, and we will be able to read the table by following the rules developed in this chapter. The table produced by the commands above is

```
Cross-tabulation:     CAPPUN     FAVOR OR OPPOSE DEATH PENALTY FOR MURDER
            By POLVIEWS   THINK OF SELF AS LIBERAL OR CONSERVATIVE

                    Count | liberal | moderate | conserva |
         POLVIEWS=> Col Pct |        |          | tive     | Row
                    |    1.00 |    2.00 |    3.00 | Total
         CAPPUN     ------ | -------- | -------- | --------
                    1.00 |   138   |   197    |   210    |   545
         FAVOR           |  67.3   |  80.1    |  81.1    |  76.8
                    | -------- | -------- | --------
                    2.00 |    67   |    49    |    49    |   165
         OPPOSE          |  32.7   |  19.9    |  18.9    |  23.2
                    | -------- | -------- | --------
                  Column     205       246       259       710
                   Total    28.9      34.6      36.5     100.0
```

```
Chi-Square  D.F.  Significance   Min. E.F.   Cells with E.F.< 5
----------  ----  ------------   ---------   ------------------
 14.47931     2      .0007        47.641         None
Number of Missing Observations = 78
```

In each cell, the count is reported first. For example, there are 138 cases in the upper left-hand cell. These are people who are liberal and are in favor of the death penalty. Below the count is the percentage of all the cases in the column in that cell. So, 67.3% of the liberals favor the death penalty, as compared with 80.1% of the moderates and 81.1% of the conservatives. These results are similar to what we expected and suggest that the variables are associated. This conclusion is reinforced by the fact that chi square is a non-zero value and, in fact, is significant at less than .05 ("Significance" is reported as .0007, to be exact). The differences in column percentages aren't very great, and the sample as a whole is fairly homogeneous on the issue. The majority of cases in each column are in favor of the death penalty, and there is not much variation in support from column to column. The relationship between support for the death penalty and political ideology is, at best, moderate in strength.

DEMONSTRATION 13.2 The Direction of Relationships: Do Attitudes Towards Sex or the Frequency of Traumatic Events Vary by Age?

Let's take a look at a positive and a negative relationship so that you can develop some experience in describing the direction as well as the strength of bivariate relationships. Both our variables have to be at least ordinal in level of measurement, so let's start with AGE as an independent variable and RECODE it into, for example, three categories, as in Demonstration 12.1:

```
RECODE AGE (LO Thru 36=1)(37 Thru 51=2)(52 Thru
HI=3).
VALUE LABELS AGE 1'36 and Younger' 2 '37 to 51'
3 '52 and Older'/.
```

If you've forgotten, I've used these unusual cutting points to divide the sample into three roughly equal categories.

Without looking at the data, I'm willing to bet that AGE will have a positive relationship with TRAUMA5 (item 70, the number of traumatic events occurring to the respondent within the past year) and a negative relationship with approval of premarital sex (PREMARSX, item 56). We need to RECODE these items in order to have tables with a reasonable number of dimensions. I used the FREQUENCIES program to find convenient cutting points and developed the following RECODE statements:

```
RECODE TRAUMA5 (0=0)(1=1)(2 thru 4=2).
VALUE LABELS TRAUMA5 0 'None' 1 'One' 2 '2 - 4'.
```

I saved myself a little trouble by specifying the two dependent variables in the same command. To do this, type both variables names in the typing window before pressing ENTER. Your command should look like this:

```
CROSSTABS /TABLES TRAUMA5 PREMARSX BY AGE/OPTIONS 4
/STATISTICS 1.
```

The bivariate table reflecting the relationship between TRAUMA5 and AGE is

```
              By AGE        AGE OF RESPONDENT

          Count  : 36 & you : 37 - 51 : 52 +     :
  AGE=>   Col Pct : nger    :         :         : Row
                  :    1.00 :    2.00 :    3.00 : Total
TRAUMA5   ------ : -------- : -------- : --------
            .00  :     64  :     47  :     34  :    145
  none           :   37.4  :   29.9  :   20.2  :   29.2
                 : -------- : -------- : ---------
           1.00  :     62  :     67  :     79  :    208
  one            :   36.3  :   42.7  :   47.0  :   41.9
                 : -------- : -------- : --------
           2.00  :     45  :     43  :     55  :    143
  2 - 4          :   26.3  :   27.4  :   32.7  :   28.8
                 : -------- : -------- : --------
          Column     171       157       168       496
          Total     34.5      31.7      33.9     100.0
```

```
Chi-Square   D.F.   Significance   Min. E.F.   Cells with E.F.< 5
----------   ----   ------------   ---------   -------------------
 12.35395      4        .0149        45.264        None
Number of Missing Observations = 292
```

Is there a relationship between these two variables? Do the conditional distributions change? Inspect the table column by column; I think you will agree that there is a relationship. Is this a positive or negative relationship? Remember that in a positive association, high scores on one variable will be associated with high scores on the other and low scores will be associated with low. In a negative relationship, high scores on one variable are associated with low scores on the other.

Now go back and look at the table. Find the single largest cell in each column and see if you can detect the pattern. For the youngest age group, the "zero trauma" (or lowest) score is most common (37.4%). For the middle and oldest groups, the middle score is most common (42.7% and 47.0%, respectively). Also note that the oldest group reports the highest percentage of cases in the bottom row, the highest level of trauma. This is a *positive* relationship, with older age groups reporting higher rates of personal trauma. The frequency of traumatic events increases with age.

What about the second relationship? The table is

```
Cross-tabulation:              PREMARSX    SEX BEFORE MARRIAGE
                          By AGE           AGE OF RESPONDENT

                   Count : 36 & you : 37 - 51 : 52 +      :
      AGE=>       Col Pct : nger     :         :          :        Row
                          :    1.00  :    2.00 :     3.00 : Total
PREMARSX         -------  : -------- : -------- : --------
                    1.00  :      32  :      39  :      68  :    139
   ALWAYS WRONG          :    18.5  :    22.2  :    41.5  :   27.1
                          : -------- : -------- : --------
                    2.00  :      16  :      16  :      24  :     56
   ALMOST ALWAYS WRG     :     9.2  :     9.1  :    14.6  :   10.9
                          : -------- : -------- : --------
                    3.00  :      37  :      37  :      29  :    103
   SOMETIMES WRONG       :    21.4  :    21.0  :    17.7  :   20.1
                          : -------- : -------- : --------
                    4.00  :      88  :      84  :      43  :    215
   NOT WRONG AT ALL      :    50.9  :    47.7  :    26.2  :   41.9
                          : -------- : -------- : --------
                  Column       173       176       164       513
                   Total      33.7      34.3      32.0     100.0
```

```
                                                             MORE
Chi-Square   D.F.  Significance   Min. E.F.   Cells with E.F.< 5
----------   ----  ------------   ---------   ------------------
  36.91395     6      .0000        17.903       None
Number of Missing Observations = 275
```

The conditional distributions change across the table, so these variables are related. To determine the direction of the relationship, proceed column by column and find the largest cell frequency for each value of the independent variable. For the youngest age group, the largest cell is in the bottom row (50.9%), which is the highest score. For the middle age group, the largest cell is also in the bottom row (47.7%), but the cell has a lower percentage of cases than the youngest age group. For the oldest group, the largest cell is in the top row (41.5%), which represents the lowest score on the row variable. As age increases, score on PREMARSX decreases and this is a negative relationship. The oldest group is the most likely to say that premarital sex is "always wrong." As age increases, approval of premarital sex decreases.

Exercises

13.1 As long as POLVIEWS has already been recoded, pick a few more "social issues" (such as GRASS, GUNLAW, or BUSING) and, with POLVIEWS as the independent variable, see if the patterns of association conform to expectations. Are "liberals" really more liberal on these issues? For each table, be sure to request OPTIONS 4 and the chi square test. For each table, write a sentence or two of interpretation.

13.2 See if you can assess the strongest determinant of attitudes on issues such as CAPPUN, GRASS, GUNLAW, and BUSING. You already have results for POLVIEWS as an independent variable. Run the same dependent or row variables against SEX, RELIG, and RACE and compare the strength and direction or pattern of the relationships with each other and with POLVIEWS. Recode either DEGREE, INCOME, or PRESTG80 as a measure of class and compare its effects with the other independent variables. Which independent has the strongest effect on these "social issue" items?

14

Association Between Variables Measured at the Nominal Level

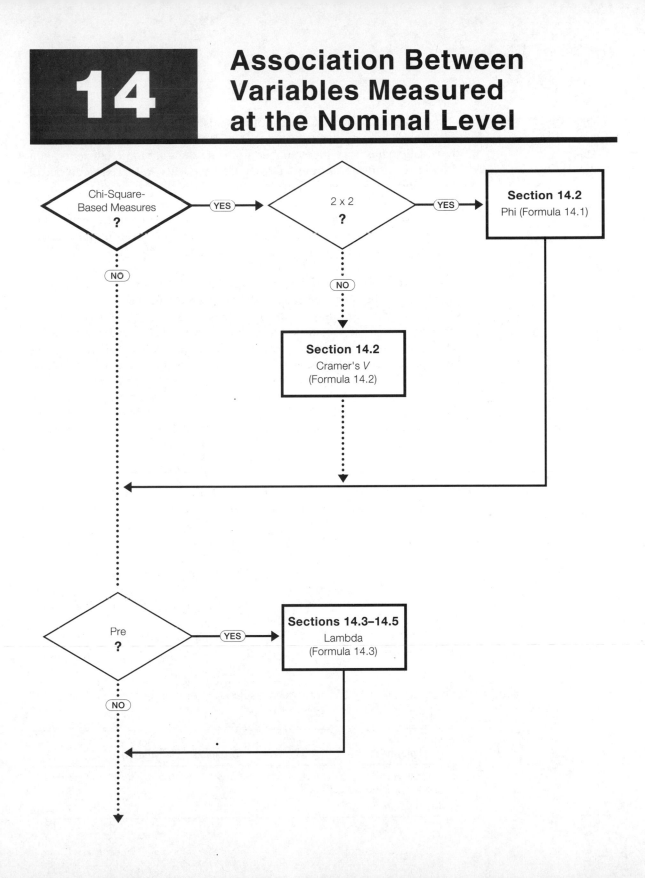

14.1 INTRODUCTION

Measures of association are descriptive statistics that summarize the overall strength of the association between two variables. Because they represent that relationship in a single number, these statistics are more efficient methods of expressing an association than the previously considered technique of calculating percentages for bivariate tables. As with any summarizing technique, however, a certain amount of information (detail and nuance) about the relationship will be lost if the researcher considers only the measure of association. Make a habit of inspecting the patterns of cell frequencies or percentages in the table along with the summary measure of association in order to maximize the amount of information you have about the relationship. This practice should be followed regardless of the level of measurement of the data or the specific measure that has been calculated.

As we shall see, there are many measures of association. In this text, I have organized these statistics according to the level of measurement for which they are most appropriate. In this chapter, we will consider measures appropriate for nominally measured variables. Several of the research situations used as examples involve variables measured at different levels (for example, one nominal-level variable and one ordinal-level variable). The general procedure in the situation of "mixed levels" is to be conservative and select measures of association appropriate for the lower of the two levels of measurement.

14.2 CHI SQUARE–BASED MEASURES OF ASSOCIATION

Over the years, statisticians have relied heavily on a group of measures of association, all based on the value of chi square. Since, ordinarily, the value of chi square will have already been obtained, these measures are quite easy to calculate and are, perhaps, most useful as "quick and dirty" indicators of the strength of a relationship.

To illustrate, let us reconsider Table 11.3, which displayed, with fictitious data, a relationship between accreditation status of undergraduate programs and employment patterns of social work majors. For the sake of convenience, this table is reproduced here as Table 14.1.

As we saw in Chapter 11, this relationship was statistically significant. The chi square test statistic was 10.78, which was significant at an alpha of

TABLE 14.1 EMPLOYMENT STATUS OF SOCIAL WORK MAJORS BY ACCREDITATION STATUS OF UNDERGRADUATE PROGRAM (Fictitious data)

Employment	Accredited	Not Accredited	
Employed as social worker	30	10	40
Not employed as social worker	25	35	60
	55	45	100

TABLE 14.2   EMPLOYMENT STATUS BY ACCREDITATION STATUS (percentages)

Employment	Accredited	Not Accredited
Employed as social worker	30%	22%
Not employed as social worker	45	78
	100%	100%
	(55)	(45)

.05. The question now concerns the strength of the association. A brief glance at Table 14.1 shows that the conditional distribution of employment status does change, so the variables are associated. To emphasize this point, it is always helpful to calculate the percentages, as in Table 14.2.

So far, we know that the relationship between these two variables is statistically significant and that there is an association of some kind between accreditation and employment. To assess the strength of the association, we will compute a **phi** (ϕ). This statistic is a frequently used chi square–based measure of association appropriate for tables with either two rows or two columns. Here is the formula for phi:

FORMULA 14.1

$$\phi = \sqrt{\frac{\chi^2}{N}}$$

Phi, being simply the square root of the value of the obtained chi square divided by N, is quite easy to calculate. For the data displayed in Table 14.1, the chi square was 10.78. Therefore, phi is

$$\phi = \sqrt{\frac{\chi^2}{N}}$$

$$\phi = \sqrt{\frac{10.78}{100}}$$

$$\phi = .33$$

For a 2 × 2 table, phi ranges in value from 0 (no association) to 1.00 (perfect association). The phi of .33 calculated above indicates a weak to moderate relationship between accreditation and employment status. The relationship between these two variables is statistically significant but not particularly strong. As for the pattern of the association, we can see from Table 14.2 that there is a tendency for graduates of accredited programs to be employed more often as social workers.

For tables with three or more columns and three or more rows, the upper limit of phi can exceed 1.00. Since this makes phi difficult to interpret, a more general form of the statistic called **Cramer's V** must be used for larger tables. Here is the formula for Cramer's V:

FORMULA 14.2

$$V = \sqrt{\frac{\chi^2}{(N)(\text{minimum of } r - 1, c - 1)}}$$

where (minimum of $r - 1$, $c - 1$) = the minimum value of $r - 1$ (number of rows minus 1) or $c - 1$ (number of columns minus 1)

To calculate V, find the square root of chi square divided by the quantity N multiplied by the number of rows minus 1 or the number of columns minus 1, whichever is the lower (or "minimum") value. Cramer's V has an upper limit of 1.00 for any size table and, like phi, can be interpreted as an index that measures the strength of the association between two variables.

To illustrate the computation of V, suppose you had gathered the data displayed in Table 14.3, which shows the relationship between membership in student organizations and academic achievement for a sample of college students. The obtained chi square for this table is 31.5, a value that is significant at the .05 level (see Chapter 11 for a detailed account of the computation of chi square). Cramer's V is

$$V = \sqrt{\frac{\chi^2}{(N)(\text{minimum of } r - 1, c - 1)}}$$

$$V = \sqrt{\frac{31.5}{(75)(2)}}$$

$$V = \sqrt{\frac{31.5}{150}}$$

$$V = \sqrt{.21}$$

$$V = .46$$

Since Table 14.3 has the same number of rows and columns, we may use either $(r - 1)$ or $(c - 1)$ in the denominator. In either case, the value of the

TABLE 14.3 ACADEMIC ACHIEVEMENT BY CLUB MEMBERSHIP

Academic Achievement	Membership			
	Fraternity or Sorority	Other Organization	No Memberships	
Low	4	4	17	25
Moderate	15	6	4	25
High	4	16	5	25
	23	26	26	75

TABLE 14.4   ACADEMIC ACHIEVEMENT BY CLUB MEMBERSHIP (percentages)

Academic Achievement	Membership		
	Fraternity or Sorority	Other Organization	No Memberships
Low	17.4	15.4	65.4
Moderate	65.2	23.1	15.4
High	17.4	61.5	19.2
	100%	100%	100%
	(23)	(26)	(26)

denominator is N multiplied by $(3 - 1)$, or 2. The computed value of V of .46 suggests an association of moderate strength between club membership and academic achievement.

The pattern of the relationship can be assessed by calculating percentages for the table, as in Table 14.4. Fraternity and sorority members tend to be moderate, members of other organizations tend to be high, and nonmembers tend to be low in academic achievement.

A major problem with both phi and Cramer's V, as you may have noticed, is the absence of a direct or meaningful interpretation for values between the extremes of 0.00 and 1.00. Both measures are indices of the strength of the association; for example, a value of .75 indicates a stronger relationship than a value of .50. The problem is that the values between 0.00 and 1.00 cannot be interpreted as anything other than an index of the relative strength of the association. Because they are easy to calculate (once the value of chi square has been obtained), perhaps the most appropriate use for these measures is as simple first indicators of the importance of an association.*

14.3 PROPORTIONAL REDUCTION IN ERROR (PRE)

In recent years, chi square–based measures of association—once very popular—have been supplanted by a group of measures that are based on a logic known as **proportional reduction in error (PRE)**. For nominal-level variables, the logic of PRE involves first attempting to guess or predict the category into which each case will fall on the dependent variable (Y) while ignoring the independent variable (X). Since we would in effect be predicting blindly in this case, we would make many errors (that is, incorrectly predict the category of many cases on the dependent variable).

---

*Two other chi square–based measures of association, T^2 and C (the contingency coefficient), are sometimes reported in the literature. Both of these measures have serious limitations. T^2 has an upper limit of 1.00 only for tables with an equal number of rows and columns, and the upper limit of C varies depending on the dimensions of the table. These characteristics make these measures more difficult to interpret and thus less useful than phi or Cramer's V.

The second step would be to predict again the category of each case on the dependent variable but this time take the independent variable into account. If the two variables are associated, the additional information supplied by the independent variable should enable us to reduce our errors of prediction (that is, misclassify fewer cases). The stronger the association between the variables, the more we will reduce our errors. In the case of a perfect association, we would make no errors at all when predicting score on Y from score on X. When there is no association between the variables, on the other hand, knowledge of the independent will not improve the accuracy of our predictions. We would make just as many errors of prediction with knowledge of the independent variable as we did without knowledge of the independent variable.

To illustrate these principles with an example, suppose you were placed in the rather unusual position of having to predict whether each of the next 100 people you meet will be less than 5 feet 9 inches in height ("short") or 5 feet 9 inches or taller ("tall"), under the condition that you would have no knowledge of these people at all (for example, the condition that you are in a sensory deprivation tank and can use none of your senses to help you with your predictions). Clearly, your predictions will be wrong quite often (you will frequently misclassify a tall person as short and vice versa).

Now assume that you must go through this ordeal twice; but, on the second round, just prior to your prediction, you are supplied with information about the sex of the person whose height you must predict. Since height is associated with sex and females are, on the average, shorter than males, the optimal strategy would be to predict that all females are short and all males are tall. Of course, you will still make errors on this second round; but, given an association between the variables, the number of errors on the second round will be less than the number of errors on the first. That is, by taking information from an associated variable into account, you will have reduced your errors of prediction. Let us see now how these unusual thoughts can be translated into a useful statistic.

14.4 A PRE MEASURE FOR NOMINAL-LEVEL VARIABLES: LAMBDA

One hundred individuals have been categorized by gender and height, and the data are displayed in Table 14.5. That the two variables are associated can be seen even without the aid of percentages. To measure the strength of this association, a PRE measure called **lambda** (symbolized by the Greek letter λ) will be calculated. Following the logic introduced in the previous section, we must find two quantities. First, the number of prediction errors made while ignoring the independent variable (gender) must be found. Then, we will find the number of prediction errors made while taking gender into account. These two sums will then be compared to derive the statistic.

First, the information given by the independent variable (gender) can be

TABLE 14.5   HEIGHT BY GENDER

Height	Gender		
	Male	Female	
Tall	44	8	52
Short	6	42	48
	50	50	100

effectively ignored by working only with the row marginals. Two different predictions can be made by using these marginals. We can predict either that all subjects are tall or that all subjects are short.* For the first prediction (all subjects are tall), 48 errors will be made. That is, for this prediction, all 100 cases would be placed in the first row. Since only 52 of the cases actually belong in this row, this prediction would result in (100 − 52), or 48, errors. If we had predicted that all subjects were short, on the other hand, we would have made 52 errors (100 − 48 = 52). We will take the lesser of these two numbers and refer to this quantity as E_1 for the number of errors made while ignoring the independent variable. So, $E_1 = 48$.

The second step in the computation of lambda is again to predict score on Y (height), this time taking X (gender) into account. Follow the same procedure as in the first step, but this time move from column to column. Since each column is a category of X, we thus take X into account in making our predictions. For the left-hand column (males), we predict that all 50 cases will be tall and make six errors (50 − 44 = 6). For the second column (females), our prediction is that all females are short, and eight errors will be made. By moving from column to column, we have taken X into account and have made a total of 14 errors of prediction, a quantity we will label E_2 ($E_2 = 6 + 8 = 14$).

Now, the logic of lambda is that, if the variables are associated, fewer errors will be made under the second procedure than under the first. Clearly, gender and height are associated, since we made fewer errors of prediction while taking gender into account ($E_2 = 14$) than while ignoring gender ($E_1 = 48$). Our errors have been reduced from 48 to only 14. To find the proportional reduction in error, use Formula 14.3:

FORMULA 14.3

$$\lambda = \frac{E_1 - E_2}{E_1}$$

---

*Other predictions are, of course, possible, but these are the only two permitted by lambda.

For the sample problem, the value of lambda would be

$$\lambda = \frac{E_1 - E_2}{E_1}$$

$$\lambda = \frac{48 - 14}{48}$$

$$\lambda = \frac{34}{48}$$

$$\lambda = .71$$

Lambda has a possible range of 0 to 1. A lambda of 0 would indicate that the information supplied by the independent variable does not improve our ability to predict the dependent and, therefore, that there is no association between the variables. A lambda of 1.00 would mean that it was possible to predict Y without error from X.

Part of what has made PRE measures popular is that values between 0.00 and 1.00, such as the lambda of .71 calculated above, have a direct and meaningful interpretation. These values can be seen as indices of the extent to which X helps us to predict (or, more loosely, understand) Y. When multiplied by 100, the value of the lambda indicates the percentage reduction in error and, therefore, the strength of the association. Thus, the lambda above would be interpreted by concluding that knowledge of gender improves our ability to predict height by a factor of 71%. Or, we are 71% better off knowing gender when attempting to predict height than we are not knowing gender.

14.5 THE COMPUTATION OF LAMBDA

Let us work through another example in order to state the computational routine for lambda in more general terms. Suppose a researcher was concerned with the relationship between religious denomination and attitude toward capital punishment and had collected the data presented in Table 14.6 from a sample of 130 respondents.

Step 1. To find E_1, the number of errors made while ignoring X (religion, in this case), subtract the largest row total from N. For Table 14.6, E_1 will be

$$E_1 = N - \text{(largest row total)}$$

$$E_1 = 130 - 50$$

$$E_1 = 80$$

Thus, we will misclassify 80 cases on attitude toward capital punishment while ignoring religion.

Step 2. To find E_2, the number of errors made when taking the independent variable into account, first subtract the largest cell frequency from the column

TABLE 14.6   ATTITUDE TOWARD CAPITAL PUNISHMENT BY RELIGIOUS DENOMINATION (fictitious data)

	Religion			
Attitude	Catholic	Protestant	Other	None	
Favors	10	9	5	14	38
Neutral	14	12	10	6	42
Opposed	11	4	25	10	50
	35	25	40	30	130

total and then add the subtotals together. For the data presented in Table 14.6:

$$
\begin{aligned}
\text{For Catholics: } 35 - 14 &= 21 \\
\text{For Protestants: } 25 - 12 &= 13 \\
\text{For "Others": } 40 - 25 &= 15 \\
\text{For "None": } 30 - 14 &= 16 \\
\hline
E_2 &= 65
\end{aligned}
$$

A total of 65 errors are made when predicting attitude on capital punishment while taking religion into account.

Step 3. In step 1, 80 errors of prediction were made as compared to 65 errors in step 2. Since the number of errors has been reduced, the variables are associated. To find the proportional reduction in error, the values for E_1 and E_2 can be directly substituted into Formula 14.3:

$$\lambda = \frac{E_1 - E_2}{E_1}$$

$$\lambda = \frac{80 - 65}{80}$$

$$\lambda = \frac{15}{80}$$

$$\lambda = .19$$

When attempting to predict attitude toward capital punishment, we would make 19% fewer errors by taking religion into account. Knowledge of a respondent's religious denomination does improve the accuracy of our predictions (the variables are associated) but not by a very large margin. The relatively low value of this lambda might indicate that factors other than religion could be associated with the dependent variable.

As a measure of association, lambda has two characteristics that should be stressed. First, lambda is asymmetric. This means that the value of the

APPLICATION 14.1

A random sample of students at a large urban university have been classified as either "traditional" (18–23 years of age and unmarried) or "nontraditional" (24 or older or married). Subjects have also been classified as "vocational," if their primary motivation for college attendance is career or job oriented, or "academic," if their motivation is to pursue knowledge for its own sake. Are these two variables associated?

	Type		
		Non-	
Motivation	Traditional	traditional	
Vocational	25	60	85
Academic	75	15	90
	100	75	175

A lambda will be computed to measure the association between these two variables.

$$E_1 = 175 - 90 = 85$$

$$E_2 = (100 - 75) + (75 - 60) = 25 + 15 = 40$$

$$\lambda = \frac{E_1 - E_2}{E_1}$$

$$\lambda = \frac{85 - 40}{85}$$

$$\lambda = \frac{45}{85}$$

$$\lambda = 0.53$$

A lambda of 0.53 indicates that we would make 53% fewer errors in predicting motivation from student type as opposed to predicting motivation while ignoring student type. The association is moderately strong, and, by inspection of the table, we can see that traditional students are more likely to have academic motivations and nontraditional types are more likely to be vocational in motivation.

statistic will vary, depending on which variable is taken as independent. For example, in Table 14.6, the value of lambda would be .14 if attitude toward capital punishment had been taken as the independent variable (verify this with your own computation). Thus, you should exercise some caution in the designation of an independent variable. If you consistently follow the convention of arraying the independent variable in the columns and compute lambda as outlined above, the asymmetry of the statistic should not be confusing.

Second, when one of the row totals is much larger than the others, lambda can take on a value of 0 even when other measures of association would not be 0 and calculating percentages for the table indicates some association between the variables. This anomaly is a function of the way lambda is calculated and suggests that great caution should be exercised in the interpretation of lambda when the row marginals are very unequal. In fact, in the case of very unequal row marginals, a chi square–based measure of association might well be preferable to lambda.

SUMMARY

1. Three measures of association—phi, Cramer's V, and lambda—were introduced. Each is used to summarize the overall strength of the association between two variables that have been organized into a bivariate table.

2. Phi and Cramer's V are chi square–based measures of association and have the advantage of being easy to compute (once the value of chi square is found). Phi is used for tables with either two rows or two columns, while Cramer's V can be used for any size table. Both indicate the strength of the relationship, but values between 0.00 and 1.00 have no direct interpretation.

3. Lambda is a PRE-based measure and provides a more direct interpretation for values between the extremes of 0.00 and 1.00. Lambda indicates the improvement in predicting the dependent variable with knowledge of the independent, compared to predicting the dependent without knowledge of the independent. Because of the meaningfulness of values between the extremes, lambda is often preferred over the more traditional chi square–based measures.

SUMMARY OF FORMULAS

Phi   14.1 $\phi = \sqrt{\dfrac{\chi^2}{N}}$

Cramer's V 14.2 $V = \sqrt{\dfrac{\chi^2}{(N)(\text{minimum of } r - 1,\ c - 1)}}$

Lambda   14.3 $\lambda = \dfrac{E_1 - E_2}{E_1}$

GLOSSARY

Cramer's V. A chi square–based measure of association. Appropriate for nominally measured variables that have been organized into a bivariate table of any number of rows and columns.

E_1. The number of errors of prediction made when predicting which category of the dependent variable cases will fall into while ignoring the independent variable.

E_2. The number of errors of prediction made when predicting which category of the dependent variable cases will fall into while taking account of the independent variable.

Lambda (λ). A measure of association appropriate for nominally measured variables that have been organized into a bivariate table. Lambda is based on the logic of proportional reduction in error (PRE).

Phi (ϕ). A chi square–based measure of association. Appropriate for nominally measured variables that have been organized into a 2 × 2 bivariate table.

Proportional reduction in error (PRE). A logic that underlies the definition and computation of lambda (and several other measures of association). Basically, the logic involves comparing the number of errors made in predicting the dependent variable while ignoring the independent variable (E_1) with the number of errors made in predicting the dependent variable while taking account of the independent (E_2). These two figures, E_1 and E_2, are then manipulated to produce the final statistic.

PROBLEMS

14.1 For each problem at the end of Chapter 13, compute a phi or Cramer's V and a lambda. Compare the value of the measure of association with your impressions of the strength of the relationships that you stated at that time.

14.2 For each table below, compute phi or Cramer's V and a lambda.

a.

Variable Y	Variable X A	Variable X B	
A	11	9	20
B	10	10	20
	21	19	40

b.

Variable Y	Variable X A	Variable X B	
A	25	10	35
B	10	25	35

c.

Variable Y	Variable X A	Variable X B	
A	15	50	65
B	10	15	25
	25	65	90

d.

Variable Y	Variable X			
	A	B	C	
A	20	10	0	30
B	0	10	20	30
	20	20	20	60

14.3 SOC St. Algebra College has a problem with attrition. A sizable number of nongraduating students do not return to classes each semester. Is attrition importantly related to race, status, or age? For each table below, how strong is the association (if any) between each of the independent variables and attrition? Does the relationship have a pattern? (Calculating percentages for the table will help to answer the second question.) Write a paragraph summarizing the results presented in these three tables.

a. Attrition by race for 532 students enrolled in fall semester

Attrition	Race			
	White	Black	Other	
Returned spring semester	260	80	55	395
Did not return	85	26	26	137
	345	106	81	532

b. Attrition by status for 532 students enrolled in fall semester

Attrition	Status		
	Part-time	Full-time	
Returned spring semester	42	353	395
Did not return	87	50	137
	129	403	532

c. Attrition by age for 532 students enrolled in fall semester

Attrition	Age		
	18 24	25 or More	
Returned spring semester	322	73	395
Did not return	57	80	137
	379	153	532

14.4 SOC/CJ A sociologist is researching public attitudes toward crime and has asked a sample of residents of his city if they think that the crime rate in their neighborhoods is rising. Is there a relationship between sex and perception of the crime rate? Between race and perception of the crime rate? What is the pattern of the relationship? Write a paragraph summarizing the information presented in these tables.

a. Perception of crime rate by sex

Crime Rate Is	Sex		
	Male	Female	
Rising	200	225	425
Stable	175	150	325
Falling	125	125	250
	500	500	1000

b. Perception of crime rate by race

Crime Rate Is	Race		
	White	Black	
Rising	300	150	450
Stable	230	85	315
Falling	170	65	235
	700	300	1000

14.5 SW The director of a shelter for battered women has noticed that many of the women who are referred to the shelter eventually return to their violent husbands even when there is every indication that the husband will continue the pattern of abuse. The director suspects that the women who return to their husbands do so because they have no place else to go—for example, no close relatives in the area with whom the women could reside. Do the data below support the director's suspicion? (Data are from the case files of former clients.)

Return to Husband?	Relatives Nearby	No Relatives Nearby	
Yes	10	23	33
No	50	17	67
	60	40	100

14.6 PS You are running for mayor of Shinbone, Kansas, and realize that, if you are to win, you must win the support of blue-collar voters. (You already have strong support in the white-collar neighborhoods.) You have a very limited advertising budget and wonder how best to reach your intended audience. An aide has found the data below, which show the relationship between social class and "main source of news" for a sample of Shinboneites. Will this information help you make a decision?

Main Source of News	Blue-Collar	White-Collar	
Television	140	200	340
Radio	25	40	65
Newspapers	85	100	185
	250	340	590

14.7 PA Traditionally, bus ridership in your town has been confined to lower-income and blue-collar patrons. As head of transportation planning for the city, you believe that ridership from white-collar, middle-income neighborhoods can be increased if bus routes linking these neighborhoods to the downtown area (where most people work) are increased. A survey is conducted and the results are displayed below. Is willingness to ride the bus related to job location? What is the pattern of the relationship (if any)?

Potential Ridership	Job Location Downtown	Other	
Would use bus	55	20	75
Would not use bus	15	21	36
	70	41	111

14.8 GER A survey of senior citizens who live in either a housing development specifically designed for retirees or an age-integrated neighborhood has been conducted. Is type of living arrangement related to sense of social isolation?

Sense of Isolation	Living Arrangement Housing Development	Integrated Neighborhood	
Low	80	30	110
High	20	120	140
	100	150	250

14.9 SOC A researcher has conducted a survey on sexual attitudes for a sample of 317 teenagers. The respondents were asked whether they considered premarital sex to be "always wrong" or "OK under certain circumstances." The tables below summarize the relationship between responses to this item and several other variables. For each table, assess the strength and pattern of the relationship and write a paragraph interpreting these results.

a. Attitudes toward premarital sex by gender

Premarital Sex	Sex Male	Female	
Always wrong	90	105	195
Not always wrong	65	57	122
	155	162	317

b. Attitudes toward premarital sex by courtship status

Premarital Sex	Ever Gone "Steady"? No	Yes	
Always wrong	148	47	195
Not always wrong	42	80	122
	190	127	317

c. Attitudes toward premarital sex by social class

Premarital Sex	Social Class Blue-Collar	White-Collar	
Always wrong	72	123	195
Not always wrong	47	75	122
	119	198	317

14.10 SOC Is there an association between the gender of college instructors and the teaching effectiveness ratings they receive from students? Write a few sentences summarizing your findings.

Teaching Effectiveness	Gender Female	Male	
High	115	241	
Low	54	113	
	169	354	

14.11 GER Community leaders are wondering if there is any relationship between the age of city residents and their attendance at city-sponsored festivities like the annual Fourth of July celebration. Does the table below indicate an association between senior citizen status and attendance at last year's celebration? Summarize your conclusions in a sentence or two.

Did You Attend Last Year's Fourth of July Celebration?	Senior Citizen?	
	Yes	No
Yes	45	178
Wanted to but couldn't	13	42
No	145	332
	203	552

14.12 SOC There is concern that suicides are motivated, in part, by imitation. Especially among young people, it may be that "epidemics" of self-destructive behaviors follow publication of suicides in local media. A number of cities have been classified by rate of suicide and by whether or not they experienced a publicized suicide within the past year. Is there an association between these two variables? Summarize your conclusions in a sentence or two.

Suicide Rate	Publicized Suicide?	
	Yes	No
Low	15	20
High	15	10
	30	30

14.13 SW Ninety male first-graders have been rated by their teachers for levels of aggressiveness (for example, hitting other children, "acting out" in the classroom), and a school social worker has rated the families of each of these children for internal conflict and stress. Is there a relationship between these two variables? Do more aggressive children tend to come from high-conflict families? Summarize your conclusions in a sentence or two.

Aggressiveness	Level of Family Conflict		
	Low	Moderate	High
Low	10	8	17
Moderate	15	10	10
High	5	12	3
	30	30	30

14.14 PS Does the American electorate really have a tendency, as is sometimes alleged, to vote for physically attractive candidates regardless of competence? A sample of adult voters were shown photographs of a physically attractive male or an unattractive male and asked if the individual pictured "seems like he would be a competent elected official." The results are presented below. Is there an association between attractiveness and electability?

Would He Be Competent?	Attractiveness	
	Attractive	Unattractive
Yes	15	12
No	17	15
Can't say	18	23
	50	50

14.15 SOC Problem 13.15 analyzed some bivariate relationships taken from the General Social Survey data set. Political ideology was used as the independent variable for five different dependent variables and, using only percentages, you were asked to characterize the relationships in terms of strength and direction. Now, with the aid of some measures of association, these characterizations should be easier to develop. If you haven't already done so for problem 14.1, compute a phi or Cramer's V and a lambda for each of the tables in problem 13.15. Compare the measures of association with your characterizations based on the percentages. Below are the same five dependent variables cross-tabulated against sex as an independent variable. Compare the strength of these relationships with POLVIEWS. Which independent variable has the stronger associations?

a. Attitudes about welfare, by sex

Spending on Welfare (item 22)	Sex	
	Male	Female
Too little	22	37
About right	41	49
Too much	95	114
	158	200

b. Attitudes about the role of the United States in world affairs, by sex

Should U.S. Be Active? (item 27)	Sex	
	Male	Female
Active	166	189
Stay out	55	89
	221	278

c. Attitudes about the national health care system

Spending on Health Care (item 23)	Sex	
	Male	Female
Too little	102	166
About right	35	26
Too much	20	12
	157	204

d. Attitudes toward pornography, by sex

Pornography Causes Breakdown. (item 59)	Sex	
	Male	Female
Yes	124	199
No	73	85
	197	284

e. Attitudes toward women's roles, by sex

Women should take care of home and family (item 69)	Sex	
	Male	Female
Strongly agree	14	18
Agree	79	77
Disagree	104	130
Strongly disagree	31	57
	228	282

SPSS/PC+ PROCEDURES FOR NOMINAL-LEVEL MEASURES OF ASSOCIATION

DEMONSTRATION 14.1 Does Support for the Death Penalty Vary by Political Ideology? Another Look

In Demonstration 13.1, we used CROSSTABS to look for an association between CAPPUN and POLVIEWS. We saw that these variables were associated and that liberals were the least supportive of the death penalty and conservatives were the most supportive. The sample was fairly homogeneous on this issue and the majority of people from each of the three political ideologies were in favor of the death penalty. The differences in conditional distributions were not great, and the relationship seemed to be, at best, "moderate" in strength. In this demonstration, we will reexamine the relationship and compute some measures of association. The SPSS/PC+ commands should repeat those in Demonstration 13.1 (remember to RECODE and specify new value labels for POLVIEWS) with the addition of a request for phi (or *V*) and lambda. Request the measures of association on the STATISTICS subcommand. The full command should read

```
CROSSTABS /TABLES CAPPUN BY POLVIEWS /OPTIONS 4
          /STATISTICS 1 2 4.
```

And the output will be

Cross-tabulation: CAPPUN   FAVOR OR OPPOSE DEATH PENALTY FOR MURDER
                  By POLVIEWS THINK OF SELF AS LIBERAL OR CONSERVATIVE

POLVIEWS->	Count Col Pct	liberal 1.00	moderate 2.00	conservative 3.00	Row Total
CAPPUN					
FAVOR	1.00	138 67.3	197 80.1	210 81.1	545 76.8
OPPOSE	2.00	67 32.7	49 19.9	49 18.9	165 23.2
	Column Total	205 28.9	246 34.6	259 36.5	710 100.0

Chi-Square	D.F.	Significance	Min. E.F.	Cells with E.F.< 5
14.47931	2	.0007	47.641	None

MORE

Statistic	Symmetric	With CAPPUN Dependent	With POLVIEWS Dependent
Lambda	.02922	.00000	.03991

Statistic	Value	Significance
Cramer's V	.14281	

Number of Missing Observations = 78

Lambda (with CAPPUN dependent) is zero and Cramer's V is .14281. Since the row totals are very unequal, we can disregard the lambda of zero (see Section 14.5). A V of .14 would be considered weak (maybe "weak to moderate") and, as we suspected, the overall association is weak. Political ideology is not an especially important cause of support for the death penalty.

DEMONSTRATION 14.2: What Variables Affect Support for the Death Penalty?

Let's see if we can do any better in "explaining" attitudes towards the death penalty with some other commonly used independent variables. As you are no doubt aware, attitudes such as CAPPUN are often associated with various measures of social location. In keeping with our focus on the nominal level of measurement, let's investigate the role of SEX, MARITAL (item 3), RACE (item 11), and religious preference, or RELIG (item 28). Incorporating a number of potential independent variables, although quite common in social science research, generates a large volume of output, and we will abbreviate the actual output from SPSS/PC+ in this demonstration.

	Percent in Favor	Significance of Chi Square[1]	Phi or V	Lambda
SEX				
Male	81.2			
Female	73.4	.01	.09	.00
RELIGION				
Protestant	76.1			
Catholic	80.0			
Jew	66.7			
None	75.8			
Other	63.6	.58	.06	.00
RACE				
White	80.9			
Black	51.8			
Other	63.9	.00	.23	.00
MARITAL[2]				
Married	80.2			
Unmarried	73.0	.02	.09	.04

1. For 2 × 2 tables, this is based on the unadjusted chi square, before Yate's correction.
2. This variable is recoded. All "nonmarried" respondents have been collapsed into the same category.

Basically, these results indicate that CAPPUN is not particularly related to any of these demographic variables. The strongest relationship is with RACE, with Blacks being less supportive of the death penalty than other groups. Since the rows are very unequal, lambda should be disregarded.

Exercises

14.1 See if you can do any better finding a variable with a strong association with CAPPUN. If necessary, use the RECODE command to reduce the number of categories in the column variable. As a suggestion, try DEGREE and INCOME as independent variables.

14.2 Run some of the tables you did for Exercises 13.1 and 13.2 again, with the /STATISTICS 1 2 4 command added. See if the measures of association help you interpret the relationship.

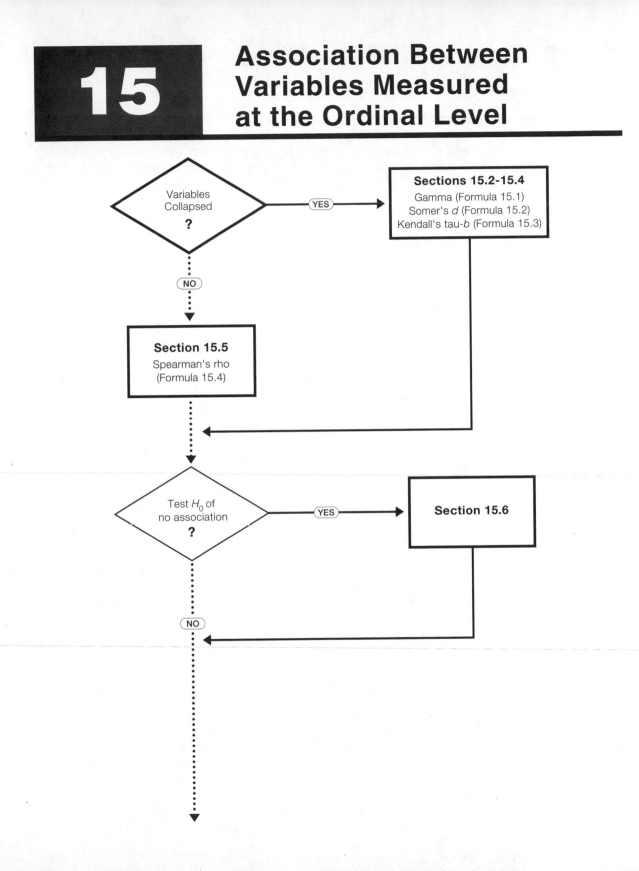

15.1 INTRODUCTION

The measurement process for ordinal-level variables can produce data of two different forms. The first form, which we shall call "continuous ordinal data," is easily recognizable by the wide range of many possible values per variable. Continuous ordinal data bear a close resemblance to interval-ratio data. Examples of such data might be found in any attitude scale that incorporated many different items and, therefore, had many possible values.

The second form, which we shall call "collapsed ordinal data," has only a few (no more than five or six) values per variable and can arise either by collecting data in collapsed form or by collapsing a continuous scale. Collapsed ordinal data would be produced, for example, by measuring the variable "social class" with only a few categories (such as upper, middle, and lower). Collapsed ordinal data could also be produced by dividing the scores on an attitude scale into just a few categories (such as high, moderate, and low).

A large variety of measures of association has been invented for use with ordinal-level variables. Rather than attempt a comprehensive coverage of all of these measures, I will concentrate on just two. I will present **gamma (G)** for situations where we have two collapsed ordinal variables and **Spearman's rho** (r_s) for measuring the association between two continuous ordinal variables. Both of these measures have limitations that make them less than ideal for all possible situations. On the other hand, they are quite popular and commonly used in the literature. In addition, two other measures (**Somer's d** and **Kendall's tau-b**) will be briefly reviewed so that you will have access to some alternative statistics for measuring association between two collapsed ordinal scales.

15.2 PROPORTIONAL REDUCTION IN ERROR (PRE)

For nominal-level variables, the logic of PRE was based on two prediction rules. First, we predicted Y while ignoring X and, second, we predicted Y while taking X into account. Lambda measured the proportional reduction in the errors of prediction made with the second rule as compared to the first.

For variables measured at the ordinal level, we have two similar prediction rules, and gamma, like lambda, measures the proportional reduction in error gained by predicting one variable while taking the other into account. The major difference in the logic of gamma and lambda lies in the way predictions are made. In the case of gamma, we predict the order of pairs of cases. That is, we predict whether one case will have a higher or lower score than the other on the variable in question. The first prediction rule is to predict the order of a pair of cases on one variable while ignoring their order on the other. The second rule is to predict the order of a pair of cases on one variable while taking their order on the other variable into account.

As an illustration, assume that a researcher is concerned about the causes of "burnout" (that is, demoralization and loss of commitment) among elementary school teachers and wonders if levels of burnout might be related to the number of years a teacher has been employed in the profession. Another way to state the problem would be to ask if teachers who rank higher on years of service would also rank higher on burnout. If we knew that teacher A had more years of service than teacher B, would we be able to predict accurately that teacher A is also higher than teacher B on burnout? That is, would knowledge of the order of this pair of cases on one variable help us predict their order on the other?

If the two variables are associated, we will reduce our errors when our predictions about one of the variables are based on knowledge of the other. Furthermore, the stronger the association, the fewer the errors we will make. With no association between the variables, gamma will be 0, and knowledge of the order of a pair of cases on one variable will not improve our ability to predict their order on the other. A gamma of ± 1 denotes a perfect relationship: the order of all pairs of cases on one variable would be predictable without error from their order on the other variable.

Ordinal-level variables, unlike variables measured at the nominal level, have direction (they range from "high" to "low"). Consequently, with ordinal variables, relationships can be either positive or negative. A positive relationship would be one where cases tended to be ranked in the same order on both variables. For example, if Case A is ranked above Case B on one variable, it would also be ranked above Case B on the second variable. The relationship suggested above between years of service and burnout would be a positive relationship. In a negative relationship, the order of the cases would be reversed between the two variables. If Case A ranked above Case B on one variable, it would tend to rank below Case B on the second variable.

In the preceding paragraphs, I have attempted to explain the logic of ordinal measures of association in the language most appropriate for an understanding of gamma. Thus, we have discussed the order of pairs of cases and how that order is predicted from variable to variable. Note, however, that we are still trying to ascertain information about the same three characteristics of an association between two variables that we introduced in Chapter 13. We are still concerned with the existence of the association, its strength, and its direction. Associations between collapsed ordinal-level variables can be (and probably should be) investigated by calculating percentages for the table and observing the conditional distributions of the dependent variable. By computing gamma (or any measure of association), we get in a single number a summary measure of the existence, strength, and direction of the relationship.

15.3 THE COMPUTATION OF GAMMA

Table 15.1 summarizes the relationship between "length of service" and "burnout" for a fictitious sample of 100 teachers. To compute gamma, two sums are needed. First, we must find the number of pairs of cases that are ranked the same on both variables (we will label this N_s) and then the number of pairs of cases ranked differently on the variables (N_d). We do so by working with the cell frequencies cell by cell.

TABLE 15.1   BURNOUT BY LENGTH OF SERVICE (fictitious data)

Burnout	Length of Service			
	Low	Moderate	High	
Low	20	6	4	30
Moderate	10	15	5	30
High	8	11	21	40
	38	32	30	100

To find the number of pairs of cases ranked the same (N_s), begin with the cell containing the cases that were ranked the lowest on both variables. In Table 15.1, this would be the upper left-hand cell. (Not all tables are constructed with values increasing from left to right and from top to bottom. When using other tables, always be certain that you have located the proper cell.) The 20 cases in the upper left-hand cell all rank low on both burnout and length of service, and we will refer to these cases as "low-lows," or "LLs."

Now, let us form a pair of cases by selecting one case from this cell and one from any other cell—for example, the middle cell in the table. All 15 cases in this cell are moderate on both variables and, following our practice above, can be labeled moderate-moderates or MMs. Any pair of cases formed between these two cells will be ranked the same on both variables. That is, all LLs are lower than all MMs on both variables (on X, low is less than moderate, and on Y, low is less than moderate). The total number of pairs of cases is given by multiplying the cell frequencies. So, the contribution of these two cells to the total N_s is (20) (15), or 300.

Note that, if we attempt to form pairs of cases between the LLs and any other cell in the top row or the left-hand column, we will find that all of these pairs are tied on either X or Y. Forming pairs of cases within the cell will result in pairs that are tied on both X and Y. Gamma ignores all ties and, therefore, the pairs of cases formed within the same row, column, or cell are not computed. Practically, this means that, in computing N_s, we will take account only of the pairs of cases that can be formed between each cell and the cells below and to the right of it.

The computational routine for finding N_s can now be stated. To find the total number of pairs of cases ranked the same on both variables, multiply the frequency in each cell by the total of all frequencies below and to the right of that cell. Repeat for each cell and add the resultant products. The total of these products is N_s. This procedure is displayed below for each cell in Table 15.1. Note that none of the cells in the bottom row or the right-hand column can contribute to N_s because they have no cells below and to the right of them. Figure 15.1 shows the direction of multiplication for each of

FIGURE 15.1   COMPUTING N_s IN A 3 × 3 TABLE

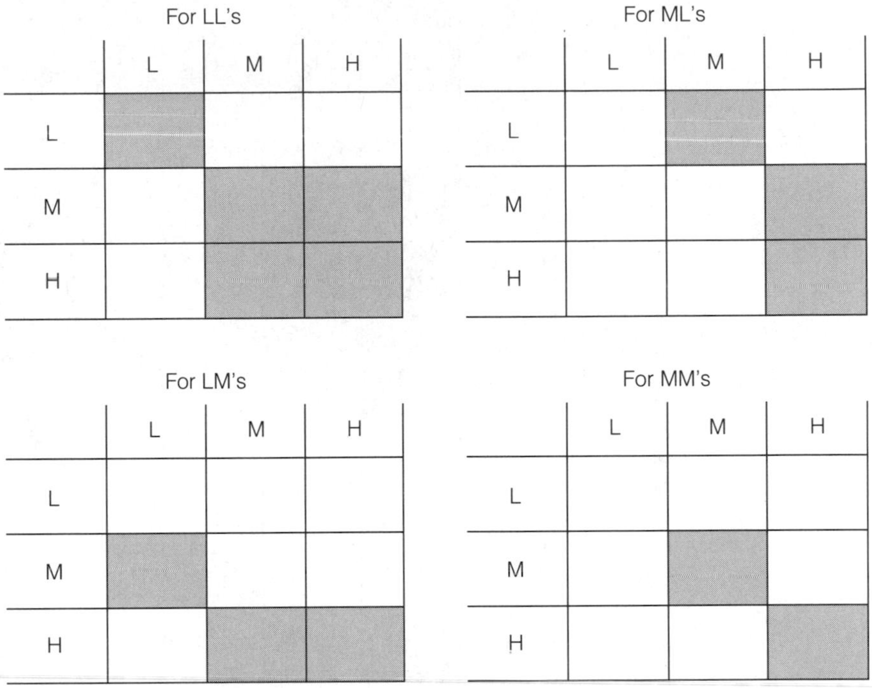

the four cells that, in a 3 × 3 table, can contribute to N_s. Computing N_s for Table 15.1:

	Contribution to N_s
For LLs, 20(15 + 5 + 11 + 21) =	1040
For MLs, 6(21 + 5) =	156
For HLs, 4(0) =	0
For LMs, 10(11 + 21) =	320
For MMs, 15(21) =	315
For HMs, 5(0) =	0
For LHs, 8(0) =	0
For MHs, 11(0) =	0
For HHs, 21(0) =	0
	$N_s = 1831$

The computations above indicate that a total of 1831 pairs of cases are ranked the same on both variables. Our next step is to find the number of pairs of cases ranked differently (N_d) on both variables. The computation of N_d follows a pattern that is the reverse of that for N_s. This time, we begin with

FIGURE 15.2 COMPUTING N_d IN A 3 × 3 TABLE

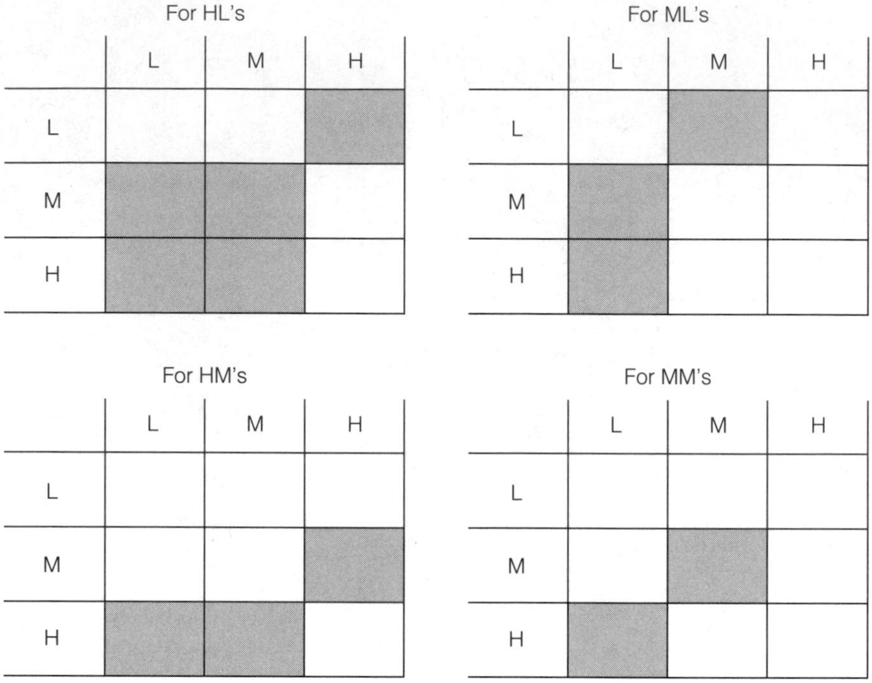

For HL's

	L	M	H
L			
M			
H			

For ML's

	L	M	H
L			
M			
H			

For HM's

	L	M	H
L			
M			
H			

For MM's

	L	M	H
L			
M			
H			

the upper right-hand cell (high-lows, or HLs) and multiply the number of cases in the cell by the total frequency of cases below and to the left. The four cases in the upper right-hand cell are low on Y and high on X; and, if a pair is formed with any case from this cell and any cell below and to the left, the cases will be ranked differently on the two variables. For example, if a pair is formed between any HL case and any case from the middle cell (moderate-moderates, or MMs), the HL case would be less than the MM case on Y ("low" is less than "moderate") but more than the MM case on X ("high" is greater than "moderate"). The computation of N_d is detailed below and shown graphically in Figure 15.2. In the computations, we have omitted cells that cannot contribute to N_d because they have no cells below and to the left of them.

Computing N_d for Table 15.1:

		Contribution to N_d
For HLs, $4(10 + 15 + 8 + 11) =$		176
For MLs, $6(10 + 8)$	$=$	108
For HMs, $5(8 + 11)$	$=$	95
For MMs, $15(8)$	$=$	120
	$N_d =$	499

APPLICATION 15.1

For local political elections, do voters turn out at higher rates when the candidates for office spend more money on media advertising? Each of 177 localities has been rated as high or low on voter turnout. Also, the total advertising budgets for all candidates in each locality have been ascertained and classified as high, moderate, or low. Are these variables associated?

Voter Turnout	Expenditure			
	Low	Moderate	High	
Low	35	32	17	84
High	23	27	43	93
	58	59	60	177

Because both variables are ordinal in level of measurement, we will compute a gamma to summarize the strength and direction of the association. The number of pairs of cases ranked in the same order on both variables (N_s) would be

$$N_s = 35(27 + 43) + 32(43) = 2450 + 1376$$
$$= 3826$$

The number of pairs of cases ranked in different order on both variables (N_d) would be

$$N_d = 17(27 + 23) + 32(23) = 850 + 736$$
$$= 1586$$

Gamma is

$$G = \frac{N_s - N_d}{N_s + N_d}$$

$$G = \frac{3826 - 1586}{3826 + 1586}$$

$$G = \frac{2240}{5412}$$

$$G = 0.41$$

A gamma of 0.41 means that, when predicting the order of pairs of cases on voter turnout, we would make 41% fewer errors by taking candidates' advertising expenditures into account, as opposed to ignoring the latter variable. There is a moderate, positive association between these two variables.

Thus, Table 15.1 has 499 pairs of cases ranked in different order and 1831 pairs of cases ranked in the same order. The formula for computing gamma is

FORMULA 15.1

$$G = \frac{N_s - N_d}{N_s + N_d}$$

where N_s = the number of pairs of cases ranked the same as both variables
N_d = the number of pairs of cases ranked differently on the two variables

For Table 15.1, the value of gamma would be

$$G = \frac{1831 - 499}{1831 + 499}$$

$$G = \frac{1332}{2330}$$

$$G = .57$$

A gamma of .57 indicates that we would make 57% fewer errors if we predicted the order of pairs of cases on one variable from the order of pairs of cases on the other—as opposed to predicting order while ignoring the other variable. Length of service is associated with degree of burnout, and the relationship is positive. Knowing the respective rankings of two teachers on length of service (Case A is higher on length of service than Case B) will help us predict their ranking on burnout (we would predict that Case A will also be higher than Case B on burnout).

To use the computational routine for gamma presented above, you must arrange the table in the manner of Table 15.1, with the column variable increasing in value as you move from left to right and the row variable increasing from top to bottom. Be careful to construct your tables according to this format; if you are working with data already in table format, you may have to rearrange the table or rethink the direction of patterns. Gamma is a symmetrical measure of association; that is, the value of gamma will be the same regardless of which variable is taken as independent.

15.4 SOMER'S *d* AND KENDALL'S TAU-*b*

The two new measures of association I will now introduce are similar to gamma in many respects. Both are appropriate for collapsed ordinal-level variables that have been arrayed in a bivariate table. Both are interpretable by the logic of PRE and, computationally, they are both based on a comparison of the number of pairs of cases ranked in the same order on both variables (N_s) with the number of pairs ranked differently (N_d).

The difference between these statistics and gamma lies in their treatment of pairs of cases that are tied. Gamma represents the proportional reduction in error in prediction for pairs of cases that are not tied. Somer's *d* and Kendall's tau-*b*, on the other hand, incorporate tied pairs of cases. Specifically, Somer's *d* includes pairs that are tied on the dependent variable but not the independent, and Kendall's tau-*b* includes pairs that are tied on either the dependent or the independent variable. Before addressing the statistics themselves, let us see how the number of tied pairs can be computed.

The number of pairs of cases tied on the dependent variable (T_y) is determined by finding the number of pairs of cases within each row, since all cases in the same row are, by definition, tied on the dependent variable. To find the contribution of each row to T_y, multiply each cell frequency by all frequencies to the right. For example, in the bottom row of Table 15.1, there are 8 cases in the far left-hand cell (LHs) and 32 cases (11 + 21) to the right. Hence, 8 × 32 or 256 pairs of cases formed from this cell are tied on the dependent variable. Similarly, from the 11 cases in the next cell to the right (MHs), we get 11 × 21, or 231, pairs of cases tied on "burnout" (not counting the pairs of cases from the first cell). The third cell in this row cannot contribute to T_y because there are no cells to the right of it. Thus, for the bottom row, the total number of pairs of cases tied on the dependent variable is 256 + 231, or 487.

Moving to the middle row, we see that there are 10 cases in the first cell and a total of $15 + 5 = 20$ cases to the right. Thus, the total number of pairs of cases tied on the dependent variable that can be formed from this row is $10(20) + 15(5)$, or 275. In similar fashion, the contribution of the top row to T_y is $20(6 + 4) + 6(4)$, or 224. Adding these subtotals from all three rows, we find that T_y equals $487 + 275 + 224$, or 986. In Table 15.1, a total of 986 pairs of cases are tied on the dependent variable ($T_y = 986$).

The number of pairs of cases tied on the independent variable ($\mathbf{T_x}$) is found in an analogous manner by working within columns instead of rows. The contribution of each cell to the total of T_x would be found by multiplying the frequency in the cell by all frequencies below the cell. Using Table 15.1 as our example, the number of pairs of cases tied on the independent variable for the left-hand column would be $20(10 + 8) + 10(8)$, or 440. The contribution of the middle column to T_x would be $6(15 + 11) + 15(11)$, or 321. Finally, the contribution of the right-hand column to T_x would be $4(5 + 21) + 5(21)$, or 209. The total number of pairs of cases tied on the independent variable would be $440 + 321 + 209$, or 970. In Table 15.1, 970 pairs of cases are tied on the independent variable ($T_x = 970$). As we will see below, the quantity T_y is used in the computation of Somer's d, while Kendall's tau-b uses both T_y and T_x.

The formula for Somer's d is

FORMULA 15.2

$$d = \frac{N_s - N_d}{N_s + N_d + T_y}$$

Note that this formula is the same as that for gamma except for the addition of the quantity T_y in the denominator. For all practical purposes, the inclusion of this quantity in the denominator of the equation means that the value of Somer's d will always be less than the value of gamma for the same table. For example, the gamma previously computed for Table 15.1 was .57, while Somer's d is .40:

$$d = \frac{N_s - N_d}{N_s + N_d + T_y}$$

$$d = \frac{1831 - 499}{1831 + 499 + 986}$$

$$d = \frac{1332}{3316}$$

$$d = .40$$

A Somer's d of .40 indicates a positive relationship of moderate strength. In predicting the order of pairs of cases on "burnout," we would make 40% fewer errors when predicting from the order of the pairs on the independent variable (length of service). Somer's d is a lower value than gamma because it includes pairs tied on the dependent variable in the pool of pairs of cases for which predictions will be made.

READING STATISTICS 7: BIVARIATE TABLES AND ASSOCIATED STATISTICS

The statistics associated with any bivariate table will usually be reported directly below the table itself. This information would include the name of the measure of association, its value, and (if relevant) its sign. Also, if the research involves a random sample, the results of the chi square test, including the value of the obtained chi square, the degrees of freedom in the table, and the significance (alpha level), will be reported. So the information will look something like this:

$$\lambda = 0.47$$
$$\chi^2 = 13.23 \ (2df, \ p < 0.05)$$

Note that the alpha level is reported in the "$p <$" format, which we discussed in Reading Statistics 5.

Besides simply reporting the value of these statistics, the researcher will also interpret them in the text of the article. The preceding statistics might be characterized by the statement: "The association between the variables is statistically significant and moderately strong." Again we see that researchers will avoid our rather wordy (but more careful) style of stating results. Where we might say that "a lambda of 0.47 indicates that we will reduce our errors of prediction by 47% when predicting the dependent variable from the independent variable, as opposed to predicting the dependent while ignoring the independent," the professional researcher will simply report the value of lambda and characterize that value in a word or two (for example, "The association is moderately strong"). The researcher assumes that his or her audience is statistically literate and can make the more detailed interpretations for themselves.

Statistics in the Professional Literature

What is the relationship between exposure to TV violence and violent behavior? A group of researchers proposed the hypothesis that " . . . exposure to TV violence will be positively related to violent behavior." (343) More than 2000 high school juniors and seniors were asked to name their favorite TV show and also asked to indicate their involvement in violent behavior. The violence level of the TV shows was rated, and the table below displays the relationship between the two variables.

TABLE 1.   Violent Behavior by Average Violence Rating of Favorite Television Shows

Average Violence Rating	Violent Behavior			
	Low	Medium	High	
Low	409 (63%)	179 (28%)	62 (10%)	650
Medium	518 (56%)	255 (28%)	147 (16%)	920
High	280 (50%)	175 (31%)	103 (19%)	558
	1,207	609	312	2,128

Chi square = 28.85   Significant at .001   Gamma = .16

Analyze this table carefully. Remember that there are no absolute conventions for presenting tables in the literature. In this case, the researchers have used a format that differs from the one used in this text. The independent variable (Average Violence Rating of the TV show) is in the rows, not the columns. Percentages are still calculated in the direction of the independent variable but, for this table, that means that they total 100% in the rows.

We can see that the conditional distributions do change (from row to row), so the variables are related. The chi square is not zero and is reported to be significant at the .001 level. Gamma is .16, indicating a weak to moderate relationship between the variables. Here's what the authors have to say about this table:

> A positive but low correlation (gamma = .16) was found between the average violence rating and violent behavior. (W)hile only 10 percent of the respondents who were low on the average violence rating were high on violent behavior, only 19 percent of those who were high on average violence rating were high on violent behavior. . . . Therefore, only minimal support for the . . . hypothesis was obtained.

Make sure you understand how the percentages in the table support this conclusion.

To help you interpret the direction of this relationship, remember that, in a positive relationship, low values on one variable occur together with low values on the other, and high values also occur together. For the table above, a positive relationship would mean that the bulk of the cases will lie along the diagonal from the upper left-hand cell ("low-lows") through the middle cell ("medium-mediums") to the lower right-hand cell ("high-highs").

Somer's d is an asymmetric measure of association and will change values, depending on which variable is taken as independent. If the row variable is taken as independent, then ties would be computed within columns instead of rows (that is, T_x becomes T_y). For Table 15.1, the values of T_x and T_y are nearly identical, and the value of d would not have changed appreciably had "burnout" been taken as the independent variable. But this similarity between T_x and T_y does not always occur, and you should exercise caution when deciding which variable to take as independent, for the value of this measure can be appreciably affected by your decision.

Kendall's tau-b differs from gamma and Somer's d in the way that ties are handled. This feature is reflected in the denominator, which incorporates not only pairs of cases tied on the dependent variable (like Somer's d) but also pairs of cases tied on the independent (T_x). The formula for Kendall's tau-b is

FORMULA 15.3

$$\text{tau-}b = \frac{N_s - N_d}{\sqrt{N_s + N_d + T_y)(N_s + N_d + T_x)}}$$

Using Table 15.1 as an illustration once again, we would have

$$\text{tau-}b = \frac{1831 - 499}{\sqrt{(1831 + 499 + 986)(1831 + 499 + 970)}}$$

$$\text{tau-}b = \frac{1332}{\sqrt{(3316)(3300)}}$$

$$\text{tau-}b = \frac{1332}{\sqrt{10,942,800}}$$

$$\text{tau-}b = \frac{1332}{3307.99}$$

$$\text{tau-}b = .40$$

A value of .40 for Kendall's tau-b indicates that we will make 40% fewer errors in predicting the order of pairs on one variable from their order on the other variable (as opposed to predicting the order of pairs while ignoring their order on the other variable). This indicates a positive association of moderate strength.

Kendall's tau-b is a symmetrical measure of association. Its major peculiarity is that it can achieve a value of ± 1 only in a square table (that is, a table where the number of rows equals the number of columns) and, therefore, tau-b would probably not be the measure of first choice in any rectangular table. Since this measure incorporates tied pairs of cases in the denominator, it will tend to be lower in value than gamma for the same table.

In summary, for the situation where we have two ordinally measured variables arrayed in table format, we have three different ways of measuring the strength of the association. The three measures differ from each other primarily in the way they treat tied pairs of cases. Gamma ignores all ties; Somer's d includes only pairs tied on the dependent variable; and tau-b includes pairs tied on either the dependent or the independent variable.

Which of these three is "best?" Unfortunately, there is no simple answer to this question. Gamma is probably the most commonly used of the three but is certainly not equally preferable in all possible situations. When a sharp distinction can be made between independent and dependent variables, Somer's d is in many ways preferable to gamma, since it allows for pairs of cases that are different on the independent variable but tied on the dependent and thus includes more information (that is, more pairs of cases). Kendall's tau-b includes the most information about the relationship (pairs tied on either X or Y) but should not be used for tables that are not square. At any rate, perhaps the most important point to be made is that, in terms of the existence, strength, and direction of an association, the three measures will generally result in the same conclusion. For Table 15.1, for example, all three measures indicate a moderate, positive association between burnout and length of service.

15.5 SPEARMAN'S RHO (r_s)   To this point, we have considered three measures of association for ordinal variables that have a limited number of categories (possible values) and are presented in tables. In many situations, however, data from ordinal-level variables will have a broad range of possible scores and thus many distinct values. Such data may be collapsed into a few broad categories (such as high, moderate, low), organized into a bivariate table, and analyzed with the measures introduced in the previous sections. Collapsing scores in this manner may be beneficial and desirable in many instances, but some important distinctions between cases may be obscured or lost as a consequence.

For example, suppose a researcher wished to test the frequently voiced claim that jogging is beneficial not only physically but also psychologically. Specifically, suppose that the researcher wanted to know if people who were deeply involved in jogging experienced an enhanced sense of self-esteem. To this end, 10 joggers are measured on two scales, the first measuring involvement in jogging and the other measuring self-esteem. The scores of the 10 joggers on the two scales are reported in Table 15.2.

These data could be collapsed and a bivariate table produced. We could, for example, dichotomize both variables to create only two values (high and low) for both variables. Although collapsing scores in this way is certainly legitimate and often necessary,* two difficulties with this practice must be noted. First, the scores seem continuous because there are no obvious or natural division points in the distribution that would allow us to distinguish, in a nonarbitrary fashion, between high scores and low ones. Second, and more important, grouping these cases into broader categories will cause us to lose information. That is, if both Wendy and Debbie are placed in the cate-

---

*For example, collapsing scores into broad categories may be advisable when the researcher is not certain that fine distinctions between scores are meaningful.

TABLE 15.2 THE SCORES OF 10 SUBJECTS ON INVOLVEMENT IN JOGGING
AND A MEASURE OF SELF-ESTEEM

Joggers	Involvement in Jogging (X)	Self-esteem (Y)
Wendy	18	15
Debbie	17	18
Phyllis	15	12
Frank	12	16
Evelyn	10	6
Tricia	9	10
Christy	8	8
Patsy	8	7
Marsha	5	5
Siegfried	1	2

gory "high" on involvement, we won't see that they had different scores on this variable. If differences like this are important and meaningful, then we should opt for a measure of association that permits the retention of as much detail and precision in the scores as possible.

Spearman's rho (r_s) is a measure of association for ordinal-level variables that have a broad range of many different scores and few ties between cases on either variable. Scores on ordinal-level variables cannot, of course, be manipulated mathematically except for judgments of "greater than" or "less than." To compute Spearman's rho, cases are first ranked from high to low on each variable and then the ranks (not the scores) are manipulated to produce the final measure. Table 15.3 displays the original scores and the rankings of the cases on both variables.

To rank the cases, first find the highest score on each variable and assign it rank 1. Wendy has the high score on X (18) and is thus ranked number 1. Debbie, on the other hand, is highest on Y and is ranked first on that variable. All other cases are then ranked in descending order of scores. If any cases have the same score on a variable, assign them the average of the ranks they would have used up had they not been tied. Christy and Patsy have identical scores of 8 on involvement. Had they not been tied, they would have used up ranks 7 and 8. The average of these two ranks is 7.5, and this average of used ranks is assigned to all tied cases. (For example, had Marsha also had a score of 8, three ranks—7, 8, and 9—would have been used and all three tied cases would have been ranked eighth.)

The formula for Spearman's rho is

FORMULA 15.4
$$r_s = 1 - \frac{6\Sigma D^2}{N(N^2 - 1)}$$

where ΣD^2 = the sum of the differences in ranks, the quantity squared

TABLE 15.3   COMPUTING SPEARMAN'S RHO

Case	Involvement (X)	Rank	Self-image (Y)	Rank	D	D²
Wendy	18	1	15	3	-2	4
Debbie	17	2	18	1	1	1
Phyllis	15	3	12	4	-1	1
Frank	12	4	16	2	2	4
Evelyn	10	5	6	8	-3	9
Tricia	9	6	10	5	1	1
Christy	8	7.5	8	6	1.5	2.25
Patsy	8	7.5	7	7	.5	.25
Marsha	5	9	5	9	0	0
Siegfried	1	10	2	10	0	0
					$\Sigma D = 0$	$\Sigma D^2 = 22.5$

To compute ΣD^2, the rank of each case on Y is subtracted from its rank on X (D is the difference between rank on Y and rank on X). A column has been provided in Table 15.3 so that these differences may be recorded on a case-by-case basis. Note that the sum of this column (ΣD) is 0. That is, the negative differences in rank are equal to the positive differences, as will always be the case. Thus, you should find the total of this column as a check on your computations to this point. If the ΣD is not equal to 0, you have made a mistake either in ranking the cases or in subtracting the differences.

In the column headed D^2, each difference is squared to eliminate negative signs. The sum of this column is ΣD^2, and this quantity is entered directly into the formula. For our sample problem:

$$r_s = 1 - \frac{6\Sigma D^2}{N(N^2 - 1)}$$

$$r_s = 1 - \frac{(6)(22.5)}{10(100 - 1)}$$

$$r_s = 1 - \frac{135}{990}$$

$$r_s = 1 - .14$$

$$r_s = .86$$

Spearman's rho is an index of the strength of association between the variables; it ranges from 0 (no association) to ±1 (perfect association). A perfect positive association ($r_s = +1.00$) would exist if there were no disagreements in ranks between the two variables (if cases were ranked in exactly the same order on both variables). A perfect negative relationship

APPLICATION 15.2

Five cities have been rated on an index that measures the quality of life. Also, the percentage of the population that has moved into each city over the past year has been determined. Have cities with higher quality-of-life scores attracted more new residents? The table below summarizes the scores, ranks, and differences in ranks for each of the five cities.

City	Quality of Life	Rank	Percentage of New Residents	Rank	D	D^2
A	30	1	17	1	0	0
B	25	2	14	3	−1	1
C	20	3	15	2	1	1
D	10	4	3	5	−1	1
E	2	5	5	4	1	1
					0	4

Spearman's rho for these variables is

$$r_s = 1 - \frac{6\Sigma D^2}{N(N^2 - 1)}$$

$$r_s = 1 - \frac{(6)(4)}{5(25 - 1)}$$

$$r_s = 1 - \frac{24}{120}$$

$$r_s = 1 - 0.20$$

$$r_s = 0.80$$

These variables have a strong, positive association. The higher the quality-of-life score, the greater the percentage of new residents. The value of r_s^2 is 0.64 ($0.80^2 = 0.64$), which indicates that we will make 64% fewer errors when predicting rank on one variable from rank on the other, as opposed to ignoring rank on the other variable.

($r_s = -1.00$) would exist if the ranks were in perfect disagreement (if the case ranked highest on one variable were lowest on the other, and so forth).

The obtained Spearman's rho above indicates a strong, positive relationship between these two variables. The respondents who were highly involved in jogging also ranked high on self-image. These results would be supportive of claims regarding the psychological benefits of jogging.

Spearman's rho is an index of the relative strength of a relationship, and values between 0 and ±1 have no direct interpretation. However, if the value of rho is squared, a PRE interpretation is possible. Rho squared (r_s^2) represents the proportional reduction in errors of prediction when predicting rank on one variable from rank on the other variable, as compared to predicting rank while ignoring the other variable. In the example above, r_s was .86 and r_s^2 would be .74. Thus, our errors of prediction would be reduced by 74% if, when predicting the rank of a subject on self-image, the rank of the subject on involvement in jogging were taken into account.

15.6 TESTING THE NULL HYPOTHESIS OF "NO ASSOCIATION" WITH GAMMA AND SPEARMAN'S RHO

Whenever a researcher is working with EPSEM, or random, samples, he or she will need to ascertain if the sample findings can be generalized to the population. In Part II of this text, we considered various ways that information taken from samples—for example, the difference between two sample means—could be generalized to the populations from which the samples were drawn. A test of the null hypothesis, regardless of the form or specific test used, asks essentially if the patterns (or differences, or relationships) that have been observed in the samples can be assumed to exist in the population.

The various measures of association can also be tested for their statistical significance. Like means and proportions, measures of association are descriptive statistics that supply the researcher with certain kinds of information about the nature of the data at hand. When data concerning the relationship between two variables have been collected from a random sample, we will not only need to measure the existence, strength, and direction of the association, but we will also want to know if we can assume that the variables are related in the population.

For nominal-level variables, the statistical significance of a relationship is usually judged by the chi square test. Chi square tests could also be conducted on tables displaying the relationship between ordinal-level variables. However, chi square tests the probability that the observed cell frequencies occurred by chance alone and is therefore not a direct test of the measure of association itself.

Below, we will consider tests of significance for gamma and for Spearman's rho. In both cases, the null hypothesis will state that there is no association between the variables in the population and that, therefore, the population values for both sample measures are 0. The population values will be denoted by the Greek letters gamma (γ) and rho (ρ_s). For both measures, the test procedures will be organized around the familiar five-step model (see Chapter 8).

To illustrate the test of significance for gamma, we will use Table 15.1, where gamma was .57.

Step 1. **Making assumptions.** When sample size is greater than 10, the sampling distribution of all possible sample gammas can be assumed to be normal in shape.

> Model: Random sampling
> Level of measurement is ordinal
> Sampling distribution is normal

Step 2. **Stating the null hypothesis.**

$$H_0: \gamma = 0.0$$
$$(H_1: \gamma \neq 0.0)$$

Step 3. **Selecting the sampling distribution and establishing the critical region.** For samples of 10 or more, the Z distribution (Appendix A) can be used to find areas under the sampling distribution:

$$\text{Sampling distribution} = Z \text{ distribution}$$
$$\text{Alpha} = .05$$
$$Z \text{ (critical)} = \pm 1.96$$

Step 4. **Computing the test statistic.**

$$Z \text{ (obtained)} = G \sqrt{\frac{N_s + N_d}{N(1 - G^2)}}$$

$$Z \text{ (obtained)} = .57 \sqrt{\frac{1831 + 499}{100(1 - .33)}}$$

$$Z \text{ (obtained)} = .57 \sqrt{\frac{2330}{(100)(.67)}}$$

$$Z \text{ (obtained)} = .57 \sqrt{34.78}$$

$$Z \text{ (obtained)} = (.57)(5.90)$$

$$Z \text{ (obtained)} = 3.36$$

Step 5. **Making a decision.** Comparing the Z (obtained) with the Z (critical):

$$Z \text{ (obtained)} = 3.36$$
$$Z \text{ (critical)} = \pm 1.96$$

We see that the null hypothesis can be rejected. The sample gamma is unlikely to have occurred by chance alone, and we may conclude that these variables are related in the population from which the sample was drawn.

Spearman's rho can be tested in similar fashion. The null hypothesis is again a statement that the population value (ρ_s) is actually 0 and, therefore, that the value of the sample Spearman's rho (r_s) is the result of mere random chance. When the number of cases in the sample is 10 or more, the sampling distribution of Spearman's rho approximates the t distribution, and we will use this distribution to conduct the test. To illustrate, the Spearman's rho computed in Section 15.5 will be used.

Step 1. **Making assumptions.**

Model: Random sampling
        Level of measurement is ordinal
        Sampling distribution is normal

Step 2. Stating the null hypothesis.

$$H_0: \rho_s = 0.00$$
$$(H_1: \rho \neq 0.00)$$

Step 3. Selecting the sampling distribution and establishing the critical region.

$$\text{Sampling distribution} = t \text{ distribution}$$
$$\text{Alpha} = .05$$
$$\text{Degrees of freedom} = N - 2 = 8$$
$$t \text{ (critical)} = \pm 2.306$$

Step 4. Computing the test statistic.

$$t \text{ (obtained)} = r_s \sqrt{\frac{N - 2}{1 - r_s^2}}$$

$$t \text{ (obtained)} = .86 \sqrt{\frac{8}{1 - .74}}$$

$$t \text{ (obtained)} = .86 \sqrt{\frac{8}{.26}}$$

$$t \text{ (obtained)} = .86\sqrt{30.77}$$

$$t \text{ (obtained)} = (.86)(5.55)$$

$$t \text{ (obtained)} = 4.77$$

Step 5. Making a decision. Comparing the test statistic with the critical region:

$$t \text{ (obtained)} = 4.77$$
$$t \text{ (critical)} = \pm 2.306$$

We see that the null hypothesis can be rejected. We may conclude, with a .05 chance of making an error, that the variables are related in the population from which the samples were drawn.

SUMMARY

1. Many statistics have been designed for measuring association between ordinally measured variables. When both variables are collapsed, gamma is probably the most commonly used measure. Somer's d and Kendall's tau-b are also frequently used measures of asso-

ciation for this situation. Spearman's rho is appropriate for measuring association between "continuous" ordinal variables. All four measures summarize the overall strength and direction of the association between the variables.

2. Gamma is a PRE-based measure that shows the improvement in our ability to predict the order of pairs of cases on one variable from the order of pairs of cases on the other variable, as opposed to ignoring the order of the pairs of cases on the other variable.

3. Gamma ignores all pairs of cases that are tied on either or both variables. Somer's *d* incorporates pairs of cases that are tied on the dependent but not on the independent. Kendall's tau-*b* incorporates all cases tied on either variable but should be computed only for "square" tables. All three measures can be interpreted by the logic of PRE.

4. Spearman's rho is an index of the strength of the association between two "continuous" ordinal variables. The scores of the cases are ranked on each variable, and the ranks are manipulated to yield the final statistic. Spearman's rho varies from zero to plus or minus one and, when squared, can be interpreted by the logic of PRE.

5. Both gamma and Spearman's rho should be tested for their statistical significance when computed for a random sample drawn from a defined population. The null hypothesis is that the variables are not related in the population, and the test can be organized by using the familiar five-step model.

SUMMARY OF FORMULAS

Gamma	15.1	$G = \dfrac{N_s - N_d}{N_s + N_d}$
Somer's *d*	15.2	$d = \dfrac{N_s - N_d}{N_s + N_d + T_y}$
Kendall's tau-*b*	15.3	

$$\text{tau-}b = \frac{N_s - N_d}{\sqrt{(N_s + N_d + T_y)(N_s + N_d + T_y)}}$$

Spearman's rho	15.4	$r_s = 1 - \dfrac{6\Sigma D^2}{N(N^2 - 1)}$

GLOSSARY

Gamma. A measure of association appropriate for ordinally measured variables organized into table format; *G* is the symbol for any sample gamma; γ is the symbol for any population gamma.

Kendall's tau-*b*. A measure of association appropriate for ordinally measured variables organized into table format. Especially appropriate for square tables.

N_s. The number of pairs of cases ranked in the same order on two variables.

N_d. The number of pairs of cases ranked in different order on two variables.

Somer's *d*. A measure of association appropriate for ordinally measured variables organized into table format.

Spearman's rho. A measure of association appropriate for ordinally measured variables that are "continuous" in form; r_s is the symbol for any sample Spearman's rho; ρ_s is the symbol for any population Spearman's rho.

T_x. The number of pairs of cases tied on the independent variable.

T_y. The number of pairs of cases tied on the dependent variable.

PROBLEMS

15.1 For each of the four tables below, compute a gamma as a measure of the strength of the association between the two variables.

a.

Variable Y	Variable X Low	High	
Low	35	0	35
High	0	35	35
	35	35	70

b.

Variable Y	Variable X Low	High	
Low	17	118	135
Mod	43	52	95
High	97	9	106
	157	179	336

c.

Variable Y	Low	Moderately Low	Moderately High	High	
Low	43	45	10	11	109
Mod	7	57	72	15	151
High	5	8	98	153	264
	55	110	180	179	524

d.

Variable Y	Variable X			
	Low	Moderate	High	
Low	15	17	10	42
High	18	15	9	42
	33	32	19	84

15.2 SOC The table below reports the scores of a random sample of 300 undergraduates on a measure of political alienation and a measure of the extent of their use of illegal drugs.

Drug Use	Political Alienation				
	Low	Moder- ately Low	Moder- ately High	High	
Nonusers	30	15	8	10	63
Rare	25	25	12	20	82
Occasional	20	20	20	25	85
Frequent	10	10	15	35	70
	85	70	55	90	300

a. Calculate percentages for this table and observe the conditional distributions of Y (drug use). Do the two variables appear to be related? How would you characterize the relationship in terms of the strength of the association? What is the direction of the association?

b. Compute gamma for this table. Do the value and sign of gamma confirm your conclusions in part a?

c. Write a paragraph of interpretation for gamma.

d. Compute Somer's d and Kendall's tau-b for this table. Compare these values with gamma. Which is lower? Why?

e. Assuming that the 300 undergraduates represent a random sample, conduct and interpret a test of significance on the gamma computed in part b.

15.3 PA All applicants for municipal jobs in Shinbone, Kansas, are given an aptitude test, but the test has never been evaluated to see if test scores are in any way related to job performance. The following table reports aptitude test scores and job performance ratings for a random sample of 75 city employees.

Efficiency Ratings	Test Scores			
	Low	Moderate	High	
Low	11	6	7	24
Moderate	9	10	9	28
High	5	9	9	23
	25	25	25	75

a. Are these two variables associated? Write a paragraph describing the strength and direction of the relationship.

b. Is the measure of association you computed in part a statistically significant?

c. Should the aptitude test continue to be administered? Why or why not?

15.4 SW Over the past few years, a large number of Indochinese refugees have settled in and around greater metropolitan Shinbone, Kansas. To better identify which of these immigrants might benefit from special programs designed to help them adjust to American society, the Social Service Bureau has surveyed a random sample of 178 refugees. As a part of the survey, each respondent was asked to judge subjectively his or her overall degree of success in coping with the new environment. The table below reports the relationship between the "subjective success score" and the number of members in the household of the respondent. Is "living arrangement" predictive of "success in adjustment?"

Problems of Adjustment	Size of Household				
	1	2	3–5	More than 5	
Few	2	12	15	7	36
Moderate	4	27	27	15	73
Difficult	7	25	10	10	52
Severe	10	4	2	1	17
	23	68	54	33	178

a. Is the relationship significant?

b. What are the strength and direction of the relationship?

c. Does the association reported above suggest a client population needing special programs?

15.5 [SOC] Some research has shown that families vary by how they socialize their children to sports, games, and other leisure-time activities. In middle-class families, such activities are carefully monitored by parents and are, in general, dominated by adults (for example, Little League Baseball). In working-class families, children more often organize and initiate such activities themselves and parents are much less involved (for example, sandlot or playground baseball games). Are the data below consistent with these findings? Summarize your conclusions in a few sentences.

As a Child, Did You Play Mostly Organized or Sandlot Sports?	Social Class Background	
	White-collar	Blue-collar
Organized	155	123
Sandlot	101	138
	256	261

15.6 [PA] Do municipal employees become more or less efficient as a function of longevity? Are the "old guard" better at their jobs than the younger, less experienced workers? A city manager has gathered the yearly efficiency ratings for all municipal employees along with data on longevity. Does the table below suggest any relationship? Summarize your conclusions in a few sentences.

Efficiency	Longevity		
	Low	High	
Low	54	25	79
Moderate	149	254	403
High	257	387	644
	460	666	1126

15.7 [SW] A sample of children has been observed and rated for symptoms of depression. Their parents have been rated for authoritarianism. Is there any relationship between these variables? Write a few sentences stating your conclusions.

Symptoms of Depression	Authoritarianism			
	Low	Moderate	High	
Few	7	8	9	24
Some	15	10	18	43
Many	8	12	3	23
	30	30	30	90

15.8 [SOC] Are prejudice and level of education related? State your conclusion in a few sentences.

Prejudice	Level of Education				
	Elementary School	High School	Some College	College Graduate	
Low	48	50	61	42	201
High	45	43	33	27	148
	93	93	94	69	349

15.9 [SOC] In a recent survey, respondents were asked to indicate how happy they were with their situations in life. Are their responses related to income level?

Happiness	Income			
	Low	Moderate	High	
Not happy	101	82	36	219
Pretty happy	40	227	100	367
Very happy	216	198	203	617
	357	507	339	1203

a. Describe the strength and direction of the relationship.
b. Is the relationship significant?

15.10 [CJ] One hundred and fifty cities have been classified as small, medium, or large by population and as high or low on crime rate. Is there a relationship between city size and crime rate?

Crime Rate	Size			
	Small	Medium	Large	
Low	21	17	8	46
High	29	33	42	104
	50	50	50	150

a. Describe the strength and direction of the relationship.

b. Is the relationship significant?

15.11 SOC Below are the scores of 15 nations on a measure of ethnic heterogeneity (or diversity) and a measure of civil strife (such as riots, demonstrations, and assassinations). Are these variables related? Do ethnically diverse nations experience more strife?

Nation	Ethnic Heterogeneity	Strife
A	.93	1.26
B	.85	1.25
C	.82	4.78
D	.80	5.02
E	.75	1.00
F	.60	.07
G	.45	.78
H	.43	.75
I	.42	.70
J	.39	1.26
K	.29	.07
L	.25	.52
M	.15	.48
N	.05	.48
O	.00	.33

15.12 SOC Twenty ethnic, racial, or national groups were rated by white and black students on a Social Distance Scale. Lower scores represent less social distance and, presumably, less prejudice. How similar are these rankings? Is the relationship statistically significant?

	Average Social Distance Scale scores	
Group	White Students	Black Students
1. White Americans	1.2	2.6
2. English	1.4	2.9
3. Canadians	1.5	3.6
4. Irish	1.6	3.6
5. Germans	1.8	3.9
6. Italians	1.9	3.3
7. Norwegians	2.0	3.8
8. American Indians	2.1	2.7

	Average Social Distance Scale scores	
Group	White Students	Black Students
9. Spanish	2.2	3.0
10. Jews	2.3	3.3
11. Poles	2.4	4.2
12. Black Americans	2.4	1.3
13. Japanese	2.8	3.5
14. Mexicans	2.9	3.4
15. Koreans	3.4	3.7
16. Russians	3.7	5.1
17. Arabs	3.9	3.9
18. Vietnamese	3.9	4.1
19. Turks	4.2	4.4
20. Iranians	5.3	5.4

15.13 SOC A random sample of eleven neighborhoods in Shinbone, Kansas, have been rated by an urban sociologist on a "quality-of-life" scale (which includes measures of affluence, availability of medical care, and recreational facilities) and a social cohesion scale. The results are presented below in scores. Higher scores indicate higher "quality of life" and greater social cohesion.

Neighborhood	Quality of Life	Social Cohesion
Happy Acres	17	8.8
North End	40	3.9
Brentwood	47	4.0
Rolling Meadows	90	3.1
Midtown	35	7.5
Bliss Park	52	3.5
Church View	23	6.3
Hidden Valley	67	1.7
College Park	65	9.2
Beaconsdale	63	3.0
Riverview 17	100	5.3

a. Are the two variables associated? What is the strength and direction of the association? Write a paragraph summarizing the relationship.

b. Conduct a test of significance for this relationship. Summarize your findings.

15.14 PA For years the city of Shinbone, Kansas, has been appropriating funds to help police officers defray the cost of college courses. Now the city council wants to know if this college experience has produced better police officers. Below, the number of college credits and efficiency ratings are reported for a random sample of 20 police officers. What can you tell the city council? (Efficiency ratings can range from a high score of 20 to a low of 0.)

Officer	Number of College Credits	Efficiency Ratings
A	124	17
B	120	19
C	100	18
D	90	16
E	89	15
F	75	20
G	70	14
H	60	12
I	60	13
J	59	11
K	45	6
L	30	5
M	24	8
N	21	9
O	18	2
P	12	0
Q	9	10
R	6	3
S	0	4
T	0	1

15.15 SW Several years ago, a job training program began and a team of social workers screened the candidates for suitability for employment. Now the screening process is being evaluated, and the actual work performance of these same individuals has been rated. Did the screening process work? Is there a relationship between the original scores and performance evaluation on the job?

Case	Original Score	Performance Evaluation
A	17	78
B	17	85
C	15	82
D	13	92
E	13	75

Case	Original Score	Performance Evaluation
F	13	72
G	11	70
H	10	75
I	10	92
J	10	70
K	9	32
L	8	55
M	7	21
N	5	45
O	2	25

15.16 SOC In problems 13.15 and 14.15, we looked at the relationships between five dependent variables and, respectively, political ideology and sex. In this exercise, we'll use income (item 12) as an independent variable and assess its relationship with this set of variables. To make income suitable for tabular analysis, we'll collapse it into three categories. For each table, calculate percentages and a measure of association. Describe the relationships in a few sentences. Data are from the 1993 General Social Survey, which is summarized in Appendix G.

a. Attitudes about welfare, by income

Spending on Welfare (item 22)	Income			
	Less than $22,500	$22,500 to $50,000	More than $50,000	
Too little	35	16	8	59
About right	45	26	19	90
Too much	74	78	57	209
	154	120	84	358

b. Attitudes about the role of the US in world affairs, by income

Should the U.S. Be Active? (item 27)	Income			
	Less than $22,500	$22,500 to $50,000	More than $50,000	
Active	129	123	103	355
Stay out	89	33	22	144
	218	156	125	499

c. Attitudes about national health care

Spending on Health Care (item 23)	Income			
	Less than $22,500	$22,500 to $50,000	More than $50,000	
Too little	116	93	59	268
About right	24	22	15	61
Too much	18	6	8	32
	158	121	82	361

d. Attitudes toward pornography, by income

Pornography Causes Breakdown. (item 59)	Income			
	Less than $22,500	$22,500 to $50,000	More than $50,000	
Yes	153	98	72	323
No	56	59	43	158
	209	157	115	481

e. Attitudes toward women's roles, by income

Women Should Take Care of Home and Family. (item 69)	Income			
	Less than $22,500	$22,500 to $50,000	More than $50,000	
Strongly agree	19	8	5	32
Agree	84	40	32	156
Disagree	92	82	60	234
Strongly disagree	31	28	29	88
	226	158	126	510

SPSS/PC+ PROCEDURES FOR ORDINAL-LEVEL MEASURES OF ASSOCIATION.

DEMONSTRATION 15.1 Do Sexual Attitudes or the Frequency of Traumatic Events Vary by Age? Another Look

SPSS/PC+ can compute five or six ordinal measures of association, including the three discussed in this chapter for collapsed ordinal scales. Spearman's rho, however, is not available in SPSS/PC+. If you need to compute this measure, it is available in SPSS[x], the mainframe version of this program. If you have access only to SPSS/PC+, you could compute the measure yourself or, perhaps, use the interval level measure of association, which will be introduced in the next chapter. As a final alternative, you could consider collapsing the variables down into a few categories and computing gamma or one of the other measures.

We will once again use the CROSSTAB program to calculate our statistics. In Demonstration 13.2, we used percentages to look at the relationships between TRAUMA5 (number of traumatic events over the past 5 years, item #50), PREMARSX (attitude toward premarital sex, item #42), and AGE. Let's reproduce these tables here and see if the measures of association clarify the relationships. AGE was the independent variable in both tables and TRAUMA5 was recoded. Follow the directions in Demonstration 13.2 for recoding TRAUMA5 and for the CROSSTABS command. Add /STATISTICS

1 6 8 9 to the CROSSTABS command. Your final command should look like this:

```
CROSSTABS /TABLES TRAUMA5 BY AGE /OPTIONS 4
           /STATISTICS 1 6 8 9.
```

And the first table will be

Cross-tabulation: TRAUMA5   TRAUMA SCALE, LAST 5 YEARS
                      By AGE   AGE OR RESPONDENT

```
                 Count :  < 36  : 37 - 51 : 52 +        :
         AGE ->  Col Pct :       :         :            : Row
                 :        1.00 :    2.00 :     3.00 : Total
 TRAUMA5         ------- : ------- : ------- : ------- :
            .00 :   64   :   47    :   34    :   145
                :  37.4  :  29.9   :  20.2   :  29.2
                : ------- : ------- : ------- :
           1.00 :   62   :   67    :   79    :   208
                :  36.3  :  42.7   :  47.0   :  41.9
                : ------- : ------- : ------- :
           2.00 :   45   :   43    :   55    :   143
     2 +        :  26.3  :  27.4   :  32.7   :  28.8
                : ------- : ------- : ------- :
         Column    171      157       168      496
         Total    34.5     31.7      33.9    100.0
```

Chi-Square	D.F.	Significance	Min. E.F.	Cells with E.F.< 5
12.35395	4	.0149	45.264	None

MORE

Statistic	Symmetric	With TRAUMA5 Dependent	With AGE Dependent
Somers' D	.11441	.11349	.11534

Statistic	Value	Significance
Kendall's Tau B	.11441	.0021
Gamma	.17208	

Number of Missing Observations = 292

The three measures are consistent with each other. The relationship between TRAUMA5 and AGE is weak and positive in direction. Older respondents report higher numbers of traumatic incidents (or: the older the respondent, the greater the number of incidents). Observe the percentages once again and see how they are consistent with the sign of the measures.

The CROSSTABS command with PREMARSX as the dependent variable, should look like this

```
CROSSTABS /TABLES PREMARSX BY AGE /OPTIONS 4 /
STATISTICS 1 6 8 9.
```

The resulting table is

```
Cross-tabulation:    PREMARSX   SEX BEFORE MARRIAGE
                     By AGE     AGE OF RESPONDENT
```

	Count	< 36	37 _ 51	52 +	
AGE ->	Col Pct				Row
		1.00	2.00	3.00	Total
PREMARSX					
1.00 ALWAYS WRONG		32 / 18.5	39 / 22.2	68 / 41.5	139 / 27.1
2.00 ALMST ALWAYS WRG		16 / 9.2	16 / 9.1	24 / 14.6	56 / 10.9
3.00 SOMETIMES WRONG		37 / 21.4	37 / 21.0	29 / 17.7	103 / 20.1
4.00 NOT WRONG AT ALL		88 / 50.9	84 / 47.7	43 / 26.2	215 / 41.9
Column Total		173 / 33.7	176 / 34.3	164 / 32.0	513 / 100.0

Chi-Square	D.F.	Significance	Min. E.F.	Cells with E.F.< 5
36.91395	6	.0000	17.903	None

MORE

Statistic	Symmetric	With PREMARSX Dependent	With AGE Dependent
Somers' D	-.20956	-.21465	-.20471

```
      Statistic              Value          Significance
      ---------              -------        ------------

Kendall's Tau B            -.20962              .0000
Gamma                      -.30219

Number of Missing Observations = 275
```

The three measures are consistent once again. This is a weak to moderate negative relationship. As age increases, support of premarital sex (the percentage who say it's "not wrong at all") decreases.

Which of the three measures should we choose? As I indicated in Section 15.4, gamma is the largest of the three measures for both tables, but not by much. The second table is not "square," so we should not use tau-*b*. For the first table, all three measures are about equally logical and appropriate. Whichever you choose to report, the conclusion will be that the association is weak to moderate in strength, positive for TRAUMA5 and AGE, and negative for PREMARSX and AGE.

DEMONSTRATION 15.2 A Note about the Direction of Ordinal Relationships

Let's pause to briefly consider the direction of relationships between ordinal-level variables. Direction for ordinal variables is a surprisingly tricky matter, mostly because of the arbitrary nature of coding at the ordinal level. We usually think of *larger* scores as indicating *more* of the quantity being measured and, for every interval-ratio level variable I can think of, this pattern will be true. For ordinal variables, however, larger scores may indicate *less* of the quantity because the codes are arbitrary. Depending on how you code the values, a high score on a scale measuring, say, prejudice might indicate great prejudice or its complete absence.

Looking at the table for PREMARSX and AGE in Demonstration 15.1, remember that a negative gamma means that *high* values on one variable are associated with *low* values on the other. This means, for example, that those who scored a 1 (the lowest possible value) on AGE tended to score a 4 (the highest possible value) on PREMARSX. You may think of a 4 on PREMARSX as representing "high support for premarital sex" or "low opposition to premarital sex." A negative gamma *always* means that the scores of the variables are inversely related, but this does *not* necessarily mean that the underlying relationship is truly negative. Always inspect tables carefully to make sure that you are interpreting the direction of the relationship properly.

DEMONSTRATION 15.3 Interpreting the Direction of Relationships: Are Prestigious Jobs More Satisfying?

What's the relationship between prestige of occupation and job satisfaction? Some would argue that high-prestige jobs are rewarding in a variety of ways (besides simply income) and that people in such jobs would express great satisfaction with their work. Others might argue an opposing point of view: because people in lower-prestige jobs have less pressure and fewer responsibilities, they will experience high levels of satisfaction. In an attempt to resolve the debate, we'll run a CROSSTABS on PRESTG80 (item 2) and SATJOB (item 49). I first found the median on PRESTG80 with the FREQUENCIES command and recoded the variable into a dichotomy. I also recoded SATJOB into two approximately equal categories. My commands were

```
RECODE PRESTG80 (Low thru 42 = 1)(43 Thru High =2).
VALUE LABELS PRESTG80 1 'Lower' 2 'Higher'.
RECODE SATJOB (1=1)(2 Thru 4=2).
VALUE LABELS SATJOB 1 'very satisfied' 2 'less
satisfied'.
```

Follow the instructions in Demonstration 15.1 for CROSSTABS. Your command should look like this:

```
CROSSTABS /TABLES SATJOB BY PRESTG80/OPTIONS 4
         /STATISTICS 1 6 8 9.
```

and the table will be

```
Cross-tabulation:    SATJOB    JOB OR HOUSEWORK
                  By PRESTG80  RS OCCUPATIONAL PRESTIGE SCORE (1980)
```

	Count Col Pct	lower	higher	Row
PRESTG80->		1.00	2.00	Total
SATJOB	-------	------	------	
1.00 very satisfied		108 35.6	160 54.6	268 45.0
2.00 less satisfied		195 64.4	133 45.4	328 55.0
Column Total		303 50.8	293 49.2	596 100.0

```
Chi-Square  D.F.  Significance  Min. E.F.  Cells with E.F.< 5
----------  ----  ------------  ---------  ------------------

  20.88783    1      .0000       131.752    None
  21.64737    1      .0000      ( Before Yates Correction )


------------------------------------------------------------------

                                With SATJOB    With PRESTG80
          Statistic   Symmetric  Dependent     Dependent
          ---------   ---------  ---------     -------------

Somers' D             -.19058     -.18964          -.19153

          Statistic          Value          Significance
          ---------          -----          ------------

Kendall's Tau B            -.19058              .0000
Gamma                      -.36950

Number of Missing Observations = 192
```

This relationship is statistically significant (the significance of the chi square is less than .05), weak to moderate in strength (the measures range from $-.19$ to $-.37$), and all three measures are negative. As prestige increases, job satisfaction decreases. Right? Wrong. Look at the codes for SAT-JOB. A *higher* score indicates a *lower* level of satisfaction. So, "high" (or a score of 2) on PRESTG80 is associated with "high" (or a score of 1) on SAT-JOB. In spite of the negative signs for all three measures, this is a *positive* relationship and job satisfaction increases with prestige. I know, it shouldn't be this complicated. Reread Demonstration 15.2 and try to make the best of it.

Exercises

15.1 Examine the relationship between recoded POLVIEWS and CON-CLERG, CONEDUC, CONFED, CONPRESS, and CONLEGIS. Run the CROSSTABS task with OPTIONS 4 and STATISTICS 1 6 8 9. Interpret the strength and direction of the relationship in each table. Which of the three political ideologies is associated with the most confidence in each institution?

15.2 Recode AGE into two or three categories and construct CROSSTAB tables with recoded AGE as the independent and the three "confidence" variables used in exercise 15.1 as dependent. Which age group has the greatest "confidence?" How strong is the relationship? Is the relationship significant? (You can make this judgment from the chi square statistic without doing the test of significance presented in Section 15.6.)

15.3 Following the procedure in Demonstration 15.3, examine the relationship between SATFIN and PRESTG80. Recode INCOME91 and DEGREE and use them as independents against SATJOB and SATFIN. Summarize the relationships (percentage patterns, significance, strength, and direction) in a paragraph. Be careful in interpreting the direction of these relationships.

16

Association Between Variables Measured at the Interval-Ratio Level

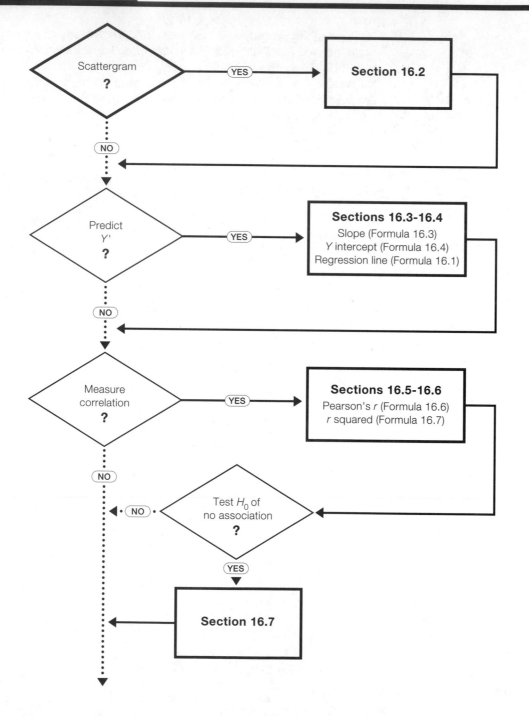

16.1 INTRODUCTION

This chapter presents a set of statistical techniques for analyzing the association (or correlation) between variables measured at the interval-ratio level.* As we shall see, these techniques are rather different in their logic and computation from those covered in Chapters 14 and 15. Let me stress at the outset, therefore, that the questions these new techniques are designed to answer are exactly the same as questions we have previously attempted to answer: Is there a relationship between the two variables? How strong is the relationship? What is the direction of the relationship? You might tend to become preoccupied by the technical details of the underlying logic and the computational routines necessary for these new techniques, but remind yourself occasionally that the goals of the analysis are identical to those of other measures of association. Ultimately, we seek to understand the nature of a bivariate relationship as a way of probing possible causal relationships and improving our ability to predict scores of a case on one variable from that case's score on another variable.

16.2 SCATTERGRAMS

Over the past several chapters, we have repeatedly noted that by properly calculating percentages for bivariate tables, the researcher can gain important information about a bivariate association. When analyzing nominal- or ordinal-level variables, you will almost always be rewarded for taking the time to examine the pattern of cell frequencies in the table in addition to computing a summary measure of association.

By the same token, the usual first step in analyzing a relationship between interval-ratio variables is to construct and examine a **scattergram**. Scattergrams are graphic display devices that are functionally analogous to bivariate tables, since they permit the researcher to quickly perceive several important features of the relationship.

To illustrate the construction and use of scattergrams, let us begin with an example. Suppose a researcher is interested in analyzing how dual wage-earner families (that is, families where both husband and wife have jobs outside the home) cope with housework. Specifically, the researcher wonders if the number of children in the family is related to the amount of time the husband contributes to housekeeping chores. The relevant data for a sample of 12 dual-wage-earner families are displayed in Table 16.1.

To construct a scattergram, begin by drawing two axes of about equal length and at right angles to each other. Array the independent (X) variable along the horizontal axis (the abscissa) and the dependent (Y) variable along the vertical axis (the ordinate). Calibrate both axes in convenient units of the original scale. Then, for each case, locate the point along the abscissa that corresponds to the score of that case on the X variable. Draw a straight line

---

*The term *correlation* is commonly used instead of *association* when discussing the relationship between interval-ratio variables. We will use the two terms interchangeably.

TABLE 16.1 NUMBER OF CHILDREN AND HUSBAND'S CONTRIBUTION TO HOUSEWORK (fictitious data)

Family	Number of Children	Hours Per Week Husband Spends on Housework
A	1	1
B	1	2
C	1	3
D	1	5
E	2	3
F	2	1
G	3	5
H	3	0
I	4	6
J	4	3
K	5	7
L	5	4

FIGURE 16.1 HUSBAND'S HOUSEWORK BY NUMBER OF CHILDREN

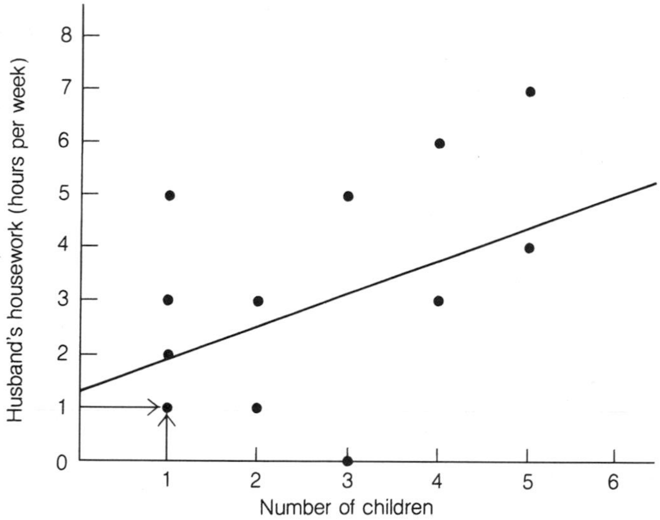

up from that point and at right angles to the axis. Then, locate the point along the ordinate that corresponds to the score of that same case on the *Y* variable. Draw a straight line out from this point and perpendicular to the ordinate. Where the line from the ordinate crosses the line from the abscissa, place a dot to represent the case. Repeat with all cases.

Figure 16.1 shows a scattergram displaying the relationship between "number of children" and "husband's housework" for the sample of 12 families presented in Table 16.1. Note that each family (case) is represented by a dot lying at the intersection of the two scores for that family. Also note that, as always, the scattergram is clearly titled and both axes are labeled.

The overall pattern of the dots (or observation points) succinctly summarizes the nature of the relationship between the two variables. The clarity of the pattern formed by the dots can be enhanced by drawing a straight line through the cluster such that the line touches every dot or comes as close to doing so as possible. In Section 16.3, a technique for fitting this line to the pattern of the dots will be explained; but, for now, an "eyeball" approximation will suffice. This summarizing line is called the **regression line** and has already been added to the scattergram.

Even a crudely drawn scattergram with a freehand regression line can be used for a variety of purposes. Scattergrams provide at least impressionistic information about the existence, strength, and direction of the relationship and can also be used to check the relationship for linearity (that is, how well the pattern of dots can be approximated with a straight line). Finally, the scattergram can be used to predict the score of a case on one variable from the score of that case on the other variable. Let us briefly examine each of these uses.

To ascertain the existence of a relationship, we can return to the basic definition of an association stated in Chapter 13. Two variables are associated if the distributions of Y change for the various conditions of X. In Figure 16.1, the scores along the abscissa (number of children) are conditions or values of X. The dots above each X value can be thought of as the conditional distributions of Y (that is, the dots represent scores on Y for each value of X). Figure 16.1 shows that these conditional distributions of Y change as X changes in that the Y scores vary across the scores of X. The existence of an association is further reinforced by the fact that the regression line lies at an angle to the X axis (the abscissa). If these two variables had not been associated, the conditional distributions of Y would not have changed, and the regression line would have been parallel to the abscissa.

The strength of the association can be judged by observing the spread of the dots around the regression line. In a perfect association, all dots would lie on the regression line, and the less the scattering of the dots around the regression line, the stronger the association.

The direction of the relationship can be detected by observing the angle of the regression line with respect to the abscissa. Figure 16.1 displays a positive relationship because cases with high scores on X also tend to have high scores on Y. If the relationship had been negative, the regression line would have sloped in the opposite direction to indicate that high scores on one variable were associated with low scores on the other.

To summarize these points about the existence, strength, and direction of the relationship, Figure 16.2 shows a perfect positive and a perfect negative relationship and a "zero relationship," or "nonrelationship" between two variables.

One key assumption underlying the statistical techniques to be intro-

FIGURE 16.2 POSITIVE, NEGATIVE, AND ZERO RELATIONSHIPS

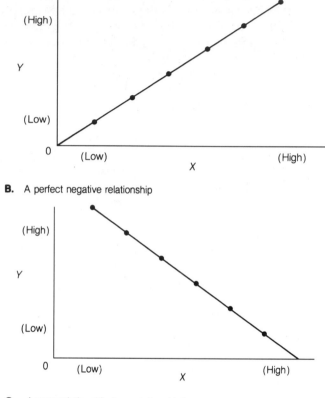

A. A perfect positive relationship

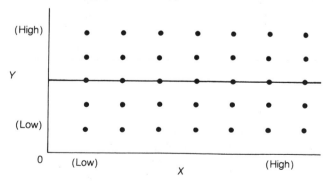

B. A perfect negative relationship

C. A zero relationship (nonrelationship)

FIGURE 16.3   SOME NONLINEAR RELATIONSHIPS

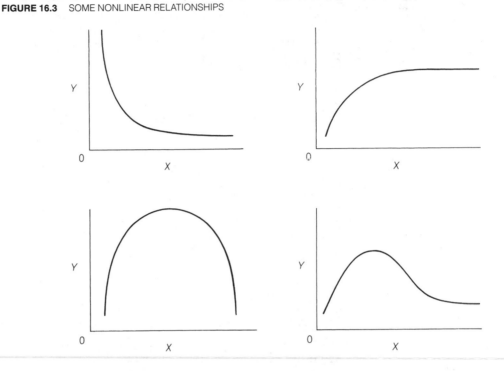

duced later in this chapter is that the two variables have an essentially **linear relationship**. In other words, the observation points or dots in the scattergram must form a pattern that can be approximated with a straight line. Significant departures from linearity would require the use of statistical techniques beyond the scope of this text. Examples of some common curvilinear relationships are presented in Figure 16.3.

If the scattergram indicates that the variables are associated in a non-linear fashion, the techniques described in this chapter should be used with great caution or, more circumspectly, not at all. Checking for the linearity of the relationship is perhaps the most important reason for constructing at least a crude, hand-drawn scattergram before proceeding with the statistical analysis. If the relationship is nonlinear, you will not meet the assumptions of the interval-ratio measure of association, and you might need to treat the variables as if they were ordinal rather than interval-ratio in level of measurement.

16.3 REGRESSION AND PREDICTION

A final use of the scattergram is to predict scores of cases on one variable from their score on the other. To illustrate, suppose that, based on the relationship between number of children and husband's housework displayed in Figure 16.1, we wish to predict the number of hours of housework a husband

FIGURE 16.4 PREDICTING HUSBAND'S HOUSEWORK

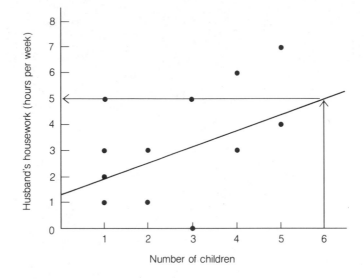

with a family of six children would do each week. The sample of 12 families had no cases of six-children families; but, if the axes and regression line in Figure 16.1 are extended to incorporate this score, a prediction can be made. Figure 16.4 reproduces the scattergram and illustrates how the prediction would be made.

The predicted score on Y, which we will symbolize as Y' to distinguish our predictions of Y from actual Y scores, is found by first locating the relevant score on X ($X = 6$ in this case) and then drawing a straight line from that point on the abscissa to the regression line. From the regression line, another straight line parallel to the abscissa is drawn across to the Y axis, or ordinate. The predicted Y score (Y') is found at the point where the line from the regression line crosses the Y axis. In our example, we would predict that, in a dual wage-earner family with six children, the husband would devote about five hours per week to housework.

Of course, this technique for predicting Y' is crude and impressionistic. The most serious limitation of this informal prediction technique is that Y' can change in value, depending on how accurately the freehand regression line is drawn. One way to eliminate this source of error would be to find the straight line that most accurately summarizes the pattern of the observation points and so best describes the relationship between the two variables. Is there such a "best-fitting" straight line? If there is, how is it defined?

Recall that our criterion for the freehand regression line was that it touch all the dots or come as close to doing so as possible. Also, recall that the dots above any given value of X can be thought of as conditional distributions of

Y. Within each conditional distribution of *Y*, we can locate a point around which the variation of the scores is minimized. This point of minimized variation is nothing other than the mean of the conditional distributions of *Y*.

In Chapter 3, we noted that the mean of any distribution of scores is the point around which the variation of the scores, as measured by squared deviations, is minimized:

$$\Sigma(X_i - \bar{X})^2 = \text{minimum}$$

Thus, if the regression line is fitted so that it touches the mean of each conditional distribution of *Y*, we would have a line coming as close to all the scores as possible. Such a line would minimize the deviations of the *Y* scores because it would contain all **conditional means of** ***Y***; and the mean of any distribution is the point of minimized variation.

Conditional means are found by summing all *Y* values for each value of *X* and then dividing by the number of cases. For example, four families had one child (*X* = 1), and the husbands of these four families devoted 1, 2, 3, and 5 hours per week to housework. Thus, for *X* = 1, *Y* = 1, 2, 3, and 5, and the conditional mean of *Y* for *X* = 1 is 2.75 (11/4 = 2.75). Husbands in families with one child worked an average of 2.75 hours per week doing housekeeping chores. Conditional means of *Y* are computed in the same way for each value of *X* displayed in Table 16.2 and plotted on a scattergram in Figure 16.5.

Let us quickly remind ourselves of the reason for these calculations. We are seeking the single best-fitting regression line for summarizing the relationship between *X* and *Y*, and we have seen that a line drawn through the conditional means of *Y* will minimize the spread of the observation points around the line. It will come as close to all the observation points as possible and will therefore be the single best-fitting regression line for these data.

Now, a line drawn through the points on Figure 16.5 (the conditional means of *Y*) will be the best-fitting line we are seeking. But, you can see from the scattergram that such a line will not be straight. In fact, only rarely will conditional means fall in a perfectly straight line (that is, in a perfect relation-

TABLE 16.2   CONDITIONAL MEANS OF *Y* (Husband's housework)
FOR VARIOUS VALUES OF *X* (Number of children)

Number of Children (*X*)	Husband's Housework (*Y*)	Conditional Means of *Y*
1	1,2,3,5	2.75
2	3,1	2.00
3	5,0	2.50
4	6,3	4.50
5	7,4	5.50

FIGURE 16.5 CONDITIONAL MEANS OF *Y*

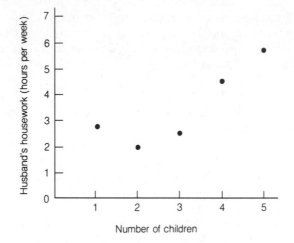

ship between *X* and *Y*). Since we still must meet the condition of linearity, let us revise our criterion and define the regression line as the unique straight line that touches all conditional means of *Y* or comes as close to doing so as possible. It can be shown that the straight line described in Formula 16.1 will conform to this criterion:

FORMULA 16.1

$$Y = a + bX$$

where *Y* = score on the dependent variable
 a = the **Y intercept**, or the point where the regression line crosses the *Y* axis
 b = the **slope** of the regression line, or the amount of change produced in *Y* by a unit change in *X*
 X = score on the independent variable

This formula describes the "least-squares" regression line, or the regression line that best fits the pattern of the data points. The formula introduces two new concepts. First, the *Y* intercept (*a*) is the point at which the regression line crosses the vertical, or *Y*, axis. Second, the slope (*b*) of the least-squares regression line is the amount of change produced in the dependent variable (*Y*) by a unit change in the independent variable (*X*). Think of the slope of the regression line as a measure of the effect of the *X* variable on the *Y* variable. If the variables have a strong association, then changes in the value of *X* will be accompanied by substantial changes in the value of *Y*, and the slope (*b*) will have a high value. The weaker the effect of *X* on *Y* (the weaker the association between the variables), the lower the value of the slope (b). If the two variables are unrelated, the least-squares regression line would be parallel to the abscissa, and b would be 0 (the line would have no slope).

With the least-squares formula (Formula 16.1), we can predict values of Y in a much less arbitrary and impressionistic way than through mere eyeballing. This will be so, remember, because the least-squares regression line as defined by Formula 16.1 touches or comes as close to the conditional means of Y as possible and is, therefore, the line that best fits the data. Before seeing how predictions of Y can be made, however, we must first calculate a and b.

16.4 THE COMPUTATION OF *a* AND *b*

Since the value of b is needed to solve for a, let us begin with the computation of the slope of the least-squares regression line. The definitional formula for the slope is

FORMULA 16.2

$$b = \frac{\Sigma(X - \bar{X})(Y - \bar{Y})}{\Sigma(X - \bar{X})^2}$$

The numerator of this formula is called the covariation of X and Y. It is a measure of how X and Y vary together, and its value will reflect both the direction and strength of the relationship. It is tedious to find the slope with this formula, and the following computational formula, which can be derived from Formula 16.2, should be used instead:

FORMULA 16.3

$$b = \frac{N\Sigma XY - (\Sigma X)(\Sigma Y)}{N\Sigma(X^2 - (\Sigma X)^2}$$

where b = the slope
N = the number of cases
ΣXY = the summation of the crossproducts of the scores
ΣX = the summation of the scores on X
ΣY = the summation of the scores on Y
ΣX^2 = the summation of the squared scores on X

Admittedly, this formula appears formidable at first glance; but it can be solved without too much difficulty if computations are organized into table format. The computing table displayed in Table 16.3 has a column for each of the four quantities needed to solve the formula. The data are from the dual wage-earner family sample (see Table 16.1).

In Table 16.3, the first two columns list the original X and Y scores for each case. The third column contains the squared scores on X, and the fourth lists the squared scores on Y. The fifth column lists the crossproducts of the scores for each case. In other words, the entries in the last column are deter-

TABLE 16.3 COMPUTATION OF THE SLOPE (*b*)

X	Y	X²	Y²*	XY
1	1	1	1	1
1	2	1	4	2
1	3	1	9	3
1	5	1	25	5
2	3	4	9	6
2	1	4	1	2
3	5	9	25	15
3	0	9	0	0
4	6	16	36	24
4	3	16	9	12
5	7	25	49	35
5	4	25	16	20
Total = 32	40	112	184	125

$$\Sigma X = 32$$
$$\Sigma Y = 40$$
$$\Sigma X^2 = 112$$
$$\Sigma Y^2 = 184$$
$$\Sigma XY = 125$$

*The quantity ΣY^2 is not used in the computation of *b*. We will need it later, however, when we compute Pearson's *r* (see Section 16.5).

mined by multiplying both scores for each case. We can now replace the symbols in Formula 16.3 with the proper sums:

$$b = \frac{N\Sigma XY - (\Sigma X)(\Sigma Y)}{N\Sigma X^2 - (\Sigma X)^2}$$

$$b = \frac{(12)(125) - (32)(40)}{(12)(112) - (32)^2}$$

$$b = \frac{1500 - 1280}{1344 - 1024}$$

$$b = \frac{220}{320}$$

$$b = .69$$

A slope of .69 indicates that, for each unit change in *X*, there is an increase of .69 units in *Y*. For the specific example under consideration, the addition of each child (an increase of one unit in *X*) results in an increase of .69 hours of housework being done by the husband (an increase of .69 units—or hours—in *Y*).

Once the slope has been calculated, finding the intercept (*a*) is relatively easy. To compute the mean of $X(\bar{X})$ and the mean of $Y(\bar{Y})$, divide the sums of columns 1 and 2 of Table 16.3 by *N* and enter these figures in Formula 16.4:

FORMULA 16.4
$$a = \bar{Y} - b\bar{X}$$

For our sample problem, the value of a would be

$$a = \bar{Y} - b\bar{X}$$
$$a = 3.33 - (.69)(2.67)$$
$$a = 3.33 - 1.84$$
$$a = 1.49$$

Thus, the least-squares regression line will cross the Y axis at the point where Y equals 1.49.

The full least squares regression line for our sample data can now be specified:

$$Y = a + bX$$
$$Y = (1.49) + (.69)\ X$$

This formula can be used to estimate or predict scores on Y for any value of X. In Section 16.3, we used the freehand regression line to predict a score on Y (husband's housework) for a family with six children ($X = 6$). Our prediction was that, in families of six children, husbands would contribute about five hours per week to housekeeping chores. Let us see with the least-squares regression line how close our impressionistic, eyeball prediction was.

$$Y' = a + bX$$
$$Y' = (1.49) + (.69)(6)$$
$$Y' = (1.49) + (4.14)$$
$$Y' = 5.63$$

Based on the least-squares regression line, we would predict that in a dual wage-earner family with six children, husbands would devote 5.63 hours a week to housework. What would our prediction of husband's housework be for a family of seven children ($X = 7$)?

Note that our predictions of Y scores are basically "educated guesses." We will be unlikely to predict values of Y exactly except in the (relatively rare) case where the bivariate relationship is perfect and perfectly linear. Note also, however, that the accuracy of our predictions will increase as relationships become stronger. That is, the less the spread of the observation points around the least-squares regression line (the stronger the relationship), the more accurate will be our predictions of values of Y from values of X.

16.5 THE CORRELATION COEFFICIENT (PEARSON'S r)

I pointed out in Section 16.4 that the slope of the least-squares regression line (b) is a measure of the effect of X on Y. Since the slope is the amount of change produced in Y by a unit change in X, b will increase in value as the relationship increases in strength. Thus, the value of b is a function of the strength of the relationship; but, because b does not vary between 0 and 1, it is awkward to use as a measure of association per se. For a measure of association for two interval-ratio variables, researchers rely heavily (almost exclusively) on a statistic called **Pearson's r**, or the correlation coefficient.

Like the ordinal measures of association discussed in Chapter 15, Pearson's r varies from 0 to ± 1, with 0 indicating no association and $+1$ and -1 indicating perfect positive and perfect negative relationships, respectively. The definitional formula for Pearson's r is

FORMULA 16.5

$$r = \frac{\Sigma(X - \bar{X})(Y - \bar{Y})}{\sqrt{[\Sigma(X - \bar{X})^2][\Sigma(Y - \bar{Y})^2]}}$$

Note that the numerator of this formula is the covariation of X and Y, as was the case with Formula 16.2. This formula is awkward to use, and the computational Formula 16.6 is usually preferred.

FORMULA 16.6

$$r = \frac{N\Sigma XY - (\Sigma X)(\Sigma Y)}{\sqrt{[N\Sigma X^2 - (\Sigma X)^2][N\Sigma Y^2 - (\Sigma Y)^2]}}$$

A computing table such as Table 16.3 is strongly recommended as a way of organizing the quantities needed to solve this equation. For our sample problem involving dual-wage-earner families, the quantities displayed in Table 16.3 can be substituted directly into Formula 16.6:

$$r = \frac{(12)(125) - (32)(40)}{\sqrt{[(12)(112) - (32)^2][(12)(184) - (40)^2]}}$$

$$r = \frac{1500 - 1280}{\sqrt{(1344 - 1024)(2208 - 1600)}}$$

$$r = \frac{220}{\sqrt{194{,}560}}$$

$$r = \frac{220}{441.09}$$

$$r = .50$$

An r value of .50 indicates a moderately strong, positive linear relationship between the variables. As the number of children in the family increases, the hourly contribution of husbands to housekeeping duties also increases.

16.6 INTERPRETING THE CORRELATION COEFFICIENT: r^2

Pearson's r is an index of the strength of the linear relationship between two variables. While a value of 0.00 indicates no linear relationship and a value of ± 1.00 indicates a perfect linear relationship, values between these extremes have no direct interpretation. We can, of course, describe relationships in terms of how closely they approach the extremes (for example, coefficients approaching 0.00 can be described as "weak" and those approaching ± 1.00 as "strong"), but this description is somewhat subjective and less than desirable.

Fortunately, a more direct interpretation is provided by calculating an additional statistic called the **coefficient of determination**. This statistic, which is simply the square of Pearson's r (r^2), can be interpreted with a logic akin to proportional reduction in error (PRE). As you recall, the logic of PRE measures of association is to predict the value of the dependent variable under two different conditions. First, Y is predicted while ignoring the information supplied by X and, second, the independent is taken into account when predicting the dependent. In previous applications of the logic of PRE, we actually counted the number of errors made under each prediction rule and manipulated the sums to derive a final measure. In the case of r^2, both the method of prediction and the construction of the final statistic are somewhat different and require the introduction of some new concepts.

When working with variables measured at the interval-ratio level, the predictions of Y under the first condition (while ignoring X) will be the mean of the Y scores ($\bar{Y}$) for every case. Given no information on X, this prediction strategy will be optimal because we know that the mean of any distribution is closer than any other point to all the scores in the distribution. I remind you of the principle of minimized variation introduced in Chapter 3 and expressed as

$$\Sigma(Y - \bar{Y})^2 = \text{minimum}$$

This statement expresses the fact that the scores of any variable vary around the mean less than they vary around any other point. If we predict the mean of Y for every case, we will make fewer errors of prediction than if we predict any other value for Y.

Of course, we will still make many errors in predicting Y even if we faithfully follow this strategy. The amount of error is represented in Figure 16.6,

FIGURE 16.6   PREDICTING Y WITHOUT X (dual-career families)

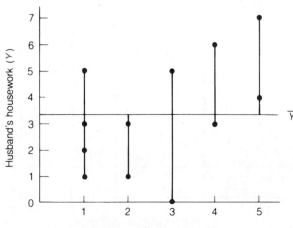

which displays the relationship between number of children and husband's housework with the mean of $Y(\bar{Y})$ noted. The vertical lines from the actual scores to the predicted score $(\bar{Y})$ represent the amount of error we would make when predicting Y while ignoring X.

We can find a more precise measure of the extent of our prediction error under the first condition by taking each actual Y score, subtracting the mean of Y from the score, and squaring and summing these deviations. The resultant figure, which can be noted as $\Sigma(Y - \bar{Y})^2$, is called the **total variation** in Y. We now have a visual representation (Figure 16.6) and a method for calculating the error we incur by predicting Y without knowledge of X. As we shall see below, we do not need to actually calculate the total variation to find the value of the coefficient of determination, r^2.

Our next step will be to determine the extent to which knowledge of X improves our ability to predict Y. If the two variables have a linear relationship, then predicting scores on Y from the least-squares regression equation will incorporate knowledge of X and reduce our errors of prediction. So, under the second condition, our predicted Y score for each value of X will be

$$Y' = a + bX$$

Figure 16.7 displays the data from the dual-career families with the regression line, as determined by the above formula, drawn in. The vertical lines from each data point to the regression line represent the amount of error in predicting Y that remains even after X has been taken into account.

As was the case under the first condition, we can develop precise ways of describing the reduction in error that results from taking X into account. Specifically, two different sums can be found and then compared with the

FIGURE 16.7 PREDICTING Y WITH X (dual-career families)

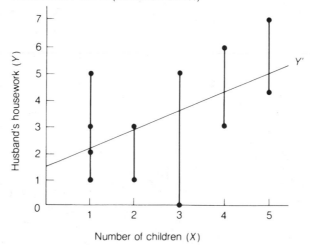

total variation of Y to construct a statistic that will indicate the improvement in prediction.

The first sum, called the **explained variation**, represents the improvement in our ability to predict Y when taking X into account. This sum is found by subtracting $\bar{Y}$ (our predicted Y score without X) from the score predicted by the regression equation (Y', or the Y score predicted with knowledge of X) for each case and then squaring and summing these differences. These operations can be summarized as $\Sigma(Y' - \bar{Y})^2$, and the resultant figure could then be compared with the total variation in Y to ascertain to what extent our knowledge of X improves our ability to predict Y. Specifically, it can be shown mathematically that

FORMULA 16.7
$$r^2 = \frac{\Sigma(Y' - \bar{Y})^2}{\Sigma(Y - \bar{Y})^2} = \frac{\text{explained variation}}{\text{total variation}}$$

Thus, the coefficient of determination, or r^2, is the proportion of the total variation in Y attributable to or explained by X. Like other PRE measures, r^2 indicates, in a precise way, to what extent knowledge of X helps us to predict, or understand, or explain Y.

Above, we refer to the improvement in predicting Y with X as the explained variation. The use of this term suggests that some of the variation in Y will be "unexplained" or not attributable to the influence of X. In fact, the vertical lines in Figure 16.7 represent the **unexplained variation**, or the difference between our best prediction of Y with X and the actual scores. The unexplained variation is thus the scattering of the actual scores around the regression line and can be found by subtracting the predicted Y scores from the actual Y scores for each case and then squaring and summing these differences. These operations can be summarized as $\Sigma(Y - Y')^2$, and the resultant sum would measure the amount of error in predicting Y that remains even after X has been taken into account. The proportion of the total variation in Y unexplained by X can be found by subtracting the value of r^2 from 1.00. Unexplained variation is usually attributed to the influence of some combination of other variables, measurement error, and random chance.

As you may have recognized by this time, the explained and unexplained variations bear a reciprocal relationship with each other. As one of these sums increases in value, the other decreases. Furthermore, the stronger the linear relationship between X and Y, the greater the value of the explained variation and the lower the unexplained variation. In the case of a perfect relationship ($r = \pm 1.00$), the unexplained variation would be 0 and r^2 would be 1.00. This would indicate that X explains or accounts for all the variation in Y and that we could predict Y from X without error. On the other hand, when X and Y are not linearly related ($r = 0.00$), the explained variation would be 0 and r^2 would be 0.00. In such a case, we would conclude that X explains none of the variation in Y and does not improve our ability to predict Y.

APPLICATION 16.1

For five cities, information has been collected on number of civil disturbances (riots, strikes, and so forth) over the past year and on unemployment rate. Are these variables associated?

The data are presented in the table below. Columns have been added for all necessary sums.

The slope (b) is

$$b = \frac{N\Sigma XY - (\Sigma X)(\Sigma Y)}{N\Sigma X^2 - (\Sigma X)^2}$$

$$b = \frac{(5)(985) - (76)(53)}{(5)(1290) - (76)^2}$$

$$b = \frac{897}{674}$$

$$b = 1.33$$

A slope of 1.33 means that for every unit change in X (for every increase of 1 in the unemployment rate) there was a change of 1.33 units in Y (the number of civil disturbances increased by 1.33).

The Y intercept (a) is

$$a = \bar{Y} - b\bar{X}$$

$$a = \left(\frac{53}{5}\right) - (1.33)\left(\frac{76}{5}\right)$$

$$a = 10.6 - (1.33)(15.20)$$

$$a = 10.6 - 20.2$$

$$a = -9.6$$

The least-squares regression equation is

$$Y = a + bX$$

$$Y = -9.6 + (1.33)X$$

The correlation coefficient is

$$r = \frac{N\Sigma XY - (\Sigma X)(\Sigma Y)}{\sqrt{[N\Sigma X^2 - (\Sigma X)^2][N\Sigma Y^2 - (\Sigma Y)^2]}}$$

$$r = \frac{(5)(985) - (76)(53)}{\sqrt{[(5)(1290) - (76)^2][(5)(919) - (53)^2]}}$$

$$r = \frac{897}{\sqrt{(674)(1786)}}$$

$$r = \frac{897}{\sqrt{1203764}}$$

$$r = \frac{897}{1097.16}$$

$$r = 0.82$$

These variables have a strong, positive association. The number of civil disturbances increases as the unemployment rate increases. The coefficient of determination, r^2, is $(0.82)^2$, or 0.67. This indicates that 67% of the variance in civil disturbances is explained by the unemployment rate.

City	Unemployment Rate (X)	Civil Disturbances (Y)	X^2	Y^2	XY
A	22	25	484	625	550
B	20	13	400	169	260
C	10	10	100	100	100
D	15	5	225	25	75
E	9	0	81	0	0
	76	53	1290	919	985

Relationships intermediate between these two extremes can be interpreted in terms of how much X increases our understanding of or ability to predict or explain Y. For the dual-career families, we calculated an r of 0.50. Squaring this value yields a coefficient of determination of 0.25 ($r^2 = 0.25$), which indicates that number of children (X) explains 25% of the total variation in husband's housework (Y). When predicting the number of hours per week that husbands in such families would devote to housework, we will make 25% fewer errors by basing the predictions on number of children and predicting from the regression line, as opposed to ignoring this variable and predicting the mean of Y for every case. Also, 75% of the variation in Y is unexplained by X and presumably due to some combination of the influence of other variables, measurement error, and random chance.

16.7 TESTING PEARSON'S r FOR SIGNIFICANCE

When the relationship measured by Pearson's r is based on data from a random sample, you will usually need to test r for its statistical significance. That is, we will need to know if a relationship between the variables can be assumed to exist in the population from which the sample was drawn. In the example below, the r of .50 from the dual wage-earner family sample is tested for its statistical significance. As was the case when testing gamma and Spearman's rho, the null hypothesis states that there is no linear association between the two variables in the population from which the sample was drawn. The population parameter is symbolized as ρ (rho), and the appropriate sampling distribution is the t distribution.

To conduct this test, we need to make a number of assumptions in step 1. Several of these assumptions should be quite familiar by now, but several others are new. First, we must assume that both variables are normal in distribution (**bivariate normal distributions**). Second, we must assume that the relationship between the two variables is roughly linear in form.

The third assumption involves a new concept: **homoscedasticity**. Basically, a homoscedastistic relationship is one where the variance of the Y scores is uniform for all values of X. That is, if the Y scores are evenly spread above and below the regression line for the entire length of the line, the relationship is homoscedastistic.

A visual inspection of the scattergram will usually be sufficient to appraise to what extent the relationship conforms to the assumptions of linearity and homoscedasticity. As a rule of thumb, if the data points fall in a roughly symmetrical, cigar-shaped pattern, whose shape can be approximated with a straight line, then it is appropriate to proceed with this test of significance. Any significant evidence of nonlinearity or marked departures from homoscedasticity may indicate the need for an alternative measure of association and thus a different test of significance. As always, of course, when not certain that the assumptions of the technique are satisfied, the researcher must demonstrate that the conclusions are reasonable and sensible.

Step 1. Making assumptions.

Model: Random sampling
Level of measurement is interval-ratio
Bivariate normal distributions
Linear relationship
Homoscedasticity
Sampling distribution is normal

Step 2. Stating the null hypothesis.

$$H_0: \rho = 0.0$$
$$(H_1: \rho \neq 0.0)$$

Step 3. Selecting the sampling distribution and establishing the critical region. With the null of "no relationship" in the population, the sampling distribution of all possible sample r's is approximated by the t distribution. Degrees of freedom are equal to $(N - 2)$.

Sampling distribution = t distribution
Alpha = .05
Degrees of freedom = $N - 2 = 10$
t (critical) = ± 2.228

Step 4. Computing the test statistic.

$$t \text{ (obtained)} = r\sqrt{\frac{N - 2}{1 - r^2}}$$

$$t \text{ (obtained)} = (.50)\sqrt{\frac{12 - 2}{1 - (.50)^2}}$$

$$t \text{ (obtained)} = (.50)\sqrt{\frac{10}{.75}}$$

$$t \text{ (obtained)} = (.50)\sqrt{13.33}$$

$$t \text{ (obtained)} = (.50)(3.65)$$

$$t \text{ (obtained)} = 1.83$$

Step 5. Making a decision. Since the test statistic does not fall in the critical region as marked by t (critical), we fail to reject the null hypothesis. Even though the variables are substantially related in the sample, we do not have sufficient evidence to conclude that the variables are also related in the population. The test indicates that the sample value of $r = .50$ could have occurred by chance alone if the null hypothesis is true and the variables are unrelated in the population.

SUMMARY

1. This summary is based on the example used throughout the chapter. We began with a question: Is the number of children in dual wage-earner families related to the number of hours per week husbands devote to housework? We presented the observations in a scattergram (Figure 16.1), and our visual impression was that the variables were associated in a positive direction. The pattern formed by the observation points in the scattergram could be approximated with a straight line; thus the relationship was roughly linear.

2. In matters of prediction, although values of Y could be predicted with the freehand regression line, the accuracy of our predictions is enhanced if we use the least-squares regression line, defined as the line that best fits the data by minimizing the variation in Y. Using the formula that defines the least-squares regression line ($Y = a + bX$), we found a slope (b) of .69, which indicates that each additional child (a unit change in X) is accompanied by an increase of .69 hours of housework per week for the husbands. We also predicted, based on this formula, that in a dual wage-earner family with six children ($X = 6$), husbands would contribute 5.63 hours of housework a week ($Y' = 5.63$ for $X = 6$).

3. Pearson's r is a statistic that measures the overall linear association between X and Y. Our impression from the scattergram of a substantial positive relationship was confirmed by the computed r of .50. We also saw that this relationship yields an r^2 of .25, which indicates that 25% of the total variation in Y (husband's housework) is accounted for or explained by X (number of children).

4. Assuming that the 12 families represented a random sample, we tested the Pearson's r for its statistical significance and found that, at the .05 level, we could not assume that these two variables were also related in the population.

5. We acquired a great deal of information about this bivariate relationship. We not only know the strength and direction of the relationship, but have also identified the regression line that best summarizes the effect of X on Y. We know the amount of change we can expect in Y for a unit change in X. In short, we have a greater volume of more precise information about this association between interval-ratio variables than we ever did about associations between ordinal or nominal variables. This is possible, of course, because the data generated by interval-ratio measurement are more precise and flexible than those produced by ordinal or nominal measurement techniques.

SUMMARY OF FORMULAS

Least-squares regression line   16.1 $Y = a + bX$

Definitional formula for the slope   16.2 $b = \dfrac{\Sigma(X - \bar{X})(Y - \bar{Y})}{\Sigma(X - \bar{X})^2}$

Computational formula for the slope   16.3 $b = \dfrac{N\Sigma XY - (\Sigma X)(\Sigma Y)}{N\Sigma X^2 - (\Sigma X)^2}$

Y intercept   16.4 $a = \bar{Y} - b\bar{X}$

Definitional formula for Pearson's r 16.5 $r = \dfrac{\Sigma(X - \bar{X})(Y - \bar{Y})}{\sqrt{[\Sigma(X - \bar{X})^2][\Sigma(Y - \bar{Y})^2]}}$

Computational formula for Pearson's r 16.6 $r = \dfrac{N\Sigma XY - (\Sigma X)(\Sigma Y)}{\sqrt{[N\Sigma X^2 - (\Sigma X)^2][N\Sigma Y^2 - (\Sigma Y)^2]}}$

Coefficient of determination   16.7 $r^2 = \dfrac{\Sigma(Y' - \bar{Y})^2}{\Sigma(Y - \bar{Y})^2} = \dfrac{\text{explained variation}}{\text{total variation}}$

GLOSSARY

Bivariate normal distributions. The model assumption in the test of significance for Pearson's r that both variables are normally distributed.

Coefficient of determination (r^2). The proportion of all variation in Y that is explained by X. Found by squaring the value of Pearson's r.

Conditional means of Y. The mean of all scores on Y for each value of X.

Explained variation. The proportion of all variation in Y that is attributed to the effect of X. Equal to $\Sigma(Y' - \bar{Y})^2$.

Homoscedasticity. The model assumption in the test of significance for Pearson's r that the variance of the Y scores is uniform across all values of X.

Linear relationship. A relationship between two variables in which the observation points (dots) in the scattergram can be approximated with a straight line.

Pearson's r (r). A measure of association for variables that have been measured at the interval-ratio level; ρ (Greek letter rho) is the symbol for the population value of Pearson's r.

Regression line. The single, best-fitting straight line that summarizes the relationship between two variables. Regression lines are fitted to the data points by the least-squares criterion whereby the line touches all conditional means of Y or comes as close to doing so as possible.

Scattergram. Graphic display device that depicts the relationship between two variables.

Slope. The amount of change in one variable per unit change in the other; b is the symbol for slope of a regression line.

Total variation. The spread of the Y scores around the mean of Y. Equal to $\Sigma(Y - \bar{Y})^2$.

Unexplained variation. The proportion of the total variation in Y that is not accounted for by X. Equal to $\Sigma(Y - Y')$.

Y intercept (a). The point where the regression line crosses the Y axis.

Y'. Symbol for predicted score on Y.

PROBLEMS

16.1   For each data set below
   a. Draw a scattergram and a freehand regression line.
   b. Compute the slope (b) and find the Y intercept (a).
   c. State the least-squares regression line.
   d. Compute r and r^2.

A.

X	Y
15	2
16	4
18	3

X	Y
19	10
25	9
30	20
35	21
40	23
42	30
50	32

B.

X	Y
7	30
8	25
9	20
10	19
15	17
16	20
17	20
18	15
20	10
25	5

C.

X	Y
10	40
15	48
20	45
30	46
50	42
75	52
80	48
85	43
90	42
100	47

16.2   SOC The table presents the scores of 20 states on each of six variables: three measures of criminal activity and three measures of population structure. For each combination of crime rates and population measure, draw a scattergram, find the least-squares regression line, and compute r and r^2. Assume that these 20 states are a random sample of all states and test the correlations for their significance. Write a paragraph interpreting the relationships between these population variables and crime.

Crime Rates per 100,000 Pop., 1986

State	Homicide	Robbery	Auto Theft	Population Growth[1]	Population Density[2]	Urbani- zation[3]
Maine	2	28	164	4.3	38	37
Massachusetts	4	193	906	1.7	745	91
New York	11	514	637	1.2	375	91
Pennsylvania	6	152	354	.2	265	85
Ohio	6	142	376	−.4	262	79
Wisconsin	3	73	254	1.7	88	67
Iowa	2	42	158	−2.2	51	43
South Dakota	4	16	99	2.5	9	28
Virginia	7	106	219	8.2	146	72
South Carolina	9	99	277	8.2	112	60
Kentucky	7	83	193	1.9	94	46
Alabama	10	112	267	4.1	80	64
Texas	14	240	714	17.3	64	81
Montana	3	20	215	4.1	6	24
Arizona	9	169	419	22.1	29	75
Utah	3	59	223	14.0	20	77
Washington	5	135	315	8.0	67	81
California	11	343	762	14.0	173	96
Arkansas	9	88	604	32.8	1	44
Hawaii	5	106	328	10.1	165	77

[1] Percentage change in population from 1980 to 1986.
[2] Population per square mile of land area, 1986.
[3] Percent of population living in metropolitan areas, 1986.
Source: United States Bureau of the Census, *Statistical Abstracts of the United States: 1988* (108th edition). Washington, D.C., 1988.

16.3 PS The city of Shinbone recently had an election for city council. For the 15 randomly selected precincts listed below, was voter turnout related to the percentage of minority-group population?

Precinct	Registered Voters Who Voted (%)	Minority-group Voters (%)
A	23	0
B	47	15
C	65	75
D	17	10
E	9	5
F	64	25
G	50	97
H	49	45
I	45	7
J	43	16

Precinct	Registered Voters Who Voted (%)	Minority-group Voters (%)
K	27	17
L	25	8
M	33	23
N	34	24
O	25	25

a. Display these data by means of a scattergram.
b. Find the least-squares regression line.
c. Compute r and r^2.
d. Is the r statistically significant?
e. Summarize the relationship in terms of its strength, direction, and statistical significance.

16.4 SW For 18 census tracts randomly selected from the greater metropolitan Shinbone area, data have been collected on rates of child abuse, average level of education, and percentage of intact (two-parent) families. Is there any relationship between the latter two variables and child abuse? Rate of child abuse is

$$\frac{\text{number of reported cases}}{\text{population}} \times 100$$

Rate of Child Abuse	Mean Years of Formal Schooling	Intact Families (%)
1.2	12.1	90
.7	12.2	86
3.5	9.2	80
6.7	11.1	75
5.8	8.5	65
4.2	11.8	76
3.8	10.5	67
1.0	12.3	75
1.0	12.7	74
.5	12.4	88
.3	13.1	85
4.7	10.1	73
4.5	9.8	72
5.3	12.0	61
6.8	11.9	64
7.1	9.0	60
9.1	11.1	63
9.3	9.2	57

a. Draw scattergrams to display the relationships between education and child abuse and between percentage of intact families and child abuse.
b. Find the least-squares regression equation for each of the independents and the rate of child abuse. What rate of child abuse would you predict for a census tract where parents had an average of 14.0 years of formal schooling? For a census tract where only 50% of the families were intact?
c. Compute r and r^2 for each of the bivariate relationships. Is either r statistically significant?
d. Summarize your findings in terms of the strength, direction, and significance of these relationships.

16.5 SOC A researcher is interested in the ways college students become "popular." One possibility

is that students who are high academic achievers also tend to be more popular. That is, students who have earned high grade-point averages have, by definition, achieved high status in terms of a central value system of the university. Do they translate this academic success into social success? From the 25 residents of a single dormitory wing, the researcher has collected GPA's and an index of popularity. Are these variables related? (Popularity is measured by a simple count of the number of times each resident was rated as popular by his neighbors.)

GPA	Popularity
1.75	10
3.01	3
3.35	5
2.00	5
1.52	1
3.50	8
3.20	11
1.67	10
2.52	5
3.10	14
2.95	10
1.99	11
2.57	22
2.80	2
3.00	19
1.30	12
3.62	21
2.00	19
3.65	18
2.12	15
3.80	12
2.82	10
3.85	15
3.51	7
3.42	14

16.6 SOC The table below presents the scores of 15 states on three variables. For each combination of variables, draw a scattergram, find the least-squares regression line, and compute r and r^2. Assume that these 15 states are a random sample of all states and test the correlations for their significance. Write a paragraph interpreting the relationship among these three variables.

State	Per Capita Expenditures on Education 1987	Average Annual Pay 1986	Population Growth 1980–1986
Arkansas	543	16,162	3.8
Colorado	713	20,275	13.1
Connecticut	830	22,516	2.6
Florida	574	17,679	19.8
Illinois	585	21,452	1.1
Kansas	712	17,934	4.1
Louisiana	576	18,290	7.0
Maryland	706	20,121	5.8
Michigan	717	22,720	-1.3
Mississippi	496	15,420	4.1
Nebraska	577	16,106	1.8
New Hampshire	529	18,303	11.5
North Carolina	565	17,001	7.7
Pennsylvania	682	19,404	.2
Wyoming	1440	18,969	8.0

Source: United States Bureau of the Census, *Statistical Abstracts of the United States: 1988* (108th edition). Washington, D.C., 1988.

16.7 SOC The basketball coach at a small local college believes that his team plays better and scores more points in front of larger crowds. The number of points scored and attendance for all home games last season are reported below. Do these data support the coach's argument?

Game	Points scored	Attendance
1	54	378
2	57	350
3	59	320
4	80	478
5	82	451
6	75	250
7	73	489
8	53	451
9	67	410
10	78	215
11	67	113
12	56	250
13	85	450
14	101	489
15	99	472

16.8 SOC The table below reports some characteristics of 20 countries in 1990. For each combination of variables, draw a scattergram, find the least-squares regression line, and compute r and r^2. Assuming that these nations are a random sample, test the correlations for their significance. Write a paragraph interpreting the relationship between these variables.

Percent Urban	Population Growth Rate	Birth Rate	Death Rate
15	1.1	45	22
83	1.3	21	9
64	2.1	25	7
29	2.9	41	10
74	.3	13	11
45	2.8	39	25
16	3.1	49	15
76	.5	12	8
60	.3	11	10
4	2.4	41	16
90	.1	14	13
85	.5	10	12
65	1.0	20	11
60	.8	21	15
44	2.2	32	8
66	2.6	35	15
64	3.6	48	10
75	.3	13	11
56	-.1	11	12
22	2.9	46	20

16.9 GER The residents of a housing development for senior citizens have completed a survey whereon they indicated how physically active they are and how many visitors they receive each week. Are these two variables related for the 10 cases reported below?

Case	Level of Activity	Number of Visitors
A	10	14
B	11	12
C	12	10
D	10	9
E	15	8
F	9	7

Case	Level of Activity	Number of Visitors
G	7	10
H	3	15
I	10	12
J	9	2

16.10 PS The variables below were collected for a random sample of 10 precincts as of the last national election. For each combination of variables, draw a scattergram, find the least-squares regression line, and compute r and r^2. Test the correlations for their significance. Write a paragraph interpreting the relationship between these variables.

Precinct	Percent Democrat	Percent Minority	Voter Turnout
A	50	10	56
B	45	12	55
C	56	8	52
D	78	15	60
E	13	5	89
F	85	20	25
G	62	18	64
H	33	9	88
I	25	0	42
J	49	9	36

16.11 SOC Twenty-five individuals were randomly selected from the 1993 General Social Survey data set and their scores on five variables are reproduced below. Is there a relationship between occupational prestige and age? Between church attendance and number of children? Between number of children and hours of TV watching? Between age and hours of TV watching? Between age and number of children? Between hours of TV watching and occupational prestige? Variables and codes are explained in detail in Appendix G.

Occupational Prestige (Item 2)	Number of Children (Item 6)	Age (Item 7)	Church Attendance (Item 29)	Hours of TV Watching per Day (Item 67)
32	3	34	3	1
50	0	41	0	3
17	0	52	7	2
69	3	67	0	5
17	0	40	0	5
52	0	22	2	3
32	3	31	0	4
50	0	23	8	4
19	9	64	1	6
37	4	55	0	2
14	3	66	5	5
51	0	22	6	0
45	0	19	3	7
44	0	21	4	1
46	4	58	2	0
20	0	22	4	5
40	0	40	6	1
26	3	41	2	5
50	2	31	3	4
52	5	70	7	4
50	0	45	6	12
46	0	24	4	1
17	4	72	3	2
50	3	40	0	7
22	2	28	1	4

SPSS/PC+ PROCEDURES FOR PEARSON'S *R*

DEMONSTRATION 16.1 What Are the Correlates of Occupational Prestige?

We'll use a program called CORRELATIONS to compute Pearson's *r*. The prestige score of the respondent's occupation (item 2) will be our primary focus, and we'll look at how this variable is associated with the prestige of the respondent's father's occupation (item 5, a measure of ascribed status), the respondent's education (item 8, a measure of level of preparation for the job market), and the respondent's age (item 7, since prestige commonly increases with longevity in the job market). We would expect that PRESTIGE would have a positive relationship with all three variables. By comparing the strength of these bivariate correlations, we may be able to make some judgment about the relative importance of these factors in determining occupational prestige. Assuming that the 1993 GSS file is active:

1. From the main menu, select 'analyze data' and then select 'correlation and regression'.

2. Select and paste CORRELATIONS. Select !/VARIABLES, press ALT-T, and type PRESTG80. Next, select WITH and !/VARIABLES again. This time, type or select PAPRES80, EDUC, and AGE. In this procedure, the variable(s) named before (to the left of) WITH are taken as dependent, and the variables named after (to the right of) WITH are the independents. You may name more than one variable in either place. The program will compute correlations for every combination of variables on either side of the WITH.

3. Select and paste /OPTIONS 5 and /STATISTICS 1. These will print a set of descriptive statistics for all variables and the exact probability (alpha) and number of cases for each correlation coefficient.

Your command should look like this:

```
CORRELATIONS /VARIABLES PRESTG80 WITH PAPRES80 EDUC AGE /OPTIONS 5 /
STATISTICS 1.
```

and the output will be

Variable	Cases	Mean	Std Dev
PRESTG80	636	43.1965	13.1365
PAPRES80	636	43.1274	12.2841
EDUC	636	13.2500	4.5315
AGE	636	46.7846	16.8357

Correlations:	PAPRES80	EDUC	AGE
PRESTG80	.2155	.3385	-.0260
	(636)	(636)	(636)
	P= .000	P= .000	P= .256

(Coefficient / (Cases) / 1-tailed Significance)
" . " is printed if a coefficient cannot be computed

The top section of the output presents some univariate descriptive statistics for each variable. Note that the number of cases is low ($N = 636$). It is low because SPSS/PC+ deletes all cases that are missing scores on any of the variables, and 121 cases are missing scores for PAPRES80.

The bottom half of the output displays Pearson's r for PRESTG80 and each of the other variables. Also reported are the number of cases and the exact alpha, or p value for each coefficient. As expected, there are positive, moderate-to-strong relationships between PRESTG80 and PAPRES80 (.2155) and EDUC (.3385). Both relationships are statistically significant at an alpha level of less than .001 ($p = .000$). EDUC has a stronger relationship with PRESTG80 than PAPRES80. This might be taken as evidence of a "meritocratic" system of achievement. In other words, for this sample, occupational prestige is more a function of preparation (EDUC) than of family background and "ascribed" factors (PAPRES80).

Unexpectedly, AGE shows virtually no relationship with PRESTG80 ($-.0260$). For this sample, older respondents are not in positions of higher prestige than younger respondents. What might explain this counterintuitive result? One possibility is that the relationship between AGE and PRESTG80 is curvilinear with younger and older (retired?) respondents holding positions of lower prestige and people in between holding positions of higher prestige. If this were the case, then a scattergram of these two variables would show an inverted U shape. See the PLOT procedure for a way to construct a scattergram.

Another possible explanation relates to the nature of the sample. Our statistics were calculated on all the respondents in the sample. We might get clearer results if we limit the sample to people who are presently employed. To limit the sample in this way, use the SELECT IF command and the WRKSTAT variable (item 1). See Demonstration 11.3 or the manual for further information on the SELECT IF command. The command should look like this:

```
SELECT IF (WRKSTAT EQ 1).
```

Next, run the CORRELATIONS task again; now the output will include only respondents who are presently employed. The output will look like this:

Variable	Cases	Mean	Std Dev
PRESTG80	339	45.8378	12.9889
PAPRES80	339	44.5221	12.7164
EDUC	339	14.0324	2.6461
AGE	339	40.9794	10.9678

Correlations:	PAPRES80	EDUC	AGE
PRESTG80	.1730	.5432	.0298
	(339)	(339)	(339)
	P= .001	P= .000	P= .292

```
(Coefficient / (Cases) / 1-tailed Significance)
" . " is printed if a coefficient cannot be computed
```

Note that the number of cases in this refined sample is reduced ($N = 339$) but that the pattern of correlations is basically the same. The refined sample is younger than the sample in the previous demonstration but is otherwise quite similar, and the relationship between AGE and PRESTG80 has not been clarified.

A third possible explanation for the nearly zero relationship between AGE and PRESTG80 is that older respondents have much lower levels of education than the younger respondents and are, therefore, lower than expected on PRESTG80 because they are low on EDUC. You can check out this possibility by running the CORRELATIONS program again for just EDUC and AGE. If older respondents are lower in prestige because they are lower in education, then EDUC and AGE should display a negative relationship. (In fact, the Pearson's r between these two variables is $-.14$.) We will return to these relationships in Demonstration 18.1.

DEMONSTRATION 16.2 Creating an SPSS/PC+ File: What are the Correlates of Crime?

It has been awhile since we looked at any data other than the 1993 GSS, so let's indulge ourselves in a little diversity and create an SPSS/PC+ file from the data presented in problem 16.2. Not incidentally, this demonstration may also suggest a way of doing your statistics homework more efficiently. The procedures for entering data and creating files are presented in Appendix F and Demonstration 2.1. We'll run some correlations on the variables and save the file for later analysis in Chapter 18. The file, with SPSS/PC+ commands and some of the data, should look like this:

```
DATA LIST FIXED /HOMICIDE 1-2 ROBBERY 3-5 CARTHEFT 6-8
      GROWTH 9-13 DENSITY 14-16 URBAN 17-18.
BEGIN DATA.
 2 28164  4.3 3838
 4193906  1.774591
11514637  1.237591
 .
 .
 .
END DATA.
```

Save the file by pressing the F9 key and answering the prompts. Let's refer to this file as the 'CRIME' file. Following the END DATA command, enter the CORRELATIONS command to request Pearson's *r* for HOMICIDE and the three demographic variables. The command would be

```
CORRELATIONS /VARIABLES HOMICIDE WITH GROWTH DENSITY
     URBAN /OPTIONS 5 /STATISTICS 1.
```

The output from this command, with descriptive statistics eliminated to conserve space, will be

```
Correlations:   GROWTH      DENSITY      URBAN

   HOMICIDE       .4559        .0358       .4343
               (   20)      (   20)     (   20)
               P= .022     P= .440     P= .028
```

For these 20 states, homicide rates have statistically significant, moderate to strong, positive relationships with both rate of population growth and percent of the population living in urban areas. There is little or no relationship between homicide rate and population density. Higher rates of homicide are associated with more urbanized states and with higher rates of population growth.

Exercises

16.1 Run the analysis in Demonstration 16.1 again with INCOME91 as the focus rather than PRESTG80. Using the same variables (PAPRES80, EDUC, and AGE) as "independent" variables, see if the patterns are similar to those identified in Demonstration 16.1. (Ignore the fact that INCOME91 is only ordinal in level of measurement.) Write up your conclusions.

16.2 Run the same analysis one more time with TVHOURS against several of the measures of ascribed and achieved status. Who watches more TV? Write up your conclusions.

16.3 Rewrite the command from Demonstration 16.2 to get correlations between robbery, auto theft, and the demographic variables. Write up your conclusions.

16.4 Use the COMPUTE command to create a scale to measure satisfaction (items 37 and 38), attitudes towards abortion (items 52 and 53), or attitudes towards gender roles (items 68 and 69). See Demonstrations 12.3, 9.3, 9.2, 4.3, or the Manual for information on the COMPUTE command. Run the CORRELATIONS program for your computed variable against PRESTG80, AGE, and EDUC. Write up your conclusions.

PART III CUMULATIVE EXERCISES

A number of research questions are stated below. Each can be answered by at least one of the techniques presented in Chapters 13 through 16. For each research situation, compute the most appropriate measure of association and write a sentence or two of response to the question. The questions are presented in random order.

In selecting a measure of association, you need to consider the number of possible values and the level of measurement of the variables. The flow charts at the beginning of each chapter may be helpful.

The research questions refer to the data base below, which is taken from the 1993 General Social Survey (GSS). The actual questions asked and the complete response codes for the GSS are presented in Appendix G. Abbreviated codes are listed below. Some variables have been recoded for this exercise.

a. Are scores on "satisfaction with family and friends" associated with income, age, or number of children? Compute a measure of association, assuming that all four variables are interval-ratio in level of measurement.

b. Is fear associated with area of residence? With marital status?

c. Is attitude about busing associated with area of residence? With marital status?

Survey Items (numbers in parentheses refer to Appendix G)

1. Marital status of respondent (recode of 3)
      1. Married
      2. Not married (includes widowed, divorced, etc.)
2. How many children have you ever had? (6) (Values are actual numbers.)
3. Age of respondent (7) (Values are actual numbers.)
4. Respondent's total family income (12) (See Appendix G for codes.)
5. Area of residence (recode of 14)
      1. Urban
      2. Suburban
      3. Rural
6. Support for busing (33)
      1. Favor
      2. Oppose
7. Fear of walking alone at night (66)
      1. Yes
      2. No

8. Satisfaction with family and friends (created by adding scores on 37 and 38). On this scale, 2 indicates the *highest* possible level of satisfaction and 14 the *lowest*.

SCORES

Case	Mari-tal Status	Nmbr of Kids	Age	Income	Area	Busing	Fear	Satis-faction
1	2	0	22	12	2	2	2	5
2	1	0	52	13	3	2	2	4
3	1	2	44	16	1	2	1	8
4	1	3	56	10	3	2	2	10
5	1	7	61	8	3	2	2	4
6	1	0	28	19	2	1	2	2
7	2	2	59	9	1	2	1	7
8	1	2	69	11	1	2	2	4
9	2	0	23	4	1	1	2	9
10	2	2	31	20	2	2	1	3
11	1	3	67	21	2	2	1	8
12	1	1	46	9	3	1	1	11
13	2	0	19	10	3	1	1	6
14	1	2	34	11	2	2	2	2
15	2	0	29	18	2	2	1	8
16	1	1	31	16	1	1	1	5
17	2	1	88	6	3	2	2	3
18	1	0	24	15	2	1	1	9
19	2	2	69	11	1	1	1	3
20	1	4	60	14	3	2	2	3
21	1	1	29	12	2	1	2	4
22	1	2	43	13	2	2	1	2
23	1	0	35	20	1	2	1	4
24	2	2	38	9	1	2	2	7
25	2	0	83	19	3	2	1	6
26	1	3	56	12	2	1	1	3
27	2	2	46	19	2	1	2	6
28	1	1	22	12	2	1	2	4
29	2	0	54	14	1	1	2	4
30	1	2	35	16	2	2	2	5
31	1	3	45	6	3	2	2	4
32	1	2	49	12	2	2	2	9
33	2	3	73	15	2	1	1	3
34	2	0	23	2	1	1	1	7
35	2	0	40	21	3	2	2	4

Part IV   Multivariate Techniques

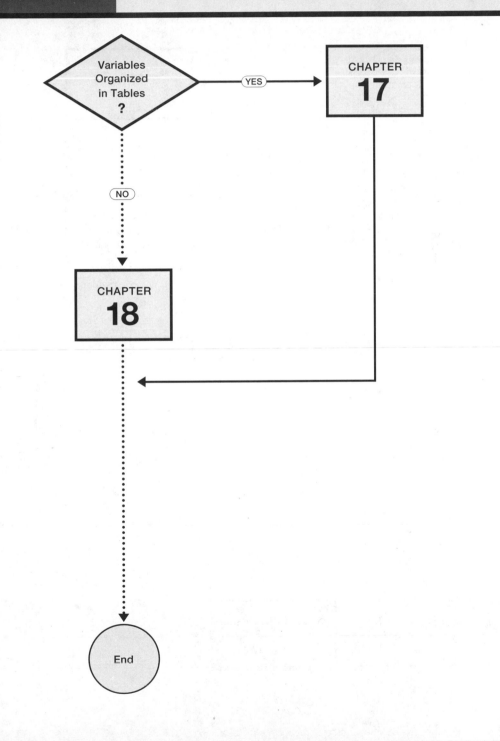

17

Elaborating Bivariate Tables

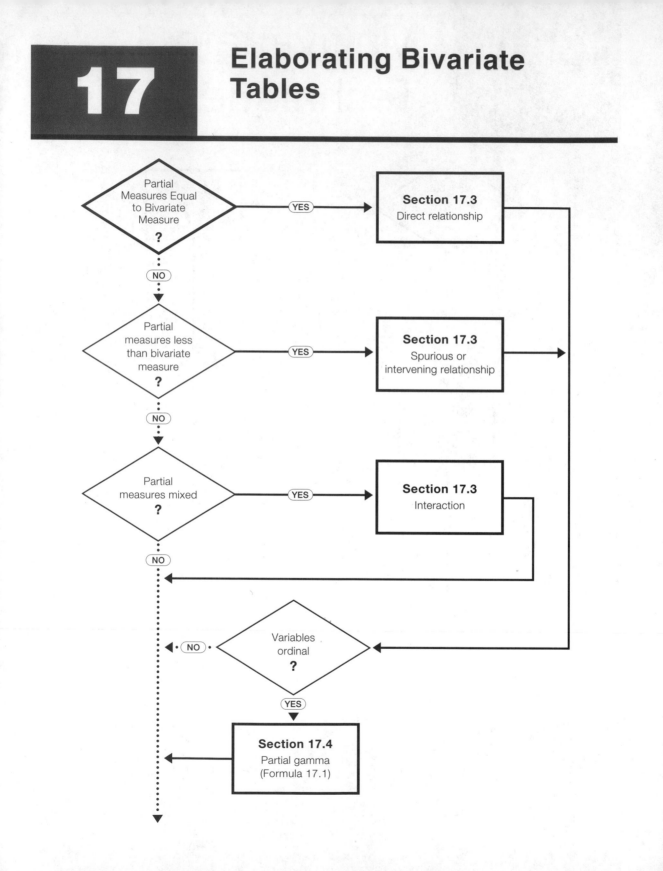

17.1 INTRODUCTION

Few research questions can be answered through a statistical analysis of only two variables. Most, if not all, research projects will require the analysis of many different variables, even when the researcher is principally concerned with a specific bivariate relationship. In Chapters 13–16, we have seen how various statistical techniques can be applied to the analysis of bivariate relationships. In this chapter and Chapter 18, we will see how some of these techniques can be utilized to probe the relationships among three or more variables. This chapter will present some multivariate techniques appropriate for variables that have been measured at the nominal or ordinal level and organized into table format. Chapter 18 will present some techniques that can be used when the variables have been measured at the interval-ratio level.

Before considering the techniques themselves, we should briefly consider why they are important and what they might be able to tell us. There are two general reasons for utilizing multivariate techniques. First, and most fundamental, is the goal of simply gathering additional information about a specific bivariate relationship by observing how that relationship is affected (if at all) by the presence of a third variable (or a fourth or a fifth variable). Multivariate techniques will increase the amount of information we have on the basic bivariate relationship and (we hope) will enhance our understanding of that relationship.

A second, and very much related, rationale for multivariate statistics involves the issue of causation. While multivariate statistical techniques cannot prove the existence of causal connections between variables, they can provide valuable evidence in support of causal arguments. Needless to say, their ability to find evidence for or against causal relationships between variables makes these techniques valuable for purposes of testing and revising theory.

17.2 CONTROLLING FOR A THIRD VARIABLE

For variables arrayed in bivariate tables, multivariate analysis proceeds by systematically observing the effects of other variables on a bivariate relationship. These effects are measured by controlling for the additional variables. The additional variable is fixed so that its values are no longer free to vary (thus, it is "controlled"), and the impact of this procedure on the bivariate relationship is then assessed.*

To illustrate, suppose that a researcher wishes to analyze the relationship between how well an individual is integrated into a group or organization and that individual's level of satisfaction with the group. The researcher has decided to focus on students and their level of satisfaction with the college as a whole. The necessary information on satisfaction (Y) is gathered from a

---

*For the sake of brevity, we will focus on the simplest application of multivariate statistical analysis, the case where the relationship between an independent (X) and dependent (Y) variable is analyzed in the presence of a single control variable (Z). Once the basic techniques are grasped, they can be readily expanded to situations involving more than one control variable.

TABLE 17.1 OVERALL SATISFACTION WITH COLLEGE BY NUMBER OF MEMBERSHIPS IN STUDENT ORGANIZATIONS (frequencies)

Satisfaction (Y)	Memberships (X)		
	None	At Least One	
Low	57	56	113
High	48	109	157
	105	165	270
	Gamma = .40		

TABLE 17.2 OVERALL SATISFACTION WITH COLLEGE BY NUMBER OF MEMBERSHIPS IN STUDENT ORGANIZATIONS (percentages)

Satisfaction (Y)	Memberships (X)	
	None	At Least One
Low	54.3	33.9
High	45.7	66.1
	100.0%	100.0%
	(105)	(165)

sample of 270 students; all are asked to list the student clubs or organizations to which they currently belong. Integration into the organization (X) is measured by dividing the students into two categories. The first category includes students who are not members of any organizations (nonmembers), and the second includes students who are members of at least one organization (members). The researcher suspects that membership and satisfaction are positively related and that students who are members of at least one organization will report higher levels of satisfaction than students who are nonmembers. The basic relationship between these two variables is displayed in Tables 17.1 and 17.2.

Inspection of these tables suggests that these two variables are associated. The conditional distributions of satisfaction (Y) change across the two conditions of membership (X), and membership in at least one organization is associated with high satisfaction while nonmembership is associated with low satisfaction. The existence and direction of the relationship is confirmed by the computation of a gamma of +.40 for this table.* In short, these tables

*We will not display the computation of gamma here. See Chapter 15 for a review of this measure of association.

FIGURE 17.1   A DIRECT RELATIONSHIP BETWEEN TWO VARIABLES

$$X \longrightarrow Y$$

strongly suggest the existence of a relationship between integration and morale, and these results can be taken as evidence for a causal or direct relationship between the two variables. The causal relationship is summarized symbolically in Figure 17.1, where the arrow represents the effect of X on Y.

The researcher recognizes, of course, that membership is not the sole factor associated with satisfaction (if it were, the gamma noted above would be $+1.00$). Other variables may well alter this relationship, and the researcher will need to consider the effects of these third variables (**Z's**) in a systematic and logical way. By doing so, the researcher will accumulate further information and detail about the bivariate relationship and can further probe the possible causal connection between X and Y.

For example, perhaps levels of satisfaction are affected by the rewards received by students from the university. Perhaps the most significant reward the university can dispense is recognition of academic achievements (that is, grades). Perhaps students who are more highly rewarded (have higher grade-point averages) are more satisfied with the university and are also more likely to participate actively in the life of the campus by joining student organizations. How can the effects of this third variable on the bivariate relationship be investigated?

Consider Table 17.1 again. This table displays the distribution of 270 cases on each of two variables and yields a variety of information about these two variables. For example, we can see that 165 students hold at least one membership, that 48 students are both nonmembers and highly satisfied, that the majority of the students are highly satisfied, and so forth. What the table cannot show us, of course, is the distribution of these students on GPA. For all we know at this point, the 109 highly satisfied members could all have high GPA's, low GPA's, or any combination of scores on this third variable. GPA is "free to vary" in this table, since the distribution of the cases on this variable is not accounted for.

We can control for the effect (if any) of third variables by fixing their distributions so that they are no longer free to vary. We do so by sorting all cases in the sample according to their score on the third variable (Z) and then observing the relationship between X and Y for each value (or score) of Z. In the example at hand, we will construct separate tables displaying the relationship between membership and satisfaction for each category of GPA. The tables so produced are called **partial tables** and are displayed in Tables 17.3 and 17.4. The students have been grouped into two categories according to whether they are "high" or "low" on GPA. There are exactly 135 students in both categories. Each subgroup has been reclassified by membership (X) and satisfaction (Y) in order to produce the two partial tables.

TABLE 17.3   SATISFACTION BY MEMBERSHIP CONTROLLING FOR GPA (frequencies)

A. High GPA

Satisfaction	Memberships		
	None	At Least One	
Low	29	28	57
High	24	54	78
	53	82	135
	Gamma = .40		

B. Low GPA

Satisfaction	Memberships		
	None	At Least One	
Low	28	28	56
High	24	55	79
	52	83	135
	Gamma = .39		

TABLE 17.4   SATISFACTION BY MEMBERSHIP CONTROLLING FOR GPA (percentages)

A. High GPA

Satisfaction	Memberships	
	None	At Least One
Low	54.7	34.2
High	45.3	65.9
	100.0%	100.1%
	(53)	(82)

B. Low GPA

Satisfaction	Memberships	
	None	At Least One
Low	53.9	33.7
High	46.2	66.3
	100.1%	100.0%
	(52)	(83)

Note that these partial tables present the original bivariate relationship in a more detailed form; that is, the partial tables elaborate the relationships displayed in Tables 17.1 and 17.2. The cell frequencies in the partial tables are, in this sense, subdivisions of the cell frequencies reported in Table 17.1. For example, if the frequencies of the cells in the partial tables are added together, the original frequencies of Table 17.1 will be reproduced. The partial tables can be combined to reproduce the original table, and you can convince yourself of this by adding all cell frequencies across the two partial tables and checking these totals against Table 17.1.*

Also note the way this method of controlling for other variables can be extended. First, there will be a partial table for each value of the control variable. In our example above, the control variables had two categories and, thus, generated two partial tables. Had there been three categories, there would have been three partial tables. Second, we can control for more than one variable at a time by sorting the cases on all scores of all control variables and producing partial tables for each combination of scores on the control variables. Thus, if we had controlled for both GPA and gender, we would have four partial tables to consider. There would have been one partial table for males with low GPA's, a second for males with high GPA's, and two partial tables for females with high and low GPA's. To summarize, for nominal and ordinal variables, multivariate analysis proceeds by first constructing partial tables—tables displaying the relationship between X and Y for each value of Z.

The next step is to trace the effect of Z by comparing the partial tables with each other and with the original bivariate table. Now that we have seen the mechanics for controlling other variables, let us see how interpreting the partial tables can lead us to understand the effects (if any) of the control variable on the original bivariate relationship.

17.3 INTERPRETING PARTIAL TABLES

When controlling for the effects of a third variable, the partial tables may display a variety of different patterns, but we will concentrate on three basic patterns, as determined by comparing the partial tables with each other and with the original bivariate table. The three basic patterns are

1. **Direct relationships** (The relationship between X and Y is the same across all partial tables and the same as in the original bivariate table.)

---

*The total of the cell frequencies in the partial tables will always equal the corresponding cell frequencies in the bivariate table except when, as often happens in "real life" research situations, the researcher is missing scores on the third variable for some cases. These cases must be deleted from the analysis and, as a consequence, the partial tables will have fewer cases than the bivariate table.

2. **Spurious relationships** or **intervening relationships** (The relationship between X and Y is the same across all partial tables but much weaker than in the bivariate table.)

3. **Interaction** (The relationships between X and Y in the partial tables are different from each other and from the bivariate table.)

Each of these three basic patterns has different implications for the causal interrelationships among the three variables and for the subsequent course of the statistical analysis. I will now describe each pattern in detail and then summarize our discussion in Table 17.9.

Direct Relationships. In this pattern, often called **replication**, the partial tables reproduce or replicate the bivariate table. The pattern of cell frequencies is the same across all partial tables and identical to the pattern in the bivariate table. Measures of association calculated on the partial tables would have the same value as the measure of association calculated for the bivariate table.

This outcome indicates that the relationship between X and Y is unaffected by the control variable, Z. Tables 17.3 and 17.4 provide an example of this outcome. In these two tables, the relationship between membership in student organizations (X) and overall satisfaction (Y) was investigated with GPA (Z) controlled. Tables 17.3A and 17.4A show the relationship for high-GPA students, and Tables 17.3B and 17.4B show the relationship for low-GPA students. The partial tables show the same conditional distributions of Y: about 45% of the nonmembers are highly satisfied, versus about 67% of the members. This same pattern was observed in the bivariate table (see Tables 17.1 and 17.2). Thus, the conditional distributions of Y are the same in each partial table as they were in the bivariate table.

You will find that it is easier to detect these patterns if you calculate appropriate measures of association. Working from the cell frequencies presented in Table 17.3, the gamma for high-GPA students is .40, and the gamma for low-GPA students is .39. The bivariate gamma (from Table 17.1) is .40, and the essential equivalence of these gammas reinforces our finding that the relationship between X and Y is essentially the same in the partial tables and the bivariate table.

This pattern of outcomes indicates that the control variable has no important impact on the bivariate relationship (if it did, the pattern in the partial tables would be different from the bivariate table) and may be ignored in any further analysis. In terms of the original research problem, the researcher may conclude that students who are members are more likely to express high satisfaction with the university regardless of their GPA. The level of rewards dispensed to students by the institution (as measured by GPA) has no effect on the relationship between membership and satisfaction. Low-GPA students who are members of at least one organization are just as likely to report high satisfaction as high-GPA students who are members of at least one organization.

Spurious or Intervening Relationships.   In this pattern, the relationship between X and Y is much weaker in the partial tables than in the bivariate table but the same across all partials. Measures of association for the partial tables are much lower in value (perhaps even dropping to 0.00) than the measure computed for the bivariate table.

This outcome is consistent with two different causal relationships among the three variables. The first is called a spurious relationship or **explanation**; in this situation, Z is conceptualized as being antecedent to both X and Y (that is, Z is thought to occur before the other two variables in time). In this pattern, Z is a common cause of both X and Y, and the original bivariate relationship is said to be spurious. The apparent bivariate relationship between X and Y is due to the effect of Z. Once Z is controlled, the association between X and Y disappears (the value of the measures for the partial tables drops to 0).

To illustrate, suppose that the researcher in our example had also controlled for class standing by dividing the sample into upperclasses (seniors and juniors) and underclasses (sophomores and freshmen). The reasoning of the researcher might be that upperclass students, as a function of simple longevity, display higher levels of satisfaction with the college than do underclass students. Self-selection processes may be operating. Students who are dissatisfied may have transferred to another college or dropped out before they attain upperclass standing. Students who have been on campus longer will be more likely to locate an organization of sufficient appeal or interest to join. This might especially be the case for organizations based on major field (such as the Accounting Club), which underclass students are less likely to join, or honorary organizations for which underclass students are unlikely to qualify.

These thoughts about the possible relationships among these variables are expressed in diagram form in Figure 17.2. The absence of an arrow from membership (X) to satisfaction (Y) indicates the possibility that these variables are not truly associated with each other but rather that both are linked to class standing.

FIGURE 17.2   A SPURIOUS RELATIONSHIP

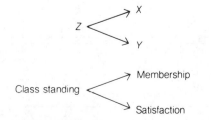

If this causal diagram is a correct description of the relationship of these three variables (if the association between X and Y is spurious), then the association between membership and satisfaction should disappear once class standing has been controlled. That is, even though the bivariate gamma was .40 (Table 17.1), the gammas computed on the partial tables will approach 0. Tables 17.5 and 17.6 display the partial tables generated by controlling for class standing.

The partial tables indicate that, once class standing is controlled, membership is no longer related to satisfaction. Upperclass students are likely to express high satisfaction and underclass students are likely to express low satisfaction regardless of their number of memberships. In the partial tables, the distributions of Y no longer vary by the conditions of X, and the gammas computed on the partial tables are virtually 0. These results indicate that the bivariate association is spurious and that X and Y have no direct relationship. Class standing (Z) is a key factor in accounting for varying levels of satisfaction, and the analysis must be reoriented with class standing as an independent variable.

This outcome (partial measures much weaker than the original measure but equal to each other) is consistent with another conception of the causal links among the variables. In addition to a causal scheme where Z is antecedent to both X and Y, Z may also intervene between the two variables. This pattern is also called **interpretation** and is illustrated in Figure 17.3, where

TABLE 17.5 SATISFACTION BY MEMBERSHIP CONTROLLING FOR CLASS STANDING (frequencies)

A. Upperclass students			
	Memberships		
Satisfaction	None	At Least One	
Low	8	32	40
High	24	97	121
	32	129	161
	Gamma = .01		

B. Underclass students			
	Memberships		
Satisfaction	None	At Least One	
Low	49	24	73
High	24	12	36
	73	36	109
	Gamma = .01		

TABLE 17.6   SATISFACTION BY MEMBERSHIP CONTROLLING FOR CLASS STANDING (percentages)

A. Upperclass students	Memberships	
Satisfaction	None	At Least One
Low	25.0	24.8
High	75.0	75.2
	100.0%	100.0%
	(32)	(129)

B. Underclass students	Memberships	
Satisfaction	None	At Least One
Low	67.1	66.7
High	32.9	33.3
	100.0%	100.0%
	(73)	(36)

FIGURE 17.3   AN INTERVENING RELATIONSHIP

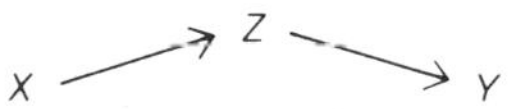

X is causally linked to Z, which is in turn linked to Y. This pattern indicates that, although X and Y are related, they are associated primarily through the control variable Z. This particular outcome does not allow the researcher to distinguish between spurious relationships (Figure 17.2) and intervening relationships (Figure 17.3). The differentiation between these two types of causal patterns may be made on temporal or theoretical grounds, but not on statistical grounds.

Interaction.   In this pattern, also called **specification**, the relationship between X and Y changes markedly, depending on the value of the control variable. The partial tables differ from each other and from the bivariate table. Interaction can be manifested in various ways in the partial tables. One possible pattern, for example, is for one partial table to display a stronger relationship between X and Y than that displayed in the bivariate table, while in a second partial table, the relationship between X and Y drops to 0. Symbolically, this outcome could be represented as in Figure 17.4, which would indicate that X and the first category of Z (Z_1) have strong effects on Y; but, for

FIGURE 17.4 AN INTERACTIVE RELATIONSHIP

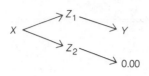

FIGURE 17.5 AN INTERACTIVE RELATIONSHIP

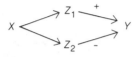

the second category of $Z(Z_2)$, there is no association between X and Y. Such a pattern might be encountered, for example, if all employees of a corporation were required to attend a program (X) designed to reduce racial prejudice (Y). Such a program would be likely to have stronger effects on white employees (Z_1) than African American employees (Z_2). That is, the program would have an effect on prejudice only for certain categories of subjects. The partial tables for white employees (Z_1) might show a strong relationship between program attendance and reduction in prejudice, while the partial tables for African American employees showed no relationship. The African American employees, being relatively unprejudiced against themselves in the first place, would be less likely to be affected by the program.

Interaction can take other forms. Let us investigate a case where the relationship between X and Y varies not only in strength but also in direction between the partial tables. This causal relationship is symbolically represented in Figure 17.5, which indicates a situation where X and Y are positively related for the first category of $Z(Z_1)$ and negatively related for the second category of $Z(Z_2)$.

The researcher investigating the relationship between club membership and satisfaction with college life establishes a final control for race and divides the sample into white and black students. The partial tables are displayed in Tables 17.7 and 17.8.

These partial tables display an interactive relationship. The relationship between membership and satisfaction is different for white students (Z_1) and for black students (Z_2), and each partial table is different from the bivariate table. For white students, the relationship is positive and stronger than in the bivariate table. White students who are also members are much more likely to express high overall satisfaction with the university, as indicated by both the percentage distribution (83% of the white students who are members are highly satisfied) and the measure of association (gamma = +.667 for white students).

TABLE 17.7   SATISFACTION BY MEMBERSHIP CONTROLLING FOR RACE
(frequencies)

A. White students			
	Memberships		
Satisfaction	None	At Least One	
Low	40	20	60
High	40	100	140
	80	120	200
	Gamma = .67		

B. Black students			
	Memberships		
Satisfaction	None	At Least One	
Low	17	36	53
High	8	9	17
	25	45	70
	Gamma = − .31		

TABLE 17.8   SATISFACTION BY MEMBERSHIP CONTROLLING FOR RACE
(percentages)

A. White students		
	Memberships	
Satisfaction	None	At Least One
Low	50.0	16.7
High	50.0	83.3
	100.0%	100.0%
	(80)	(120)

B. Black students		
	Memberships	
Satisfaction	None	At Least One
Low	68.0	80.0
High	32.0	20.0
	100.0%	100.0%
	(25)	(45)

APPLICATION 17.1

Seventy-eight juvenile males in a sample have been classified as high or low on a scale that measures involvement in delinquency. Also, each subject has been classified, using school records, as either a good or poor student. The following table displays a strong relationship between these two variables for this sample ($G = -.69$).

Delinquency	Academic Record		
	Poor	Good	
Low	13	20	33
High	35	10	45
	48	30	78
	Gamma = $-.69$		

Juvenile males with poor academic records seem to be especially prone to delinquency.

Is this relationship between delinquency and academic record affected by whether the subject resides in an urban or nonurban area?

Urban areas

Delinquency	Academic Record		
	Poor	Good	
Low	10	3	13
High	26	7	33
	36	10	46
	Gamma = $-.05$		

Nonurban areas

Delinquency	Academic Record		
	Poor	Good	
Low	3	17	20
High	9	3	12
	12	20	32
	Gamma = $-.89$		

For urban juvenile males, the relationship between academic record and delinquency disappears ($G = -.05$). For this group, delinquency is not associated with experience in school.

For nonurban males, there is a very strong relationship between the two variables ($G = -.89$), with poor students being especially prone to delinquency.

Comparing the partial tables with each other and with the bivariate table reveals an interactive relationship among these three variables. Although urban juvenile males are more delinquent than nonurban males (33 out of the 46, or 71.74%, of the urban males were highly delinquent, as compared to 12 out of 32, or 37.50%, of the nonurban males), their delinquency is not associated with academic record. For nonurban males, academic record is very strongly associated with delinquency. Urban males are more delinquent than nonurban males but not because of their experience in school. Nonurban males who are also poor students are especially likely to become involved in delinquency.

For black students, the relationship between membership and satisfaction is very different—nearly the reverse of the pattern evinced by white students. The great majority (80%) of the black students who are members report low satisfaction, and the gamma for this partial table is negative ($-.31$). Thus, for white students, satisfaction increases with membership, while, for black students, satisfaction decreases with membership. Based on these results, one might conclude that the social meanings and implications of joining student clubs and organizations vary by race and that the function of joining has different effects for black students. It may be, for example, that black students join different kinds of organizations than do white students. If the black students belonged primarily to a black student association that had an antagonistic relationship with the university, then belonging to such an organization could increase dissatisfaction with the university as it increased awareness of racial problems.

In closing this section, let me stress that, for the sake of clarity, I have presented unusually "clean" examples of these three possible outcomes. In any given research project, the results of controlling for third variables will probably be considerably more ambiguous and open to varying interpretation than are the examples presented here. In the case of spurious relationships, for example, the measures of association computed for the partial tables will probably not actually drop to zero, even though they may be dramatically lower than the bivariate measure. (The pattern where the partial measures are roughly equivalent and much lower than the bivariate measure, but not zero, is sometimes referred to as attenuation.) It is probably best to consider the examples above as ideal types against which any given empirical result can be compared.

Table 17.9 summarizes this discussion by outlining guidelines for decision making for each of the three outcomes discussed in this section. Since your own results will probably be more ambiguous than the ideal types presented here, you should regard this table as a set of suggestions and not as a substitute for your own creativity and sensitivity to the problem under consideration.

17.4 PARTIAL GAMMA (G_p)

When the results of controlling for a third variable indicate a direct, spurious, or intervening relationship, it is often useful to compute an additional measure that indicates the overall strength of the association between X and Y after the effects of the control variable (Z) have been removed. This statistic, called **partial gamma (G_p)**, is somewhat easier to compare with the bivariate gamma than are the gammas computed on the partial tables separately. G_p is computed across all partial tables by Formula 17.1:

FORMULA 17.1
$$G_p = \frac{\Sigma N_s - \Sigma N_d}{\Sigma N_s + \Sigma N_d}$$

where ΣN_s = the number of pairs of cases ranked the same across all partial tables
ΣN_d = the number of pairs of cases ranked differently across all partial tables

In words, ΣN_s is the total of all N_s's from the partial tables, and ΣN_d is the total of all N_d's from all partial tables. (See Chapter 15 to review the computation of N_s and N_d.)

To illustrate the computation of G_p, let us return to Tables 17.3 and 17.5. In Table 17.3, the relationship between satisfaction and membership while controlling for GPA was displayed. The gammas computed on the partial ta-

TABLE 17.9 A SUMMARY OF THE POSSIBLE RESULTS OF CONTROLLING FOR THIRD VARIABLES

Partial Tables (compared with bivariate table) Show	Pattern	Implications for Further Analysis	Likely Next Step in Statistical Analysis	Theoretical Implications
Same relationship between X and Y	Direct relationship, replication	Disregard Z	Select another control variable to test further the directness of the relationship	Theory that X causes Y is supported
Weaker relationship between X and Y	Spurious relationship	Incorporate Z	Focus on relationship between Z and Y	Theory that X causes Y is not supported
	Intervening relationship	Incorporate Z	Focus on relationship between X, Z, and Y	Theory that X causes Y is partially supported but must be revised to take Z into account
Mixed	Interaction	Incorporate Z	Analyze subgroups (categories of Z) separately	Theory that X causes Y partially supported but must be revised to take Z into account

bles were essentially equal to each other and to the bivariate gamma. Our conclusion was that the control variable had no effect on the relationship, and this conclusion can be confirmed by also computing partial gamma.

From Table 17.3A (High GPA)	From Table 17.3B (Low GPA)
$N_s = (29)(54) = 1566$	$N_s = (28)(55) = 1540$
$N_d = (28)(24) = 672$	$N_d = (28)(24) = 672$

$$\Sigma N_s = 1566 + 1540 = 3106$$

$$\Sigma N_d = 672 + 672 = 1344$$

$$G_p = \frac{\Sigma N_s - \Sigma N_d}{\Sigma N_s + \Sigma N_d}$$

$$G_p = \frac{3106 - 1344}{3106 + 1344}$$

$$G_p = \frac{1762}{4450}$$

$$G_p = .40$$

The partial gamma measures the strength of the association between X and Y once the effects of Z have been removed. In this instance, the partial gamma is the same value as the bivariate gamma ($G_p = G = .40$) and indicates that GPA has no effect on the relationship between satisfaction and membership.

When class standing was controlled (Table 17.5), clear evidence of a spurious relationship was found, since the gammas computed on the partial tables dropped almost to zero. Let us see what the value of partial gamma (G_p) would be for this second control:

From Table 17.5A (Upperclasses)	From Table 17.5B (Underclasses)
$N_s = (8)(97) = 776$	$N_s = (49)(12) = 588$
$N_d = (32)(24) = 768$	$N_d = (24)(24) = 576$

$$\Sigma N_s = 776 + 588 = 1364$$
$$\Sigma N_d = 768 + 576 = 1344$$

$$G_p = \frac{\Sigma N_s - \Sigma N_d}{\Sigma N_s + \Sigma N_d}$$

$$G_p = \frac{1364 - 1344}{1364 + 1344}$$

$$G_p = \frac{20}{2708}$$

$$G_p = .01$$

Once the effects of the control variable are removed, there is no relationship between X and Y. The very low value of G_p confirms our previous conclusion that the bivariate relationship between membership and satisfaction is spurious and actually due to the effects of class standing.

In a sense, G_p tells us no more about the relationships than we can see for ourselves from a careful analysis of the percentage distributions of Y in the partial tables or by a comparison of the measures of association computed on the partial tables. The advantage of G_p is that it tells us, in a single number (that is, in a compact and convenient way), the precise effects of Z on the relationship between X and Y. While G_p is no substitute for the analysis of the partial tables per se, it is a convenient way of stating our results and conclusions when working with direct or spurious relationships.

Although G_p can be calculated in cases of interactive relationships (see Table 17.7), it is rather difficult to interpret in these instances. If substantial interaction is found in the partial tables, this pattern indicates that the control variable has a profound effect on the bivariate relationship. Thus, the analyst should not attempt to separate the effects of Z from the bivariate relationship and, since G_p involves exactly this kind of separation, it should not be computed.

17.5 WHERE DO CONTROL VARIABLES COME FROM?

In one sense, this question is quite easy to answer. Although control variables can arise from various sources, they arise especially from theory. Social research proceeds in many different ways and is begun in response to a variety of problems. However, virtually all research projects are guided by a more-or-less-explicit theory or by some question about the relationship between two or more variables. The ultimate goal of social research is to develop defensible generalizations that improve our understanding of the variables under consideration and link back to theory at some level.

Thus, research projects are anchored in theory; and the concepts of interest, which will later be operationalized as variables, are first identified and their interrelationships first probed at the theoretical level. Since the social world is exceedingly complex, very few theories attempt to encompass that world in only two variables. Theories are, almost by definition, multivariate even when they focus on a bivariate relationship. Thus, to the extent that a research project is anchored in theory, the theory itself will suggest the control variables that need to be incorporated into the analysis. In the example used throughout this chapter, I tried to suggest that any researcher attempting to probe the relationship between involvement in an organization and satisfaction would, in the course of thinking over the possibilities, identify a number of additional variables that needed to be explicitly incorporated into the analysis.

Of course, textbook descriptions of the research process are oversimplified. They tend to imply that research flows smoothly from conceptualization to operationalization to quantification to generalization. In reality, research is commonly characterized by surprises, unexpected outcomes, and unanticipated results. Research in "real life" is typically more loosely structured and requires more imagination and creativity than textbooks can fully convey. My point is that the control variables that might be appropriate to incorporate in the data-analysis phase will be suggested or implied in the theoretical backdrop of the research project. They will flow from the researcher's imagination and sensitivity to the problem being addressed as much as from any other source.

These considerations have taken us well beyond the narrow realm of statistics and back to the planning stages of the research project. At this early time the researcher must make decisions about which variables to measure during the data-gathering phase and, thus, which variables might be incorporated as potential controls. Careful thinking and an extended consideration of possible outcomes at the planning stage will pay significant dividends during the data-analysis phase. Ideally, all relevant control variables will be incorporated and readily available for statistical analysis. Thus, control variables come from the theory underlying the research project and from creative and imaginative thinking and planning during the early phases of the project. Nonetheless, it is not unheard of for a researcher to realize during data analysis that the control variable now so obviously relevant was never measured during data gathering and is thus unavailable for statistical analysis.

17.6 THE LIMITATIONS OF ELABORATING BIVARIATE TABLES

The basic limitation of this technique involves sample size. Elaboration is a relatively inefficient technique for multivariate analysis because it requires that the researcher divide the sample into a series of partial tables. If the control variable has more than two or three possible values, or if we attempt to control for more than one variable at a time, many partial tables will be produced. The greater the number of partial tables, the more likely we are to run out of cases to fill all the cells of each partial table. Empty or small cells, in turn, can create serious problems in terms of generalizability and confidence in our findings.

To illustrate, the example used throughout this chapter began with two dichotomized variables (Table 17.1) and a four-cell table. Each of the control variables was also dichotomized, and we never confronted more than two partial tables with four cells each, or eight cells for each control variable. If we had used a control variable with three values, we would have had 12 cells to fill up, and if we had attempted to control for two dichotomized variables, we would have had 16 cells in four different partial tables. Clearly, as control variables become more elaborate and/or as the process of controlling becomes more complex, the phenomenon of empty or small cells will increasingly become a problem.

Two potential solutions to this dilemma immediately suggest themselves. The easy solution is to reduce the number of cells in the partial tables by collapsing categories within variables. If all variables are dichotomized, for example, the number of cells will be kept to a minimum. The best solution is to work with only very large samples. Unfortunately, the easy solution will often violate common sense (since the more categories are collapsed, the more likely that dissimilar elements will be grouped under a common label); and the best solution is not always feasible (mundane matters of time and money rear their ugly heads).

A third solution to the problem of empty cells requires the (sometimes risky) assumption that the variables of interest are measured at the interval-ratio level. At that level the techniques of partial and multiple correlation and regression, to be introduced in Chapter 18, are available. These multivariate techniques are more efficient than elaboration because they utilize all cases simultaneously and do not require that the sample be divided among the various partial tables.

SUMMARY

1. Most research questions require the analysis of the interrelationship among many variables, even when the researcher is primarily concerned with a specific bivariate relationship. Multivariate statistical techniques provide the researcher with a set of tools by which additional information can be gathered about the variables of interest and by which causal interrelationships can be probed.

2. When variables have been organized in bivariate tables, multivariate analysis proceeds by controlling for a third variable. Partial tables are constructed and compared with each other and with the original bivariate table. Comparisons are made easier if appropriate measures of association are computed for all tables.

3. A direct relationship exists between the independent (X) and dependent (Y) variables if, after controlling for the third variable (Z), the relationship between X and Y is the same across all partial tables and the same as in the bivariate table. This pattern suggests a causal relationship between X and Y.

4. If the relationship between X and Y is the same across all partial tables but much weaker than in the bivariate table, the relationship is either spurious (Z causes both X and Y) or intervening (X causes Z, which causes Y). Either pattern suggests that Z must be explicitly incorporated into the analysis.

5. Interaction exists if the relationship between X and Y varies across the partial tables and between each partial and the bivariate table. This pattern suggests that no simple or direct causal relationship exists between X and Y and that Z must be explicitly incorporated into the causal scheme.

6. Partial gamma (G_p) is a useful summary statistic that measures the strength of the association between X and Y after the effects of the control variable (Z) have been removed. Partial gamma should not be computed when an analysis of the partial tables shows substantial interaction.

7. Potential control variables must be identified before the data-gathering phase of the research project. The theoretical backdrop of the research project, along with creative thinking and some imagination, will suggest the variables that should be controlled for and measured.

8. Controlling for third variables by constructing partial tables is inefficient in that the cases must be spread out across many cells. If the variables have many categories and/or the researcher attempts to control for more than one variable simultaneously, "empty cells" may become a problem. It may be possible to deal with this problem by either collapsing categories or gathering very large samples. If interval-ratio level of measurement can be assumed, the multivariate techniques presented in the next chapter will be preferred, since they do not require the partitioning of the sample.

SUMMARY OF FORMULAS

Partial gamma $\qquad$ 17.1 $\quad G_p = \dfrac{\Sigma N_s - \Sigma N_d}{\Sigma N_s + \Sigma N_d}$

GLOSSARY

Direct relationship. A multivariate relationship where the control variable has no effect on the bivariate relationship.

Explanation. See **spurious relationship.**

Interaction. A multivariate relationship where a bivariate relationship changes across the categories of the control variable.

Interpretation. See **intervening relationship.**

Intervening relationship. A multivariate relationship where a bivariate relationship becomes substantially weaker after a third variable is controlled for. The independent and dependent variables are linked primarily through the control variable.

Partial gamma (G_p). A statistic that indicates the strength of the association between two variables after the effects of a third variable have been removed.

Partial tables. Tables produced when controlling for a third variable.

Replication. See **direct relationship.**

Specification. See **interaction.**

Spurious relationship. A multivariate relationship in which a bivariate relationship becomes substantially weaker after a third variable is controlled for. The independent and dependent variables are not causally linked. Rather, both are caused by the control variable.

Z. Symbol for any control variable.

PROBLEMS

17.1 [SOC] Data on suicide rates, age structure, and unemployment rates have been gathered for 100 census tracts. Suicide and unemployment rates have been dichotomized at the median so that each tract could be rated as high or low. Age structure is measured in terms of the percentage of the population age 65 and older. This variable has also been dichotomized and tracts have been rated as high or low. The tables below display the bivariate relationship between suicide rate and age structure and the same relationship controlling for unemployment.

Suicide Rate	Population 65 and Older (%)		
	Low	High	
Low	45	20	65
High	10	25	35
	55	45	100

a. Calculate percentages for the table so that the effect of age (X) on suicide rate (Y) is displayed. Is there a relationship between these two variables? Do the conditional distributions of Y change?

b. Compute and interpret gamma.

c. Describe the bivariate relationship in terms of strength and direction.

Suicide rate by age, controlling for unemployment:

A. High unemployment

Suicide Rate	Population 65 and Older (%)		
	Low	High	
Low	23	10	33
High	5	12	17
	28	22	50

B. Low unemployment

Suicide Rate	Population 65 and Older (%)		
	Low	High	
Low	22	10	32
High	5	13	18
	27	23	50

d. Calculate percentages for the partial tables so that the effect of age on suicide rates is displayed.

e. Compute gamma for both partial tables. Compare these gammas with the bivariate gamma.

f. Compute partial gamma (G_p).

g. Summarize the results of the control. Does unemployment rate have any effect on the relationship between age and suicide rate? Describe the effect of the control variable in terms of the pattern of percentages, the value of the gammas, and the possible causal relationships among these variables.

17.2 [SOC] Is there a relationship between attitudes on sexuality and age? Are older people more conservative with respect to questions of sexual morality? A national sample of 925 respondents has been questioned about attitudes on premarital sex. Responses have been collapsed into two categories: those who believe that premarital sex is "always wrong" and those who believe it is not wrong under certain conditions ("sometimes wrong"). These responses have been crosstabulated by age, and the results are reported below:

Premarital Sex Is	Age		
	Less than 35	35 or More	
Always wrong	90	235	325
Sometimes wrong	420	180	600
	510	415	925

a. Calculate percentages for the table so that the effect of age (X) on attitude (Y) is displayed. Is there a relationship between these two variables? Do the conditional distributions of Y change?

b. Compute and interpret gamma.

c. Describe the bivariate relationship in terms of its strength and direction. Below, the bivariate relationship is reproduced after controlling for the sex of the respondent. Does gender have any effect on the relationship?

Attitude toward premarital sex by age, controlling for gender:

A. Male respondents

Premarital Sex Is	Age		
	Less than 35	35 or More	
Always wrong	70	55	125
Sometimes wrong	190	80	270
	260	135	395

B. Female respondents

Premarital Sex Is	Age		
	Less than 35	35 or More	
Always wrong	20	180	200
Sometimes wrong	230	100	330
	250	280	530

d. Calculate percentages for the partial tables so that the effect of age on attitude is displayed.
e. Compute gamma for both partial tables. Compare these gammas with the bivariate gamma.
f. Compute partial gamma (G_p).
g. Summarize the results of this control. Does gender have any effect on the relationship between age and attitude? If so, describe the effect of the control variable in terms of the pattern of percentages, the value of the gammas, and the possible causal interrelationships among these variables.

17.3 SOC A job-training center is trying to justify its existence to its funding agency. To this end, data on four variables have been collected for each of the 403 trainees served over the past three years: (1) whether or not the trainee completed the program, (2) whether or not the trainee got and held a job for at least a year after training, (3) the sex of the trainee, and (4) the race of the trainee. Is employment related to completion of the program? Is the relationship between completion and employment affected by race or sex?

Employment by training:

Held Job for at Least One Year?	Training Completed?		
	Yes	No	
Yes	145	60	205
No	72	126	198
	217	186	403

Employment by training, controlling for race:

A. Whites

Held Job for at Least One Year?	Training Completed?		
	Yes	No	
Yes	85	33	118
No	38	47	85
	123	80	203

B. Blacks

Held Job for at Least One Year?	Training Completed?		
	Yes	No	
Yes	60	27	87
No	34	79	113
	94	106	200

Employment by training, controlling for sex:

C. Males

Held Job for at Least One Year?	Training Completed?		
	Yes	No	
Yes	73	30	103
No	36	62	98
	109	92	201

D. Females

Held Job for at Least One Year?	Training Completed?		
	Yes	No	
Yes	72	30	102
No	36	64	100
	108	94	202

17.4 SOC In a survey, 247 respondents were asked to indicate their overall level of satisfaction with their lot in life. Is satisfaction related to social class? Is the bivariate relationship affected by gender? By marital status?

Satisfaction by social class:

	Social Class		
	Working and	Upper and	
Satisfaction	Lower	Middle	
Low	19	12	31
Moderate	80	54	134
High	27	55	82
	126	121	247

a. Calculate percentages for the table so that the effect of social class (X) on satisfaction (Y) is displayed.
b. Compute and interpret gamma for this table.
c. Describe the bivariate relationship in terms of strength and direction.

Satisfaction by social class, controlling for gender:

A. Male respondents

	Social Class		
	Working and	Upper and	
Satisfaction	Lower	Middle	
Low	10	8	18
Moderate	28	20	48
High	12	29	41
	50	57	107

B. Female respondents

	Social Class		
	Working and	Upper and	
Satisfaction	Lower	Middle	
Low	9	4	13
Moderate	52	34	86
High	15	26	41
	76	64	140

d. Calculate percentages for the partial tables so that the effect of social class on satisfaction is displayed.
e. Compute gammas for both partial tables. Compare these with the gamma for the bivariate table.

f. Compute partial gamma (G_P).

Satisfaction by social class, controlling for marital status:

A. Married respondents

	Social Class		
	Working and	Upper and	
Satisfaction	Lower	Middle	
Low	8	8	16
Moderate	42	38	80
High	20	35	55
	70	81	151

B. Nonmarried respondents

	Social Class		
	Working and	Upper and	
Satisfaction	Lower	Middle	
Low	11	4	15
Moderate	38	16	54
High	7	20	27
	56	40	96

g. Calculate percentages for the partial tables so that the effect of social class on satisfaction is displayed.
h. Compute gammas for both partial tables. Compare these with the gamma for the bivariate table.
i. Compute partial gamma (G_P).
j. Summarize the results of the two controls. Are satisfaction and social class related? Does gender or marital status have any effect on the relationship?

17.5 SOC What are the social sources of support for the environmental movement? A recent survey gathered information on level of concern for such issues as global warming, acid rain, and related issues. Is concern for the environment related to level of education? What effects do the control variables have? Write a paragraph summarizing your conclusions.

Concern for the environment by level of education:

Concern for the Environment	Level of Education		
	Low	High	
Low	27	35	62
High	22	48	70
	49	83	132

Concern for the environment by level of education, controlling for gender:

A. Males

Concern for the Environment	Level of Education		
	Low	High	
Low	14	17	31
High	11	22	33
	25	39	64

B. Females

Concern for the Environment	Level of Education		
	Low	High	
Low	13	18	31
High	11	26	37
	24	44	68

Concern for the environment by level of education, controlling for "level of trust in the nation's leadership":

A. Low levels of trust

Concern for the Environment	Level of Education		
	Low	High	
Low	6	22	28
High	10	40	50
	16	62	78

B. High levels of trust

Concern for the Environment	Level of Education		
	Low	High	
Low	21	13	34
High	12	8	20
	33	21	54

Concern for the environment by level of education, controlling for race:

A. White

Concern for the Environment	Level of Education		
	Low	High	
Low	19	11	30
High	18	44	62
	37	55	92

B. Black

Concern for the Environment	Level of Education		
	Low	High	
Low	8	24	32
High	4	4	8
	12	28	40

17.6 [SW] Do long-term patients in mental health facilities become more withdrawn and reclusive over time? A sample of 608 institutionalized patients was rated by a standard "reality orientation scale." Is there a relationship with length of institutionalization? Does gender have any effect on the relationship?

Reality orientation by length of institutionalization:

Reality Orientation	Length of Institutionalization		
	Less than 5 Years	More than 5 Years	
Low	200	213	413
High	117	78	195
	317	291	608

Reality orientation by length of institutionalization, controlling for gender:

A. Females

Reality Orientation	Length of Institutionalization		
	Less than 5 Years	More than 5 Years	
Low	95	120	215
High	60	37	97
	155	157	312

B. Males

Reality Orientation	Length of Institutionalization		
	Less than 5 Years	More than 5 Years	
Low	105	93	198
High	57	41	98
	162	134	296

17.7 SOC In problem 15.16, we used the 1993 General Social Survey data set to investigate the relationships between income and five dependent variables. Let's return to two of those relationships and see if sex has any effect on the bivariate relationships. The partial tables are presented below, and bivariate gammas were computed in the previous exercise. Compute percentages and gammas for each of the partial tables and state your conclusions. Does controlling for sex have any effect?

a. Attitudes about the role of the U.S. in world affairs by income, controlling for sex:

A. Males

Should U.S. be active? (item 27)	Income		
	Less than $22,500	$22,500 to $50,000	More than $50,000
Active	49	59	58	166
Stay out	33	11	11	55
	82	70	69	221

B. Females

Should U.S. be active? (item 27)	Income		
	Less than $22,500	$22,500 to $50,000	More than $50,000
Active	80	64	45	189
Stay out	56	22	11	89
	136	86	56	278

b. Attitudes about pornography, by income, controlling for sex:

A. Males

Pornography Leads to Breakdown. (item 59)	Income		
	Less than $22,500	$22,500 to $50,000	More than $50,000
Yes	56	30	38	124
No	28	25	20	73
	84	55	58	197

B. Females

Pornography Leads to Breakdown. (item 59)	Income		
	Less than $22,500	$22,500 to $50,000	More than $50,000
Yes	107	58	34	199
No	37	25	23	85
	144	83	57	284

17.8 SOC Does support for traditional views of women vary by social class? Is this relationship affected by sex? By political ideology? By age? We looked at this relationship in problem 15.16e using FEFAM (item 69) as the dependent variable. The original version of the variable has four possible scores. We will need to reduce the number of values (and, thus, the number of cells in the tables) to conduct multivariate analysis. The bivariate table, with FEFAM collapsed, is presented below.

Attitude towards women's roles by income:

Women Should Take Care of Home and Family. (item 69)	Income			
	Less than $22,500	$22,500 to $50,000	More than $50,000	
Agree	113	38	37	188
Disagree	136	97	89	322
	249	135	126	510

Attitudes towards women's roles by income, controlling for gender:

A. Males

Women Should Take Care of Home and Family. (item 69)	Income			
	Less than $22,500	$22,500 to $50,000	More than $50,000	
Agree	54	19	20	93
Disagree	48	38	49	135
	102	57	69	228

B. Females

Women Should Take Care of Home and Family. (item 69)	Income			
	Less than $22,500	$22,500 to $50,000	More than $50,000	
Agree	59	19	17	95
Disagree	88	59	40	187
	147	78	57	282

Attitudes towards women's roles by income, controlling for political ideology:*

A. Liberals

Women Should Take Care of Home and Family. (item 69)	Income			
	Less than $22,500	$22,500 to $50,000	More than $50,000	
Agree	27	8	8	43
Disagree	37	34	23	94
	64	42	31	137

B. Moderates

Women Should Take Care of Home and Family. (item 69)	Income			
	Less than $22,500	$22,500 to $50,000	More than $50,000	
Agree	36	17	7	60
Disagree	61	38	32	131
	97	55	39	191

C. Conservatives

Women Should Take Care of Home and Family. (item 69)	Income			
	Less than $22,500	$22,500 to $50,000	More than $50,000	
Agree	42	13	20	75
Disagree	34	25	34	93
	76	38	54	168

Attitudes towards women's roles by income, controlling for age:

A. Younger (less than 40)

Women Should Take Care of Home and Family. (item 69)	Income			
	Less than $22,500	$22,500 to $50,000	More than $50,000	
Agree	23	9	13	45
Disagree	79	47	39	165
	102	56	52	210

B. Older (40 and older)

Women Should Take Care of Home and Family. (item 69)	Income			
	Less than $22,500	$22,500 to $50,000	More than $50,000	
Agree	90	29	24	143
Disagree	57	50	50	157
	147	79	74	300

*Partial tables have fewer cases than the bivariate table because of missing scores. Not all respondents answered all questions.

SPSS/PC+ PROCEDURES FOR ELABORATING BIVARIATE TABLES

DEMONSTRATION 17.1 Analyzing the Effects of AGE and SEX on Sexual Attitudes and Number of Traumatic Events

Since we have been concerned with bivariate tables in this chapter, it will come as no surprise to find that we will once again make use of the CROSS-TABS procedure. We have used this task often in previous chapters so I'll dispense with the usual step-by-step directions and simply tell you that, to control for a third variable with CROSSTABS, all you have to do is include another 'BY' in the /TABLES specification and name the control variable after this second BY.

To illustrate, in Demonstration 13.2 and again in Demonstration 15.1, we looked at the relationships between recoded TRAUMA5, PREMARSX and AGE (see Demonstrations 13.2 and 15.1 for the recoding scheme). We found that AGE had weak to moderate, positive relationships with the other variables. Now, let's see if SEX has any effect on these bivariate relationships. Are older women any more likely than older men to suffer traumatic events? Are the sexes likely to express different attitudes about premarital sex as a function of age?

Recode AGE and TRAUMA5 as in Demonstration 13.2 and PREMARSX as in Demonstration 15.1. Our command will be:

```
CROSSTABS /TABLES TRAUMA5 PREMARSX BY AGE BY SEX/
OPTIONS 4 /STATISTICS 8.
```

and the output would consist of the two partial tables, one for males and one for females, and gammas computed for each of these partial tables.

The bivariate relationships and statistics were displayed in Demonstration 15.1. At that point, we found a bivariate gamma of .17 for TRAUMA5 and AGE (older respondents were more likely to report traumatic events) and a bivariate gamma of −.30 between PREMARSX and AGE (older respondents were more opposed to premarital sex). To conserve space, we will concentrate on the gammas and not reproduce the partial tables.

After controlling for SEX, the relationship between TRAUMA5 and AGE is very similar across the partial tables, and each of these is similar to the bivariate relationship. The gamma for males is .11 and, for females, the statistic is .22. About a third of the oldest males and females reported the highest level of traumatic events. Almost exactly the same percentage of the sample as a whole reported this number of traumas. This looks like a direct relationship between TRAUMA5 and AGE. Gender has no important effect on the relationship between these two variables. Females and males show about the same relationship between age and the number of traumas.

The partial tables for PREMARSX and AGE also show that SEX has no effect. The partial gammas are about the same value as the bivariate gamma. For both males and females, disapproval of premarital sex increases with age.

DEMONSTRATION 17.2 Are Prestigious Jobs More Satisfying? Another Look

In Demonstration 15.3, we looked at the relationship between SATJOB and PRESTG80. We found a bivariate gamma of $-.37$, indicating that people with higher-prestige jobs reported higher levels of satisfaction. While not exactly surprising, we might be able to learn a little more about the relationship with some additional controls. One immediate control that might occur to you is employment status. We included everyone in the sample—students, house-wives, the retired, and so forth—in the original table. So, how can we limit the sample to only the employed?

We could control for the effect of employment status by running separate partial tables for the various conditions of WRKSTAT (item 1 in Appendix G). A moment's reflection should convince you that, in this case, such a control makes little sense, and it would be more appropriate to simply eliminate re-spondents who are not currently employed. Following the procedures from Demonstration 16.1, this can be done with the following command:

```
SELECT IF (WRKSTAT EQ 1).
```

Let's control for SEX and for DEGREE. Does the relationship between satisfaction and prestige vary by sex? Is the job satisfaction of the more edu-cated ("at least some college") differently affected by level of prestige than the job satisfaction of the less educated? We will have to recode DEGREE into just two categories:

```
RECODE DEGREE (0,1=1)(2 Thru 4=2).
VALUE LABELS DEGREE 1'HS or less' 2 'At Least Some
College'/.
```

Copy the other recode schemes from Demonstration 15.3 and add the control variables to the CROSSTABS command:

```
CROSSTABS /TABLES SATJOB BY PRESTG80 BY SEX DEGREE/
OPTIONS 4 /STATISTICS 8.
```

The bivariate table in Demonstration 15.3 showed a moderate relation-ship (gamma was $-.37$) between job satisfaction and prestige. Confining our attention only to the respondents who are presently employed, we find essentially the same relationship. Even though we have over 200 fewer cases, the gamma and the percentage patterns are essentially the same as the previous table.

```
Cross-tabulation:      SATJOB       JOB OR HOUSEWORK
                    By PRESTG80     RS OCCUPATIONAL
                                    PRESTIGE SCORE (1980)


               Count ¦ lower      ¦ higher    ¦
PRESTG80=>  Col Pct ¦            ¦           ¦ ¦ Row
                     ¦     1.00   ¦    2.00 ¦ Total
SATJOB         ------- ¦ ---------- ¦ -------- ¦
               1.00 ¦    52      ¦   125     ¦   177
   very sat         ¦    32.3    ¦   53.6    ¦  44.9
                     ¦ ---------- ¦ -------- ¦
               2.00 ¦   109      ¦   108     ¦   217
   less sat         ¦    67.7    ¦   46.4    ¦  55.1
                     ¦ ---------- ¦ -------- ¦
            Column     161          233         394
             Total     40.9         59.1       100.0

         Statistic                 Value
         ---------                 -----

Gamma                            -.41625

Number of Missing Observations =        0
```

The control for gender reveals some interaction by gender. The gammas show that the relationship is stronger for men than for women. Prestige has a more marked effect on how men rate their jobs. For example, in the top row of the partial tables, comparing higher- and lower-prestige men, there is a 26 point difference in the percentage who claim to be "very satisfied." The equivalent difference for women is only about 16 percentage points. Thus, prestige seems to have more of an effect on job satisfaction for men than for women.

```
Cross-tabulation:      SATJOB       JOB OR HOUSEWORK
                    By PRESTG80     RS OCCUPATIONAL
                                    PRESTIGE SCORE (1980)
          Controlling for SEX      RESPONDENTS SEX

                                      = 1.00 MALE
```

```
                   Count │ lower   │ higher  │
PRESTG80=>  Col Pct │         │         │      Row
                    │    1.00 │    2.00 │ Total
SATJOB       ------- │ --------│ --------│
                1.00 │    27   │    70   │    97
      very sat       │   31.4  │   57.4  │   46.6
                     │ --------│ --------│
                2.00 │    59   │    52   │   111
      less sat       │   68.6  │   42.6  │   53.4
                     │ --------│ --------│
              Column       86       122      208
              Total       41.3      58.7    100.0


        Statistic                   Value
        ---------                   -----

Gamma                              -.49259
```

```
Cross-tabulation:       SATJOB      JOB OR HOUSEWORK
                   By  PRESTG80      RS OCCUPATIONAL
                                     PRESTIGE SCORE (1980)
          Controlling for SEX        RESPONDENTS SEX

                                      = 2.00 FEMALE

                   Count │ lower   │ higher  │
PRESTG80=>  Col Pct │         │         │      Row
                    │    1.00 │    2.00 │ Total
SATJOB       ------- │ --------│ --------│
                1.00 │    25   │    55   │    80
      very sat       │   33.3  │   49.5  │   43.0
                     │ --------│ --------│
                2.00 │    50   │    56   │   106
      less sat       │   66.7  │   50.5  │   57.0
                     │ --------│ --------│
              Column       75       111      186
              Total       40.3      59.7    100.0


        Statistic                   Value
        ---------                   -----

Gamma                              -.32530


Number of Missing Observations =          0
```

The control for education also reveals an interactive relationship. For the respondents with lower level of educational attainment, job satisfaction is related to prestige in a pattern that is almost identical to the bivariate table. For respondents with "at least some college," however, the relationship is much stronger. Inspecting the column percentages reveals that better-educated respondents with higher-prestige jobs are about as satisfied as the less well educated respondents. The difference between the two partial tables lies in the left-hand column. Well-educated repondents in lower prestige jobs are especially likely to be less satisfied. Thus, there is interaction between education, prestige, and job satisfaction.

```
Cross-tabulation:        SATJOB      JOB OR HOUSEWORK
                     By PRESTG80    RS OCCUPATIONAL
                                    PRESTIGE SCORE (1980)
        Controlling for DEGREE      RS HIGHEST DEGREE

                                    = 1.00 hs or less

                 Count │ lower   │ higher  │
PRESTG80=>       Col Pct │       │         │       Row
                       │   1.00 │    2.00 │ Total
SATJOB           ------- │ ------- │ -------- │
                  1.00 │    45   │    58   │   103
        very sat        │  33.8  │  52.3  │  42.2
                        │ -------- │ -------- │
                  2.00 │    88   │    53   │   141
        less sat        │  66.2  │  47.7  │  57.8
                        │ -------- │ -------- │
                 Column    133      111      244
                 Total     54.5     45.5    100.0


           Statistic               Value
           ---------               -----

Gamma                            -.36307
Cross-tabulation:        SATJOB      JOB OR HOUSEWORK
                     By PRESTG80    RS OCCUPATIONAL
                                    PRESTIGE SCORE (1980)
        Controlling for DEGREE      RS HIGHEST DEGREE

                                    = 2.00 at least some
                                    college

                 Count │ lower   │ higher  │
PRESTG80=>       Col Pct │       │         │       Row
                       │   1.00 │    2.00 │ Total
SATJOB           ------- │ ------- │ -------- │
                  1.00 │     7   │    67   │    74
        very sat        │  25.0  │  54.9  │  49.3
                        │ -------- │ -------- │
                  2.00 │    21   │    55   │    76
        less sat        │  75.0  │  45.1  │  50.7
                        │ -------- │ -------- │
                 Column     28      122      150
                 Total     18.7     81.3    100.0
```

```
     Statistic                    Value
     ---------                    -----

Gamma                           -.57031
```

Exercises

17.1 Use Demonstration 17.1 as a guide and analyze the relationship of re-coded AGE on CAPPUN, GRASS, and GUNLAW, using SEX as the control variable. Summarize the results in a paragraph or two. Be sure to characterize the relationships as direct, spurious, or interactive.

17.2 Dredge up our analysis of the old maxim that money can't buy happiness (see Demonstrations 10.2 and 12.3). Analyze the relationship between HAPPY and INCOME91 (be sure to recode as in the previous demonstrations) while controlling for SEX, AGE, and then DEGREE (be sure to recode the latter two). Summarize the results of your investigation.

17.3 Is prejudice a regional phenomenon? Recode REGION into two categories: the south* (values 5, 6, and 7) and the nonsouth (all other values) and analyze the bivariate relationships with RACSEG[†]. Use a SELECT IF statement to limit the analysis to white respondents only. Since region is nominal in level of measurement, use phi or lambda as the measure of association. Be sure to analyze the percentage patterns in the table.

Now run the CROSSTABS procedure again, this time controlling for SEX, DEGREE, and AGE. Be sure to recode the latter two variables into dichotomies. After comparing the partial tables (percentages as well as the measure of association) with each other and the original bivariate table, characterize these multivariate relationships and answer the original question. Is prejudice related to region?

---

*The recoding scheme for region doesn't quite match conventional notions of "the South." The South Atlantic states (code 5) include three states that were not a part of the Confederacy (Delaware, Maryland, and West Virginia) and the District of Columbia.

[†]You will probably need to recode RACSEG into two categories: agree (codes 1 and 2) and disagree (codes 3 and 4).

18 Partial Correlation and Multiple Regression and Correlation

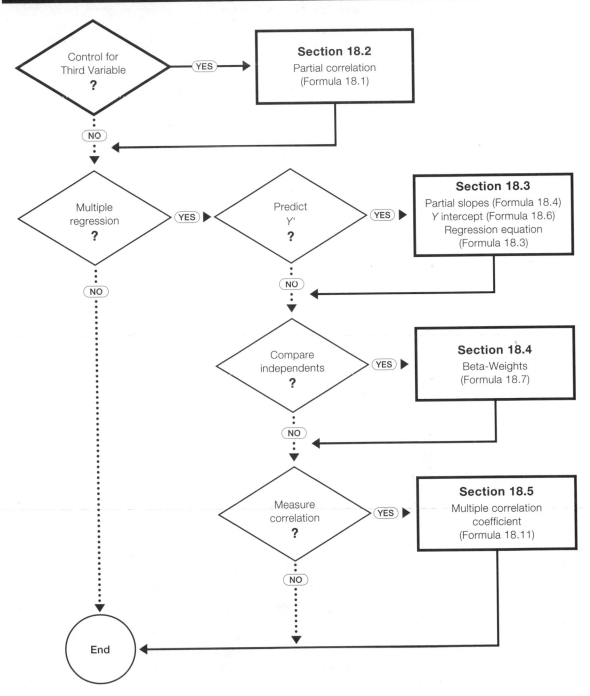

18.1 INTRODUCTION

In Chapter 17, we saw that additional information about a bivariate relationship could be gathered by observing what happens to that relationship when a third variable is brought into the analysis. Not only is our understanding of the bivariate relationship enhanced by the techniques discussed in Chapter 17, but we can also gather information with respect to the possible causal relationships among the variables.

In this chapter, I will introduce some techniques for analyzing the interrelationships of three or more variables that have been measured at the interval-ratio level. These techniques are based on Pearson's r (see Chapter 16) and, in general, are more flexible and produce more information than the techniques presented in Chapter 17. That is, these multivariate techniques provide a wider variety of ways of disentangling the interrelationships among the variables and also provide more detail with respect to the nature of these interrelationships.

The first technique to be introduced will be partial correlation analysis, a technique analogous to controlling for a third variable by constructing partial tables. In fact, the partial correlation coefficient provides much the same kind of information as a partial gamma does. The second technique involves multiple regression and correlation and allows the researcher to assess the effects, separately and in combination, of more than one independent variable on the dependent variable.

Throughout this chapter, we will focus our attention on research situations involving only three variables. This is the least complex application of these techniques, but extensions to situations involving four or more variables are relatively straightforward. To deal efficiently with the computations required by the more complex applications, I refer you to any of the computerized statistical packages probably available at your local computer center.

18.2 PARTIAL CORRELATION

The technique of **partial correlation** can be used when a researcher wishes to observe how a specific bivariate relationship behaves in the presence of a third variable. By observing the partial correlation coefficients, spurious and intervening relationships can be detected, and the causal relationships among the variables can be probed. As opposed to the techniques considered in Chapter 17, however, partial correlation analysis does not involve the inspection of the bivariate relationship for each category of the control variable. For this reason, partial correlation techniques will not reveal interactive relationships among the variables (that is, relationships where the nature of the bivariate relationship changes across the categories of the control variable).

Before dealing with partial correlation per se, let me introduce some new terminology. Since we will be dealing with more than two variables, we will use subscripts to denote exactly which variables are being correlated. Thus, the symbol r_{yx} will refer to the correlation coefficient between variable Y and variable X. By the same token, r_{yz} will refer to the correlation coefficient between Y and Z, and r_{xz} to the correlation coefficient between X and Z. Correlation coefficients calculated for bivariate relationships are often referred to as **zero-order correlations.**

Partial correlation coefficients, when controlling for a single variable, are called first-order partials and can be symbolized as $r_{yx \cdot z}$. The variable to the right of the dot is the control variable. Thus, $r_{yx \cdot z}$ refers to the partial correlation coefficient that measures the relationship between variables X and Y while controlling for variable Z. The formula for the first-order partial is

FORMULA 18.1

$$r_{yx \cdot z} = \frac{r_{yx} - (r_{yz})(r_{xz})}{\sqrt{1 - r_{yz}^2}\ \sqrt{1 - r_{xz}^2}}$$

To solve this formula, first calculate the zero-order coefficients between all possible pairs of variables (variables X and Y, X and Z, and Y and Z).

To illustrate the computation of a first-order partial, let us return to the problem introduced in Chapter 16 that examined the relationship between number of children (X) and husband's contribution to housework (Y) for 12 dual-career families. The zero-order r between these two variables ($r_{yx} = .50$) indicated a moderate, positive relationship. Suppose the researcher wished to investigate the possible effects of socioeconomic status (SES) on the bivariate relationship. The original data (from Table 16.1) and the scores of the 12 families on the new variable of SES (measured by the years of education completed by the husband) are presented in Table 18.1.

The three correlations indicate that the husband's contribution to housework is related to number of children ($r_{yx} = .50$), that husbands in higher SES families tend to do less housework ($r_{yz} = -.30$), and that higher-SES families

TABLE 18.1 SCORES ON THREE VARIABLES FOR 12 DUAL WAGE-EARNER FAMILIES AND ZERO-ORDER CORRELATIONS

Family	Husband's Housework (Y)	Number of Children (X)	Husband's Years of Education (Z)
A	1	1	12
B	2	1	14
C	3	1	16
D	5	1	16
E	3	2	18
F	1	2	16
G	5	3	12
H	0	3	12
I	6	4	10
J	3	4	12
K	7	5	10
L	4	5	16
	$r_{yx} = .50$	$r_{yz} = -.30$	$r_{xz} = -.47$

have fewer children ($r_{xz} = -.47$). Is the relationship between husband's housework and number of children affected by SES? Substituting the above correlations into Formula 18.1, we would have

$$r_{yx \cdot z} = \frac{r_{yx} - (r_{yz})(r_{xz})}{\sqrt{1 - r_{yz}^2}\,\sqrt{1 - r_{xz}^2}}$$

$$r_{yx \cdot z} = \frac{.50 - (-.30)(-.47)}{\sqrt{1 - (-.30)^2}\,\sqrt{1 - (-.47)^2}}$$

$$r_{yx \cdot z} = \frac{.50 - (.14)}{\sqrt{1 - .09}\,\sqrt{1 - .22}}$$

$$r_{yx \cdot z} = \frac{.36}{\sqrt{.91}\,\sqrt{.78}}$$

$$r_{yx \cdot z} = \frac{.36}{(.95)(.88)}$$

$$r_{yx \cdot z} = \frac{.36}{.84}$$

$$r_{yx \cdot z} = .43$$

The first-order partial ($r_{yx \cdot z} = .43$) is lower in value than the zero-order coefficient ($r_{yx} = .50$), but the difference in the two values is not great. This result suggests a direct relationship between variables X and Y (see Section 17.3). That is, when controlling for SES, the statistical relationship between husband's housework and number of children is essentially unchanged. Regardless of SES, the hours devoted to housework by husbands increase with the number of children. As was the case when a direct relationship was found in Chapter 17, the next step in statistical analysis would probably be to discard this control variable (SES) and select another (see Table 17.9). The more the bivariate relationship retains its strength across a series of controls for the third variables (Z's), the stronger the evidence for a direct relationship between X and Y.

In addition to the pattern discussed above, two other possible relationships between the partial and zero-order correlation coefficients exist. If the partial is much lower in value than the zero-order coefficient ($r_{yx \cdot z} < r_{yx}$), the bivariate relationship between X and Y may be spurious, or Z may intervene between X and Y. The interpretation of this outcome follows the lines established in Chapter 17; the probable next step in the statistical analysis would be to incorporate the control variable into the analysis as an independent (see Table 17.9). Let me stress that partial correlation techniques cannot distinguish between the case where the control variable is antecedent (where Z "causes" both X and Y) and the case where the control variable is intervening (when X "causes" Z, which in turn "causes" Y). Judgments about possible causal relationships must be made on the grounds of temporal ordering among the variables and/or theory.

FIGURE 18.1   A POSSIBLE CAUSAL RELATIONSHIP AMONG THREE VARIABLES

A third possible outcome of the application of partial correlation is for the partial correlation coefficient to be greater in value than the zero-order coefficient ($r_{xy.z} > r_{yw}$). This outcome would be consistent with a causal model where the variable taken as independent and the control variable each had a separate effect on the dependent variable and were uncorrelated with each other. This relationship is depicted in Figure 18.1. The absence of an arrow between X and Z indicates that they have no mutual relationship.

Upon finding this pattern, a researcher would conclude that both X and Z were independents, and the next step in the statistical analysis would probably involve multiple correlation and regression. As we shall see in Sections 18.3 and 18.4, these techniques enable the researcher to isolate the separate effects of several independent variables on the dependent variable and thus to make judgments about which independent has the stronger effect on the dependent.

18.3 MULTIPLE REGRESSION: PREDICTING THE DEPENDENT VARIABLE

In Chapter 16, the least-squares regression line was introduced as a way of describing the overall linear relationship between two interval-ratio variables and of predicting scores on Y from scores on X. This line was the best-fitting line to summarize the bivariate relationship and was defined by the formula

FORMULA 18.2

$$Y = a + bX$$

where a = the Y intercept
b = the slope

The least-squares regression line can be modified to include (theoretically) any number of independent variables. This technique is called **multiple regression**. For ease of explication, we will confine our attention to the case involving two independent variables. The least-squares multiple regression equation is

FORMULA 18.3

$$Y = a + b_1 X_1 + b_2 X_2$$

where b_1 = the partial slope of the linear relationship between the first independent variable and Y
b_2 = the partial slope of the linear relationship between the second independent variable and Y

Some new notation and some new concepts are introduced in this formula. First, while the dependent variable is still symbolized as Y, the independent variables are differentiated by subscripts. Thus, X_1 identifies the first independent variable and X_2 the second. The symbol for the slope (b) is also subscripted to identify the independent with which it is associated.

A major difference between the multiple and bivariate regression equations concerns the slopes (b's). In the multiple regression case, the b's are actually **partial slopes**, since they show the amount of change in Y for a unit change in the independent while controlling for the effects of the other independents in the equation. The partial slopes are thus analogous to partial correlation coefficients and represent the direct effect of the associated independent variable on Y.

The partial slopes for the independent variables are determined by Formula 18.4 and Formula 18.5:*

FORMULA 18.4
$$b_1 = \left(\frac{s_y}{s_1}\right)\left(\frac{r_{y1} - r_{y2}r_{12}}{1 - r_{12}^2}\right)$$

FORMULA 18.5
$$b_2 = \left(\frac{s_y}{s_2}\right)\left(\frac{r_{y2} - r_{y1}r_{12}}{1 - r_{12}^2}\right)$$

where b_1 = the partial slope of X_1 on Y
b_2 = the partial slope of X_2 on Y
s_y = the standard deviation of Y
s_1 = the standard deviation of the first independent variable (X_1)
s_2 = the standard deviation of the second independent variable (X_2)
r_{y1} = the bivariate correlation between Y and X_1
r_{y2} = the bivariate correlation between Y and X_2
r_{12} = the bivariate correlation between X_1 and X_2

To illustrate the computation of the partial slopes, we will assess the combined effects of number of children (X_1) and SES (X_2) on husband's contribution to housework. All the relevant information can be calculated from Table 18.1 and is reproduced below:

Husband's Housework	Number of Children	SES
$\bar{Y} = 3.3$	$\bar{X}_1 = 2.7$	$\bar{X}_2 = 13.7$
$s_y = 2.1$	$s_1 = 1.5$	$s_2 = 2.6$

Zero-order correlations
$r_{y1} = .50$
$r_{y2} = -.30$
$r_{12} = -.47$

---

*Partial slopes can be computed from zero-order slopes, but Formulas 18.4 and 18.5 are somewhat easier to use.

The partial slope for the first independent variable (X_1) is

$$b_1 = \left(\frac{s_y}{s_1}\right)\left(\frac{r_{y1} - r_{y2}r_{12}}{1 - r_{12}^2}\right)$$

$$b_1 = \left(\frac{2.1}{1.5}\right)\left(\frac{.50 - (-.30)(-.47)}{1 - (-.47)^2}\right)$$

$$b_1 = (1.4)\left(\frac{.50 - .14}{1 - .22}\right)$$

$$b_1 = (1.4)\left(\frac{.36}{.78}\right)$$

$$b_1 = (1.4)(.46)$$

$$b_1 = .65$$

For the second independent variable, SES or X_2, the partial slope is

$$b_2 = \left(\frac{s_y}{s_2}\right)\left(\frac{r_{y2} - r_{y1}r_{12}}{1 - r_{12}^2}\right)$$

$$b_2 = \left(\frac{2.1}{2.6}\right)\left(\frac{-.30 - (-.24)}{1 - .22}\right)$$

$$b_2 = (.81)\left(\frac{-.30 + .24}{.78}\right)$$

$$b_2 = (.81)\left(\frac{-.06}{.78}\right)$$

$$b_2 = (.81)(-.08)$$

$$b_2 = -.07$$

Now that partial slopes have been determined for both independent variables, the final unknown in the least-squares multiple regression equation can be found. Note that the Y intercept (a) is calculated from the mean of the dependent variable (symbolized as $\bar{Y}$) and the means of the two independent variables ($\bar{X}_1$ and $\bar{X}_2$).

FORMULA 18.6
$$a = \bar{Y} - b_1\bar{X}_1 - b_2\bar{X}_2$$

Substituting in the proper values for the example problem at hand, we would have

$$a = \bar{Y} - b_1\bar{X}_1 - b_2\bar{X}_2$$
$$a = 3.3 - (.65)(2.7) - (-.07)(13.7)$$
$$a = 3.3 - (1.8) - (-1.0)$$
$$a = 3.3 - 1.8 + 1.0$$
$$a = 2.5$$

For our example problem, the full least-squares multiple regression equation would be

$$Y = a + b_1X_1 + b_2X_2$$
$$Y = 2.5 + (.65)X_1 + (-.07)X_2$$

As was the case with the bivariate regression line, this formula can be used to predict scores on the dependent variable from scores on the independent variables. For example, what would be our best prediction of husband's housework (Y') for, say, a family of four children ($X_1 = 4$) where the husband had completed 11 years of schooling ($X_2 = 11$)? Substituting these values into the least-squares formula, we would have

$$Y' = 2.5 + (.65)(4) + (-.07)(11)$$
$$Y' = 2.5 + 2.6 - .8$$
$$Y' = 4.3$$

Our prediction would be that, in a family of four children when the husband had 11 years of education, the husband would contribute 4.3 hours per week to housework. This prediction is, of course, a kind of "educated guess." While our predictions of Y are unlikely to be perfectly accurate, we will make fewer errors of prediction using the least-squares line (and, thus, incorporating information from the independent variables) than we would using any other method of prediction (assuming, of course, that a linear association exists between the independents and the dependent).

18.4 MULTIPLE REGRESSION: ASSESSING THE EFFECTS OF THE INDEPENDENT VARIABLES

The least-squares multiple regression equation (Formula 18.3) is quite useful for purposes of isolating the separate effects of the independents and for predicting scores on the dependent variable. However, in many situations, using this formula to determine the relative importance of the various independents will be awkward—especially when the independent variables differ in terms of units of measurement. In our example problem, the first independent is number of children, while the second is years of education. In situations where the units of measurement differ, we will not necessarily be able to tell from the partial slopes which independent has the strongest effect on the dependent and is thus the most important of the independent variables. Comparing the partial slopes of variables that differ in units of measurement is a little like comparing apples and oranges.

The independent variables can be made more comparable by converting all of the variables in the equation to a common scale and thereby eliminating variations in the values of the partial slopes that are solely a function of differences in units of measurement. We can, for example, standardize all distributions by converting all scores into Z scores. Each distribution of scores for each variable would then have a mean of 0 and a standard deviation of 1, and comparisons between the standardized independent variables would become much more meaningful. This standardization could be accomplished by converting all scores for each variable into the equivalent Z scores (that is, subtracting the mean of the distribution from each score and dividing by the standard deviation of the distribution) and then recomputing the slopes and the Y intercept by using the Z scores. This procedure requires a good deal of work and, fortunately, a shortcut is available for directly computing the slopes of the standardized scores. These **standardized partial slopes** are called **beta-weights** and are symbolized b^*. The beta-weights will show the amount of change in the standardized scores of Y for a one-unit change in the standardized scores of the independent while controlling for the effects of the other independents. For the case where we have two independents, the beta-weight for each is found by using Formula 18.7 and Formula 18.8:

FORMULA 18.7

$$b_1^* = b_1 \left(\frac{s_1}{s_y} \right)$$

FORMULA 18.8

$$b_2^* = b_2 \left(\frac{s_2}{s_y} \right)$$

where b_1^* = the standardized partial slope of X_1 on Y
b_2^* = the standardized partial slope of X_2 on Y

Using standardized scores, the least-squares regression equation can be written as

FORMULA 18.9

$$Z_y = a_z + b_1^* Z_1 + b_2^* Z_2$$

where the symbol Z indicates that all scores have been standardized to the normal curve

The standardizes regression equation can be further simplified by dropping the term for the Y intercept, since this term will always be zero when scores have been standardized. This value is the point where the regression line crosses the Y axis and is equal to the mean of Y when all independents equal 0. This relationship can be seen by substituting 0 for all independent variables in Formula 18.6:

$$a = \bar{Y} - b_1 \bar{X}_1 - b_2 \bar{X}_2$$
$$a = \bar{Y} - b_1(0) - b_2(0)$$
$$a = \bar{Y}$$

Since the mean of any standardized distribution of scores is zero, the mean of the standardized Y scores will be zero and the Y intercept will also be zero ($a = \bar{Y} = 0$). Thus, Formula 18.9 simplifies to

FORMULA 18.10

$$Z_y = b_1^* Z_1 + b_2^* Z_2$$

We can now compute the beta-weights for our sample problem to see which of the two independents has the stronger effect on the dependent. For the first independent variable, number of children (X_1):

$$b_1{}^* = b_1 \left(\frac{s_1}{s_y} \right)$$

$$b_1{}^* = (.65) \left(\frac{1.5}{2.1} \right)$$

$$b_1{}^* = (.65)(.71)$$

$$b_1{}^* = .46$$

For the second independent variable, SES (X_2):

$$b_2{}^* = b_2 \left(\frac{s_2}{s_y} \right)$$

$$b_2{}^* = (-.07) \left(\frac{2.6}{2.1} \right)$$

$$b_2{}^* = (-.07)(1.24)$$

$$b_2{}^* = -.09$$

Thus, the standardized regression equation, with beta-weights noted, would be

$$Z_y = (.46)Z_1 + (-.09)Z_2$$

and it is immediately obvious that the first independent variable has a much stronger direct effect on Y than the second independent variable.

In summary, multiple regression analysis permits the researcher to summarize the linear relationship among two or more independents and a dependent variable. The unstandardized regression equation (Formula 18.2) permits values of Y to be predicted from the independent variables in the original units of the variables. The standardized regression equation (Formula 18.10) allows the researcher to assess the relative importance of the various independents more easily by noting which of the beta-weights is greatest in value.

18.5 MULTIPLE CORRELATION

We use the multiple regression equations to disentangle the separate direct effects of each independent variable on the dependent. Using **multiple correlation** techniques, we can also ascertain the combined effects of all independents on the dependent variable. We do so by computing the **multiple correlation coefficient (R)** and the **coefficient of multiple determination (R^2).** The value of the latter statistic represents the proportion of the variance in Y that is explained by all the independent variables combined. In terms of zero-order correlation, we have seen that "number of children" (X_1) explains a proportion of .25 of the variance in Y ($r_{y1}^2 = (.50)^2 = .25$) by itself and that SES explains a proportion of .09 of the variance in Y ($r_{y2}^2 = (-.30)^2 = .09$). The two zero-order correlations cannot be simply added together to ascertain their combined effect on Y, because the two independents are also correlated with each other and, therefore, will "overlap" in their effects on Y and explain some of the same variance. This overlap is eliminated in Formula 18.11:

FORMULA 18.11

$$R^2 = r_{y1}^2 + r_{y2\cdot1}^2 (1 - r_{y1}^2)$$

where R^2 = the multiple correlation coefficient
r_{y1}^2 = the zero-order correlation between Y and X_1, the quantity squared
$r_{y2\cdot1}^2$ = the partial correlation of Y and X_2 while controlling for X_1, the quantity squared
r_{y1}^2 = the zero-order correlation between Y and X_1, the quantity squared

The first term in this formula (r_{y1}^2) is the coefficient of determination for the relationship between Y and X_1. It represents the amount of variation in Y explained by X_1 by itself. To this quantity we add the amount of the variation remaining in Y (given by $1 - r_{y1}^2$) that can be explained by X_2 after the effect of X_1 is controlled ($r_{y2\cdot1}^2$). Basically, Formula 18.11 allows X_1 to explain as much of Y as it can and then adds in the effect of X_2 after X_1 is controlled (thus eliminating the "overlap" in the variance of Y that X_1 and X_2 have in common).

To observe the combined effects of number of children (X_1) and SES (X_2) on husband's housework (Y), we need two quantities. The correlation between X_1 and Y ($r_{y1} = .50$) has already been found; but, before we can solve Formula 18.11, we must first calculate the partial correlation of Y and X_2 while controlling for X_1 ($r_{y2\cdot1}$):

$$r_{y2\cdot1} = \frac{r_{y2} - (r_{y1})(r_{12})}{\sqrt{1 - r_{y1}^2}\ \sqrt{1 - r_{12}^2}}$$

$$r_{y2\cdot1} = \frac{(-.30) - (.50)(-.47)}{\sqrt{1 - (.50)^2}\ \sqrt{1 - (-.47)^2}}$$

$$r_{y2\cdot1} = \frac{-.30 - (-.235)}{\sqrt{.75}\ \sqrt{.78}}$$

$$r_{y2\cdot1} = \frac{-.06}{(.87)(.88)}$$

$$r_{y2\cdot1} = \frac{-.06}{.77}$$

$$r_{y2\cdot1} = -.08$$

Formula 18.11 can now be solved for our sample problem:

$$R^2 = r_{y1}^2 + r_{y2 \cdot 1}^2 (1 - r_{y1}^2)$$
$$R^2 = (.50)^2 + (-.08)^2(1 - .50^2)$$
$$R^2 = (.25) + (.006)(1 - .25)$$
$$R^2 = (.25) + (.006)(.75)$$
$$R^2 = .25 + .005$$
$$R^2 = .255$$

The first independent variable (X_1), number of children, explains 25% of the variance in Y by itself. To this total, the second independent (X_2), SES, adds only a half a percent, for a total explained variance of 25.5%. In combination, the two independents explain a total of 25.5% of the variation in the dependent variable.

18.6 A NON-COMPUTATIONAL EXAMPLE

You have been presented with much technical material over the past several sections, so let us pause for a time and consider what all this means. I will present an example of how these techniques might be used and interpreted in a statistical investigation. I will consciously minimize matters of calculation and equations so that we may concentrate on meaning. I will also take this opportunity, the final "sample problem" of the text, to indulge myself in some thinly disguised statistical moralizing.

A researcher wonders about the causes of "success" in statistics courses. Why do some students achieve higher grades than others? Is it simply a matter of aptitude or level of preparation for college-level mathematics? What are the effects, if any, of motivation and willingness to work hard? Can level of success in statistics be reasonably predicted? With the appropriate data and the help of the techniques introduced in this chapter, we can answer these questions.

To begin, assume that the researcher has three pieces of information about a large number of students who have completed courses in statistics: final grade in the course (success in statistics), score on the college board mathematics test (a measure of level of preparation for college-level statistics courses), and the average number of hours each student spent studying statistics each week (a measure of motivation or willingness to work hard). Statistics grade will be taken as the dependent variable (Y) and the other two as independents $(X_1$ and $X_2)$.

The first step in the analysis would probably involve the careful inspection of the three possible zero-order relationships. Scattergrams would be produced, regression equations found, and correlation coefficients calculated. Let us suppose that the bivariate relationships are all roughly linear and that the "preparation" variable (college board score) has a stronger bivariate relationship with statistics grade than does the "motivation" variable (hours studying). We would have a preliminary indication that success in statistics was more closely related to preparation than to motivation. We theorize that

TABLE 18.2   ZERO-ORDER CORRELATION COEFFICIENTS (fictitious data)

	College Board Score (Preparation)	Hours of Study (Motivation)
Statistics grade (success)	.60	.32
College board score (preparation)		− .15

successful students achieve higher grades mainly because they are better pre-pared. The fictitious zero-order relationships are presented in Table 18.2.

These zero-order relationships display the effects of each independent, one at a time, on the dependent variable. To assess the combined effects of both independent variables and the direct effects of each while controlling for the other, we must use the techniques presented in the previous sections.

To begin with matters of prediction, we can find the least-squares mul-tiple regression equation as defined by Formula 18.3. After calculating the partial slopes and the Y intercept, we can predict a student's final grade in statistics. If the partial slopes were .09 for preparation ($b_1 = .09$) and 1.17 for motivation ($b_2 = 1.17$), with the Y intercept equal to 12 ($a = 12$), the predic-tion equation would be

$$Y' = 12 + (.09)X_1 + (1.17)X_2$$

and the researcher would be ready to make some educated guesses about statistics grades. For example, how well would a very well prepared student ($X_1 = 650$) do in the course if he or she did not study ($X_2 = 1$)? The multiple regression equation can help us answer this question by providing a pre-dicted grade for such a student:

$$Y' = 12 + (.09)(650) + (1.17)(1)$$
$$Y' = 12 + 58.5 + 1.17$$
$$Y' = 71.67$$

At the very best, a grade in the low 70s would be average. The prediction would be that a well-prepared student might be able to slide by without working but would be extremely unlikely to do well in the course. How about a not-so-well-prepared student ($X_1 = 400$) who compensates by work-ing hard and studying an average of 40 hours per week ($X_2 = 40$)? Again, the regression equation can supply a prediction:

$$Y' = 12 + (.09)(400) + (1.17)(40)$$
$$Y' = 12 + 36 + 46.8$$
$$Y' = 94.8$$

A score in the middle 90s would indicate a high level of success and would suggest that any lack of preparation could indeed be overcome by hard work.

APPLICATION 18.1

Five recently divorced men have been asked to rate subjectively the success of their adjustment to single life on a scale ranging from 5 (very successful adjustment) to 1 (very poor adjustment). Is adjustment related to the length of time married? Is adjustment related to socioeconomic status as measured by yearly income?

Case	Adjustment (Y)	Years Married (X_1)	Income (dollars) (X_2)
A	5	5	30,000
B	4	7	45,000
C	4	10	25,000
D	3	2	27,000
E	1	15	17,000

$$\bar{Y} = 3.4 \qquad \bar{X}_1 = 7.8 \quad \bar{X}_2 = 28,800.00$$
$$s = 1.4 \qquad s = 4.5 \quad s = \ 9173.88$$

The zero-order correlations among these three variables are

	Years Married (X_1)	Income (X_2)
Adjustment (Y)	− .62	.62
Years married (X_1)		− .49

These results suggest strong but opposite relationships between each independent and adjustment. Adjustment decreases as years married increases, and increases as income increases.

To find the multiple regression equation, we must find the partial slopes.

For years married (X_1):

$$b_1 = \left(\frac{s_y}{s_1}\right)\left(\frac{r_{y1} - r_{y2}r_{12}}{1 - r_{12}^2}\right)$$

$$b_1 = \left(\frac{1.4}{4.5}\right)\left(\frac{(-.62) - (.62)(-.49)}{1 - (-.49)^2}\right)$$

$$b_1 = (.31)\left(\frac{(-.62) - (-.30)}{1 - .24}\right)$$

$$b_1 = (.31)\left(\frac{-.32}{.76}\right)$$

$$b_1 = (.31)(-.42)$$

$$b_1 = -.13$$

For income (X_2):

$$b_2 = \left(\frac{s_y}{s_2}\right)\left(\frac{r_{y2} - r_{y1}r_{12}}{1 - r_{12}^2}\right)$$

$$b_2 = \left(\frac{1.4}{9173.88}\right)\left(\frac{.62 - (-.62)(-.49)}{1 - (-.49)^2}\right)$$

$$b_2 = (.00015)\left(\frac{.62 - .30}{1 - .24}\right)$$

$$b_2 = (.00015)\left(\frac{.32}{.76}\right)$$

$$b_2 = (.00015)(.42)$$

$$b_2 = .000063$$

Finally, what would our prediction be for a student of average preparation $(X_1 = 500)$ who studied an average number of hours per week $(X_2 = 15)$?

$$Y' = 12 + (.09)(500) + (1.17)(15)$$
$$Y' = 12 + 45 + 17.6$$
$$Y' = 74.6$$

The Y intercept would be

$$a = \bar{Y} - b_1\bar{X}_1 - b_2\bar{X}_2$$
$$a = 3.4 - (-.13)(7.8) - (.000063)(28,800.00)$$
$$a = 3.4 - (-1.01) - (1.81)$$
$$a = 3.4 + 1.01 - 1.81$$
$$a = 2.60$$

The multiple regression equation is

$$Y = a + b_1X_1 + b_2X_2$$
$$Y = 2.60 + (-.13)X_1 + (.000063)X_2$$

What adjustment score could we predict for a male who had been married 30 years ($X_1 = 30$) and had an income of \$50,000 ($X_2 = 50,000$)?

$$Y' = 2.60 + (-.13)(30) + (.000063)(50,000)$$
$$Y' = 2.60 + (-3.9) + (3.15)$$
$$Y' = 1.85$$

To assess which of the two independents has the stronger effect on adjustment, the standardized partial slopes must be computed. For years married (X_1):

$$b_1^* = b_1\left(\frac{s_1}{s_y}\right)$$

$$b_1^* = (-.13)\left(\frac{4.5}{1.4}\right)$$

$$b_1^* = -0.42$$

For income (X_2):

$$b_2^* = b_2\left(\frac{s_2}{s_y}\right)$$

$$b_2^* = (.000063)\left(\frac{9173.88}{1.4}\right)$$

$$b_2^* = 0.41$$

The standardized regression equation is

$$Z_y = b_1^*Z_1 + b_2^*Z_2$$
$$Z_y = (-0.42)Z_1 + (0.41)Z_2$$

and the independents have nearly equal but opposite effects on adjustment. To assess the combined effects of the two independents on adjustment, the coefficient of multiple determination must be computed.

$$R^2 = r_{y1}^2 + r_{y2\cdot1}^2 (1 - r_{y1}^2)$$

$$R^2 = (-.62)^2 + (-.46)^2(1 - (-.62)^2)$$

$$R^2 = .38 + (.21)(1 - .38)$$

$$R^2 = .38 + (.21)(.62)$$

$$R^2 = .38 + .13$$

$$R^2 = .51$$

The first independent, years married, explains 38% of the variation in adjustment by itself. To this quantity, income explains an additional 13% of the variation in adjustment. Taken together, the two independents explain a total of 51% of the variation in adjustment.

An average student with average motivation would be likely to end the course with an average grade.

Let us stress several points about these predictions. First, we can make no claim to fortune-telling. In the absence of perfect relationships, we will be unlikely to predict final grade perfectly. On the other hand, the stronger the relationships between the independents and the dependent, the more accurate our predictions will be.

Second, we know intuitively that other factors besides the two identified in our regression equation have an effect on grades. For example, the efficiency of study time is probably as important as the amount, and there are affective dimensions (for example, whether or not the student likes statistics) that should be explored as well. The accuracy of our predictions would be enhanced if we measured these other variables and incorporated them into the equation as additional independents. As a general rule, increasing the number of independents will increase the amount of variation explained in the dependent. However, to the extent that the independents are correlated with each other, they will overlap in the portion of the variation that they explain in the dependent. Each additional independent variable will likely explain a smaller and smaller portion of the variation in the dependent variable, and the researcher will reach a point of diminishing returns.

Third, although the unstandardized partial slopes permit predictions of the dependent in the original units of measurement (final average grade), they do not permit an assessment of the relative importance of the two independents. Since the independent variables have different scales (college board scores and hours per week), the unstandardized partial slopes cannot be directly compared to assess which has the stronger effect.

To deal with this issue, we could compute the beta-weights (standardized partial slopes) for each independent. Since the beta-weights are based on standardizing all distributions to Z scores, they eliminate the difficulties that arise from differences in scale among the independents. Suppose that the standardized regression equation for the present relationship was (based on Formula 18.10)

$$Z_y = b_1^* Z_1 + b_2^* Z_2$$
$$Z_y = (.64)Z_1 + (.42)Z_2$$

These results would indicate that both independents had fairly strong direct effects on statistics grade and that level of preparation (Z_1) was the stronger of the two.

As a final step in the analysis, we would no doubt want to assess the combined effects of these two independents on final grade in statistics. This can be accomplished by computing the coefficient of multiple determination (R^2) as defined in Formula 18.11:

$$R^2 = r_{y1}^2 + r_{y2\cdot1}^2 (1 - r_{y1}^2)$$
$$R^2 = .60^2 + (.44^2)(1 - .60^2)$$
$$R^2 = .36 + (.19)(.64)$$
$$R^2 = .36 + .12$$
$$R^2 = .48$$

The first independent variable, level of preparation, explains 36% of the variation in statistics grade by itself. To this quantity, the second independent, study time, adds another 12%. In combination, the two independents explain almost half of the variation in the dependent variable. About half of the variation remains unexplained, presumably because of some combination of the other, unmeasured variables mentioned above (efficiency of study time, affective factors), measurement error, and random chance.

To summarize in terms of our original questions, we have seen that both independents have important relationships with final grade in statistics. The unstandardized regression equation was used to predict grades for various combinations of preparation and motivation. The standardized regression equation showed that both independents had substantial direct effects on the dependent and that the preparation variable had a stronger effect on final grade than study time. Finally, we saw that, in combination, the two independents account for about half of the variation in final grade.

The analysis shows that, although well-prepared students may be able to slide by without too much effort, students who are not well prepared can compensate and even achieve high levels of success by increased study time. High levels of preparation do not guarantee high levels of success in the absence of a willingness to work hard. Regardless of background, highly motivated students can acquit themselves with distinction.

18.7 THE LIMITATIONS OF MULTIPLE REGRESSION AND CORRELATION

Multiple regression and correlation are very powerful tools for analyzing the interrelationships among three or more variables. The techniques presented in this chapter permit the researcher to predict scores on one variable from two or more other variables, to distinguish between independent variables in terms of the importance of their direct effects on a dependent, and to ascertain the total effect of a set of independent variables on a dependent variable. In terms of the flexibility of the techniques and the volume of information they can supply, multiple regression and correlation represent some of the most powerful statistical techniques available to the researcher.

Such powerful tools are not, of course, cheap. They demand high-quality data; and, for many concepts and variables, measurement at the interval-ratio level may be very difficult considering the present stage of development of the social sciences. Furthermore, these techniques assume that the interrelationships among the variables follow a particular form. First, they assume that each independent variable is related to the dependent in a linear fashion. How well a given set of variables meets this assumption can be quickly checked by use of scattergrams.

Second, the techniques presented in this chapter assume that there is no interaction among the variables in the equation. If there is interaction among the variables, it will not be possible to accurately estimate or predict the dependent variable by simply adding the effects of the independents. There are techniques for handling interaction among the variables in the set, but these techniques are beyond the scope of this text.

READING STATISTICS 8: REGRESSION AND CORRELATION

Research projects that analyze the interrelationships among many variables are particularly likely to employ regression and correlation as central statistical techniques. The results of these projects will typically be presented in summary tables that report only the zero-order correlations, multiple correlations, slopes, and, if applicable, the significance of the results. The zero-order correlations are often presented in the form of a matrix that displays the value of Pearson's r for every possible bivariate relationship in the data set. Sometimes, for clarity, r's that are considered weak or unimportant (for example, r's weaker than $\pm .20$) will be omitted from the matrix. Also, r's that are statistically significant at some selected level of alpha (usually the 0.05 level) may be identified by some symbol (often, an asterisk, or "*"). These matrices have variable names for column and row headings, and the body of the table reports the value of Pearson's r for each pair of variables. An example of such a matrix can be found in Section 18.6.

Tables may also be used to summarize the results of a multiple regression and correlation analysis, especially when the research involves more than one dependent variable. At a minimum, these summary tables will report the value of R^2 and the slope for each independent variable in the regression equation. The independent variables will usually be ranked by the strength of their direct effect on the dependent. An example of this kind of summary table would look like this:

Independents	Multiple R^2	Beta-weights
X_1	.17	.47
X_2	.23	.32
X_3	.27	.16

This table reports that the first independent variable, X_1, has the strongest direct relationship with the dependent variable and explains 17% of the variance in the dependent by itself ($R^2 = 0.17$). The second independent, X_2, adds 6% to the explained variance ($R^2 = 0.23$ after X_2 is entered into the equation). The third independent, X_3, adds 4% to the explained variance of the dependent ($R^2 = 0.27$ after X_3 is entered into the equation). A table such as this would be included for each dependent variable.

Statistics in the Professional Literature

Besides the obvious burden of the academic workload, what are the sources of stress for students? Professor Anne Fortune asked this question of a sample of social work graduate students. She was particularly interested in the possibility that playing multiple social roles (for example, parent, spouse, and worker) might increase stress and decrease psychological well-being for these students. One theory of role conflict, which she calls the "scarcity model," suggests that multiple roles will overload the individual. All of the roles will be played poorly, and stress and feelings of inner conflict will increase.

On the other hand, Professor Fortune points to an alternative theory, which she labels the "expansion model," which ". . . suggests that more roles increase the individual's sense of identity, privileges from role-statuses, and interpersonal relationships, and consequently increases self-esteem and decreases stress" (82). The table below presents some of her results for second-year social work graduate students. The entries in the table are beta-weights.

	Well-being $(N = 255)$	General Stress $(N = 260)$	Stress as Student $(N = 283)$
Variable:			
Married	.054	−.074	−.007
Children	.066	−.001	−.013
Working	.017	.009	−.161*
Age	−.096	−.107	−.051
Male	−.047	.068	−.057
Part-time Student	.065	.004	.026
Hours Study	−.026	.053	−.006
Years of Soc. Wk. Exper.	−.012	.031	.010
Debt	−.001	.000	−.049
Locus of Control	.288*	−.368*	−.329*
ADJUSTED R^2	.06*	.15*	.11*

*Probability less than or equal to .05.

Note that the author uses three dependent variables to measure stress and well-being and a total of ten independent variables. In the data-analysis phase of the project, Professor Fortune finds support for the "expansion" rather than the "scarcity" model. Looking specifically at the results of her regression analysis, she notes that ". . . only one of the role-status variables was significant: working . . . students experienced less stress. . . . Overall, only one factor, Locus of Control, was consistently related to the dependent variables. . . . Students who have additional roles . . . are not at greater risk than 'traditional' students" (87). Strong betas for the variables that measure multiple roles would have supported the "scarcity" model. The evidence is that stress and well-being are not related to multiple roles. Professor Fortune goes on to suggest that factors such as support groups, quality of role experience, and the like are more important than the number of roles.

Anne E. Fortune: 1987. "Multiple Roles, Stress and Well-Being Among MSW Students." *Journal of Social Work Education*, 23:3, pp.81–90. Reprinted by permission.

Third, the techniques of multiple regression and correlation assume that the independent variables are uncorrelated with each other. Strictly speaking, this condition means that the zero-order correlation among all pairs of independents should be zero; but, practically, we act as if this assumption has been met if the intercorrelations among the independents are low.

To the extent that these assumptions are violated, the regression coefficients (especially partial and standardized slopes) and the coefficient of multiple determination (R^2) become less and less trustworthy and the techniques less and less useful. If the assumptions of the model cannot be met, the alternative might be to turn to the multivariate techniques described in the previous chapter. Unfortunately those techniques, in general, supply a lower volume of less precise information about the interrelationships among the variables.

Finally, we should note that, in this chapter, we have been concerned only with the simplest applications of the techniques of partial correlation and multiple regression and correlation. In terms of logic and interpretation, the extensions to situations involving more than one control variable or more than two independent variables are relatively straightforward. However, the computational routines required to handle such situations can become extremely complex. If you are faced with a situation involving more than three variables, turn to the services offered by your local computer center. Several excellent preprogrammed statistical packages are commonly available, and all can be successfully utilized with minimal computer literacy and can handle these complex calculations in, literally, the blink of an eye. Efficient use of these packages will enable you to avoid drudgery and will also virtually guarantee the accuracy of your results. Furthermore, these statistical packages will free you to do what social scientists everywhere enjoy doing most: pondering the meaning of your results and, by extension, the nature of social life in general.

SUMMARY

1. Partial correlation involves controlling for third variables in a manner analogous to that introduced in the previous chapter. Partial correlations permit the detection of direct and spurious or intervening relationships between X and Y.

2. Multiple regression includes statistical techniques by which predictions of the dependent variable from more than one independent variable can be made (by partial slopes and the multiple regression equation) and by which we can disentangle the relative importance of the independents (by standardized partial slopes).

3. The multiple correlation coefficient (R^2) summarizes the combined effects of all independents on the dependent variable in terms of the proportion of the total variation in Y that is explained by all of the independents.

4. Partial correlation and multiple regression and correlation are some of the most powerful tools available to the researcher and demand high-quality measurement and relationships among the variables that are linear and noninteractive. Further, correlations among the independents must be low (preferably zero). Although the price is high, these techniques pay considerable dividends in the volume of precise and detailed information they generate about the interrelationships among the variables.

SUMMARY OF FORMULAS

Partial correlation coefficient	18.1	$r_{yx \cdot z} = \dfrac{r_{yx} - (r_{yz})(r_{xz})}{\sqrt{1 - r_{yz}^2}\sqrt{1 - r_{xz}^2}}$
Least-squares regression line (bivariate)	18.2	$Y = a + bX$
Least-squares multiple regression line	18.3	$Y = a + b_1 X_1 + b_2 X_2$
Partial slope for X_1	18.4	$b_1 = \left(\dfrac{s_y}{s_1}\right)\left(\dfrac{r_{y1} - r_{y2} r_{12}}{1 - r_{12}^2}\right)$
Partial slope for X_2	18.5	$b_2 = \left(\dfrac{s_y}{s_2}\right)\left(\dfrac{r_{y2} - r_{y1} r_{12}}{1 - r_{12}^2}\right)$
Y intercept	18.6	$a = \bar{Y} - b_1 \bar{X}_1 - b_2 \bar{X}_2$
Standardized partial slope (beta-weight) for X_1	18.7	$b_1^* = b_1\left(\dfrac{s_1}{s_y}\right)$
Standardized partial slope (beta-weight) for X_2	18.8	$b_2^* = b_2\left(\dfrac{s_2}{s_y}\right)$
Standardized least-squares regression line	18.9	$Z_y = a_z + b_1^* Z_1 + b_2^* Z_2$
Standardized least-squares regression line (simplified)	18.10	$Z_y = b_1^* Z_1 + b_2^* Z_2$
Coefficient of multiple determination	18.11	$R^2 = r_{y1}^2 + r_{y2 \cdot 1}^2 (1 - r_{y1}^2)$

GLOSSARY

Beta-weights (b^*). Standardized partial slopes.

Coefficient of multiple determination (R^2). A statistic that equals the total variation explained in the dependent variable by all independent variables combined.

Multiple correlation. A multivariate technique for examining the combined effects of more than one independent variable on a dependent variable.

Multiple correlation coefficient (R). A statistic that indicates the strength of the correlation between a dependent variable and two or more independent variables.

Multiple regression. A multivariate technique that breaks down the separate effects of the independent variables on the dependent variable; used to make predictions of the dependent variable.

Partial correlation. A multivariate technique for examining a bivariate relationship while controlling for other variables.

Partial correlation coefficient. A statistic that shows the relationship between two variables while controlling for other variables; $r_{yx \cdot z}$ is the symbol for the partial correlation coefficient when controlling for one variable.

Partial slopes. In a multiple regression equation, the slope of the relationship between a particular independent variable and the dependent variable while controlling for all other independents in the equation.

Standardized partial slopes (beta-weights). The slope of the relationship between a particular independent variable and the dependent when all scores have been normalized.

PROBLEMS

18.1 In problem 16.2, crime and population data were presented for each of 20 states. For each of the three crime variables:

    a. Compute the partial correlation coefficient with population growth rate while controlling for urbanization.

    b. Compute the partial correlation coefficient with urbanization while controlling for population growth rate.

c. Compute the partial correlation coefficient with population density while controlling for population growth rate.

d. Find the multiple regression equations (unstandardized) with growth and urbanization as the independents.

e. Make a prediction for each dependent variable for a state with a 5% growth rate and a population that is 90% urbanized.

f. Compute beta-weights for each independent in each equation and compare their relative effect on each dependent.

g. Compute the multiple correlation coefficient for each crime variable, using the population variables as independents.

h. Write a paragraph summarizing your findings in terms of the effects of the population variables on these measures of crime.

18.2 In problem 16.4, child-abuse rates were presented along with information on education and family structure for each of 18 census tracts. Taking child-abuse rates as the dependent variable:

a. Compute the partial correlation coefficient with education (mean years of schooling), controlling for percentage of intact families.

b. Compute the partial correlation coefficient with percentage of intact families, controlling for education.

c. Find the multiple regression equation and make a prediction for child-abuse rate for a census tract with 70% of the families intact and an average years of schooling at 14.

d. Compute beta-weights for both independents and compare their relative effects on the dependent.

e. Compute the multiple correlation coefficient.

f. Write a paragraph summarizing your findings.

18.3 Problem 16.6 presented per capita expenditures on education for 15 states, along with information on average annual pay and population growth. Taking educational expenditures as the dependent variable:

a. Compute the partial correlation coefficient with average annual pay while controlling for growth.

b. Compute the partial correlation coefficient with growth while controlling for average annual pay.

c. Find the multiple regression equation and make a prediction for educational expenditures for a state with an average annual pay of 16,000 and a growth rate of 5.0.

d. Compute beta-weights for both independents and compare their relative effects on the dependent.

e. Compute the multiple correlation coefficient.

f. Write a paragraph summarizing your findings.

18.4 Data on civil strife (number of incidents), unemployment, and urbanization have been gathered for 10 nations. Take civil strife as the dependent variable and do a full multivariate analysis of the interrelationships among these three variables. Compute both partial and multiple correlation coefficients and find both the unstandardized and standardized regression equations. Write a paragraph summarizing your findings.

Number of Incidents of Civil Strife	Unemployment Rate	Percentage of Population Living in Urban Areas
0	5.3	60
1	1.0	65
5	2.7	55
7	2.8	68
10	3.0	69
23	2.5	70
25	6.0	45
26	15.2	40
30	7.8	75
53	9.2	80

18.5 Problem 16.8 presented some data for 20 nations. Take "percent urban" as the dependent variable and do a full regression analysis with birthrate and death rates as the independents. Summarize your conclusions in a paragraph.

18.6 For the data presented in problem 16.8, take death rate as the dependent and do a full regression analysis with percent urban and growth rate as the independents. Summarize your conclusions in a paragraph.

18.7 Problem 16.10 presented some data on 10 precincts. Take voter turnout as the dependent variable and do a full regression analysis with percent Democrat and percent minority as the independents. Write a paragraph summarizing your findings.

18.8 SW Twelve families have been referred to a counselor and she has rated each of them on a cohesiveness scale. Also, she has information on family income and number of children currently living at home. Take family cohesion as the dependent and do a full regression analysis of the relationships among these variables. Write a paragraph summarizing your findings.

Family	Cohesion Score	Family Income	Number of Children
A	10	30,000	5
B	10	70,000	4
C	9	35,000	4
D	5	25,000	0
E	1	55,000	3
F	7	40,000	0
G	2	60,000	2
H	5	30,000	3
I	8	50,000	5
J	3	25,000	4
K	2	45,000	3
L	4	50,000	0

18.9 SOC A scale measuring support for increases in the national defense budget has been administered to a sample. The respondents have also been asked to indicate how many years of school they have completed and how many years, if any, they served in the military. Taking "support" as the dependent variable, do a full regression analysis of the relationships among these variables. Write a paragraph summarizing your findings.

Case	Support	Years of School	Years of Service
A	20	12	2
B	15	12	4
C	20	16	20
D	10	10	10
E	10	16	20
F	5	8	0
G	8	14	2
H	20	12	20
I	10	10	4
J	20	16	0

18.10 SOC The fatality rate for motor vehicle accidents is reported below for 25 states along with information on some potential independent variables. Take fatality rate as the dependent variable and do a full regression analysis of the relationship among these variables. Write a paragraph summarizing your conclusions.

State	Motor Vehicle Fatality Rate[1]	Urbanization[2]	Age of Population[3]
Maine	1.92	36.1	25.8
Mass.	1.31	90.8	23.0
Rhode Isl.	1.28	92.5	23.3
New York	1.20	90.5	24.6
Penna.	1.53	84.6	24.0
Indiana	1.90	68.0	26.8
Wisconsin	1.63	66.5	26.6
N. Dakota	1.71	46.9	27.9
Nebraska	1.61	46.9	26.7
Kansas	2.04	49.3	26.3
Maryland	1.74	92.9	24.9
Virginia	1.79	71.5	24.9
S. Carolina	2.83	60.2	27.8
Florida	2.61	90.9	22.5
Tennessee	2.56	68.8	2.60
Alabama	2.50	64.1	27.5
Louisiana	2.25	69.1	29.9
Texas	2.32	80.7	29.5
Wyoming	2.84	28.8	30.5
Colorado	1.94	81.6	26.5
Arizona	2.95	75.4	27.4
Utah	1.81	77.0	37.2
Washington	1.78	81.0	25.9
California	1.98	95.7	26.3
Alaska	2.38	44.0	32.1

[1]Number of fatalities from motor vehicle accidents per one million population.
[2]Percentage of the population living in metropolitan areas.
[3]Percentage of the population less than 18 years of age.

Source: United States Bureau of the Census, *Statistical Abstracts of the United States: 1988* (108th edition). Washington, D.C., 1988.

18.11 SOC Thirty individuals were randomly selected from the 1993 General Social Survey data set, and their scores on four variables are reported below. Take "Hours of TV Watching per Day" as the dependent variable and do a full regression analysis, first with occupational prestige and age as independents, and then with number of children and age as independents. See Appendix G for the full wording of the questions. Write a paragraph summarizing your conclusions.

Hours of TV Watching per Day (Item 67)	Occupational Prestige (Item 2)	Number of Children (Item 6)	Age (Item 7)
4	50	2	43
3	36	3	58
3	36	1	34
4	50	2	42
2	45	2	27
3	50	5	60
4	50	0	28
7	40	3	55
1	57	2	46
3	33	2	65
1	46	3	56
3	31	1	29

Hours of TV Watching per Day (Item 67)	Occupational Prestige (Item 2)	Number of Children (Item 6)	Age (Item 7)
1	19	2	41
0	52	0	50
2	48	1	62
4	36	1	24
3	48	0	25
1	62	1	87
5	50	0	45
1	27	3	62
5	40	7	61
3	50	6	34
4	52	5	70
6	45	3	44
1	50	2	40
6	60	3	32
4	17	5	50
8	45	4	31
2	17	1	32
2	29	2	50

SPSS/PC+ PROCEDURES FOR REGRESSION ANALYSIS

DEMONSTRATION 18.1 What Are the Correlates of Occupational Prestige? Another Look

In Demonstration 16.1, we used the CORRELATIONS procedure to calculate zero-order correlation coefficients between PRESTG80 and three independent variables. In this section, we will use a significantly more complex and flexible program called REGRESSION to analyze the effects of these same three independents on PRESTG80. This procedure gives the user control over many aspects of building the regression formula, and it can produce a much greater volume of output than the CORRELATIONS program. Among other things, REGRESSION displays the slope (b) and the Y intercept (a) so we can use this procedure to find least-squares regression lines. REGRESSION can be run for simple bivariate relationships as well as complex multivariate relationships. The task described below represents a very sparing use of the power of this program and an extremely economical use of all the options available. I urge you to consult the Manual and explore some of the variations and capabilities of this powerful data-analysis command.

To run the program after using the GET command to make the 1993 GSS data the active file:

1. Select 'analyze data' from the main menu and then choose 'correlation and regression'.
2. Select and paste REGRESSION and !/VARIABLES and then use ALT-T to list all variables in the equation.
3. To get a set of summary statistics for each variable, select DESCRIPTIVES and DEFAULT.
4. Select !/DEPENDENT and then use ALT-T to name PRESTG80.
5. Select !/METHOD and ENTER.

Your completed command should look like this:

```
REGRESSION /VARIABLES PRESTG80 PAPRES80 EDUC AGE
/DESCRIPTIVES DEFAULT /DEPENDENT PRESTIGE /METHOD
ENTER.
```

and your results should be

```
* * * * MULTIPLE REGRESSION * * * *

Listwise Deletion of Missing Data

               Mean Std Dev Label

PRESTG80 43.197  13.136 RS OCCUPATIONAL PRESTIGE SCORE
                        (1980)
PAPRES80 43.127  12.284 FATHERS OCCUPATIONAL PRESTIGE
                        SCORE (1980)
EDUC     13.250   4.531 HIGHEST YEAR OF SCHOOL
                        COMPLETED
AGE      46.785  16.836 AGE OF RESPONDENT

N of Cases = 636

Correlation:

              PRESTG80      PAPRES80      EDUC      AGE

PRESTG80       1.000          .215       .338    -.026
PAPRES80        .215         1.000       .148    -.151
EDUC            .338          .148      1.000    -.176
AGE            -.026         -.151      -.176    1.000
--------------------------------------------------------
```

```
* * * * MULTIPLE REGRESSION * * * *

Equation Number 1  Dependent Variable..   PRESTG80   RS
OCCUPATIONAL PRESTIGE

Variable(s) Entered on Step Number
     1..    AGE        AGE OF RESPONDENT
     2..    PAPRES80   FATHERS OCCUPATIONAL PRESTIGE
                       SCORE (198
     3..    EDUC       HIGHEST YEAR OF SCHOOL COMPLETED

Multiple R             .38170
R Square               .14570
Adjusted R Square      .14164
Standard Error      12.17066

Analysis of Variance

                 DF       Sum of Squares      Mean Square
Regression        3          15965.42066      5321.80689
Residual        632          93615.01173       148.12502

F =       35.92781  Signif F =  .0000
-----------------------------------------------------------
              * * * * MULTIPLE REGRESSION * * * *

Equation Number 1  Dependent Variable..   PRESTG80   RS
OCCUPATIONAL PRESTIGE

------------- Variables in the Equation -------------

Variable               B      SE B       Beta      T Sig T

AGE                .04472    .02938     .05732   1.522  .1285
PAPRES80           .18871    .04008     .17646   4.708  .0000
EDUC               .93496    .10913     .32252   8.568  .0000
(Constant)       20.57759  2.79512               7.362  .0000
-----------------------------------------------------------
```

The output begins with summary statistics on all the variables in the equation and a correlation matrix showing the zero-order correlations between all pairs of variables. This section essentially repeats the output of the CORRELATIONS task we used in Demonstration 16.1. The next section, along with other information, lists the variables entered into the regression equation, the multiple R (.38170), R Square (.14570), and an ANOVA test of significance (Signif F = .0000). So far, we know that the three independent variables explain about 15% of the variance in PRESTG80 and that this result is statistically significant.

In the last section of output, we see the slopes (*B*) of the three independent variables on PRESTG80, the standardized partial slopes (beta), and the *Y* intercept (reported as a constant of 20.57759). From this information, we can build a regression equation to predict scores on PRESTG80, and assess the relative importance of the three independent variables. The betas show that EDUC has the strongest effect (beta = .32252) and that AGE has a very weak relationship with PRESTG80 (beta = .05732). With a beta of .18, the effect of PAPRES80 is much less than that of EDUC.

What does all this mean? At least for this sample, occupational status (PRESTG80) is the most important correlate of level of educational achievement (EDUC). With the effect of the other variables controlled, the social class of one's father (PAPRES80) has very little effect on PRESTG80, and PRESTG80 essentially has no relationship with AGE. The higher the level of education, the more prestigious the job of the respondent. Neither the social class of one's father nor age have very much effect on occupational prestige.

DEMONSTRATION 18.2 Analyzing Homicide Rates

As an additional example, let's do part of problem 18.1. If you've already done this problem by hand, you will be impressed by how the computer speeds things up. If you haven't done the problem yet, take my word for it— the computer will do these complex calculations at a rate that will make your hand calculator tremble in embarrassment. (Of course, you still have to enter the data and write the SPSS/PC+ program—but by this time in the course, we are seasoned veterans, unafraid of these minor tasks. Right?)

We used this data in Demonstration 16.2. I suggested at that point that you save the file under the name CRIME. If you did this, retrieve the file by pressing F3 and answering the prompts. If you didn't, you'll have to enter the data again, following the instructions in Demonstration 16.2. Once the file has been loaded, make sure that any old commands have been deleted and, to analyze the homicide rate, add the REGRESSION commands below to the end of the file. Follow the instructions in Demonstration 18.1 to create the commands.

```
REGRESSION /HOMICIDE GROWTH DENSITY URBAN
/DESCRIPTIVES DEFAULT /DEPENDENT HOMICIDE /METHOD
ENTER.
```

Although I won't reproduce the output here to conserve space, it should display the descriptive statistics, a correlation matrix for all variables, and the regression information (e.g., an R Square of .37573). The betas show that URBAN and GROWTH have nearly equal, strong positive effects on HOMI-

CIDE and that DENSITY has a weaker negative effect. The greater the extent to which the population is urbanized and growing, the higher the homicide rate.

DEMONSTRATION 18.3 Who Watches TV?

Item 67 in Appendix G (TVHOURS) measures the average hours of TV viewing per day. What type of person would tend to be a heavy viewer? My guesses were that the well-educated would watch less (or, at least, claim to watch less), and older respondents would watch more. I also thought that people who express fear of going out at night would be more likely to stay home and watch TV. I used FEAR (item 66) to determine level of fearfulness. So, I would predict positive relationships between TVHOURS and AGE and negative relations with TVHOURS and EDUC and FEAR. Note that FEAR has only two categories. We're violating some assumptions here so we have to be extra cautious in interpreting results.

To produce all of the relevant statistics, the command would be

```
REGRESSION /TVHOURS EDUC AGE FEAR/DESCRIPTIVES DEFAULT
/DEPENDENT TVHOURS/METHOD ENTER.
```

To conserve space, I will not reproduce the output here. The correlation matrix is consistent with my predictions even though the relationships are very weak. The R Square is .03, indicating that these three variables account for only 3% of the variance in the dependent variable. The 'Signif F' is significant at less than .01.

The values from the regression equation again are consistent with the predictions. The betas show that FEAR and EDUC have weak effects and that TVHOURS increases with age.

Exercises

18.1 Conduct the analysis in Demonstration 18.1 again with INCOME91 as the dependent variable. Compare your conclusions with those you made in Demonstration 16.1. Write a paragraph summarizing the relationships.

18.2 Use ROBBERY and CARTHEFT in place of HOMICIDE and conduct the analysis in Demonstration 18.2 again. Write a paragraph comparing the effects of the demographic variables on the various crime rates.

18.3 Use the COMPUTE command to create a summary scale for attitudes on abortion (items 52 and 53). Use INCOME91, EDUC, and AGE as independents and conduct a full regression analysis. Write up your conclusions in a paragraph.

18.4 Not afraid to beat a dead horse? Willing to run the risk of violating some level-of-measurement assumptions? Take HAPPY as a dependent variable and INCOME91, CHILDS, and EDUC as independents and conduct a full regression analysis. See Demonstrations 12.3, 11.2, and 10.2 for previous episodes. As an alternative, COMPUTE a satisfaction scale from items 37 and 38 in Appendix G for a dependent variable (see Demonstration 12.3).

PART IV CUMULATIVE EXERCISES

A number of research questions are stated below. Each can be answered by one of the techniques presented in Chapters 17 and 18. For each research situation, choose either the elaboration technique or regression analysis. The level of measurement of the variables should have a great deal of influence on your decision.

   The research questions refer to the data base below, which is taken from the General Social Survey (GSS). The actual questions asked and the complete response codes for the GSS are presented in Appendix G. Abbreviated codes are listed below. Some variables have been recoded for this exercise.

a. Is there a relationship between rate of church attendance and anomia? Take anomia as the dependent variable. Anomia is a condition characterized by a sense of normlessness, apathy, and meaninglessness. It is measured by two items (47 and 48) on the General Social Survey, and the variable below is the addition of those two scores. Is the relationship affected by social class (measured by occupational prestige) or by the number of traumatic events experienced recently?

b. Is attitude on pornography associated with age? Is the relationship affected by sex or by education?

Survey Items (numbers in parentheses refer to Appendix G)

1. Occupational prestige (2)
2. Age of respondent (7)
      1. 35 and younger
      2. 36 and older
3. Education (recode of 9)
      1. HS or less
      2. At least some college
4. Sex of respondent (10)
      1. Male
      2. Female
5. How often do you attend church? (29)
      0. Never
      1. Less than once a year
      2. Once or twice a year
      3. Several times a year
      4. About once a month
      5. 2–3 times a month
      6. Nearly every week

7. Every week
8. Several times a week
6. Anomia (created by adding scores on 47 and 48). On this scale, 2 indicates the *highest* possible level of anomia and 4 the *lowest*.
7. Sexual materials lead to a breakdown of morals. (59)
1. Yes
2. No
8. Number of traumatic events over the last five years (70) (Scores are actual numbers.)

SCORES

Case	Prestige	Age	Educa-tion	Sex	Church Attend.	Anomia	Attitude on Porno.	Number of Traumas
1	39	2	1	1	0	4	2	1
2	47	1	2	2	3	2	2	1
3	29	2	1	2	4	3	1	3
4	50	2	1	1	2	3	1	2
5	37	1	1	1	4	2	2	1
6	50	1	1	1	7	2	1	2
7	29	1	1	1	0	3	2	0
8	36	2	1	2	5	2	1	1
9	17	2	1	1	0	2	1	1
10	34	2	1	2	7	3	1	2
11	50	2	1	2	7	2	1	4
12	50	1	2	1	2	3	2	0
13	50	2	2	2	7	4	1	4
14	40	2	2	1	4	2	2	2
15	50	1	2	2	5	2	1	3
16	19	2	1	2	8	2	1	3
17	35	1	1	1	1	4	2	0
18	48	2	2	2	3	4	2	1
19	60	2	2	2	3	4	2	1
20	36	2	2	2	1	3	2	0
21	52	2	2	1	0	4	2	1
22	48	2	1	1	7	2	1	0
23	48	2	2	1	0	4	1	4
24	27	2	1	2	5	4	1	3
25	25	2	1	2	1	4	1	0
26	48	1	2	2	2	4	2	0
27	36	1	2	2	2	4	1	0
28	45	2	1	2	7	3	2	1
29	45	1	2	1	7	3	1	3
30	36	1	2	2	2	4	1	0
31	45	2	1	2	7	3	2	1

Case	Prestige	Age	Educa-tion	Sex	Church Attend.	Anomia	Attitude on Porno.	Number of Traumas
32	45	1	2	1	7	3	1	3
33	36	2	1	1	5	2	1	1
34	18	2	1	2	3	2	1	1
35	55	2	2	2	6	2	2	2
36	15	2	2	1	0	4	2	2
37	26	1	1	2	1	2	2	0
38	45	2	1	2	1	2	2	0
39	36	1	1	2	7	2	1	0
40	36	2	1	2	7	2	1	3

Appendix A Area Under the Normal Curve

Column (a) lists Z scores from 0.00 to 4.00. Only positive scores are displayed, but, since the normal curve is symmetrical, the areas for negative scores will be exactly the same as areas for positive scores. Column (b) lists the proportion of the total area between the Z score and the mean. Figure A.1 displays areas of this type. Column (c) lists the proportion of the area beyond the Z score, and Figure A.2 displays this type of area.

FIGURE A.1 AREA BETWEEN MEAN AND Z

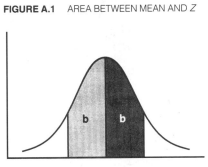

FIGURE A.2 AREA BEYOND Z

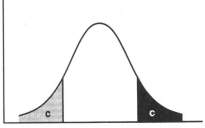

(a) Z	(b) Area Between Mean and Z	(c) Area Beyond Z	(a) Z	(b) Area Between Mean and Z	(c) Area Beyond Z
0.00	0.0000	0.5000	0.21	0.0832	0.4168
0.01	0.0040	0.4960	0.22	0.0871	0.4129
0.02	0.0080	0.4920	0.23	0.0910	0.4090
0.03	0.0120	0.4880	0.24	0.0948	0.4052
0.04	0.0160	0.4840	0.25	0.0987	0.4013
0.05	0.0199	0.4801	0.26	0.1026	0.3974
0.06	0.0239	0.4761	0.27	0.1064	0.3936
0.07	0.0279	0.4721	0.28	0.1103	0.3897
0.08	0.0319	0.4681	0.29	0.1141	0.3859
0.09	0.0359	0.4641	0.30	0.1179	0.3821
0.10	0.0398	0.4602			
0.11	0.0438	0.4562	0.31	0.1217	0.3783
0.12	0.0478	0.4522	0.32	0.1255	0.3745
0.13	0.0517	0.4483	0.33	0.1293	0.3707
0.14	0.0557	0.4443	0.34	0.1331	0.3669
0.15	0.0596	0.4404	0.35	0.1368	0.3632
0.16	0.0636	0.4364	0.36	0.1406	0.3594
0.17	0.0675	0.4325	0.37	0.1443	0.3557
0.18	0.0714	0.4286	0.38	0.1480	0.3520
0.19	0.0753	0.4247	0.39	0.1517	0.3483
0.20	0.0793	0.4207	0.40	0.1554	0.3446

(a) Z	(b) Area Between Mean and Z	(c) Area Beyond Z	(a) Z	(b) Area Between Mean and Z	(c) Area Beyond Z
0.41	0.1591	0.3409	0.91	0.3186	0.1814
0.42	0.1628	0.3372	0.92	0.3212	0.1788
0.43	0.1664	0.3336	0.93	0.3238	0.1762
0.44	0.1700	0.3300	0.94	0.3264	0.1736
0.45	0.1736	0.3264	0.95	0.3289	0.1711
0.46	0.1772	0.3228	0.96	0.3315	0.1685
0.47	0.1808	0.3192	0.97	0.3340	0.1660
0.48	0.1844	0.3156	0.98	0.3365	0.1635
0.49	0.1879	0.3121	0.99	0.3389	0.1611
0.50	0.1915	0.3085	1.00	0.3413	0.1587
0.51	0.1950	0.3050	1.01	0.3438	0.1562
0.52	0.1985	0.3015	1.02	0.3461	0.1539
0.53	0.2019	0.2981	1.03	0.3485	0.1515
0.54	0.2054	0.2946	1.04	0.3508	0.1492
0.55	0.2088	0.2912	1.05	0.3531	0.1469
0.56	0.2123	0.2877	1.06	0.3554	0.1446
0.57	0.2157	0.2843	1.07	0.3577	0.1423
0.58	0.2190	0.2810	1.08	0.3599	0.1401
0.59	0.2224	0.2776	1.09	0.3621	0.1379
0.60	0.2257	0.2743	1.10	0.3643	0.1357
0.61	0.2291	0.2709	1.11	0.3665	0.1335
0.62	0.2324	0.2676	1.12	0.3686	0.1314
0.63	0.2357	0.2643	1.13	0.3708	0.1292
0.64	0.2389	0.2611	1.14	0.3729	0.1271
0.65	0.2422	0.2578	1.15	0.3749	0.1251
0.66	0.2454	0.2546	1.16	0.3770	0.1230
0.67	0.2486	0.2514	1.17	0.3790	0.1210
0.68	0.2517	0.2483	1.18	0.3810	0.1190
0.69	0.2549	0.2451	1.19	0.3830	0.1170
0.70	0.2580	0.2420	1.20	0.3849	0.1151
0.71	0.2611	0.2389	1.21	0.3869	0.1131
0.72	0.2642	0.2358	1.22	0.3888	0.1112
0.73	0.2673	0.2327	1.23	0.3907	0.1093
0.74	0.2703	0.2297	1.24	0.3925	0.1075
0.75	0.2734	0.2266	1.25	0.3944	0.1056
0.76	0.2764	0.2236	1.26	0.3962	0.1038
0.77	0.2794	0.2206	1.27	0.3980	0.1020
0.78	0.2823	0.2177	1.28	0.3997	0.1003
0.79	0.2852	0.2148	1.29	0.4015	0.0985
0.80	0.2881	0.2119	1.30	0.4032	0.0968
0.81	0.2910	0.2090	1.31	0.4049	0.0951
0.82	0.2939	0.2061	1.32	0.4066	0.0934
0.83	0.2967	0.2033	1.33	0.4082	0.0918
0.84	0.2995	0.2005	1.34	0.4099	0.0901
0.85	0.3023	0.1977	1.35	0.4115	0.0885
0.86	0.3051	0.1949	1.36	0.4131	0.0869
0.87	0.3078	0.1922	1.37	0.4147	0.0853
0.88	0.3106	0.1894	1.38	0.4162	0.0838
0.89	0.3133	0.1867	1.39	0.4177	0.0823
0.90	0.3159	0.1841	1.40	0.4192	0.0808

(a)	(b)	(c)		(a)	(b)	(c)
Z	Area Between Mean and Z	Area Beyond Z		Z	Area Between Mean and Z	Area Beyond Z
1.41	0.4207	0.0793		1.91	0.4719	0.0281
1.42	0.4222	0.0778		1.92	0.4726	0.0274
1.43	0.4236	0.0764		1.93	0.4732	0.0268
1.44	0.4251	0.0749		1.94	0.4738	0.0262
1.45	0.4265	0.0735		1.95	0.4744	0.0256
1.46	0.4279	0.0721		1.96	0.4750	0.0250
1.47	0.4292	0.0708		1.97	0.4756	0.0244
1.48	0.4306	0.0694		1.98	0.4761	0.0239
1.49	0.4319	0.0681		1.99	0.4767	0.0233
1.50	0.4332	0.0668		2.00	0.4772	0.0228
1.51	0.4345	0.0655		2.01	0.4778	0.0222
1.52	0.4357	0.0643		2.02	0.4783	0.0217
1.53	0.4370	0.0630		2.03	0.4788	0.0212
1.54	0.4382	0.0618		2.04	0.4793	0.0207
1.55	0.4394	0.0606		2.05	0.4798	0.0202
1.56	0.4406	0.0594		2.06	0.4803	0.0197
1.57	0.4418	0.0582		2.07	0.4808	0.0192
1.58	0.4429	0.0571		2.08	0.4812	0.0188
1.59	0.4441	0.0559		2.09	0.4817	0.0183
1.60	0.4452	0.0548		2.10	0.4821	0.0179
1.61	0.4463	0.0537		2.11	0.4826	0.0174
1.62	0.4474	0.0526		2.12	0.4830	0.0170
1.63	0.4484	0.0516		2.13	0.4834	0.0166
1.64	0.4495	0.0505		2.14	0.4838	0.0162
1.65	0.4505	0.0495		2.15	0.4842	0.0158
1.66	0.4515	0.0485		2.16	0.4846	0.0154
1.67	0.4525	0.0475		2.17	0.4850	0.0150
1.68	0.4535	0.0465		2.18	0.4854	0.0146
1.69	0.4545	0.0455		2.19	0.4857	0.0143
1.70	0.4554	0.0446		2.20	0.4861	0.0139
1.71	0.4564	0.0436		2.21	0.4864	0.0136
1.72	0.4573	0.0427		2.22	0.4868	0.0132
1.73	0.4582	0.0418		2.23	0.4871	0.0129
1.74	0.4591	0.0409		2.24	0.4875	0.0125
1.75	0.4599	0.0401		2.25	0.4878	0.0122
1.76	0.4608	0.0392		2.26	0.4881	0.0119
1.77	0.4616	0.0384		2.27	0.4884	0.0116
1.78	0.4625	0.0375		2.28	0.4887	0.0113
1.79	0.4633	0.0367		2.29	0.4890	0.0110
1.80	0.4641	0.0359		2.30	0.4893	0.0107
1.81	0.4649	0.0351		2.31	0.4896	0.0104
1.82	0.4656	0.0344		2.32	0.4898	0.0102
1.83	0.4664	0.0336		2.33	0.4901	0.0099
1.84	0.4671	0.0329		2.34	0.4904	0.0096
1.85	0.4678	0.0322		2.35	0.4906	0.0094
1.86	0.4686	0.0314		2.36	0.4909	0.0091
1.87	0.4693	0.0307		2.37	0.4911	0.0089
1.88	0.4699	0.0301		2.38	0.4913	0.0087
1.89	0.4706	0.0294		2.39	0.4916	0.0084
1.90	0.4713	0.0287		2.40	0.4918	0.0082

(a)	(b)	(c)	(a)	(b)	(c)
Z	Area Between Mean and Z	Area Beyond Z	Z	Area Between Mean and Z	Area Beyond Z
2.41	0.4920	0.0080	2.91	0.4982	0.0018
2.42	0.4922	0.0078	2.92	0.4982	0.0018
2.43	0.4925	0.0075	2.93	0.4983	0.0017
2.44	0.4927	0.0073	2.94	0.4984	0.0016
2.45	0.4929	0.0071	2.95	0.4984	0.0016
2.46	0.4931	0.0069	2.96	0.4985	0.0015
2.47	0.4932	0.0068	2.97	0.4985	0.0015
2.48	0.4934	0.0066	2.98	0.4986	0.0014
2.49	0.4936	0.0064	2.99	0.4986	0.0014
2.50	0.4938	0.0062	3.00	0.4986	0.0014
2.51	0.4940	0.0060	3.01	0.4987	0.0013
2.52	0.4941	0.0059	3.02	0.4987	0.0013
2.53	0.4943	0.0057	3.03	0.4988	0.0012
2.54	0.4945	0.0055	3.04	0.4988	0.0012
2.55	0.4946	0.0054	3.05	0.4989	0.0011
2.56	0.4948	0.0052	3.06	0.4989	0.0011
2.57	0.4949	0.0051	3.07	0.4989	0.0011
2.58	0.4951	0.0049	3.08	0.4990	0.0010
2.59	0.4952	0.0048	3.09	0.4990	0.0010
2.60	0.4953	0.0047	3.10	0.4990	0.0010
2.61	0.4955	0.0045	3.11	0.4991	0.0009
2.62	0.4956	0.0044	3.12	0.4991	0.0009
2.63	0.4957	0.0043	3.13	0.4991	0.0009
2.64	0.4959	0.0041	3.14	0.4992	0.0008
2.65	0.4960	0.0040	3.15	0.4992	0.0008
2.66	0.4961	0.0039	3.16	0.4992	0.0008
2.67	0.4962	0.0038	3.17	0.4992	0.0008
2.68	0.4963	0.0037	3.18	0.4993	0.0007
2.69	0.4964	0.0036	3.19	0.4993	0.0007
2.70	0.4965	0.0035	3.20	0.4993	0.0007
2.71	0.4966	0.0034	3.21	0.4993	0.0007
2.72	0.4967	0.0033	3.22	0.4994	0.0006
2.73	0.4968	0.0032	3.23	0.4994	0.0006
2.74	0.4969	0.0031	3.24	0.4994	0.0006
2.75	0.4970	0.0030	3.25	0.4994	0.0006
2.76	0.4971	0.0029	3.26	0.4994	0.0006
2.77	0.4972	0.0028	3.27	0.4995	0.0005
2.78	0.4973	0.0027	3.28	0.4995	0.0005
2.79	0.4974	0.0026	3.29	0.4995	0.0005
2.80	0.4974	0.0026	3.30	0.4995	0.0005
2.81	0.4975	0.0025	3.31	0.4995	0.0005
2.82	0.4976	0.0024	3.32	0.4995	0.0005
2.83	0.4977	0.0023	3.33	0.4996	0.0004
2.84	0.4977	0.0023	3.34	0.4996	0.0004
2.85	0.4978	0.0022	3.35	0.4996	0.0004
2.86	0.4979	0.0021	3.36	0.4996	0.0004
2.87	0.4979	0.0021	3.37	0.4996	0.0004
2.88	0.4980	0.0020	3.38	0.4996	0.0004
2.89	0.4981	0.0019	3.39	0.4997	0.0003
2.90	0.4981	0.0019	3.40	0.4997	0.0003

(a)	(b)	(c)	(a)	(b)	(c)
	Area Between	Area Beyond		Area Between	Area Beyond
Z	Mean and Z	Z	Z	Mean and Z	Z
3.41	0.4997	0.0003	3.60	0.4998	0.0002
3.42	0.4997	0.0003			
3.43	0.4997	0.0003	3.70	0.4999	0.0001
3.44	0.4997	0.0003			
3.45	0.4997	0.0003	3.80	0.4999	0.0001
3.46	0.4997	0.0003			
3.47	0.4997	0.0003	3.90	0.4999	<0.0001
3.48	0.4997	0.0003			
3.49	0.4998	0.0002	4.00	0.4999	<0.0001
3.50	0.4998	0.0002			

Appendix B   Distribution of *t*

Degrees of freedom (df)	Level of significance for one-tailed test					
	.10	.05	.025	.01	.005	.0005
	Level of significance for two-tailed test					
	.20	.10	.05	.02	.01	.001
1	3.078	6.314	12.706	31.821	63.657	636.619
2	1.886	2.920	4.303	6.965	9.925	31.598
3	1.638	2.353	3.182	4.541	5.841	12.941
4	1.533	2.132	2.776	3.747	4.604	8.610
5	1.476	2.015	2.571	3.365	4.032	6.859
6	1.440	1.943	2.447	3.143	3.707	5.959
7	1.415	1.895	2.365	2.998	3.499	5.405
8	1.397	1.860	2.306	2.896	3.355	5.041
9	1.383	1.833	2.262	2.821	3.250	4.781
10	1.372	1.812	2.228	2.764	3.169	4.587
11	1.363	1.796	2.201	2.718	3.106	4.437
12	1.356	1.782	2.179	2.681	3.055	4.318
13	1.350	1.771	2.160	2.650	3.012	4.221
14	1.345	1.761	2.145	2.624	2.977	4.140
15	1.341	1.753	2.131	2.602	2.947	4.073
16	1.337	1.746	2.120	2.583	2.921	4.015
17	1.333	1.740	2.110	2.567	2.898	3.965
18	1.330	1.734	2.101	2.552	2.878	3.922
19	1.328	1.729	2.093	2.539	2.861	3.883
20	1.325	1.725	2.086	2.528	2.845	3.850
21	1.323	1.721	2.080	2.518	2.831	3.819
22	1.321	1.717	2.074	2.508	2.819	3.792
23	1.319	1.714	2.069	2.500	2.807	3.767
24	1.318	1.711	2.064	2.492	2.797	3.745
25	1.316	1.708	2.060	2.485	2.787	3.725
26	1.315	1.706	2.056	2.479	2.779	3.707
27	1.314	1.703	2.052	2.473	2.771	3.690
28	1.313	1.701	2.048	2.467	2.763	3.674
29	1.311	1.699	2.045	2.462	2.756	3.659
30	1.310	1.697	2.042	2.457	2.750	3.646
40	1.303	1.684	2.021	2.423	2.704	3.551
60	1.296	1.671	2.000	2.390	2.660	3.460
120	1.289	1.658	1.980	2.358	2.617	3.373
∞	1.282	1.645	1.960	2.326	2.576	3.291

For 50, you would look up 40

Source: Table III of Fisher & Yates: *Statistical Tables for Biological, Agricultural and Medical Research*, published by Longman Group Ltd., London (1974), 6th edition (previously published by Oliver & Boyd Ltd., Edinburgh). By permission of the authors and publishers.

Appendix C   Distribution of Chi Square

df	.99	.98	.95	.90	.80	.70	.50	.30	.20	.10	.05	.02	.01	.001
1	$.0^3157$	$.0^3628$	.00393	.0158	.0642	.148	.455	1.074	1.642	2.706	3.841	5.412	6.635	10.827
2	.0201	.0404	.103	.211	.446	.713	1.386	2.408	3.219	4.605	5.991	7.824	9.210	13.815
3	.115	.185	.352	.584	1.005	1.424	2.366	3.665	4.642	6.251	7.815	9.837	11.341	16.268
4	.297	.429	.711	1.064	1.649	2.195	3.357	4.878	5.989	7.779	9.488	11.668	13.277	18.465
5	.554	.752	1.145	1.610	2.343	3.000	4.351	6.064	7.289	9.236	11.070	13.388	15.086	20.517
6	.872	1.134	1.635	2.204	3.070	3.828	5.348	7.231	8.558	10.645	12.592	15.033	16.812	22.457
7	1.239	1.564	2.167	2.833	3.822	4.671	6.346	8.383	9.803	12.017	14.067	16.622	18.475	24.322
8	1.646	2.032	2.733	3.490	4.594	5.527	7.344	9.524	11.030	13.362	15.507	18.168	20.090	26.125
9	2.088	2.532	3.325	4.168	5.380	6.393	8.343	10.656	12.242	14.604	16.919	19.670	21.666	27.877
10	2.558	3.059	3.940	4.865	6.179	7.267	9.342	11.781	13.442	15.987	18.307	21.161	23.209	29.588
11	3.053	3.609	4.575	5.578	6.989	8.148	10.341	12.899	14.631	17.275	19.675	22.618	24.725	31.264
12	3.571	4.178	5.226	6.304	7.807	9.034	11.340	14.011	15.812	18.549	21.026	24.054	26.217	32.909
13	4.107	4.765	5.892	7.042	8.634	9.926	12.340	15.119	16.985	19.812	22.362	25.472	27.688	34.528
14	4.660	5.368	6.571	7.790	9.467	10.821	13.339	16.222	18.151	21.064	23.685	26.873	29.141	36.123
15	5.229	5.985	7.261	8.547	10.307	11.721	14.339	17.322	19.311	22.307	24.996	28.259	30.578	37.697
16	5.812	6.614	7.962	9.312	11.152	12.624	15.338	18.418	20.465	23.542	26.296	29.633	32.000	39.252
17	6.408	7.255	8.672	10.085	12.002	13.531	16.338	19.511	21.615	24.769	27.587	30.995	33.409	40.790
18	7.015	7.906	9.390	10.865	12.857	14.440	17.338	20.601	22.760	25.989	28.869	32.346	34.805	42.312
19	7.633	8.567	10.117	11.651	13.716	15.352	18.338	21.689	23.900	27.204	30.144	33.687	36.191	43.820
20	8.260	9.237	10.851	12.443	14.578	16.266	19.337	22.775	25.038	28.412	31.410	35.020	37.566	45.315
21	8.897	9.915	11.591	13.240	15.445	17.182	20.337	23.858	26.171	29.615	32.671	36.343	38.932	46.797
22	9.542	10.600	12.338	14.041	16.314	18.101	21.337	24.939	27.301	30.813	33.924	37.659	40.289	48.268
23	10.196	11.293	13.091	14.848	17.187	19.021	22.337	26.018	28.429	32.007	35.172	38.968	41.638	49.728
24	10.856	11.992	13.848	15.659	18.062	19.943	23.337	27.096	29.553	33.196	36.415	40.270	42.980	51.179
25	11.524	12.697	14.611	16.473	18.940	20.867	24.337	28.172	30.675	34.382	37.652	41.566	44.314	52.620
26	12.198	13.409	15.379	17.292	19.820	21.792	25.336	29.246	31.795	35.563	38.885	42.856	45.642	54.052
27	12.879	14.125	16.151	18.114	20.703	22.719	26.336	30.319	32.912	36.741	40.113	44.140	46.963	55.476
28	13.565	14.847	16.928	18.939	21.588	23.647	27.336	31.391	34.027	37.916	41.337	45.419	48.278	56.893
29	14.256	15.574	17.708	19.768	22.475	24.577	28.336	32.461	35.139	39.087	42.557	46.693	49.588	58.302
30	14.953	16.306	18.493	20.599	23.364	25.508	29.336	33.530	36.250	40.256	43.773	47.962	50.892	59.703

Source: Table IV of Fisher & Yates: *Statistical Tables for Biological, Agricultural and Medical Research*, published by Longman Group Ltd., London (1974), 6th edition (previously published by Oliver & Boyd Ltd., Edinburgh). By permission of the authors and publishers.

Appendix D   Distribution of *F*

$p = .05$

n_1 / n_2	1	2	3	4	5	6	8	12	24	∞
1	161.4	199.5	215.7	224.6	230.2	234.0	238.9	243.9	249.0	254.3
2	18.51	19.00	19.16	19.25	19.30	19.33	19.37	19.41	19.45	19.50
3	10.13	9.55	9.28	9.12	9.01	8.94	8.84	8.74	8.64	8.53
4	7.71	6.94	6.59	6.39	6.26	6.16	6.04	5.91	5.77	5.63
5	6.61	5.79	5.41	5.19	5.05	4.95	4.82	4.68	4.53	4.36
6	5.99	5.14	4.76	4.53	4.39	4.28	4.15	4.00	3.84	3.67
7	5.59	4.74	4.35	4.12	3.97	3.87	3.73	3.57	3.41	3.23
8	5.32	4.46	4.07	3.84	3.69	3.58	3.44	3.28	3.12	2.93
9	5.12	4.26	3.86	3.63	3.48	3.37	3.23	3.07	2.90	2.71
10	4.96	4.10	3.71	3.48	3.33	3.22	3.07	2.91	2.74	2.54
11	4.84	3.98	3.59	3.36	3.20	3.09	2.95	2.79	2.61	2.40
12	4.75	3.88	3.49	3.26	3.11	3.00	2.85	2.69	2.50	2.30
13	4.67	3.80	3.41	3.18	3.02	2.92	2.77	2.60	2.42	2.21
14	4.60	3.74	3.34	3.11	2.96	2.85	2.70	2.53	2.35	2.13
15	4.54	3.68	3.29	3.06	2.90	2.79	2.64	2.48	2.29	2.07
16	4.49	3.63	3.24	3.01	2.85	2.74	2.59	2.42	2.24	2.01
17	4.45	3.59	3.20	2.96	2.81	2.70	2.55	2.38	2.19	1.96
18	4.41	3.55	3.16	2.93	2.77	2.66	2.51	2.34	2.15	1.92
19	4.38	3.52	3.13	2.90	2.74	2.63	2.48	2.31	2.11	1.88
20	4.35	3.49	3.10	2.87	2.71	2.60	2.45	2.28	2.08	1.84
21	4.32	3.47	3.07	2.84	2.68	2.57	2.42	2.25	2.05	1.81
22	4.30	3.44	3.05	2.82	2.66	2.55	2.40	2.23	2.03	1.78
23	4.28	3.42	3.03	2.80	2.64	2.53	2.38	2.20	2.00	1.76
24	4.26	3.40	3.01	2.78	2.62	2.51	2.36	2.18	1.98	1.73
25	4.24	3.38	2.99	2.76	2.60	2.49	2.34	2.16	1.96	1.71
26	4.22	3.37	2.98	2.74	2.59	2.47	2.32	2.15	1.95	1.69
27	4.21	3.35	2.96	2.73	2.57	2.46	2.30	2.13	1.93	1.67
28	4.20	3.34	2.95	2.71	2.56	2.44	2.29	2.12	1.91	1.65
29	4.18	3.33	2.93	2.70	2.54	2.43	2.28	2.10	1.90	1.64
30	4.17	3.32	2.92	2.69	2.53	2.42	2.27	2.09	1.89	1.62
40	4.08	3.23	2.84	2.61	2.45	2.34	2.18	2.00	1.79	1.51
60	4.00	3.15	2.76	2.52	2.37	2.25	2.10	1.92	1.70	1.39
120	3.92	3.07	2.68	2.45	2.29	2.17	2.02	1.83	1.61	1.25
∞	3.84	2.99	2.60	2.37	2.21	2.09	1.94	1.75	1.52	1.00

Values of n_1 and n_2 represent the degrees of freedom associated with the between and within estimates of variance respectively.

Source: Table V of Fisher and Yates: *Statistical Tables for Biological, Agricultural and Medical Research*, published by Longman Group Ltd., London (1974), 6th edition (previously published by Oliver and Boyd Ltd., Edinburgh). By permission of the authors and publishers.

$p = .01$

n_1 / n_2	1	2	3	4	5	6	8	12	24	∞
1	4052	4999	5403	5625	5764	5859	5981	6106	6234	6366
2	98.49	99.01	99.17	99.25	99.30	99.33	99.36	99.42	99.46	99.50
3	34.12	30.81	29.46	28.71	28.24	27.91	27.49	27.05	26.60	26.12
4	21.20	18.00	16.69	15.98	15.52	15.21	14.80	14.37	13.93	13.46
5	16.26	13.27	12.06	11.39	10.97	10.67	10.27	9.89	9.47	9.02
6	13.74	10.92	9.78	9.15	8.75	8.47	8.10	7.72	7.31	6.88
7	12.25	9.55	8.45	7.85	7.46	7.19	6.84	6.47	6.07	5.65
8	11.26	8.65	7.59	7.01	6.63	6.37	6.03	5.67	5.28	4.86
9	10.56	8.02	6.99	6.42	6.06	5.80	5.47	5.11	4.73	4.31
10	10.04	7.56	6.55	5.99	5.64	5.39	5.06	4.71	4.33	3.91
11	9.65	7.20	6.22	5.67	5.32	5.07	4.74	4.40	4.02	3.60
12	9.33	6.93	5.95	5.41	5.06	4.82	4.50	4.16	3.78	3.36
13	9.07	6.70	5.74	5.20	4.86	4.62	4.30	3.96	3.59	3.16
14	8.86	6.51	5.56	5.03	4.69	4.46	4.14	3.80	3.43	3.00
15	8.68	6.36	5.42	4.89	4.56	4.32	4.00	3.67	3.29	2.87
16	8.53	6.23	5.29	4.77	4.44	4.20	3.89	3.55	3.18	2.75
17	8.40	6.11	5.18	4.67	4.34	4.10	3.79	3.45	3.08	2.65
18	8.28	6.01	5.09	4.58	4.25	4.01	3.71	3.37	3.00	2.57
19	8.18	5.93	5.01	4.50	4.17	3.94	3.63	3.30	2.92	2.49
20	8.10	5.85	4.94	4.43	4.10	3.87	3.56	3.23	2.86	2.42
21	8.02	5.78	4.87	4.37	4.04	3.81	3.51	3.17	2.80	2.36
22	7.94	5.72	4.82	4.31	3.99	3.76	3.45	3.12	2.75	2.31
23	7.88	5.66	4.76	4.26	3.94	3.71	3.41	3.07	2.70	2.26
24	7.82	5.61	4.72	4.22	3.90	3.67	3.36	3.03	2.66	2.21
25	7.77	5.57	4.68	4.18	3.86	3.63	3.32	2.99	2.62	2.17
26	7.72	5.53	4.64	4.14	3.82	3.59	3.29	2.96	2.58	2.13
27	7.68	5.49	4.60	4.11	3.78	3.56	3.26	2.93	2.55	2.10
28	7.64	5.45	4.57	4.07	3.75	3.53	3.23	2.90	2.52	2.06
29	7.60	5.42	4.54	4.04	3.73	3.50	3.20	2.87	2.49	2.03
30	7.56	5.39	4.51	4.02	3.70	3.47	3.17	2.84	2.47	2.01
40	7.31	5.18	4.31	3.83	3.51	3.29	2.99	2.66	2.29	1.80
60	7.08	4.98	4.13	3.65	3.34	3.12	2.82	2.50	2.12	1.60
120	6.85	4.79	3.95	3.48	3.17	2.96	2.66	2.34	1.95	1.38
∞	6.64	4.60	3.78	3.32	3.02	2.80	2.51	2.18	1.79	1.00

Values of n_1 and n_2 represent the degrees of freedom associated with the between and within estimates of variance respectively.

Appendix E A Table of Random Numbers

The numbers in this table were generated by a random process and have no pattern or order to them. In the social sciences, tables such as this are used mainly in the selection of random samples (see Chapter 6). To enhance readability, the numbers have been organized into 14 columns of five digits each and then subdivided into groups of five rows. This format is used solely to clarify the presentation and make the table easier to read. Columns and rows have no other significance with respect to the randomness of the digits.

To select a random sample from a numbered population list, begin anywhere in the table and select groups of digits the same length as the largest number on your population list. That is, if the list numbers in the 1000s, select digits in groups of four. If the list numbers in the 100s, select in groups of three, and so on. Every time a number selected from the table corresponds to the identification number of a case on the population list, select that case for the sample. Ignore repeated numbers and numbers that are not on the population list. Continue the process until you have achieved your desired sample size.

	1	2	3	4	5	6	7	8	9	10	11	12	13	14
1	10480	15011	01536	02011	81647	91646	69179	14194	62590	36207	20969	99570	91291	90700
	22368	46573	25595	85393	30995	89198	27982	53402	93965	34095	52666	19174	39615	99505
	24130	48360	22527	97265	76393	64809	15179	24830	49340	32081	30680	19655	63348	58629
	42167	93093	06243	61680	07856	16376	39440	53537	71341	57004	00849	74917	97758	16379
5	37570	39975	81837	16656	06121	91782	60468	81305	49684	60672	14110	06927	01263	54613
	77921	06907	11008	42751	27756	53498	18602	70659	90655	15053	21916	81825	44394	42880
	99562	72905	56420	69994	98872	31016	71194	18738	44013	48840	63213	21069	10634	12952
	96301	91977	05463	07972	18876	20922	94595	56869	60014	60045	18425	84903	42508	32307
	89579	14342	63661	10281	17453	18103	57740	84378	25331	12566	58678	44947	05585	56941
10	85475	36857	53342	53988	53060	59533	38867	62300	08158	17983	16439	11458	18593	64952
	28918	69578	88231	33276	70997	79936	56865	05859	90106	31595	01547	85590	91610	78188
	63553	40961	48235	03427	49626	69445	18663	72695	52180	20847	12234	90511	33703	90322
	09429	93969	52636	92737	88974	33488	36320	17617	30015	08272	84115	27156	30613	74952
	10365	61129	87529	85689	48237	52267	67689	93394	01511	26358	85104	20285	29975	89868
15	07119	97336	71048	08178	77233	13916	47564	81056	97735	85977	29372	74461	28551	90707
	51085	12765	51821	51259	77452	16308	60756	92144	49442	53900	70960	63990	75601	40719
	02368	21382	52404	60268	89368	19885	55322	44819	01188	65255	64835	44919	05944	55157
	01011	54092	33362	94904	31273	04146	18594	29852	71585	85030	51132	01915	92747	64951
	52162	53916	46369	58586	23216	14513	83149	98736	23495	64350	94738	17752	35156	35749
20	07056	97628	33787	09998	42698	06691	76988	13602	51851	46104	88916	19509	25625	58104
	48663	91245	85828	14346	09172	30168	90229	04734	59193	22178	30421	61666	99904	32812
	54164	58492	22421	74103	47070	25306	76468	26384	58151	06646	21524	15227	96909	44592
	32639	32363	05597	24200	13363	38005	94342	28728	35806	06912	17012	64161	18296	22851
	29334	27001	87637	87308	58731	00256	45834	15398	46557	41135	10367	07684	36188	18510
25	02488	33062	28834	07351	19731	92420	60952	61280	50001	67658	32586	86679	50720	94953
	81525	72295	04839	96423	24878	82651	66566	14778	76797	14780	13300	87074	79666	95725
	29676	20591	68086	26432	46901	20849	89768	81536	86645	12659	92259	57102	80428	25280
	00742	57392	39064	66432	84673	40027	32832	61362	98947	96067	64760	64584	96096	98253
	05366	04213	25669	26422	44407	44048	37937	63904	45766	66134	75470	66520	34693	90449
30	91921	26418	64117	94305	26766	25940	39972	22209	71500	64568	91402	42416	07844	69618
	00582	04711	87917	77341	42206	35126	74087	99547	81817	42607	43808	76655	62028	76630
	00725	69884	62797	56170	86324	88072	76222	36086	84637	93161	76038	65855	77919	88006
	69011	65795	95876	55293	18988	27354	26575	08625	40801	59920	29841	80150	12777	48501
	25976	57948	29888	88604	67917	48708	18912	82271	65424	69774	33611	54262	85963	03547
35	09763	83473	73577	12908	30883	18317	28290	35797	05998	41688	34952	37888	38917	88050
	91567	42595	27958	30134	04024	86385	29880	99730	55536	84855	29080	09250	79656	73211
	17955	56349	90999	49127	20044	59931	06115	20542	18059	02008	73708	83517	36103	42791
	46503	18584	18845	49618	02304	51038	20655	58727	28168	15475	56942	53389	20562	87338
	92157	89634	94824	78171	84610	82834	09922	25417	44137	48413	25555	21246	35509	20468
40	14577	62765	35605	81263	39667	47358	56873	56307	61607	49518	89656	20103	77490	18062
	98427	07523	33362	64270	01638	92477	66969	98420	04880	45585	46565	04102	46880	45709
	34914	63976	88720	82765	34476	17032	87589	40836	32427	70002	70663	88863	77775	69348
	70060	28277	39475	46473	23219	53416	94970	25832	69975	94884	19661	72828	00102	66794
	53976	54914	06990	67245	68350	82948	11398	42878	80287	88267	47363	46634	06541	97809
45	76072	29515	40980	07391	58745	25774	22987	80059	39911	96189	41151	14222	60697	59583
	90725	52210	83974	29992	65831	38857	50490	83765	55657	14361	31720	57375	56228	41546
	64364	67412	33339	31926	14883	24413	59744	92351	97473	89286	35931	04110	23726	51900
	08962	00358	31662	25388	61642	34072	81249	35648	56891	69352	48373	45578	78547	81788
	95012	68379	93526	70765	10592	04542	76463	54328	02349	17247	28865	14777	62730	92277
50	15664	10493	20492	38391	91132	21999	59516	81652	27195	48223	46751	22923	32261	85653

	15	16	17	18	19	20	21	22	23	24	25	26	27	28
51	16408	81899	04153	53381	79401	21438	83035	92350	36693	31238	59649	91754	72772	02338
	18629	81953	05520	91962	04739	13092	97662	24822	94730	06496	35090	04822	86774	98289
	73115	35101	47498	86737	99016	71060	88824	71013	18735	20286	23153	72924	35165	43040
	57491	16703	23167	49323	45021	33132	12544	41035	80780	45393	44812	12515	98931	91202
55	30405	83946	23792	14422	15059	45799	22716	19792	09983	74353	68668	30429	70735	25499
	16631	35006	85900	98275	32388	52390	16815	69298	82732	38480	73817	32523	41961	44437
	96773	20206	42559	78985	05300	22164	24369	54224	35083	19687	11052	91491	60383	19746
	38935	64202	14349	82674	66523	44133	00697	35552	35970	19124	63318	29686	03387	59846
	31624	76384	17403	53363	44167	64486	64758	75366	76554	31601	12614	33072	60332	92325
60	78919	19474	23632	27889	47914	02584	37680	20801	72152	39339	34806	08930	85001	87820
	03931	33309	57047	74211	63445	17361	62825	39908	05607	91284	68833	25570	38818	46920
	74426	33278	43972	10119	89917	15665	52872	73823	73144	88662	88970	74492	51805	93378
	09066	00903	20795	95452	92648	45454	09552	88815	16553	51125	79375	97596	16296	66092
	42238	12426	87025	14267	20979	04508	64535	31355	86064	29472	47689	05974	52468	16834
65	16153	08002	26504	41744	81959	65642	74240	56302	00033	67107	77510	70625	28725	34191
	21457	40742	29820	96783	29400	21840	15035	34537	33310	06116	95240	15957	16572	06004
	21581	57802	02050	89728	17937	37621	47075	42080	97403	48626	68995	43805	33386	21597
	55612	78095	83197	33732	05810	24813	86902	60397	16489	03264	88525	42786	05269	92532
	44657	66999	99324	51281	84463	60563	79312	93454	68876	25471	93911	25650	12682	73572
70	91340	84979	46949	81973	37949	61023	43997	15263	80644	43942	89203	71795	99533	50501
	91227	21199	31935	27022	84067	05462	35216	14486	29891	68607	41867	14951	91696	85065
	50001	38140	66321	19924	72163	09538	12151	06878	91903	18749	34405	56087	82790	70925
	65390	05224	72958	28609	81406	39147	25549	48542	42627	45233	57202	94617	23772	07896
	27504	96131	83944	41575	10573	08619	64482	73923	36152	05184	94142	25299	84387	34925
75	37169	94851	39117	89632	00959	16487	65536	49071	39782	17095	02330	74301	00275	48280
	11508	70225	51111	38351	19444	66499	71945	05422	13442	78675	84081	66938	93654	59894
	37449	30362	06694	54690	04052	53115	62757	95348	78662	11163	81651	50245	34971	52924
	46515	70331	85922	38329	57015	15765	97161	17869	45349	61796	66345	81073	49106	79860
	30986	81223	42416	58353	21532	30502	32305	86482	05174	07901	54339	58861	74818	46942
80	63798	64995	46583	09785	44160	78128	83991	42865	92520	83531	80377	35909	81250	54238
	82486	84846	99254	67632	43218	50076	21361	64816	51202	88124	41870	52689	51275	83556
	21885	32906	92431	09060	64297	51674	64126	62570	26123	05155	59194	52799	28225	85762
	60336	98782	07408	53458	13564	59089	26445	29789	85205	41001	12535	12133	14645	23541
	43937	46891	24010	25560	86355	33941	25786	54990	71899	15475	95434	98227	21824	19585
85	97656	63175	89303	16275	07100	92063	21942	18611	47348	20203	18534	03862	78095	50136
	03299	01221	05418	38982	55758	92237	26759	86367	21216	98442	08303	56613	91511	75928
	79626	06486	03574	17668	07785	76020	79924	25651	88325	88428	85076	72811	22717	50585
	85636	68335	47539	03129	65651	11977	02510	26113	99447	68645	34327	15152	55230	93448
	18039	14367	61337	06177	12143	46609	32989	74014	64708	00533	35398	58408	13261	47908
90	08362	15656	60627	36478	65648	16764	53412	09013	07832	41574	17639	82163	60859	75567
	79556	29068	04142	16268	15387	12856	66227	38358	22478	73373	88732	09443	82558	05250
	92608	82674	27072	32534	17075	27698	98204	63863	11951	34648	88022	56148	34925	57031
	23982	25835	40055	67006	12293	02753	14827	23235	35071	99704	37543	11601	35503	85171
	09915	96306	05908	97901	28395	14186	00821	80703	70426	75647	76310	88717	37890	40129
95	59037	33300	26695	62247	69927	76123	50842	43834	86654	70959	79725	93872	28117	19233
	42488	78077	69882	61657	34136	79180	97526	43092	04098	73571	80799	76536	71255	64239
	46764	86273	63003	93017	31204	36692	40202	35275	57306	55543	53203	18098	47625	88684
	03237	45430	55417	63282	90816	17349	88298	90183	36600	78406	06216	95787	42579	90730
	86591	81482	52667	61582	14972	90053	89534	76036	49199	43716	97543	04379	46370	28672
100	38534	01715	94964	87288	65680	43772	39560	12918	86537	62738	19636	51132	25739	56947

Abridged from *Handbook of Tables for Probability and Statistics*, Second Edition, edited by William H. Beyer (Cleveland: The Chemical Rubber Company, 1968).

Appendix F   An Introduction to SPSS/PC+

You are all aware of the profound impact computers have had on everyday life over the past twenty or thirty years. The impact might be most familiar to you in the form of Nintendo games or microwave ovens, but, as you might expect, the computer revolution has been especially significant in such areas as research and statistical analysis. Of particular importance to us is the fact that computing technology is now available that is relatively easy to use and that will perform all of the (often tedious and dull) arithmetical tasks required in this text. Even though you can do all of the end-of-chapter problems in this text with nothing more than a simple hand calculator (in fact, you can do most of them with just paper and pencil), your time will be much more efficiently spent if you learn how to use the computer to perform the routine tasks of mere calculation.

These days, the necessary skills are quite accessible even for people with little or no computer experience. There are a number of relatively easy-to-use software packages available that compute statistics and manipulate data. This appendix will introduce you to a program called SPSS/PC+, one of the leading statistical programs available today. This short guide is not intended to substitute for the SPSS/PC+ Manuals and other relevant documentation. My intent is to provide an overview and a kind of "road map" that will make SPSS/PC+ easier for you to use and will place this powerful statistical tool in the context of this text.

SPSS stands for "Statistical Package for the Social Sciences" and the "PC+" identifies the version of this package that has been designed for personal computers. A **statistical package** (or "statpak") is a set of computer programs that work with data and compute statistics. Once you have entered the data for a particular group of observations, you can easily and quickly (not to mention accurately) produce an abundance of statistical information without doing any mathematical computation yourself.

Let me illustrate the power of this technology with an example. Suppose you wanted to know the average age of a sample of 1500 respondents. How long do you think it should take you to add 1500 two-digit numbers with a calculator? If you entered the scores at a (pretty fast) rate of about 1 age per second, or 60 scores a minute, it would take you almost half an hour to enter all of the ages (assuming you made no mistakes, which is not very realistic). To accomplish the same result with a statpak like SPSS/PC+, only a few seconds would be needed (assuming that the data have already been entered). Further, the procedures for generating statistics are relatively simple. In SPSS/PC+, you would identify the statistic(s) you desire (usually by typing or selecting a word or two) and the variables you want processed (again by typing or selecting a few words). So, in a few keystrokes, you could produce statistical information that might take you minutes, hours, or even days to produce with your hand calculator.

Clearly, this is a powerful technology that is worth mastering. If you learn to use this software, you will be able to spend more time analyzing the meanings and implications of the data and less time computing and ciphering. In addition, familiarity with SPSS/PC+ will give you the ability to execute research projects using large sets of data. This capability might be very handy indeed in your senior-level courses, the work place, or graduate school.

F.1 GETTING STARTED—Data Bases and Computer Files

Before we can perform any statistical analysis, we must first supply SPSS/PC+ with information to process. SPSS/PC+ can work with numbers, letters, or both, but usually researchers use numbers to represent information. For example, suppose you had a survey that recorded gender by asking respondents to check either MALE or FEMALE. When it comes time to organize the information for computer analysis, it would be quite possible to keep track of gender by actually typing 'MALE' or 'FEMALE' as appropriate. You would save a considerable amount of time, however, if you punch a 1 for every male (a single keystroke instead of four) and a 2 for every female (a savings of five keystrokes). Because of this convenience, and for a number of other reasons, the information to be analyzed by a statpak is almost always recorded in the form of numbers.

A **data base** is an organized collection of related information. The responses to a survey might, for example, be referred to as a data base. For purposes of computer analysis, data bases are organized into files. A **file** is a collection of information that is stored under the same name in the memory of the computer or, perhaps, on a diskette. Words as well as numbers can be saved in files. If you've ever used a word processing program to type a letter or term paper, you probably saved your work in a file so that you could keep a copy or update your work or otherwise make corrections. As a first step in analyzing a data base, we would create a file that contained the responses to all of the survey items for everyone in the sample. We would also include in the file some information and instructions that SPSS/PC+ needs in order to locate and process the data and compute statistics.

To illustrate the procedures we are about to describe, let's use the data presented in Table 2.4 on page 27. This is an unrealistically small data base, but this is actually an advantage because you will be able to see the entire data base at once. Later, we'll switch to a more realistic (that is, larger) data base. Note that the information in Table 2.4 is partly in the form of numbers and partly in the form of words. For purposes of convenience, we will change the scores for gender and marital status to a numerical code. We will also use numbers for student identification. For sex, let's use a 1 to represent male and a 2 for female. For marital status, a 1 will indicate single, a 2 will mean married, and divorce will be represented as a 3. The revised data base, with number codes replacing words and letters, would look as shown in Table F.1.

TABLE F.1   DATA FROM COUNSELING CENTER SURVEY (see Table 2.4)

Student	Sex	Marital Status	Satisfaction	Age
1	1	1	4	18
2	1	2	2	19
3	2	1	4	18
4	2	1	2	19
5	1	2	1	20
6	1	1	3	20
7	2	2	4	18
8	2	1	3	21
9	1	1	3	19
10	2	3	3	23
11	2	1	3	24
12	1	2	3	18
13	2	1	1	22
14	2	2	3	26
15	1	1	3	18
16	1	2	4	19
17	2	2	2	19
18	1	3	1	19
19	2	3	3	21
20	1	1	2	20

Now, how do we get this information into a form that can be read by SPSS/PC+? The data we will submit to SPSS/PC+ will look like Table F.1 except that the spacing between columns will be eliminated. The spacing in Table F.1 is to improve legibility for human eyes and is quite unnecessary for the computer. So, we will eventually write a file with data arranged in a rectangular block of numbers, with the vertical dimension (columns) corresponding to the variables (sex, age, etc.) and the horizontal dimension (rows) corresponding to the cases (student 1, student 2, etc.).

As we enter the data, we must set aside a column for every possible digit of each score. Note that the scores on three of our variables (sex, marital status, and satisfaction) are expressed as single-digit numbers but that student identification number and age are both two-digit numbers. Thus, our data file will be seven columns across (2 columns for identification number plus 1 for sex, 1 for marital status, 1 for satisfaction, and 2 for age) and will be 20 rows long (one row per student). The first five lines of data would look like this:

```
0111418
0212219
0321418
0421219
0512120
```

Each line in this block corresponds exactly to a row in Table F.1, which in turn corresponds to the scores for a specific student. Note that the identification numbers are always recorded as two-digit numbers (e.g., 01 rather than 1). This is to leave space for the two-digit ID numbers that we will encounter later. If we had a sample with more than 99 cases, we would need to allocate three columns for the identification numbers, and the first case would have been numbered 001. Note also that a particular score is always in exactly the same column. Thus, the marital status code is always in the fourth column from the left and age is always in the last two columns. As you will see below, the column location of the variables is a very important piece of information that will enable SPSS/PC+ to read our data correctly.

F.2 DATA DEFINITION COMMANDS AND SAVING AND RETRIEVING FILES

Of course, it's not enough to supply only data to SPSS/PC+. We must also define and describe the nature of the data set. We need to tell SPSS/PC+ how to locate the variables, read the scores, and label them appropriately so that we (and others) can read the output. In this section, I will introduce some SPSS/PC+ data definition commands and, in the next section, we'll see how we can enter the data from Table 2.4 and begin to analyze the data base by using SPSS/PC+.

Below, four data definition commands are briefly summarized. The treatment of these commands is quite minimal, and you should make it a habit to consult the SPSS/PC+ Manual for further explanation and examples.

DATA LIST. The first and most important of the data definition commands is the DATA LIST command. This command performs a number of crucial functions. It tells SPSS/PC+ where to find the data set, what each of the variables is named, and the column location for each of the variables. The general format for this command is

```
1
DATA LIST FIXED/
        variable_name_1    column_numbers
        variable_name_2    column_numbers
        variable_name_3    column_numbers
        variable_name_4    column_numbers
        . . .                              .
```

Like all other SPSS/PC+ commands, this command begins in column 1. That is, the first letter of the command ("D" in DATA) is as far left as possible. In all illustrations in this appendix, the capitalized words will indicate commands to SPSS/PC+, or **keywords**, and lowercase words will indicate information that you, as the user, supply to the program. A keyword is a command that SPSS/PC+ recognizes automatically. When the program reads the words DATA LIST, for example, it knows that certain types of information about a data set will follow. The program will know from this command that the data are included in the same file as the commands. This organizational format is called "inline" data and is the **default** option. This means that, unless we specify otherwise, SPSS/PC+ will assume that data and commands are stored together. Since we said nothing at all about the location of the data, the default option will be in force.

The keyword **FIXED** tells the program that scores are always in the same column location, and the slash (/) tells SPSS/PC+ that the keyword portion of the command is completed. It is very important to spell, space, and locate all keywords correctly. If you make a mistake, you will get an **error message** from SPSS/PC+ (which is not really so terrible—statpak users get error messages all the time).

User-supplied information, the second part of the DATA LIST command, includes instructions or descriptions that you must supply yourself and that are unique to a given data set or situation. For example, we have to tell SPSS/PC+ what to call the variables we are using and which column(s) contain the scores on each variable. What we choose to call each variable (the variable names) and the exact column locations will vary from data set to data set.

Please notice that a period, or dot, is the last element in an SPSS/PC+ command. This is called the **command terminator** and it is the signal to the program that a particular command is finished. If you omit the period, SSPS/PC+ will look for continuing instructions regarding the current command and will almost certainly give you an error message (get used to error messages—you're going to see a lot of them).

When we write a DATA LIST command for the counseling center data set, we would type (or select) the keywords and a slash and then name our variables and specify their column locations, ending our command with a period. Variable names can be as many as eight characters long, but the first character must be alphabetical. Commands can continue for more than one line. It is good practice to use abbreviations and mnemonic devices to name your variables. For our data set the command might look like this:

```
DATA LIST FIXED/ IDNO 1-2 SEX 3 MARITAL 4 SATIS 5 AGE
6-7,
```

After reading this command, SSPS/PC+ would know that the data are included in the same file as the commands (the default option) and that scores are always in the same column location (from the keyword FIXED). Further, the command specifies that the numbers in columns 1 and 2 represent the scores of the cases on a variable to be called IDNO, the numbers in column 3 go with a variable to be called SEX, scores on a variable to be called MARITAL are to be found in column 4, scores on SATIS are in column 5, and AGE is in columns 6 and 7.

Please note that we need to be absolutely sure that these column specifications on the DATA LIST command are accurate. If we make a mistake,

SPSS/PC+ might read the ages of the sample as scores on marital status or vice versa. This might be a good time to note one of the most important limitations of computers: they try to do exactly what you tell them to do. If you mistakenly tell SPSS/PC+ that the scores in column 5 represent sex rather than satisfaction, the program has no way to independently check the accuracy of the instruction. SPSS/PC+ will believe what you tell it and only what you tell it.

VARIABLE LABELS. The DATA LIST command is actually the only required data definition command for SPSS/PC+, but it is almost always necessary to include three more commands to document output and handle missing scores. The VARIABLE LABELS command associates a long descriptive label with a variable name. The purpose of this command is to improve the readability of the output you will eventually generate. This command does not affect the processing of data or calculation of statistics. The descriptive label should be no longer than forty characters and must be enclosed in single quote marks ('). Whatever labels you specify will be printed every time the variable name is printed. The general format for this command is

```
VARIABLE LABELS   variable_name_1 'label'
                  variable_name_2 'label'
                  variable_name_3 'label'
                  ...                     .
```

For our data set, this command might look like this:

```
VARIABLE LABELS MARITAL 'Marital Status'
   SATIS   'Satisfaction with Counseling Center
   Services'.
```

Note that we do not need to label every variable. Variables whose meanings are clear from the variable name (like SEX and AGE) do not require further documentation.

The variable names used on this command must *exactly* match the variable names specified on the DATA LIST command. If you change the spelling of a variable name (e.g., MARTAL instead of MARITAL), SPSS/PC+ will not recognize the misspelling and will give you an error message. The labels themselves (the words between the quote marks) can be any combination of words, symbols, or numbers that you think will aid in reading output. *Be sure to terminate the command with a period.*

VALUE LABELS. The VALUE LABELS command associates a label of up to 20 characters with every value or score of a variable. Like the previous command, the purpose of this command is to document and improve the readability of output. The general format is

```
VALUE LABELS   variable_name_1   value_1 'label_1'
               value_2 'label_2'  value_3 'label_3'
               ... / variable_name_2  value_1
               'label_1' value_2 'label_2'  value_3
               'label_3' ... /
               ....                        .
```

For the variable MARITAL in our data set, the labels might look like this:

```
VALUE LABELS MARITAL 1 'Single' 2 'Married' 3
'Divorced'/.
```

This command would tell SPSS/PC+ to print the label 'Single' every time it printed the value 1 for the variable MARITAL, to print 'Married' next to a score of 2, and 'Divorced' beside a score of 3. Variables whose values have an exact, nonarbitrary meaning (like AGE) do not need labels.

MISSING VALUES.   When respondents do not answer all the questions in a survey (a very common phenomenon), we face the problem of missing data. That is, something must be recorded for the missing scores or those spaces must be left blank. In either case, we certainly would not want these values or blanks to be mistaken for actual data. In SPSS/PC+, this problem is handled by the MISSING VALUES command. We first record a code (usually some impossible score) wherever data is missing and then we declare that score to be "missing." The program will ignore these values while computing statistics.

None of our twenty cases have any missing scores but, for the sake of illustration, the command below would instruct SPSS/PC+ to treat a score of nine as missing for MARITAL (i.e., to disregard the score in the processing of the data):

```
MISSING VALUES   MARITAL (9).
```

Note that, if any scores at all are missing, the MISSING VALUES command is not really optional.

Processing Multiple Variables Together.   On the commands summarized above (with the exception of DATA LIST) and for most other SPSS commands, adjacent variables may be processed together by using the "inclusive TO" convention. For example, if several variables in a file had "9" as a missing value code, then the variables could all be processed together:

```
MISSING VALUES   SEX TO SATIS (9).
```

This command would instruct the program to treat 9 as missing for every variable between (and including) SEX and SATIS. The "TO" convention can save a considerable amount of time, effort, and space when preparing commands.

Saving and Retrieving Files.   One of the greatest advantages of using computers to analyze data, besides simply the speed of processing, is the ability to **save** files so that you can resume your analysis at some future time without having to re-enter the data. When a file is saved, it is stored either on the hard disk drive, which is internal to your computer, or on a "floppy" disk, which you may carry away with you. *Make sure you understand how to save files on your particular system before you do any actual data entry.* You may need to get additional information from your instructor or from computer center personnel.

To save and retrieve files with SPSS/PC+, you must use the **function keys**. These are labeled F1 through at least F10 and are located either in a line at the top of the keyboard or in a group on the left-hand side. To save a file, press the key labeled F9 (*not* the F key followed by the 9 key) and answer the prompts. Be sure you give the file a name that you will remember. To retrieve a file that has been stored, touch the F3 key and name the file. Further instructions on saving and retrieving files are presented below.

Summary.   The commands presented in this section will tell SPSS/PC+ how to read the data, label the output, and deal with missing scores. A visual representation of this process may be helpful. You will prepare a data base, or an input file, that includes data definition commands, the actual data, and (eventually) some commands that will calculate and display statistics. To generate output, we will submit our data base, or input file, to the SPSS/PC+ program. The software will activate the programs that will calculate the statistics we request for our data. The results of the calculations and manipulations will be your output and will be displayed on the screen. The whole process might be diagrammed as shown in Figure F.1.

In addition to the screen, you may direct the output to your printer or to another file if you wish. One way to get a printed copy of your results would be to select, from the main menu, 'session control and info', and then select SET, output, printer, and ON. This setting will cause SPSS/PC+ to print all output directly to your printer. Check with your instructor or computer center personnel if you are not familiar with the printers at your installation. Consult the SPSS/PC+ Manual for further instructions.

FIGURE F.1   A REPRESENTATION OF THE PROCESS OF SUBMITTING DATA TO SPSS/PC+ FOR ANALYSIS

```
INPUT                                              OUTPUT
Commands ----------> SPSS/PC+  ---------- > Statistics
and data                PROGRAM                    and other
                                                   information
```

F.3 IMPLEMENTING COMMANDS WITH SPSS/PC+

In this section, I will show you how to use SPSS/PC+ to implement the commands described above. Specifically, we will use the program to create a file that includes the counseling center data set and associated data definition commands. When we finish this session, we will save the file and we will be ready to conduct some statistical analysis on this data base.

Since the details about starting the SPSS/PC+ program will vary from place to place, I'll just assume that you have the program running and start from there. You should be looking at a screen with a **menu** (a list of options) in the top left-hand corner (Main Menu), some explanatory information on the right, and a blank area in the bottom half of the screen.

Since many people do not type, SPSS/PC+ is set up so that you can do most of your programming by simply selecting items from the menu system. You move through the menus by using the **arrow keys** (usually located at the bottom right of your keyboard). The highlighted area (or **cursor**) will move up and down as you touch the up and down arrows. For items on the menu that have an arrow pointing to the right, you can access additional submenus by touching the right arrow key (→). To move back to a previous menu, touch the left arrow key (←). When you have highlighted a menu option that you wish to include in your file, you "select" that item by pressing the ENTER key. When selected, the item will appear in (or be "pasted to") the bottom half of the screen.

You may also type commands and edit your file in the bottom window. To move back and forth between the two windows, type ALT-E. (That is, touch 'e' while holding the key labeled ALT down.) Note that the cursor will change shape when you switch back and forth. We will refer to the top half of the screen as MENU MODE and the bottom half of the screen as EDIT MODE. We will use both frequently.

The DATA LIST Command.   The first command we need is the DATA LIST command. To find and select this command:

1. From the Main Menu, select "read or write data." That is, move the cursor with the down arrow until this option is highlighted and then press the right arrow key.

2. From the "read or write data" menu, select and paste DATA LIST. That is, move the cursor to highlight this option and then touch ENTER. The keywords DATA LIST will appear in the bottom half of the screen (the EDIT MODE section). You are now in the "data list" submenu.

3. From the DATA LIST menu, highlight the keyword FIXED/ and touch the ENTER key. FIXED is the command that informs SPSS/PC+ that our data are arranged with the scores in the same column for all cases, and the slash (/) separates the keywords from the user-supplied information. The keyword FIXED/ will appear in the bottom half of the screen. You are now in the FIXED/ submenu.

4. The cursor is highlighting the "numerical variables" specification. This is the option we want, so type ALT-T (type "T" while holding down the ALT key) and a rectangular typing window will appear in the middle of the screen. Note that the cursor has changed shape and is now a blinking underscore. (This means an underscore symbol that blinks, not an 'expletive deleted' underscore. Speaking of expletives, many may occur to you as we move through this process. It's OK to mutter them under your breath or even aloud as long as you don't hurt the machines.) Type the names and column locations of each of the variables. In our case, we want a line that looks like this in the typing window:

```
IDNO 1-2 SEX 3 MARITAL 4 SATIS 5 AGE 6-7
```

Should you make a mistake while typing this line, use the BACK-SPACE or DELETE key to erase and the left and right arrow keys to move the cursor without erasing. When you have finished typing this line in the window, touch ENTER, and whatever you typed in the window will appear in the bottom half of the screen. Note that SPSS/PC+ adds the period to the end of the command automatically. Now, you will be in the "FIXED/" menu. Move back to the "read or write data" menu by touching the left arrow key twice.

Entering Data.   Now we need to enter our data. To begin this process, select BEGIN DATA from the menu by highlighting this command and pressing ENTER. Note that both the BEGIN DATA and END DATA keywords appear in the bottom part of the screen and that there is a blank row between them. Note also that the cursor has changed shape to a blinking underscore to let you know that you are in EDIT MODE and that the cursor is located between the two sets of keywords. The program is now ready for you to enter the data. Begin with student 1 in Table F.1 and type those scores as the first line of data. Touch ENTER when you are finished with the first student, and the cursor will move to the second line, ready to receive more data. As you finish each line of data, touch ENTER, and the cursor will move to the next line, ready to enter the scores on the next case.

When you have entered scores for all 20 cases, save your work to this point by touching the F9 key. SPSS/PC+ will ask if you wish to write (or save) the whole file. Since you do, touch ENTER and you will be asked for a file name. SPSS/PC+ might suggest the file name 'scratch.pad.' You can erase this name from the window by touching the BACKSPACE key, and you can now write your own name for the file. File names can be up to eight characters. Let's call this file CCSURVEY for Counseling Center Survey. Touch ALT-E to return to the main menu.

Variable Labels. Although technically optional, we usually want to include this command so as to make our output as readable as possible. To include this command, select "read and write data" from the main menu, and then

1. From the "read or write data" menu, highlight "labels and formatting," touch the right arrow key, and then select VARIABLE LABELS by touching ENTER while that command is highlighted. You are now in the "VARIABLE LABELS" menu and the cursor is highlighting the option "!variable(s)." The exclamation point indicates a *required* part of a command (that is, SPSS/PC+ will not run without this element).

2. Type ALT-T and the typing window will appear. Type the name of the first variable that needs to be labeled. This will be MARITAL in our case. Touch ENTER when you are finished and you will be back in the VARIABLE LABELS submenu.

3. Select "!' '." The typing window will appear in the middle of the screen, and the quote marks will be transferred to the bottom half of the screen. Type a label or a description of the variable. Something like 'Marital Status' should do it. When you touch ENTER, whatever you have typed in the window will appear between the quote marks in the EDIT window. Remember that the content of the label is up to you alone but that you are limited to 40 characters maximum.

4. Move the cursor over "!variables," type ALT-T, and type the name of the next variable that needs a label. In our case, that would be SATIS. Then, select "!' '" and type the label.

5. Continue this cycle until all variables are labeled. Since we need to label only two variables, our command will be fairly short:

```
VARIABLE LABELS MARITAL 'Marital Status' SATIS
'Satisfaction with Counseling Center Services'.
```

Double-check to make sure that you have spelled the variable names consistently with the DATA LIST command and that you have a period at the end of the command. You can make corrections by switching to EDIT MODE (touch ALT-E) and using the arrow keys to move around the file. Consult the SPSS/PC+ Manual for additional instructions on editing files.

6. Move back to the "labels and formatting" menu by touching the left arrow key.

Value Labels. Our next step is to prepare VALUE LABELS. This command will write a descriptive label beside the scores or values of a variable. Again, this is an optional command, the purpose of which is to document our output (that is, make the output easier to read).

1. From the "labels and formatting" menu, highlight VALUE LABELS and touch ENTER. The cursor will highlight "!variables."

2. Type ALT-T and type in the name of the first variable whose values you wish to label. This would be SEX in our case. Press Enter.

3. Select "!value," type ALT-T, and type in the first value in the typing window. Press ENTER and the value will appear in the bottom, or EDIT, window.

4. Select "' '" and, in the typing window, type the label you wish to have associated with that value. In this case, the label would be 'Male.' Labels can be as long as 20 characters. Touch ENTER when you are done and the label will be pasted to the EDIT window.

5. Select "!value" and type the second value (2), and then select "' '" and type the second label (Female), continuing the cycle until all values are labeled.

6. When you have finished labeling all of the values of a variable, select the slash (/). This is the signal to SPSS/PC+ that you are done with one variable. If you do not include the slash at the end of each set of value labels, you will receive an error message or a warning.

7. Repeat steps 2–5 until all variables have their values labeled.
   Your final VALUE LABELS command should look something like this:

```
VALUE LABELS  SEX 1 'Male' 2 'Female'/
              MARITAL 1 'Single' 2 'Married'
              3 'Divorced'/
              SATIS 1 'Very Dissatisfied'
                2 'Dissatisfied'
                3 'Satisfied' 4 'Very
              Satisfied'/.
```

Please make a special note of the slashes (/) that separate lists of labels and, again, the command terminator.

It is possible that you found this method of preparing the VALUE LABELS command to be tedious. If so, you can switch to EDIT MODE (touch ALT-E) and type the commands directly in the bottom window. As you gain confidence and experience with SPSS/PC+, you will often find the EDIT MODE to be significantly faster than the MENU MODE.

8. When finished, move back to the "labels and formatting" menu by touching the left arrow key.

Missing Values.   Our counseling center data has no missing scores. This is not the usual situation, and to illustrate the process, let's declare 9 to be the missing value code for sex, marital status, and satisfaction. For MISSING VALUES:

1. From the "labels and formatting" menu, select MISSING VALUE.
2. Type ALT-T and type the name of the first variable in the window.
3. Select "( )" and type 9, the value to be treated as missing, in the window. SPSS/PC+ permits only one value to be declared missing for each variable.
4. Repeat the cycle until all variables have been processed.

Your final command should look like this:

```
MISSING VALUES SEX (9) MARITAL (9) SATIS (9).
```

Since we are using the same missing value code for all variables, we could have used the inclusive TO in the EDIT mode to prepare this command:

```
MISSING VALUES SEX TO SATIS (9).
```

The latter command is identical in its effects to the former.

The SPSS/PC+ File.   The final SPSS/PC+ file, with both commands and data included, should look like this:

```
DATA LIST FIXED / IDNO 1-2 SEX 3 MARITAL 4 SATIS 5 AGE
6-7.
BEGIN DATA.
0111418
0212219
0321418
0421219
0512120
0611320
0722418
0821321
0911319
1023323
1121324
1212318
1321122
1422326
1511318
```

```
1612419
1722219
1813119
1923321
2011220
END DATA.
VARIABLE LABELS MARITAL 'Marital Status'
    SATIS 'Satisfaction with Counseling Center
    Services'.
VALUE LABELS  SEX 1 'Male' 2 'Female'/
              MARITAL 1 'Single' 2 'Married'
              3 'Divorced'/
              SATIS 1 'Very Dissatisfied'
                  2 'Dissatisfied'
                  3 'Satisfied' 4 'Very Satisfied'/.
MISSING VALUES  SEX TO AGE (9).
```

Saving, Checking, and Listing Your File. When you are finished typing these commands, save them again by pressing F9 and answering the prompts. If you want to check your work to this point for errors, place the cursor on the DATA LIST line (press ALT-E to switch to EDIT MODE and move the cursor with the arrow keys), and press F10. The program will give you two options: RUN FROM CURSOR or EXIT TO PROMPT, with the former highlighted. This is the option you want, so simply touch ENTER and your program will run. You won't receive any output yet, but SPSS/PC+ will process the commands and let you know if you have made any errors. (See Section F4 on Error Messages.)

A normal step in the research process at this point would be to proofread your data file for typographical and other errors. A useful facility to help with error checking and quality control is the LIST command. This command will list the scores on any or all variables for any or all cases. If we use it to print the scores on our variables for all 20 cases, we will produce output that will echo Table F.1. We can then compare the two and search for errors.

To get a LIST, from the Main Menu:

1. Highlight "analyze data" and then highlight "reports and tables" and touch the right arrow key. From this menu, select LIST and touch ENTER. The LIST command should appear in the bottom, or EDIT, box and you are now in the LIST menu. From this menu, you can limit the number of cases or specify the exact variables you want the program to list. Since we want to list all values for all cases, the default option, we don't need to provide any further specifications.

2. Switch to EDIT mode (type ALT-E) and place the cursor on the DATA LIST line. Now press 'F10' to submit the program to SPSS/PC+.
   Your output should look like this:

IDNO	GENDER	MARITAL	SATIS	AGE
1	1	1	4	18
2	1	2	2	19
3	2	1	4	18
4	2	1	2	19
5	1	2	1	20
6	1	1	3	20
7	2	2	4	18
8	2	1	3	21
9	1	1	3	19
10	2	3	3	23
11	2	1	3	24
12	1	2	3	18
13	2	1	1	22
14	2	2	3	26
15	1	1	3	18
16	1	2	4	19
17	2	2	2	19
18	1	3	1	19
19	2	3	3	21
20	1	1	2	20

The listing format is very convenient for comparing with Table F.1 and checking for typographical errors. If you detect any errors, load the original file (use the F3 key to retrieve the file) and type ALT-E to get into EDIT MODE. Use the arrow keys to move around the file and correct any mistakes.

When you finish this session, remember to save your file by pressing F9 and answering the prompts. Remember the file name you used so that you can retrieve this file later. We will use this file in some of the end-of-chapter exercises.

Exiting from SPSS/PC+.   To exit from SPSS/PC+, select FINISH from the MAIN MENU and press the F10 key. SPSS/PC+ will present two choices: 'Run from Cursor' (this one will be highlighted) or 'Exit to Prompt'. The first option is what you want, so just press ENTER and you will exit from SPSS/PC+.

F.4 EDITING YOUR FILES AND DEALING WITH ERROR MESSAGES

The easiest way to edit (correct, update, and/or revise) a file is to use the EDIT mode. Load the file by touching F3 and naming the file (e.g., CCSURVEY). Now, touch ALT-E and you can move around in the file with the arrow keys and correct mistakes, add new data, change or delete old commands, or add new commands. Once the changes have been made, save the revised file by touching F9. If you save by using the same file name (e.g., CCSURVEY), the old file will be erased and replaced with the revised file. If you wish to save both the old and the revised files, use a new file name in answer to the prompt. For example, if we worked on our file CCSURVEY and wanted to save both the original and the revised files, we could call the latter CCSURVEY2 and we would have two versions of the file (CCSURVEY and CCSURVEY2) available to us.

In particular, you will need to edit your file in response to error messages. If you submit a file to SPSS/PC+ with words or instructions it doesn't recognize or can't figure out, you will see an error message or a warning on the screen. DO NOT BE CONCERNED OR ALARMED. Error messages and warnings are all part of the process. They are normal and expected, and everybody gets them. You can almost always figure out what the problem is from the message SPSS/PC+ displays. Remember that the most common mistakes are just misspellings or missing command terminators. Check to make sure that you have spelled all keywords correctly and that you have spelled your variable names consistently. On the latter point, remember that the spelling you use for variable names on the DATA LIST command is what SPSS/PC+ expects to see in all commands that follow.

Incidentally, the difference between a **warning** and an **error message** has to do with how confused SPSS/PC+ gets when trying to decipher your instructions. If the program can still function and produce some output in spite of its confusion, it will do so and give you a warning that it has detected a problem. An example of a mistake that would result in a warning would be if you omitted the slash that signals the end of a set of labels on the VALUE LABELS command. SPSS/PC+ would continue to look for labeling information until it finds a slash, and it will certainly get confused and let you know about it. In Figure F.2, I have reproduced an incorrect VALUE LABELS command and the warning I received from SPSS/PC+ when I ran the file with this mistake.

First, note the mistake I made. There is no slash at the end of the list of labels for SEX. The correct syntax would be

```
VALUE LABELS
SEX 1 'Male' 2 'Female'/
MARITAL 1 'Single' 2 'Married' 3 'Divorced'/.
```

Now, look at the WARNING. It says that the word MARITAL is an "unrecognized symbol." Because it had not yet found a slash, the program expected

FIGURE F.2   AN SPSS/PC+ WARNING

```
VALUE LABELS
SEX 1 'MALE' 2 'FEMALE'
MARITAL 1 'SINGLE' 2 'MARRIED' 3 'DIVORCED'/
WARNING 255, Text: MARITAL
INVALID SYMBOL ON VALUE LABELS COMMAND_The VALUE
LABELS command contains an unrecognized symbol where a
value is expected. All labels up to the next slash are
ignored.
```

FIGURE F.3   AN SPSS/PC+ ERROR MESSAGE

```
DATA LIST FIXED/ IDNO 1-2 SEX 3 MARITAL 4 SATIS AGE
6-7.
VALUE LABELS
 SAX 1 'Male' 2 'Female'/.
ERROR 443, Text: SAX
UNDEFINED VARIABLE NAME_Check for a misspelled name.
This command not executed.
```

additional *values* (scores or numbers) and got confused when it found alphabetic characters instead. Since it can't figure out what's going on, the program decided to simply ignore all values "up to the next slash." If we had put the slash where SPSS/PC+ expected to find it, it would have known that the value labels list was completed and would have looked for a variable name (like MARITAL) or the command terminator (the period) next.

Even though your program will still execute with a warning, it is usually worthwhile to correct the mistake and resubmit your input file. That way, when you get clean output (no warnings at all), you know that the results are what you want.

An error, as distinct from a warning, is more serious and will terminate all processing. An error means that SPSS/PC+ was not able to figure out what you wanted it to do. Remember that computers are extremely literal and that they will try to do exactly what you say. When SPSS/PC+ is very unsure about your instructions, it will simply shut down and report on screen what the problem is. To illustrate, I deliberately misspelled a variable name on a VALUE LABELS command. In this case, I spelled the variable as SEX on the DATA LIST command and as SAX on the VALUE LABELS command. Figure F.3 shows what the file looked like and what SPSS/PC+ said when I submitted the file.

The program identified the cause of its confusion ('SAX'), diagnosed the problem as an "UNDEFINED VARIABLE NAME" (or a variable name that had not been included on the DATA LIST command), and even offered a suggestion for dealing with the problem ("Check for a misspelled name"). To correct the error, retrieve the file (use the F3 key), make the necessary corrections, and be sure to save the corrected file (use the F9 key).

F.5 SYSTEM FILES AND THE GENERAL SOCIAL SURVEY

When you have written and debugged a file, you may want to save the command and data files in what is called a **system file**. A system file saves the data base in a format that is convenient to access and that permits rapid processing. System files are especially convenient for large data bases that have many variables and many commands. When a system file processes, SPSS/PC+ does not display the commands and data on the screen. Therefore, the program can produce output at a considerably higher rate of speed than the "inline" file we created in the previous sections.

Before creating an SPSS/PC+ system file, make certain that your file is arranged and documented to your satisfaction. Then, select and paste from the 'read or write data' menu SAVE and OUTFILE. Type a name for the file in the typing window. Later, to access a system file, select GET from the same menu and name the file. Touch F10 to make the file active. You may then type or paste additional commands in the EDIT window to generate output, update the file, revise the file contents, or otherwise work with your data base.

Many of the end-of-chapter exercises in this text refer to an SPSS/PC+ system file that contains a data base called the General Social Survey. This is a "public opinion poll" that is conducted every year by the National Opinion Research Council. A shortened version of the General Social Survey for 1993 (we will refer to this as the GSS93) is supplied with this text and described in detail in Appendix G. Appendix G lists the variables in the data base and the values associated with each variable. To examine the contents of the file, select GET from the "read and write data menu" and type GSS93 when prompted. Touch F10, and the GSS93 data base will become the active file. See the end-of-chapter exercises for additional instructions and data-analysis projects.

GLOSSARY

Arrow keys. The keys used to move the cursor around the screen.

Command terminator. All SPSS/PC+ commands must end with a period, or dot. This symbol is called the command terminator.

Cursor. The symbol, either a blinking underscore or a rectangle, which marks where you are on the screen.

Data base. An organized collection of related information.

Default. A procedure, or routine, that SPSS/PC+ executes automatically ("by default") unless explicitly instructed to do otherwise.

Error message. When SPSS/PC+ cannot decipher instructions, it will cease processing and print an error message on the screen.

File. A data base (or any other information) that is stored under the same name in the memory of the computer or on a floppy disk.

FIXED. The part of the DATA LIST command that informs SPSS/PC+ that scores are always recorded in the same column(s).

Keywords. Commands that SPSS/PC+ recognizes automatically.

Menu. A list of options in a statistical package.

SPSS/PC+. A statistical package designed for the analysis of social science data. The version used in this text runs on personal computers.

Statistical package, or statpak. A set of computer programs designed to manipulate and statistically analyze data.

System file. A special organizational format for SPSS/PC+ files.

Warning. When SPSS/PC+ detects a minor difficulty in a file, it will print a warning message on the screen.

LIST OF SPSS/PC+ DEMONSTRATIONS

	Major New SPSS/PC+ Procedures and Commands	Page
2.1 Producing Frequency Distributions for the Counseling Center Data	FREQUENCIES	54
2.2 Producing Frequency Distributions for the 1993 General Social Survey (GSS) Data		57

Appendix G Code Library for the General Social Survey, 1993

The General Social Survey is a public opinion poll conducted yearly by the National Opinion Research Council. The data set supplied with this text includes 78 variables for a randomly selected subsample of about half of the 1606 original respondents

This code library lists each item in the data set. The abbreviated variable names in capital letters are those used in the SPSS/PC+ system file. The questions have been reproduced exactly as they were asked (with a few exceptions in order to conserve space), and the numbers beside each response alternative correspond to the scores recorded in the data file.

The data set includes variables that measure demographic or background characteristics of the respondents, including sex, age, race, religion and a variety of indicators of socioeconomic status. Also included are items that measure public opinion on such current and controversial topics as abortion, capital punishment, and homosexuality.

A common problem in social science research is that respondents may not answer all questions. In addition, many items on the General Social Survey were not presented to all respondents. Usually, researchers will note why a particular piece of information is missing for a particular respondent. In the full version of the General Social Survey, for example, different codes are used to distinguish "Don't Know" (DK) responses (where the respondent does not have the requested information) from "No Answer," or NA (where the respondent might have the requested information but chooses not to report it). SPSS/PC+ permits only one code for missing values and all nonresponses, regardless of the reason, are indicated by a single score.

WRKSTAT
1. Last week, were you working full time, part time, going to school, keeping house, or what?
   1. Working full time
   2. Working part time
   3. With a job but not at work (ill, vacation, etc.)
   4. Unemployed
   5. Retired
   6. In school
   7. Keeping house
   8. Other

PRESTG80
2. Prestige score for respondent's occupation.
   0. Not applicable, no answer, don't know
   17–86. Actual score

MARITAL   3. Are you currently married, widowed, divorced, separated, or have you never been married?
- 1. Married
- 2. Widowed
- 3. Divorced
- 4. Separated
- 5. Never married

AGEWED   4. How old were you when you first married?
- 14–50. Scores are actual ages
- 0. Not applicable

PAPRES80   5. Prestige score for respondent's father's occupation (Same scoring as PRESTG80, item 2)

CHILDS   6. How many children have you ever had? Please count all that were born alive at any time, (including any from a previous marriage).
- 0–7. Actual number
- 8. Eight or more
- 9. NA

AGE   7. Age of respondent.
- 18–89. Actual age in years
- 0. NA or DK

EDUC   8. Highest year of school completed
- 0–20. Actual number of years
- 98. DK or NA

DEGREE   9. Respondent's highest degree
- 0. Less than HS
- 1. High school
- 2. Assoc./ jr. coll.
- 3. Bachelor's
- 4. Graduate
- 8. DK or NA

SEX   10. Respondent's gender
- 1. Male
- 2. Female

RACE   11. Race of respondent
- 1. White
- 2. Black
- 3. Other

INCOME91   12. Respondent's total family income from all sources for 1991
- 1. Less than 1000
- 2. 1000 to 2999
- 3. 3000 to 3999
- 4. 4000 to 4999

5. 5000 to 5999	14. 22,500 to 24,999
6. 6000 to 6999	15. 25,000 to 29,999
7. 7000 to 7999	16. 30,000 to 34,999
8. 8000 to 9999	17. 35,000 to 39,999
9. 10,000 to 12,499	18. 40,000 to 49,999
10. 12,500 to 14,999	19. 50,000 to 59,999
11. 15,000 to 17,499	20. 60,000 to 74,999
12. 17,500 to 19,999	21. 75,000 or over
13. 20,000 to 22,499	0. Not applicable

REGION   13 Region of interview
1. New England
2. Mid-Atlantic
3. East N. Cent.
4. West N. Cent.
5. So. Atlantic
6. East So. Cent.
7. West So. Cent.
8. Mountain
9. Pacific
0. Not assigned

SRCBELT   14. Belt code
1. Central city, 12 largest SMSAs
2. Central city, remaining 100 largest SMSAs
3. Suburbs, 12 largest SMSAs
4. Suburbs, remaining largest 100 SMSAs
5. Other urban (counties with towns of 10,000 or more)
6. Other rural (counties with no towns of 10,000 or more)
0. Not assigned

SIZE   15. Size of place in thousands
Population figures from U.S. Census. Add three zeros to code for actual values.
−1. Not assigned

PARTYID   16. Generally speaking, do you usually think of yourself as a Republican, Democrat, Independent, or what?
0. Strong Democrat
1. Not very strong Demo.
2. Indep. close to Demo.
3. Independent
4. Indep. close to Rep.
5. Not very strong Rep.
6. Strong Republican
7. Other party
9. DK

PRES88   17. In 1988, you remember that Dukakis ran for president on the Democratic ticket against Bush for the Republicans. Did you vote for Dukakis or Bush? (Includes only those who said they voted in this election)
1. Dukakis
2. Bush
3. Other
5. No presidential vote
0. Not applicable, DK, NA

PRES92   18. In 1992, you remember that Clinton ran for president on the Democratic ticket against Bush for the Republicans and Perot as an Independent. Did you vote for Clinton, Bush, or Perot? (Includes only those who said they voted in this election)
1. Clinton
2. Bush
3. Perot
4. Other
6. No presidential vote
0. Not applicable, DK, NA

POLVIEWS   19. I'm going to show you a seven-point scale on which the political views that people might hold are arranged from extremely liberal—point 1—to extremely conservative—point 7. Where would you place yourself on this scale?
1. Extremely liberal
2. Liberal
3. Slightly liberal
4. Moderate
5. Slightly conservative
6. Conservative
7. Extremely conservative
0. Not applicable, DK, NA

NATAID   20. Are we spending too much money on foreign aid, too little money, or about the right amount? (see pp. 114 and 115 of the 1993 GSS Cumulative Codebook for exact wording of items 19–22)
1. Too little
2. About right
3. Too much
0. Not applicable, DK, NA

NATCRIME   21. Are we spending too much money on halting the rising crime rate, too little money, or about the right amount?
(Same scoring as NATAID, item 20)

NATFARE   22. Are we spending too much money on welfare, too little money, or about the right amount?
(Same scoring as NATAID, item 20)

NATHEAL   23. Are we spending too much money on improving and protecting the nation's health, too little money, or about the right amount?
(Same scoring as NATAID, item 20)

CAPPUN   24. Do you favor or oppose the death penalty for persons convicted of murder?
1. Favor
2. Oppose
0. Not applicable, DK, NA

GUNLAW   25. Would you favor or oppose a law which requires a person to obtain a police permit before he or she could buy a gun?
1. Favor
2. Oppose
0. Not applicable, DK, NA

GRASS   26. Do you think the use of marijuana should be made legal or not?
1. Should
2. Should not
9. Not applicable, DK, NA

USINTL   27. Do you think it would be best for the future of the country if we take an active part in world affairs or if we stay out of world affairs?
1. Active part
2. Stay out
0. Not applicable, DK, NA

RELIG   28. What is your religious preference? Is it Protestant, Catholic, Jewish, some other religion, or no religion?
1. Protestant
2. Catholic
3. Jewish
4. None
5. Other
9. No answer

ATTEND   29. How often do you attend religious services?
0. Never
1. Less than once per year
2. Once or twice a year
3. Sev. times per year
4. About once a month
5. 2–3 times a month
6. Nearly every week
7. Every week
8. Several times a week
9. DK or no answer

RELITEN

30. Would you call yourself a strong (PREFERENCE NAMED IN ITEM 28) or a not very strong (PREFERENCE NAMED IN ITEM 28)?
   1. Strong
   2. Not very strong
   3. Somewhat strong (volunteered)
   0. Not applicable, DK, NA

Here are some opinions other people have expressed in connection with black–white relations. Which statement . . . comes closest to how you yourself feel?

RACSEG

31. White people have a right to keep blacks out of their neighborhoods if they want to, and blacks should respect that right.
   1. Agree strongly
   2. Agree slightly
   3. Disagree slightly
   4. Disagree strongly
   0. Not applicable, DK, NA

RACLIVE

32. Are there any blacks living in this neighborhood now?
   1. Yes
   2. No
   0. Not applicable, DK, NA

BUSING

33. In general, do you favor or oppose the busing of black and white school children from one school district to another?
   1. Favor
   2. Oppose
   0. Not applicable, DK, NA

HAPPY

34. Taken all together, how would you say things are these days—would you say that you are very happy, pretty happy, or not too happy?
   1. Very happy
   2. Pretty happy
   3. Not too happy
   0. Not applicable, DK, NA

HAPMAR

35. Taking things all together, how would you describe your marriage? Would you say that your marriage is very happy, pretty happy, or not too happy?
   1. Very happy
   2. Pretty happy
   3. Not too happy
   0. Not applicable, DK, NA

HELPFUL

36. Would you say that most of the time people try to be helpful or that they are mostly just looking out for themselves?

    1. Try to be helpful
    2. Just look out for themselves
    3. Depends (volunteered)
    0. Not applicable, DK, NA

For each area of life . . . tell me . . . how much satisfaction you get from . . .

SATFAM

37. Your family life
    1. A very great deal
    2. A great deal
    3. Quite a bit
    4. A fair amount
    5. Some
    6. A little
    7. None
    0. Not applicable, DK, NA

SATFRND
(34)

38. Your friendships
    1. A very great deal
    2. A great deal
    3. Quite a bit
    4. A fair amount
    5. Some
    6. A little
    7. None
    0. Not applicable, DK, NA

I'm going to name some institutions in this country. As far as the people running these institutions are concerned, would you say you have a great deal of confidence, only some confidence, or hardly any confidence at all in them?

CONCLERG

39. Organized religion
    1. A great deal
    2. Only some
    3. Hardly any
    0. Not applicable, DK, NA

CONEDUC

40. Education
(Same scoring as CONCLERG, item 39)

CONFED

41. Executive branch of the federal government
(Same scoring as CONCLERG, item 39)

CONLEGIS

42. Congress
(Same scoring as CONCLERG, item 39)

CONPRESS

43. Press
(Same scoring as CONCLERG, item 39)

CONTV 44. TV
(Same scoring as CONCLERG, item 39)

DRINK 45. Do you ever have occasion to use any alcoholic beverages such as liquor, wine, or beer, or are you a total abstainer?
1. Use alcohol
2. Total abstainer
0. Not applicable, DK, NA

SMOKE 46. Do you smoke?
1. Yes
2. No
0. Not applicable, DK, NA

ANOMIA5 47. In spite of what some people say, the lot of the average man is getting worse, not better.
1. Agree
2. Disagree
0. Not applicable, DK, NA

ANOMIA6 48. It's hardly fair to bring a child into the world with the way things look for the future.
1. Agree
2. Disagree
0. Not applicable, DK, NA

SATJOB 49. On the whole, how satisfied are you with the work you do—would you say that you are very satisfied, moderately satisfied, a little dissatisfied, or very dissatisfied?
1. Very satisfied
2. Moderately satisfied
3. A little dissatisfied
4. Very dissatisfied
0. Not applicable, DK, NA

SATFIN 50. So far as you and your family are concerned, would you say that you are pretty well satisfied with your present financial situation, more or less satisfied, or not satisfied at all?
1. Pretty well satisfied
2. More or less satisfied
3. Not satisfied at all
0. Not applicable, DK, NA

GETAHEAD 51. Some people think that people get ahead by their own hard work; others say that lucky breaks or help from other people are more important. Which do you think is most important?
1. Hard work, most important
2. Hard work, luck equally important

3. Luck most important
0. Not applicable, DK, NA

Please tell me whether or not *you* think it should be possible for a pregnant woman to get a *legal* abortion if . . .

ABNOMORE   52. She is married and does not want any more children.
1. Yes
2. No
0. Not applicable, DK, NA

ABPOOR   53. The family has a very low income and cannot afford any more children. (Same scoring as ABNOMORE, item 39)

PILLOK   54. Do you strongly agree, agree, disagree, or strongly disagree that methods of birth control should be available to teenagers between the ages of 14 and 16 if their parents do not approve?
1. Strongly agree
2. Agree
3. Disagree
4. Strongly disagree
0. Not applicable, DK, NA

SEXEDUC   55. Would you be for or against sex education in your public schools?
1. Favor
2. Oppose
0. Not applicable, DK, NA

PREMARSX   56. There's been a lot of discussion about the way morals and attitudes about sex are changing in this country. If a man and a woman have sex relations before marriage, do you think it is always wrong, almost always wrong, wrong only sometimes, or not wrong at all?
1. Always wrong
2. Almost always wrong
3. Wrong only sometimes
4. Not wrong at all
0. Not applicable, DK, NA

XMARSEX   57. What is your opinion about a married person having sexual relations with someone other than the marriage partner—is it always wrong, almost always wrong, wrong only sometimes, or not wrong at all?
1. Always wrong
2. Almost always wrong
3. Wrong only sometimes
4. Not wrong at all
0. Not applicable, DK, NA

HOMOSEX

58. What about relations between two adults of the same sex: Do you think it is always wrong, almost always wrong, wrong only sometimes, or not wrong at all?
   1. Always wrong
   2. Almost always wrong
   3. Wrong only sometimes
   4. Not wrong at all
   0. Not applicable, DK, NA

The next question [is] about pornography—books, movies, magazines, or photographs that show or describe sex activities. I'm going to read some opinions about the effects of looking at or reading such sexual materials. As I read . . . please tell me if you think sexual materials do or do not have that effect.

PORNMORL

59. Sexual materials lead to a breakdown of morals.
   1. Yes
   2. No
   0. Not applicable, DK, NA

SPANKING

60. Do you strongly agree, agree, disagree, or strongly disagree that it is sometimes necessary to discipline a child with a good, hard spanking?
   1. Strongly agree
   2. Agree
   3. Disagree
   4. Strongly disagree
   0. Not applicable, DK, NA

LETDIE1

61. When a person has a disease that cannot be cured, do you think doctors should be allowed by law to end the patient's life by some painless means if the patient and his family request it?
   1. Yes
   2. No
   0. Not applicable, DK, NA

SUICIDE1

62. Do you think a person has the right to end his or her life if this person has an incurable disease?
   1. Yes
   2. No
   0. Not applicable, DK, NA

HIT

63. Have you ever been punched or beaten by another person?
   1. Yes
   2. No
   0. Not applicable, DK, NA

HITOK

64. Are there any situations that you can imagine in which you would approve of a man punching an adult male stranger?

1. Yes
2. No
0. Not applicable, DK, NA

POLHITOK   65. Are there any situations you can imagine in which you would approve of
                    a policeman striking an adult male citizen?
                    1. Yes
                    2. No
                    0. Not applicable, DK, NA

FEAR   66. Is there any area right around here—that is, within a mile—where you
                    would be afraid to walk alone at night?
                    1. Yes
                    2. No
                    0. Not applicable, DK, NA

TVHOURS   67. On the average day, about how many hours do you personally watch
                    television?
                    00–16. Actual hours
                    −1. Not applicable, DK, NA

FEHELP   68. It is more important for a wife to help her husband's career than to have
                    one herself.
                    1. Strongly agree
                    2. Agree
                    3. Disagree
                    4. Strongly disagree
                    0. Not applicable, DK, NA

FEFAM   69. It is much better for everyone involved if the man is the achiever outside
                    the home and the woman takes care of the home and family.
                    1. Strongly agree
                    2. Agree
                    3. Disagree
                    4. Strongly disagree
                    0. Not applicable, DK, NA

TRAUMA5   70. Number of traumatic events (deaths, divorces, unemployments, and hos-
                    pitalizations or disabilities) happening to respondent during last five
                    years.
                    0–4. Actual number of traumatic events
                    −1. Not applicable, DK, NA

I'm going to list some types of music. For each type, can you tell me if you
like it very much, like it, have mixed feelings, dislike it, dislike it very much,
or is this a type of music that you don't know much about?

RAP

71. Rap music
    1. Like it very much
    2. Like it
    3. Mixed feelings
    4. Dislike it
    5. Dislike it very much
    0. Not applicable, DK, NA

CONROCK

72. Contemporary pop/rock
    1. Like it very much
    2. Like it
    3. Mixed feelings
    4. Dislike it
    5. Dislike it very much
    0. Not applicable, DK, NA

TVNEWS

73. How often do you watch world or national news on television? (For exact wording see p. 496, 1993 GSS Cumulative Codebook)
    1. Every day
    2. Several times a week
    3. Several times a month
    4. Rarely
    5. Never
    0. Not applicable, DK, NA

ENGLISH

74. It is better for everyone if English is the only language used in the public schools.
    1. Strongly agree
    2. Agree
    3. Disagree
    4. Strongly disagree
    0. Not applicable, DK, NA

AMTESTS

75. It is right to use animals for medical testing if it might save human lives.
    1. Strongly agree
    2. Agree
    3. Neither agree or disagree
    4. Disgree
    5. Strongly disagree
    0. Not applicable, DK, NA

GRNSOL

76. How willing would you be to accept cuts in your standard of living in order to protect the environment?
    1. Very willing
    2. Fairly willing
    3. Neither willing nor unwilling
    4. Not very willing

5. Not at all willing
0. Not applicable, DK, NA

SEXFREQ   77. About how many times did you have sex during the last 12 months?
   0. Not at all
   1. Once or twice
   2. About once a month
   3. 2 or 3 times a month
   4. About once a week
   5. 2 or 3 times a week
   6. More than 3 times a week
   −1. Not applicable, DK, NA

CLASS   78. Subjective class identification
   1. Lower class
   2. Working class
   3. Middle class
   4. Upper class
   0. Not applicable, DK, NA

Appendix H   Basic Mathematics Review

H.1 OF ARITHMETIC, CALCULATORS, AND COMPUTERS

You will probably be relieved to hear that first courses in statistics are not particularly mathematical and do not stress computation per se. While you will encounter many numbers to work with and numerous formulas to use, the major emphasis will be on understanding the role of statistics in research and the logic by which we attempt to answer research questions empirically. You will also find that, at least in this text, the example problems and many of the homework problems have been intentionally simplified so that the computations will not unduly distract you from the task of understanding the statistics themselves.

On the other hand, you will probably regret to hear that there is, inevitably, some arithmetic that you simply cannot avoid if you want to master this material. It is likely that some of you haven't had any math in a long time, others have convinced themselves that they just cannot do math under any circumstances, and still others are just rusty and out of practice. All of you will find that even the most complex and intimidating operations and formulas can be broken down into simpler steps. If you have forgotten how to cope with some of these simpler steps or are unfamiliar with these operations, this appendix is designed to ease you into the skills you will need to do all of the computation in this textbook.

Before we begin the review, let me point out that a calculator is a virtual necessity for this text. While you could do all the arithmetic by hand, the calculator will save you time and effort and is definitely worth the small investment. Incidentally, you do not need to invest in any of the super-sophisticated, more expensive models you might see for sale. A square-root function is the only extra feature you really need. Of course, if you want the additional features, such as memories and preprogrammed functions, by all means spend the extra money and learn how to use them. However, a simple, inexpensive calculator will work fine for all of the problems in this text.

Along the same lines, many of you probably have access to computers and statistical packages. If so, take the time now to learn how to use them, because they will eventually save you time and effort (not to mention that they will guarantee the accuracy of your answers). This text includes a guide to a statistical package called SPSS/PC+, but many other programs are available that will accomplish the goals of saving time and avoiding drudgery while generating precise and accurate results.

In summary, you should find a way at the beginning of this course—with a calculator, a computer, or both—to minimize the tedium and hassle of mere computing. This will permit you to devote maximum effort to the truly important goal of increasing your understanding of the meaning of statistics in particular and social research in general.

H.2 VARIABLES AND SYMBOLS

Statistics are a set of techniques by which we can describe, analyze, and manipulate **variables**. A variable is any attribute or trait that can change value from case to case or from time to time. Examples of variables would include height, weight, level of prejudice, and political party preference. The possible values associated with a given variable might be numerous (for example, income) or relatively few (for example, gender). I will often use symbols, usually the letter X, to refer to variables in general or to a specific variable.

Sometimes we will need to refer to a specific value or set of values of a variable. This is usually done with the aid of subscripts. So, the symbol X_1 (read "X-sub-one") would refer to the first score in a set of scores, X_2 ("X-sub-two") to the second score, and so forth. On occasion, we will use the subscript i to refer to all the scores in a set. Thus, the symbol X_i ("X-sub-eye") would be taken to mean all of the scores associated with a given variable (for example, the test grades of a particular class).

H.3 OPERATIONS

You are all familiar with the four basic mathematical operations of addition, subtraction, multiplication, and division and the standard symbols ($+$, $-$, $\times$, $\div$) used to denote them. Some of you may not be aware, however, that the latter two operations can be symbolized in a variety of ways. For example, the operation of multiplying some number a by some number b may be symbolized in (at least) six different ways:

$$
\begin{array}{ll}
a \times b & ab \\
a \cdot b & a(b) \\
a * b & (a)\,(b)
\end{array}
$$

In this text, we will commonly use the "adjacent symbols" format (that is, ab), the conventional times sign ($\times$), or adjacent parentheses to indicate multiplication. On most calculators and computers, the asterisk (*) is the symbol for multiplication.

The operation of division is also indicated by multiple symbols:

$$a \div b$$

$$a/b$$

$$\frac{a}{b}$$

In this text, the latter two symbols are used almost exclusively.

Several of the formulas with which we will be working require us to find the square of a number. To do this, simply multiply the number by itself. This operation is symbolized as X^2 (read "X squared"), which is the same thing as $(X)(X)$. If X has a value of 4, then

$$X^2 = (X)(X) = (4)(4) = 16$$

or we could say that "4 squared is 16."

The square root of a number is the value that, when multiplied by itself, results in the original number. So the square root of 16 is 4 because $(4)(4)$ is 16. The operation of finding the square root of a number is symbolized as

$$\sqrt{X}$$

Be sure that you have access to a calculator with a built-in square root function.

A final operation with which you should be familiar is summation, or the addition of the scores associated with a particular variable. When a formula requires the addition of a series of scores, this operation is usually symbolized as ΣX_i, where Σ is the uppercase Greek letter sigma and stands for "the summation of." So the combination of symbols ΣX_i means "the summation of all the scores" and directs us to add the value of all the scores for that variable. If four people had family sizes of 2, 4, 5, and 7, then the summation of these four scores for this variable could be symbolized as

$$\Sigma X_i = 2 + 4 + 5 + 7 = 18$$

The symbol Σ is an operator, just like the $+$ or $\times$ signs. It directs us to add all of the scores on the variable indicated by the X symbol.

There are two other common uses of the summation sign and, unfortunately, the symbols denoting these uses are not, at first glance, sharply different from each other or from the symbol used above. A little practice and some careful attention to these various meanings should minimize the confusion.

Beginning with Chapter 4, you will frequently encounter the set of symbols ΣX_i^2, which means "the sum of the squared scores." This quantity is found by *first squaring* each of the scores and *then adding* the squared scores together. A second common set of symbols will be $(\Sigma X_i)^2$, which means "the sum of the scores, squared." This quantity is found by *first summing* the scores and *then squaring* the total.

I know that these distinctions are confusing at first, so let's see if an example helps to clarify the situation. Suppose we had a set of three scores: 10, 12, and 13. So,

$$X_i = 10, 12, 13$$

The sum of these scores would be indicated as

$$\Sigma X_i = 10 + 12 + 13 = 35$$

The sum of the squared scores would be

$$\Sigma X_i^2 = (10)^2 + (12)^2 + (13)^2 = 100 + 144 + 169 = 413$$

Take careful note of the order of operations here. First, the scores are squared one at a time and then the squared scores are added. This is a completely different operation from squaring the sum of the scores:

$$(\Sigma X_i)^2 = (10 + 12 + 13)^2 = (35)^2 = 1225$$

To find this quantity, first the scores are summed and then the total of all the scores is squared. The value of the sum of the scores, squared (1225) is not at all the same as the value of the sum of the squared scores (413). In summary, the table below indicates the operations associated with each set of symbols.

Symbols	Operations
ΣX_i	Add the scores.
ΣX_i^2	First square the scores and then add the squared scores.
$(\Sigma X_i)^2$	First add the scores and then square the total.

H.4 OPERATIONS WITH NEGATIVE NUMBERS

A number can be either positive (if it is preceded by a + sign or by no sign at all) or negative (if it is preceded by a − sign). Positive numbers are greater than zero, and negative numbers are less than zero. It is very important to keep track of signs because they will affect the outcome of virtually every mathematical operation. In this section I will briefly summarize the relevant rules for dealing with negative numbers. First, adding a negative number is the same as subtraction. For example,

$$3 + (-1) + 4 = 3 - 1 + 4 = 2 + 4 = 6$$

Second, subtraction changes the sign of a negative number:

$$3 - (-1) - 4 = 3 + 1 - 4 = 4 - 4 = 0$$

Note the importance of keeping track of signs here. It would be relatively easy for an inexperienced person to forget to change the sign of the negative number in the expression above and arrive at the wrong answer.

For multiplication and division, there are various combinations of negative and positive numbers you should be aware of. For purposes of this text, you will rarely have to multiply or divide more than two numbers at a time, and we will confine our attention to this situation. Ignoring the case of all positive numbers, this leaves several possible combinations. A negative number times a positive number results in a negative value:

$$(-3)\,(4) = -12$$
$$\text{or}$$
$$(3)\,(-4) = -12$$

A negative number times a negative number is always positive:

$$(-3)\,(-4)\,=\,12$$

Division follows the same patterns. If there is a single negative number in the calculations, the answer will be negative. If both numbers are negative, the answer will be positive. So,

$$(-4)/(2)\,=\,-2$$
and
$$(4)/(-2)\,=\,-2$$
but
$$(-4)/(-2)\,=\,2$$

Negative numbers do not have square roots, since multiplying a number by itself cannot result in a negative value. Squaring a negative number always results in a positive value (see the multiplication rules above).

H.5 ACCURACY AND ROUNDING OFF

A possible source of confusion in computation involves the issues of accuracy and rounding off. For various reasons, different people will work at different levels of accuracy and precision and, for this reason alone, may arrive at different answers to problems. This is important because, if you work at one level of precision and I (or your instructor or your study partner) work at another, we will often arrive at solutions that are at least slightly different. You may sometimes think you've gotten the wrong answer when all you've really done is round off at a different place in the calculations or in a different way.

There are two issues here: when to round off and how to round off. In this text, I have followed the convention of working in as much accuracy as my calculator or statistics package will allow and then rounding off to two places of accuracy (two places beyond the decimal point) at the very end. If a set of calculations is lengthy and requires the reporting of intermediate sums or subtotals, I will round the subtotals off to two places also.

In terms of how to round off, begin by looking at the digit immediately to the right of the last digit you want to retain. If you want to round off to 100ths (two places beyond the decimal point), look at the digit in the 1000ths place (three places beyond the decimal point). If that digit is greater than 5, round up. For example, 23.346 would round off to 23.35. If the digit to the right is less than 5, round down. So, 23.343 would become 23.34. If the digit to the right is 5, round up if the digit immediately to the left is even and round down if the digit is odd. So, 23.345 would become 23.35 and 23.355 would round to 23.35. I tell you this so that, if we both follow these conventions, we will get the same answers on homework and example problems. I hope this will help you avoid mistaking a rounding-off difference for an error.

Let's look at some more examples of how to follow the rounding rules stated above. If you are calculating the mean value of a set of test scores and your calculator shows a final value of 83.459067, and you want to round off to two places beyond the decimal point, look at the digit three places beyond the decimal point. In this case the value is 9 (greater than 5), so we would round the second digit beyond the decimal point up and report the mean as 83.46. If the value had been 83.453067, we would have reported our final answer as 83.45. A value of 83.455067 would round to 83.45, and a value of 83.445067 would be 83.45.

H.6 FORMULAS, COMPLEX OPERATIONS, AND THE ORDER OF OPERATIONS

A mathematical formula is a set of directions, stated in general symbols, for calculating a particular statistic. To "solve a formula" means that you must replace the symbols with the proper values and then manipulate the values through a series of calculations. Even the most complex formula can be rendered manageable if it is broken down into smaller steps. Working through these steps requires some knowledge of general procedure and the rules of precedence of mathematical operations. This is because the order in which you perform calculations may affect your final answer. Consider the following expression:

$$2 + 3(4)$$

Note that if you do the addition first, you will evaluate the expression as

$$5(4) = 20$$

but if you do the multiplication first, the expression becomes

$$2 + 12 = 14$$

Obviously, it is crucial to complete the steps of a calculation in the correct order.

The basic rules of precedence are to find all squares and square roots first, then do all multiplication and division, and finally complete all addition and subtraction. So the following expression:

$$8 + 2 \times 2^2/2$$

would be evaluated as

$$8 + 2 \times 4/2$$
$$= 8 + 8/2$$
$$= 8 + 4$$
$$= 12$$

The rules of precedence may be overridden when an expression contains parentheses. Solve all expressions within parentheses before applying the rules stated above. For most of the complex formulas in this text, the order of calculations will be controlled by the parentheses. Consider the following expression:

$$(8 + 2) - 4(3)^2/(8 - 6)$$

Resolving the parenthetical expressions first, we would have

$$(10) - 4 \times 9/(2)$$
$$= 10 - 36/2$$
$$= 10 - 18$$
$$= -8$$

Without the parentheses, the same expression would be evaluated as

$$8 + 2 - 4 \times 3^2/8 - 6$$
$$= 8 + 2 - 4 \times 9/8 - 6$$
$$= 8 + 2 - 36/8 - 6$$
$$= 8 + 2 - 4.5 - 6$$
$$= 10 - 10.5$$
$$= -.5$$

A final operation you will encounter in some formulas in this text involves denominators of fractions that themselves contain fractions. In this situation, solve the fraction in the denominator first and then complete the division. For example,

$$\frac{15 - 9}{6/2}$$

would become

$$\frac{15 - 9}{3} = 6/3 = 2$$

When you are confronted with complex expressions such as these, don't be intimidated. If you're patient with yourself and work through them step by step, beginning with the parenthetical expression, even the most imposing formulas can be managed.

H.7 EXERCISES

You can use the problems below as a "self-test" on the material presented in this review. If you can handle these problems, you're ready to do all of the arithmetic in this text. If you have difficulty with any of these problems, please review the appropriate section of this appendix. You might also want to use this section as an opportunity to become more familiar with your calculator. Answers are given on the next page, along with some commentary and some reminders.

1. Complete each of the following:

a. $17 \times 3 =$ _____

b. $17(3) =$ _____

c. $(17)(3) =$ _____

d. $17/3 =$ _____

e. $(42)^2 =$ _____

f. $\sqrt{113} =$ _____

2. For the set of scores (X_i) of 50, 55, 60, 65, and 70, evaluate each of the expressions below:

ΣX_i = _____

ΣX_i^2 = _____

$(\Sigma X_i)^2$ = _____

3. Complete each of the following:

a. $17 + (-3) + (4) + (-2) =$ _____

b. $15 - 3 - (-5) + 2 =$ _____

c. $(-27)(54) =$ _____

d. $(113)(-2) =$ _____

e. $(-14)(-100) =$ _____

f. $-34/-2 =$ _____

g. $322/-11 =$ _____

h. $\sqrt{-2} =$ _____

i. $(-17)^2 =$ _____

4. Round off each of the following to two places beyond the decimal point:

a. 17.17532 _____

b. 43.119 _____

c. 1076.77337 _____

d. 32.4651152301 _____

e. 32.4751152301 _____

5. Evaluate each of the following:

a. $(3 + 7)/10 =$ _____

b. $3 + 7/10 =$ _____

c. $((4 - 3) + (7 + 2))/((4 + 5)(10)) =$ _____

d. $\sqrt{(7(5 - 3)^2)/((17/3)(4))} =$ _____

e. $\dfrac{22 + 44}{15/3} =$ _____

H.8 ANSWERS

1. **a.** 51

b. 51

c. 51
(The obvious purpose of these first three problems is to remind you that there are several different ways of expressing multiplication.)

d. 5.67 (Note the rounding off.)

e. 1764

f. 10.63

2. The first expression translates to "the sum of the scores," so this operation would be

$\Sigma X_i = 50 + 55 + 60 + 65 + 70 = 300$

The second expression is the "sum of the squared scores." So

$\Sigma X_i^2 = (50)^2 + (55)^2 + (60)^2 + (65)^2 + (70)^2$

$\Sigma X_i^2 = 2500 + 3025 + 3600 + 4225 + 4900$

$\Sigma X_i^2 = 18{,}250$

The third expression is "the sum of the scores, squared":

$(\Sigma X_i)^2 = (50 + 55 + 60 + 65 + 70)^2$

$(\Sigma X_i)^2 = (300)^2$

$(\Sigma X_i)^2 = 90{,}000$

Remember that ΣX_i^2 and $(\Sigma X_i)^2$ are two completely different expressions with very different values.

3. **a.** 16

b. 19 (Remember to change the sign of -5.)

c. -1458

d. -226

e. 1400 (A negative number times a negative number results in a positive number.)

f. 17

g. -29.27

h. Your calculator probably gave you some sort of error message for this problem, since negative numbers do not have square roots.

i. 289

4. a. 17.17

b. 43.12

c. 1076.77

d. 32.47

e. 32.47

5. a. 1

b. 3.7 (Note again the importance of parentheses.)

c. 0.11

d. Whoa!—this looks complicated and should be worked out carefully, step by step. Note that complex expressions such as this are solved from the "inside out." That is, we begin within the innermost parentheses, follow the rules of precedence until we have resolved the numbers under the square root sign into a single number, and the *final* thing we do is take the square root:

$$\sqrt{(7(5-3)^2)/((17/3)\,(4))}$$

$$=\sqrt{7(2)^2/(5.6666667)\,(4)}$$

$$=\sqrt{7(4)/22.666667}$$

$$=\sqrt{28/22.666667}$$

$$=\sqrt{1.2352941}$$

$$=1.11143790 \text{ or, rounded off}$$

$$=1.11$$

e. 13.2

2.11 a. Complex A (5/20) × 100 = 25.00%
Complex B (10/20) × 100 = 50.00%

   b. Complex A 4 : 5 = 0.80
Complex B 6 : 10 = 0.60

   c. Complex A (0/20) = 0.00
Complex B (1/20) = 0.05

   d. (6/(4 + 6)) = (6/10) = 60.00%

   e. Complex A 8 : 5 = 1.60
Complex B 2 : 10 = 0.20

Chapter 3

3.1   For **problem 2.3**, the mode is 7, the median is the score associated with the 13th case, or 5, and the mean is 150/25, or 6. The mean is higher in value than the median, indicating a slight positive skew. The third quartile is the score associated with the 19th case, or 7. The fourth decile is the score associated with the 10th case, or 3. The 23rd percentile is the score associated with the sixth case, or 2. For **problem 2.5**, the mode is 17. The median is the score halfway between the scores of the 25th and 26th cases. Both cases have scores of 47, so the median is also 47. The mean is 2280/50, or 45.60. Since the mean is lower than the median, this distribution has a negative skew. The third quartile is 66, the fourth decile is 29, and the 23rd percentile is 19.

3.3   A total of 50 cases were tested, so the median is the score halfway between the scores of the 25th and 26th cases. For the "release date" test, both of these cases have a score of 22, so the median is 22. The mean for this first test is 1198/50, or 23.96. The median and mean for the second test are 27.5 and 28.58, respectively. The average scores are higher for the second test, indicating that the sample as a whole experienced an improvement in level of functioning and competence.

3.5   For the full sample, the mean is 22.75 and the median is 20.25. This distribution has a positive skew. By inspection, we can see that New York's score is much higher than any other score in the sample, and this would seem to account for the skew. Sure enough, if you remove New York and recalculate, the median falls to 19.20 and the mean falls to 20.89. There is still a skew in the distribution, but it is less pronounced (as indicated by the fact that

the difference between the median and mean is less with New York removed).

3.7   The mean is 19.44 and the median is 15.3. The distribution has a positive skew (there were a few highly urbanized nations in 1980).

3.9   The question posed could be answered with either the median or mean, but the latter is preferable since the data (number of cases) are clearly interval-ratio. In 1983, the average case load was 54.25 (1085/20 = 54.25). By 1993, the mean had risen to 60.30 (1206/20 = 60.30). The case load had increased by an average of about six cases per worker. The medians for the two years, incidentally, are 53.5 and 60.

3.13 In Table 2.4, there are 10 males and 10 females, so sex is bimodal. The mode of "marital status" is "single," with 10 cases. The median of "satisfaction" is 3, and the mean age is 20.05.

3.15 For the pretest, the mean is 9.33 and the median is 10. For the posttest, the mean is 12.93 and the median is 12.

3.17 Attitude and opinion scales almost always generate ordinal-level data, so the appropriate measure of central tendency would be the median. For the students, the median is 9 and, for the neighbors, the median is 2. Incidentally, the means are 7.80 and 4.00, respectively.

3.19 Sex is nominal in level of measurement so the mode is the appropriate measure of central tendency. There are 14 males and 11 females in the group so the mode is "male," or a score of 1. Support for gun control is ordinal (in the sense that "in favor" represents *more* support than "opposed") so the median is appropriate. There are 25 cases in the group so the median is the score associated with the 13th case. Twenty cases had a score of 1 and 5 had scores of 2. With only two scores, ranking the cases from high to low is not really necessary. The 13th case is one of the twenty with scores of 1 so the median is 1. Education is also an ordinal variable (the categories are unequal in size) so the median would be appropriate for this variable as well. The median would be the score associated with the 13th case, or a score of 1.

Answers to Odd-Numbered Computational Problems

In this answer section, I've also suggested some problem-solving strategies and provided some examples of how to interpret the numerical answers. You should try to solve and interpret the problems on your own before consulting this section.

In solving these problems, I let my calculator or computer do most of the work. I worked with whatever level of precision these devices permitted. I didn't round off until the end or until I had to record an intermediate sum. I always rounded off to two places of accuracy (or, two places beyond the decimal point, or to 100ths). I tell you this so that, if you follow the same conventions, your answers will match mine. However, I cannot guarantee that our answers will be exactly the same all the time because you are unlikely to solve these problems in exactly the same way I did. Usually, differences will be quite small but I want you to be aware that these small discrepancies might occur and that they are almost always trivial. If the difference between your answer and mine doesn't seem trivial, you should double-check to make sure you haven't made an error or solve the problem again using a greater degree of precision.

Finally, allow me a brief disclaimer about mathematical errors in this section. Let me assure you, first of all, that I know how important this section is for most students and that I worked hard to be certain that these answers are correct. Human fallibility being what it is, however, I know that I cannot make absolute guarantees. Should you find any errors, please let me know so I can make corrections in the future.

Chapter 2

2.1 a. To solve this problem, you must first find the total number of social science majors (this will give you the denominator of the fraction). This sum will be $97 + 132$, or 229. Since 97 of the social science majors are male, the percentage will be found by dividing 97 by 229 and multiplying by 100:

$$(97/229) \times 100 = (.4236) \times 100 = 42.36\%$$

b. Follow the same strategy as in 2.1a. First find the total number of business majors (295) and divide that into the number of female business majors (139). Since the problem asks for proportions, do not multiply by 100:

$$139/295 = 0.47$$

c. The two categories being compared in this problem are male humanities majors (117) and female humanities majors (83). The ratio will be

$$117/83 = 1.41$$

For every female, there are 1.41 male humanities majors.

d. $(475/899) \times 100 = 52.84\%$
e. $475/424 = 1.12$
f. $3/38 = 0.08$
g. $(229/899) \times 100 = 25.47\%$
h. $200/295 = 0.68$
i. $139/35 = 3.97$
j. $30/475 = 0.06$

3.21 For age, the mean is 42.17. For income the median is 14. For region, the mode is a score of 5 (So. Atlantic), which occurred 6 times. For political party, there were 10 scores of 1 ('not very strong Democrat'). Since this is the most common score, the mode is 1.

Chapter 4

4.1 The IQV is .89 for complex A, .99 for complex B, .71 for complex C, and .74 for complex D. Complex B is the most heterogeneous and complex C is the least.

4.3 $s = 14.62$

4.5 For **problem 2.3**, the range is $21 - 0$, or 21. Q_1 is the score of the $(.25)(25) = 6.25$th or, rounding off, the 6th case. This value is 2. Q_3 is the score of the $(.75)(25) = 18.75$, or 19th case, or 7. $Q = Q_3 - Q_1 = 7 - 2 = 5$. The standard deviation is 5.49 and the variance is $(5.49)^2$, or 30.14. For **problem 2.5**, the range is 87, Q is 46, $s = 27.24$, and $s^2 = 742.02$.

4.7 For 1975, the standard deviation is 4.02, and for 1991, the standard deviation is 4.10.

4.9 The range is 56.3 and the standard deviation is 13.80. New York City has the most extreme score in the sample, so if you eliminate this case and recalculate, the value of s will decrease.

4.11 $s = 14.50$

4.13 For 1983, $s = 7.27$
For 1993, $s = 9.09$

4.15 The issue of equalization of work load would essentially be a matter of dispersion. If all workers had the same work load (as measured by number of cases), any measure of dispersion would be zero, since there would be no difference in case load from worker to worker. To find the division that comes closest to this ideal, calculate s:

For A, $s = 5.32$
For B, $s = 2.37$
For C, $s = 1.03$
For D, $s = 7.88$

Division C has the most nearly equalized work load and Division D the least.

4.17 For town A, the IQV is 0.88. For town B, the IQV is also 0.88.

4.19 The standard deviations are 3.33 and 4.15 for students and neighbors, respectively.

4.21 For males, the mean is 77.30 and the standard deviation is 2.57. For females, the mean is 56.90 and the standard deviation is 4.08. Females participate at lower rates than males and are more heterogeneous.

4.23 The standard deviation for the age of these 30 cases is 17.05. The mean age (from problem 3.21) is 42.17.

Chapter 5

5.1

X_i	Z Score	% Area Above	% Area Below
650	1.50	6.68	93.32
400	−1.00	84.13	15.87
375	−1.25	89.44	10.56
586	.86	19.49	80.51
437	− .63	73.57	26.43
526	.26	39.74	60.26
621	1.21	11.31	88.69
498	− .02	50.80	49.20
517	.17	43.25	56.75
398	−1.02	84.61	15.39

5.3

X_i	Z Score	Number Above	Number Below
60	−2.00	195	5
57	−2.50	199	1
55	−2.83	199	1
67	− .83	159	41
70	− .33	126	74
72	0.00	100	100
78	1.00	32	168
82	1.67	9	191
90	3.00	1	199
95	3.83	1	199

5.5 The Z-score equivalent of the first score is $+0.53$, while the Z-score equivalent of the second score is +0.67. The second test score is the better performance.

5.7 a.

X_i	Z Score	% Below	% Above
19	.67	74.86	25.14
10	−2.33	00.99	99.01
14	−1.00	15.87	84.13
15	−.67	25.14	74.86
18	.33	62.93	37.07
20	1.00	84.13	15.87
22	1.67	95.25	4.75
23	2.00	97.72	2.28

b.

Scores	Z Scores	% Area Between
8 and 12	−3.00 and −1.67	4.61
9 and 13	−2.67 and −1.33	8.80
11 and 17	−2.00 and 0.00	47.72
15 and 19	−.67 and .67	49.72
16 and 20	−.33 and 1.00	47.06
17 and 23	0.00 and 2.00	47.72
18 and 19	.33 and .67	11.93
19 and 22	.67 and 1.67	20.39

c.

Scores	Z Scores	Probability
less than 17	0.00	.5000
less than 24	2.33	.9901
less than 10	−2.33	.0099
less than 8	−3.00	.0014
between 8 and 12	−3.00 and −1.67	.0461
between 11 and 17	−2.00 and 0.00	.4772
between 16 and 18	−.33 and .33	.2586
between 20 and 24	1.00 and 2.33	.1488
more than 24	2.33	.0099
more than 20	1.00	.1587
more than 15	−.67	.7486
more than 9	−2.67	.9962

5.9 a. .5328
b. .0638
c. .8944
d. .3085
e. .0062
f. .9332

5.11 a. .1112
b. .4129
c. .2177
d. .0375
e. .6711
f. .3694
g. .1174

5.13 a. 1.39%
b. 96.41%
c. 54.34%
d. 20.93%
e. 88.49%
f. 21.19%

5.15

Math and Science			Language Skills and Writing Ability			History and Social Sciences		
X_i	Z	% Area Below	X_i	Z	% Area Below	X_i	Z	% Area Below
55	−1.00	15.87	44	1.79	96.33	64	−1.07	14.23
52	−1.77	3.84	45	2.15	98.42	60	−1.46	7.21
49	−2.54	.55	41	0.70	75.80	58	−1.66	4.85
48	−2.80	.26	43	1.43	92.36	55	−1.96	2.50
45	−3.56	.02	42	1.07	85.77	50	−2.45	.71
44	−3.82	.01	38	−0.38	35.20	46	−2.85	.22
43	−4.08	.00	35	−1.47	7.08	40	−3.44	.03
47	−3.05	.11	50	3.96	99.99	75	0.02	50.80
58	−0.23	40.90	40	0.34	63.31	77	0.22	58.71
60	0.28	61.03	37	−0.75	22.66	80	0.52	69.85
61	0.54	70.54	39	−0.02	49.20	81	0.62	73.24
65	1.56	94.06	34	−1.83	3.36	89	1.41	92.07
67	2.08	98.12	33	−2.20	1.39	90	1.51	93.45
70	2.85	99.78	30	−3.28	.05	94	1.90	97.13
71	3.10	99.90	29	−3.65	.01	88	1.31	90.49

5.17 a. 33.58%
b. 14.53%
c. 30.15%
d. 20.90%
e. 37.93%
f. 18.18%
g. 50.80%
h. 58.71%
i. 69.85%
j. 84.38%

Part I Cumulative Exercises

1. Determine the level of measurement of each variable first. This will help you to set up the frequency distributions and determine the appropriate measures of central tendency and dispersion. There are two nominal variables (sex and marital status), two ordinal variables (attitude on interracial dating and area), and two interval-ratio-level variables (age and years of school). For the nominal-level variables, count the number of cases in each category of the variable. With only 30 cases, it may be sensible to reduce the number of categories for marital status to just two (married and unmarried, with the latter including separated, divorced, and widowed). The largest category of each variable is the mode. 'Female' (17 cases) is the modal category for sex, and 'Married' (16 cases) is the mode for marital status.

The index of qualitative variation (IQV) is the only measure of dispersion presented in this text for nominal-level variables. For sex, the IQV is .98, and for marital status, the IQV is .82. These are high values and indicate a great deal of dispersion in both variables. The IQV for sex approaches the maximum possible value, indicating that the two categories are roughly equal in size.

For this small sample, it may be desirable to collapse the two ordinal-level variables for the frequency distributions. First, however, use the uncollapsed versions of the variables to determine the median. Remember that, with an even number of cases, the median is the average of the scores of the two middle cases. For this sample, the middle cases will be the 15th case and 16th case. On the "attitude" variable, both middle cases have scores of 2, so the median is 2. For "area," both middle cases have scores of 3, so the median is 3.

Choosing a measure of dispersion for the two ordinal-level variables is something of a problem. Both variables have only five values so the range and interquartile range are not very meaningful, and no other measures of dispersion relevant for ordinal-level variables are presented in the text. This leaves the IQV and the standard deviation as possibilities. The former is .88 for "attitude" and .99 for "area" while the standard deviations are, respectively, 1.27 and 1.40. It is not uncommon in the social science literature for the standard deviation to be reported for ordinal-level variables, so we would probably report this statistic rather than the IQV.

For age, the mean is 37.37 and the standard deviation is 14.25. The range is (70 − 17), or 53. For education, the mean is 13.53 and the standard deviation is 2.55. The range is (18 − 8), or 10.

Chapter 7

7.1 **a.** 5.2 ± 0.12
　　 b. 100.1 ± 0.77
　　 c. 20 ± 0.40
　　 d. 1023 ± 5.74
　　 e. 7.3 ± 0.24
　　 f. 33 ± 0.82

7.5 **a.** 0.40 ± 0.10
　　 b. 0.40 ± 0.03
　　 c. 0.40 ± 0.01

7.7 10.3 ± 0.53

7.9 0.23 ± 0.08
At the 95% confidence level, the estimate would be that between 240 (15%) and 496 (31%) of the 1600 freshmen would be extremely interested.

7.11 2.30 ± 0.04

7.13 0.30 ± 0.07

7.15 2.37 ± 0.08

7.17 178.23 ± 1.97
The estimate is that students spent between $176.26 and $180.20 on books.

7.19 0.53 ± 0.10
Between 43% and 63% of the patients have been readmitted at least once.

7.21 **a.** $43.06 \pm .95$
　　 b. $2.80 \pm .14$
　　 c. $1.86 \pm .12$
　　 d. $.24 \pm .04$ (Catholics make up about 25% of the population.)
　　 e. $.19 \pm .04$ (About 23% of the population has never married.)

f. .42 ± .04 (About 43% of the electorate voted for Clinton.)

g. .37 ± .04

Chapter 8

8.3 Z (obtained) = 1.51.

At the 95% confidence level, we would fail to reject the null hypothesis. The difference is not significant. The local police are not different from the police nationally in percentage of robberies cleared by arrest.

8.5 a. Z (obtained) = − 41.00
b. Z (obtained) = 29.09

8.7 t (obtained) = − 1.14

8.9 Z (obtained) = 3.06

8.11 Z (obtained) = 2.83

8.13 a. t (obtained) = 0.67
b. t (obtained) = −7.09
The null would be rejected in the second test but not in the first. The program had a significant effect on absenteeism but not on GPA.

8.15 t (obtained) = 4.50

8.17 Z (obtained) = − 1.48

8.19 a. Z (obtained) = − 2.70
b. Z (obtained) = 1.17
c. Z (obtained) = −6.55
d. Z (obtained) = 1.55
e. Z (obtained) = −3.59

Chapter 9

9.1 a. $\sigma_{\bar{x}-\bar{x}} = 2.57$ t (obtained) = − 1.36
b. $\sigma_{\bar{x}-\bar{x}} = 1.61$ Z (obtained) = 2.49
c. $P_u = 0.19$ $\sigma_{p-p} = 0.05$
Z (obtained) = −0.60
d. $P_u = 0.61$ $\sigma_{p-p} = 0.03$
Z (obtained) = 0.67

9.3 $\sigma_{\bar{x}-\bar{x}} = 0.04$ Z (obtained) = − 4.00
With an alpha of .05, we would reject the null.

There is a significant difference in the way blue- and white-collar elderly relate to kin.

9.5 $P_u = 0.45$ $\sigma_{p-p} = 0.06$
Z (obtained) = 0.67
With alpha set at .05, we would fail to reject the null in this test. The course had no significant effect on efficiency.

9.7 $\sigma_{\bar{x}-\bar{x}} = 0.45$ Z (obtained) = − 0.67

9.9 a. $P_u = 0.46$ $\sigma_{p-p} = 0.06$
Z (obtained) = 2.17
b. $P_u = 0.80$ $\sigma_{p-p} = 0.07$
Z (obtained) = 1.43
c. $P_u = 0.72$ $\sigma_{p-p} = 0.08$
Z (obtained) = 0.75

9.11 t (obtained) = 2.33

9.13 $\sigma_{\bar{x}-\bar{x}} = 0.08$ Z (obtained) = 11.25

9.15 $P_u = 0.30$ $\sigma_{p-p} = 0.05$
Z (obtained) = 2.00

9.17 a. $P_u = .83$ $\sigma_{p-p} = .03$
Z (obtained) = − 6.00
b. $P_u = .27$ $\sigma_{p-p} = .04$
Z (obtained) = − 0.75
c. $P_u = .42$ $\sigma_{p-p} = .04$
Z (obtained) = − 2.75
d. $\sigma_{\bar{x}-\bar{x}} = .98$ Z (obtained) = 1.55
e. $\sigma_{\bar{x}-\bar{x}} = .20$ Z (obtained) = − 3.25
f. $\sigma_{\bar{x}-\bar{x}} = .12$ Z (obtained) = − 1.42

Three of these relationships are significant at the .05 level. Females are significantly more in favor of gun control, were more likely to have voted for President Clinton, and attend church at a significantly higher rate. Males and females are not significantly different in their approval of premarital sex, their level of occupational prestige, or their number of children.

Chapter 10

10.1 Mann-Whitney U test: $U = 122$
Z (obtained) = − 2.11
Runs test: $R = 6$ Z (obtained) = −4.81

10.3 Mann-Whitney U test: $U = 106.5$
Z (obtained) $= -2.11$
Runs test: $r = 12$ Z (obtained) $= -2.59$

10.5 $U = 168$ Z (obtained) $= -0.87$

10.7 $U = 107$ Z (obtained) $= -0.23$
The data have too many ties to justify the runs test.

10.9 $U = 25.5$ Z (obtained) $= -2.69$

Depending on how you account for the one tied score across the samples, the number of runs is either 6 or 8. For 6 runs, Z (obtained) is -2.92 and for 8 runs, Z (obtained) is -2.08. In either case, the null would be rejected at alpha $= 0.05$.

10.11 a. $U = 163$ Z(obtained) $= -1.00$
b. $U = 169.5$ Z(obtained) $= -0.83$
c. $U = 189$ Z(obtained) $= -0.31$
d. $U = 196.5$ Z(obtained) $= -0.10$
e. $U = 161$ Z(obtained) $= -1.07$

None of these relationships is statistically significant at the .05 level.

Chapter 11

11.1 a. 1.11
b. 0.00
c. 1.52
d. 1.46

11.3 a. 101.94
b. 11.43
c. 69.79

11.5 There are 4 degrees of freedom in a 3 × 3 table so, with alpha set at .05, the critical value for the chi square would be 9.448. The obtained chi square is 12.59 so we may reject the null hypothesis of independence between the variables. There is a statistically significant relationship between living arrangement and GPA.

11.7 With 3 degrees of freedom and alpha set at .05, the critical region will begin at 7.815. The obtained chi square of 19.34 is well within this area, and the null can be rejected. There is a statisti-

cally significant relationship between region of residence and support for the legalization of marijuana.

11.9 Expected frequencies would be found by dividing the total N equally across all four categories. So, $f_e = 580/4 = 145$ for all classes. The obtained chi square is 33.79.

11.11 The obtained chi square is 4.42.

11.13 The obtained chi square is 0.04.

11.15 The obtained chi square is 25.73.

Chapter 12

12.1 The overall mean is 12.17, SST = 230.69, SSB = 173.19, and SSW = 57.5. The F ratio is 13.55.

12.3 SST = 10609.31
SSB = 3342.99
SSW = 7265.32
F = 7.59

12.5 The overall mean is 4.12, SST = 132.64, SSB = 71.28, and SSW = 61.36. The obtained F ratio is 12.77. With alpha set at .05 and 2 and 22 degrees of freedom, the critical F ratio would be 3.44, so the null may be rejected. Decision making does vary significantly by type of relationship. By inspection of the group means, it would appear that the "cohabitational" category accounts for most of the differences.

12.7 SST = 342.99
SSB = 19.50
SSW = 323.50
F = .81

12.9 SST = 194.56
SSB = 38.22
SSW = 156.34
F = 3.30

12.11 a. SST = 3590.00
SSB = 825.80
SSW = 2764.20
F = 4.03

b. SST = 69.37
SSB = 23.27
SSW = 46.10
F = 6.81

c. SST = 162.67
SSB = 35.47
SSW = 127.20
F = 3.76

d. SST = 229.37
SSB = 1.07
SSW = 228.30
F = 0.06

e. SST = 301.47
SSB = 4.47
SSW = 297.00
F = 0.20

With alpha set at .05, the first three of these tests are significant and the last two are not. There are statistically significant differences by type of residence for prestige, number of children, and income but not for church attendance and TV watching. For prestige, the largest differences in category means is between suburban and rural dwellers. For number of children, the largest difference is between urban and rural dwellers. For income, the largest difference is between urban and suburban dwellers.

Part II Cumulative Exercises

a. One of the continuing challenges of using statistics in a reasonable way concerns test selection. As you are now aware, there are many statistical tests available, and researchers may sometimes be confused about how to select a test for a specific situation. In case it's any comfort, skilled researchers with years of experience get confused about this too. If you approach the decision systematically and consider the selection criteria carefully, the confusion and ambiguity can be minimized. Let's use this first problem of the exercise to consider some ways in which reasonable decisions can be made.

Begin by considering this research situation carefully. The situation calls for a test of hypothesis (is there a . . . significant difference?), so our choice of procedure will be limited to Chapters 8–12. Next, determine the types of variables you are working with. TV hours seems like an interval-ratio level

variable and race is definitely nominal. Note that, for this data set, race has only two categories. Income would ordinarily be treated as an interval-ratio variable but, in this case, the variable has been recoded into two categories. So, we are asked to see if an interval-level variable (TV hours) varies significantly between the categories of income and race, both of which have two categories. Which test should we use?

Since we are comparing two samples or categories (higher vs. lower on income and white vs. black on race), the techniques in Chapter 8 and Chapter 12 are irrelevant. Chi square (Chapter 11) won't work unless we collapse the scores on TV hours. This leaves Chapter 9. A test of sample means fits the situation and we have a small sample ($N = 45$) so it looks like we're going to wind up in Section 9.3.

Another way to approach test selection would be to use the flowcharts at the beginning of PART II and each chapter. The PART II flowchart would quickly lead to Chapter 9, since we are conducting a test of hypotheses with two samples and are using means. Turn to the flowchart at the beginning of Chapter 9, respond to the questions (YES, we want to test for the significance of the difference between population means; NO, the samples are not matched; NO, sample size is not large), and you should wind up again in Section 9.3. Using the formulas presented in Section 9.3 to test for the difference in TV hours by income level:

Lower Income	Higher Income
$\bar{X}_1 = 4.32$	$\bar{X}_2 = 2.00$
$s_1 = 2.70$	$s_2 = 1.25$
$N_1 = 22$	$N_2 = 23$

$\sigma_{\bar{X}-\bar{X}} = 0.64$ Z (obtained) = 3.64

Since some classes may omit Section 9.3, I will also mention the results using large-sample procedures (Section 9.2):

$\sigma_{\bar{X}-\bar{X}} = 0.65$ Z (obtained) = 3.57

In both cases, the difference is statistically significant. Lower-income respondents watch significantly more TV.

Testing for the significance of the difference by race:

Whites	Blacks
$\bar{X}_1 = 3.03$	$\bar{X}_2 = 3.71$
$s_1 = 2.23$	$s_2 = 3.06$
$N_1 = 38$	$N_2 = 7$

Using the small-sample formulas from Section 9.3:

$$\sigma_{\bar{X}-\bar{X}} = 0.98 \qquad Z \text{ (obtained)} = -0.69$$

With the large-sample formulas (Section 9.2)

$$\sigma_{\bar{X}-\bar{X}} = 1.21 \qquad Z \text{ (obtained)} = -0.56$$

In both cases, the difference is not statistically significant. White and black respondents do not differ significantly in hours spent watching TV. Because of the extremely small number of black respondents, we should be very cautious in interpreting these results.

b. One way to deal with this problem would be to construct a bivariate table and conduct the chi square test. The variables have only a few values and would fit this format rather well. The obtained chi square is 1.10, which is not significant at the .05 level. There is no statistically significant relationship between these variables. Older and younger respondents are not significantly different in happiness.

Can you think of other ways to deal with this research situation? For example, could you test the proportion of younger respondents who are "very happy" against the proportion of older respondents who are "very happy"?

c. The key words here are "estimate" (Chapter 7) and "average" (Section 7.4). Actually, the sample is too small to justify the use of these procedures, and we should present our results with great caution. At an alpha level of .05, the confidence interval would be 3.13 ± 0.71.

d. The research question focuses on the difference between a single sample and a population, so Chapter 8 is relevant. The population standard deviation is unknown and we have a small sample, so the flowcharts will direct us to Section 8.6. The sample statistics are

$$\bar{X} = 2.33$$
$$s = 1.91$$
$$N = 45$$

and Z (obtained) is 0.10. The difference is not significant. The sample is not significantly different from the population.

e. Education is coded so that it has only three categories, so the chi square test is appropriate. The obtained chi square is 4.72, which is not significant at the .05 level. Protestants and Catholics are not significantly different in level of education. What other tests might work in this situation?

f. Happiness has three categories, and TV hours is an interval-ratio level variable. If you follow the PART II flowchart, these characteristics will lead you to the ANOVA test. The sample information is

Very Happy	Pretty Happy	Not Too Happy
$\bar{X} = 3.41$	$\bar{X} = 2.43$	$\bar{X} = 3.56$
$s = 2.57$	$s = 1.40$	$s = 2.87$
$N = 22$	$N = 14$	$N = 9$

$$\text{SST} = 257.20$$
$$\text{SSB} = 10.25$$
$$\text{SSW} = 246.95$$
$$F = .87$$

The differences are not significant. The level of TV watching does not vary by level of happiness.

g. There are 7 blacks in the sample, so the sample proportion is (7/45), or .16. Setting alpha at .05, the confidence interval will be

$$.16 \pm .15$$

This is a very wide interval because the sample is so small. Expressed as percentages, we would estimate the population proportion to be between 1% and 31%. The actual parameter is about 12%.

Chapter 13

Because of space constraints, I've provided answers to only two problems here.

13.5

GPA	Attractiveness		
	Low	Moderate	High
Low	28.00	26.67	33.33
Moderate	40.00	33.33	35.56
High	32.00	40.00	31.11
	100.00	100.00	100.00

The conditional distributions change, so there is a relationship between the variables. The change from column to column is quite minimal (for example, roughly a third of the cases in each column are in each cell), and the relationship is weak. There is a slight tendency (I think—it's hard to tell without a measure of association to use as a guide) for GPA to increase with attractiveness. These trends are too weak to support the hypothesis.

13.7

Involvement	Involvement by Coverage		
	Coverage		
	None	Moderate	Extensive
None	3	4	0
	(50.0%)	(36.4%)	(0.0%)
Some	2	4	3
	(33.3%)	(36.4%)	(37.5%)
High	1	3	5
	(16.7%)	(27.3%)	(62.5%)
	6	11	8
	(100.0%)	(100.1%)	(100.0%)

There is a positive association between the variables. As press coverage increases, so does student involvement. Fad behavior was rare on campuses with no coverage (16.67%) and common on campuses with extensive coverage (62.50%). Note that we cannot tell the direction of the causal relationship from the table. The pattern is consistent with the idea that increased coverage motivates increased involvement AND the idea that extensive involvement attracts increased press coverage. We would need additional information on the timing of events before we could come to any conclusions about causation.

Chapter 14

14.3 **a.** $\phi = 0.06$ $\lambda = 0.00$
     b. $\phi = 0.54$ $\lambda = 0.33$
     c. $\phi = 0.39$ $\lambda = 0.05$

14.5 $\phi = 0.43$ $\lambda = 0.18$

14.7 $\phi = 0.31$ $\lambda = 0.03$

14.9 **a.** $\phi = .07$ $\lambda = .00$

b. $\phi = .41$ $\lambda = .27$
c. $\phi = .02$ $\lambda = .00$
     The only relationship of interest here is in Table b. Attitude toward premarital sex is related to courtship status: those who have "gone steady" are less likely to express disapproval. Attitude toward premarital sex is not associated with either gender or social class.

14.11 $V = 0.11$ Lambda $= 0.00$

14.13 $V = 0.36$ Lambda $= 0.16$

14.15 Measures of association for the tables presented in problem 13.15 are as follows:
     a. Cramer's $V = .10$
       Lambda $= .00$
     b. Cramer's $V = .08$
       Lambda $= .00$
     c. Cramer's $V = .15$
       Lambda $= .00$
     d. Cramer's $V = .20$
       Lambda $= .00$
     e. Cramer's $V = .13$
       Lambda $= .00$

   Lambda is zero for all tables. This is a result of unequal row totals, and we should probably select Cramer's V as the most appropriate measure of association. Although the Vs are nonzero, none of the associations are very strong. For this sample, political ideology is at best a moderately important correlate of a person's position on these issues.
   For the tables presented in problem 14.15:
a. Cramer's $V = .06$
   Lambda $= .00$
b. Phi $= .08$
   Lambda $= .00$
c. Cramer's $V = .19$
   Lambda $= .00$
d. Phi $= .08$
   Lambda $= .00$
e. Cramer's $V = .10$
   Lambda $= .00$

   The measures are not particularly strong and suggest generally weak relationships between the variables. The only relationship that seems at all interesting is that between sex and attitudes about health care.

Comparing these tables with those in problem 13.15, political ideology has (at least slightly) stronger relationships with these variables than sex.

Chapter 15

15.1 **a.** $G = +1.00$
b. $G = -0.88$
c. $G = +0.80$
d. $G = -0.11$

15.3 $G = 0.22$ $Z \text{(obtained)} = 0.93$

15.5 $G = -0.27$

15.7 $G = -0.17$

15.9 $G = 0.08$ $Z \text{(obtained)} = 1.26$
The relationship between the two variables is very weak and statistically insignificant. Apparently, happiness and income are not related in any important way.

15.11 $r_s = 0.70$
There is a strong, positive relationship between ethnic heterogeneity and strife. The greater the diversity, the greater the strife.

15.13 $r_s = -0.46$ $t \text{(obtained)} = -1.55$

15.15 $r_s = 0.78$

15.17 The Spearman's rho between prestige and ideology is -0.03. The Spearman's rho between prestige and attendance is 0.18. The Spearman's rho between ideology and attendance is 0.15.

Chapter 16

16.1 **a.** $b = 0.89$ $a = -10.26$
$Y = (-10.26) + (0.89)X$
$r = 0.97$ $r^2 = 0.94$
b. $b = -1.08$ $a = 33.72$
$Y = (33.72) + (-1.08)X$
$r = -0.90$ $r^2 = 0.81$
c. $b = 0.02$ $a = 44.09$
$Y = (44.09) + (0.02)X$
$r = 0.21$ $r^2 = 0.04$

16.3 **b.** $Y = 27.4 + (.37)X$
c. $r = 0.61$ $r^2 = 0.37$
d. $t \text{(obtained)} = 2.77$

16.5 $r = 0.20$ $r^2 = 0.04$

16.7 $b = 0.05$ $a = 53.18$
$r = 0.40$ $r^2 = 0.16$

16.9 $b = -0.26$ $a = 12.13$
$r = 0.31$ $r^2 = 0.10$

16.11 Pearson's r between occupational prestige and age is -0.18. This is a weak to moderate negative relationship. As age increases, prestige decreases. The relationship between church attendance and number of children is moderate in strength and negative ($r = -0.33$). As number of children increases, church attendance decreases.

The relationship between number of children and hours of TV watching is very weak and positive ($r = 0.06$), as is the relationship between age and TV watching ($r = 0.08$).

Pearson's r between age and number of children is .68. This is a strong, positive relationship. The older the respondent, the greater the number of children. The relationship between TV watching and prestige is very weak ($r = 0.02$).

Part III Cumulative Exercises

a. The PART III flowchart shows that the choice of a measure of association is largely dependent on the level of measurement of the variables involved. Assuming interval-ratio level measurement, the flowchart leads to Chapter 16 and Pearson's r. The values for this measure of association are

	Pearson's r
Satisfaction and:	
Income	$-.16$
Age	$-.15$
Number of Children	$-.15$

These values indicate weak to moderate relationships. In terms of direction, all three measures are negative. This gets a little tricky, however, because satisfaction is coded so that *low* scores indicate *high* satisfaction. Satisfaction actually increases with income, age, and number of children. In other words, higher incomes, older ages, and larger numbers of children are associated with more satisfaction. Drawing a scattergram of the relationships may help to clarify this point.

b. These variables have only a few possible values so it is appropriate to use bivariate tables to analyze the relationship. Given the way the question is asked, "fear" will be the dependent variable and should be placed in the rows. Marital status is a nominal-level variable, so we should use Cramer's V or lambda to measure the strength of the association. Area of residence could be considered ordinal (in the sense that size of place or population density would increase from rural to suburban to urban areas), and we will look at gamma as well as the nominal measures of association. Be careful when interpreting direction, however, because the "area" variable is coded with urban areas as the lowest score and rural areas as the highest.

For "fear" and "area," the conditional distributions change, so there is a relationship between these two variables. By inspection of the bivariate table, the majority (60%) of urban residents express fear, while the majority of both suburban and rural residents express no fear. Cramer's V for this table is .23 and lambda is .13. Gamma is .37. This is a weak to moderate relationship. The gamma tells us that the relationship is positive in direction. The larger the size of place, the greater the fear. (Take note of the percentage patterns to help you understand the direction of this relationship.)

For fear and marital status, the conditional distributions also change. A majority (60%) of the married respondents express no fear, and a slight majority (53.3%) of the nonmarried respondents do express fear. Lambda is .06 and phi is .13. This indicates a weak relationship between these two variables.

c. "Attitude about busing" has only two possible scores, so bivariate tables are a logical format for investigating these relationships. Taking the relationship with area first, the conditional distributions change, so there is a relationship. Urban dwellers are split, a slight majority of suburbanites (53.3%) are opposed, and a large majority of rural residents (80%) are opposed. For this relationship, lambda is

zero, Cramer's V is .26, and gamma is .38. Disregarding the lambda because of the unequal row totals, we can say that there is a moderate relationship between these variables. As size of place decreases, support for busing decreases. Remember that area is coded so that rural areas have the highest score and urban areas the lowest.

The majority of both married and nonmarried respondents are opposed to busing, with a higher percentage of the married opposed. The conditional distributions change, but the relationship is weak. Lambda is zero and phi is only .12.

Chapter 17

17.1 **b.** $G = 0.70$
e. $G = 0.69$ (high unemployment)
$G = 0.70$ (low unemployment)
f. $G_p = 0.70$

17.3 $G = -0.62$ (bivariate table)
$G = -0.52$ (whites)
$G = -0.68$ (blacks)
$G_P = -0.60$ (controlling for race)
$G = -0.62$ (males)
$G = -0.62$ (females)
$G_P = -0.62$ (controlling for sex)

The bivariate relationship is strong and negative. Completion of the training program is closely associated with holding a job for at least one year.

There is some interaction with race. The training has less impact for whites than for blacks. Blacks who did not complete the training were less likely to have held a job for at least one year. Gender has no impact at all on the relationship. Overall, there is a direct relationship between completion of the training and holding a job for at least a year, though there is some interaction with race.

17.5 $G = 0.26$ (bivariate table)
$G = 0.24$ (males)
$G = 0.26$ (females)
$G_p = 0.25$ (controlling for sex)
$G = 0.04$ (low trust)
$G = 0.04$ (high trust)
$G_p = 0.04$ (controlling for trust)

$G = 0.62$ (whites)
$G = -0.50$ (blacks)

17.7 a. Atttitudes about the role of the U.S. in world affairs by income, controlling for sex.

$G = -0.41$ (bivariate gamma from problem 15.16)
$G - -0.45$ (males)
$G = -0.36$ (females)
$G_p = -.39$ (controlling for sex)

This is a direct relationship (replication). The gammas for the partial tables are quite similar to the bivariate gamma and quite similar to each other. Observe the percentage patterns and note that a higher percentage of males than females support an active role for the U.S. at each income level. Furthermore, about 75% of all males support an active role vs. about 68% of all females (see the row totals). However, the relationship between the two variables is basically the same for both males and females. For both sexes, the higher the income level, the greater the support for an active role.

b. Attitude about pornography, controlling for sex.

$G = 0.18$ (bivariate gamma from problem 15.16)
$G = .05$ (males)
$G = 0.21$ (females)

The control for sex reveals an interactive relationship. For females, the belief that pornography leads to moral breakdown increases as income increases, and the relationship between these two variables is slightly stronger than in the bivariate table. For males, the variables have a very different relationship. The gamma is very weak, and the belief that pornography leads to moral breakdown is most common among *both* high- and low-income men.

Chapter 18

18.1 a. With population growth (X) while controlling for urbanization (Z):
For homicide, $r_{yx \cdot z} = .48$
For robbery, $r_{yx \cdot z} = .01$
For auto theft, $r_{yx \cdot z} = .41$

d. Independents are urbanization (X_1) and population growth rate (X_2).
For homicide, $Y = 1.0 + (.06) X_1 + (.17) X_2$
For robbery, $Y = -121.26 + (3.89) X_1 + (.13) X_2$
For auto theft, $Y = -113.68 + (6.47) X_1 + (8.01) X_2$

e. If $X_1 = 90$ and $X_2 = 5.0$:
For homicide: $Y = 1.00 + (.06)(90) + (.17)(5.0)$
$$Y = 1.00 + 5.40 + 0.85$$
$$Y = 7.25$$

f. For homicide, $Z_y = (.41) Z_1 + (.43) Z_2$
For robbery, $Z_y = (.71) Z_1 + (.01) Z_2$
For auto theft, $Z_y = (.61) Z_1 + (.31) Z_2$

g. With population growth (X_1) and urbanization (X_2):
For homicide, $R^2 = 0.37$
For robbery, $R^2 = 0.51$
For auto theft, $R^2 = 0.50$

18.3 a. $r_{yx \cdot z} = .32$
c. With growth as X_1 and annual pay as X_2:

$Y = 12.77 + (3.18)X_1 + (.04)X_2$
$Y = 12.77 + (3.18)(5.00) + (.04)(16000)$
$Y = 668.67$

d. $Z_y = (.08)Z_1 + (.34)Z_2$
e. $R^2 = 0.11$

18.5 With "percent urban" as Y, death rate as X_1, and birth rate as X_2,

$Y = 100.14 + (-.97)X_1 + (-1.20)X_2$
$Z_y = (-.18)Z_1 + (-.68)Z_2$
$R^2 = 0.63$

18.7 With "turnout" as Y, percent minority as X_1, and percent Democratic as X_2,

$Y = 83.80 + (2.88)X_1 + (-1.16)X_2$
$Z_y = (.84)Z_1 + (-1.27)Z_2$
$R^2 = 0.51$

18.9 With "support" as Y, years of service as X_1, and years of school as X_2,

$Y = 1.51 + (.07)X_1 + (.93)X_2$
$Z_y = (.10)Z_1 + (.45)Z_2$
$R^2 = 0.25$

18.11 With prestige (X_1) and age (X_2) as independents:

$$Y = 2.57 + (.03)X_1 + (-.01)X_2$$
$$Z_y = (.20)Z_1 + (-.10)Z_2$$
$$R^2 = .05$$

These two variables account for very little of the variation in TV watching $(R^2 = .05)$. Prestige has a stronger relationship than age. The former has a positive relationship and the latter a negative relationship with hours of TV watching. That is, hours of TV viewing increase with prestige and decrease with age.

With the number of children (X_1) and age (X_2) as independents:

$$Y = 3.46 + (.35)X_1 + (-.02)X_2$$
$$Z_y = (.33)Z_1 + (-.18)Z_2$$
$$R^2 = .11$$

These two variables account for about twice as much of the variation in TV viewing $(R^2 = .11)$ as the combination of prestige and age. Number of children has a stronger direct effect. Hours of TV viewing increase with number of children and decrease with age.

Part IV Cumulative Exercises

a. The choice of multivariate procedures will depend on level of measurement and the number of possible scores for each variable. We need interval-ratio, continuous variables to fully justify the use of the powerful and efficient regression techniques presented in Chapter 18. In this situation, we have a total of four variables: anomia, church attendance, prestige, and number of traumatic events. Only the last of these meets the level-of-measurement criteria for regression analysis, but it has only five possible values (0 through 4). The other variables are all probably ordinal, but prestige has a wide range of values. This is obviously not an ideal situation for multiple regression techniques.

As an alternative to regression analysis, we could collapse the variables into fewer categories and apply the elaboration techniques presented in Chapter 17. Elaboration is a less powerful technique (that is, it generates less information and is less flexible than regression analysis) and, in addition, we would lose information by collapsing the variables.

To resolve the ambiguity, let's proceed with regression analysis. It is not unusual for social science researchers to stray beyond strict level-of-measurement criteria, and we would, of course, present our findings with appropriate cautions to the reader.

Anomia is the dependent variable and our first task is to assess its relationship with church attendance and prestige. The zero-order correlations between all variables are

	Attendance	Prestige
Anomia	−0.42	0.08
Attendance		0.23

Anomia has a moderate relationship with church attendance and a weak relationship with prestige. The sign of the correlation coefficient for the anomia/attendance relationship is negative, but remember that anomia is coded so that higher scores indicate lower levels of anomia, the reverse of true interval-ratio variables. The underlying relationship is actually positive, and anomia increases as church attendance increases. This result seems counter-intuitive and raises some interesting causal questions. For example, is church attendance actually the result of anomia? Do people with higher levels of anomia seek resolution for their feelings of meaninglessness in church? Unfortunately, with a small sample and some cavalier treatment of the assumptions for regression analysis, we are not in a position to pursue these questions very far.

The anomia/prestige relationship is positive, with anomia increasing with prestige. The results of the regression analysis with attendance (X_1) and prestige (X_2) as independents:

$$Y = 2.90 + (-.15)X_1 + (.01)X_2$$
$$Z_y = (-.47)Z_1 + (.19)Z_2$$
$$R^2 = .21$$

The beta-weights indicate that attendance has a stronger direct effect on anomia than prestige.

For the second part of the problem, number of traumatic events replaces prestige in the analysis. The zero-order correlations between all variables are

	Attendance	Traumas
Anomia	−0.42	−0.05
Attendance		0.46

Anomia has a weak relationship with number of traumas. The sign of the correlation coefficient for the anomia/traumas relationship is negative, but let me remind you that anomia is coded so that higher scores indicate lower levels of anomia. The underlying relationship is actually positive and anomia increases as number of traumas increase. The results of the regression analysis with attendance (X_1) and number of traumatic events (X_2) as independents:

$$Y = 3.34 + (-.16)X_1 + (.12)X_2$$
$$Z_y = (-.51)Z_1 + (.18)Z_2$$
$$R^2 = .21$$

The beta-weights indicate that the direct effect of attendance is greater than the direct effect of number of traumatic events.

b. In this situation, we have a nominal-level variable (sex), and the remaining three variables have only two categories each. Age and education could be measured at the interval-ratio level but have been collapsed in this exercise. Attitude towards pornography is probably ordinal and is also a dichotomy. With so little variation in the variables, it is probably best to choose the elaboration technique to analyze these relationships.

The bivariate table shows that the belief that pornography is a threat to morality increases with age. Of the older respondents, 58% endorse the "yes" position vs. 50% of the younger respondents. The gamma for the bivariate table is .15.

Controlling for sex reveals some interaction between the three variables. The gamma for males is .25 and, for females, −.04. The percentage of males who view pornography as a threat is much higher in the older age group than in the younger (56% vs. 43%), while females do not differ very much in the percentage that sees pornography as a threat (57% vs. 59%). Age makes a difference for males but not for females.

The control for education is also consistent with the interaction pattern. The less educated, older respondents are more likely to see pornography as a threat (77% vs. 33% of the less educated, younger respondents), and the gamma for the partial table is .73. This is much greater than the bivariate gamma of .15. For the more educated, the relationship switches signs (−.71), and the more educated, older respondents are *less* likely to see pornography as a threat (22% vs. 63%).

Taken together, these results strongly suggest that both age and education are important correlates of attitude towards pornography and that the analysis should be re-oriented with these variables as independents. Age does have an effect on attitude towards pornography, but the relationship is strongly affected by the other variables.

Glossary

Each entry includes a brief definition and notes the chapter in which the term was introduced.

Alpha (α). In inferential statistics, the probability of error. (1) In estimation, the probability that a confidence interval does not contain the population value. Chapter 7 (2) In hypothesis testing, the proportion of the area under the sampling distribution that contains unlikely sample outcomes if the null is true. The probability of Type I error. Chapter 8

Alpha error. *See* **Type I error.** Chapter 8

Analysis of variance. A test of significance appropriate when we are concerned with the differences among more than two sample means. Chapter 12

ANOVA. *See* **Analysis of variance.** Chapter 12

Arrow keys. When using SPSS/PC+, the keys used to move the cursor around the screen. Appendix F

Association. The relationship between two (or more) variables. Two variables are said to be associated if the distribution of one variable changes for the various categories or scores of the other variable. Chapter 13

Average deviation. The average of the absolute deviations around the mean. Chapter 4

Bar chart. A graphic display device for nominal- and ordinal-level variables. Chapter 2

Beta error. *See* **Type II error.** Chapter 8

Beta-weights (β^*). Standardized partial slopes. Chapter 18

Bias. A criterion used to select sample statistics for estimation procedures. A statistic is unbiased if the mean of its sampling distribution is equal to the population value of interest. Chapter 7

Bivariate normal distributions. The model assumption in the test of significance for Pearson's r that both variables are normally distributed. Chapter 16

Bivariate table. A table that displays the joint frequency distribution of two variables. Chapter 2

Cells. The cross-classification categories of the variables in a bivariate table. Chapter 11

Central Limit Theorem. A theorem that specifies the mean, standard deviation, and shape of the sampling distribution, given that the sample is large. Chapter 6

χ^2 (critical). The chi square score that marks the beginnings of the critical region of the chi square sampling distribution. Chapter 11

χ^2 (obtained). The test statistic computed in step 4 of the five-step model. Chapter 11

Chi square test. A nonparametric test of hypothesis for variables organized in a bivariate table. Chapter 11

Class intervals. The categories used in frequency distributions for interval-ratio level variables. Chapter 2

Cluster sampling. A method of EPSEM sampling that is based on selecting geographical areas rather than cases for a list of the population. Chapter 6

Coefficient of determination (r^2). The proportion of all variation in Y that is explained by X. Chapter 16

Coefficient of multiple determination (R^2). A statistic that equals the total variation explained in the dependent variables by all independent variables combined. Chapter 18

Column. The vertical dimension of a table. Chapter 2

Command terminator. All SPSS/PC+ commands must end with a period, or dot. This symbol is called the command terminator. Appendix F

Conditional distribution of Y. The distribution of scores on the dependent variable for a specific score or category of the independent variable. Chapter 13

Conditional means of Y. The mean of all scores on Y for each value of X. Chapter 16

Confidence interval. An estimate of a population value in which a range of values is specified. Chapter 7

Confidence level. An alternative way to express alpha, the probability that a confidence interval will not contain the population value. Chapter 7

Continuous variable. A variable with a unit of measurement that can be subdivided infinitely. Chapter 1

Cramer's $V(V)$. A chi square–based measure of association. Chapter 14

Critical region (region of rejection). The area under the sampling distribution that includes all unlikely sample outcomes. Chapter 8

Cumulative frequency. A column in a frequency distribution that displays the number of cases in an interval and all preceding intervals. Chapter 2

Cumulative percentage. A column in a frequency distribution that displays the percentage of cases in an interval and all preceding intervals. Chapter 2

Cursor. When using SPSS/PC+, this symbol, either a blinking underscore or rectangle, marks where you are on the screen. Appendix F

Data. In social science research, information which is represented by numbers. Chapter 1

Data base. An organized collection of related information. Appendix F

Data reduction. Summarizing many scores with a few statistics. A major goal of descriptive statistics. Chapter 2

Deciles. The points that divide a distribution of scores into tenths. Chapter 3

Default option. A procedure, or routine, that SPSS/PC+ executes automatically ("by default") unless explicitly instructed to do otherwise. Appendix F

Dependent variable. A variable that is identified as an effect, result, or outcome variable. The dependent variable is thought to be caused by the independent variable. Chapter 1

Descriptive statistics. The branch of statistics concerned with (1) summarizing the distribution of a single variable or (2) measuring the relationship between two or more variables. Chapter 1

Deviations. The distances between the scores and the mean. Chapter 4

Direct relationship. A multivariate relationship in which a control variable has no effect on the bivariate relationship. Chapter 17

Discrete variable. A variable with a basic unit of measurement that cannot be subdivided infinitely. Chapter 1

Dispersion. The amount of variety or heterogeneity in a distribution of scores. Chapter 4

E_1. For lambda, the number of errors of prediction made when predicting which category of the dependent variable cases will fall into while ignoring the independent variable. Chapter 14

E_2. For lambda, the number of errors of prediction made when predicting which category of the dependent variable cases will fall into while taking the independent variable into account. Chapter 14

Efficiency. The extent to which sample outcomes are clustered around the mean of the sampling distribution. Chapter 7

EPSEM. Equal probability of selection method. A technique for selecting samples in which every element or case in the population has an equal probability of being selected for the sample. Chapter 6

Error message. When SPSS/PC+ cannot decipher instructions, it will cease processing and print an error message on the screen. Appendix F

Expected frequency (f_e). The cell frequencies that would be expected in a bivariate table if the variables were independent. Chapter 11

Explained variation. The proportion of all variation in Y that is attributed to the effect of X. Chapter 16

Explanation. *See* **spurious relationship.** Chapter 17

F **ratio.** For the analysis of variance, the test statistic computed in step 4 of the five-step model. Chapter 12

File. A data base (or any other information) that is stored under the same name in the memory of the computer or on a floppy disk. Appendix F

Five-step model. A step-by-step guideline for conducting tests of hypothesis. Chapter 8

FIXED. The part of the DATA LIST command that informs SPSS/PC+ that scores are always recorded in the same column(s). Appendix F

Frequency distribution. A table that displays the number of cases in each category of a variable. Chapter 2

Frequency polygon. A graphic display device for interval-ratio variables. Chapter 2

Gamma (G). A measure of association for ordinal variables organized in table format. Chapter 15

Goodness-of-fit test. A chi square test to see if a variable is randomly distributed across a series of categories. Chapter 11

Histogram. A graphic display device for interval-ratio variables. Chapter 2

Homoscedasticity. The model assumption in the test of significance for Pearson's r that the variance of the Y scores is uniform across all values of X. Chapter 16

Hypothesis. A statement about the relationship between variables that is derived from a theory. Hypotheses are more specific than theories, and all terms and concepts are fully defined. Chapter 1

Independence. The null hypothesis in the chi square test. Two variables are independent if the classification of a case on one variable has no effect on the probability that the case will be classified in any particular category of the second variable. Chapter 11

Independent random samples. Random samples gathered so that the selection of a case for one sample has no effect on the probability that any particular case will be selected for the other samples. Chapter 9

Independent variable. A variable that is identified as a causal variable. The independent variable is thought to cause the dependent variable. Chapter 1

Index of qualitative variation (IQV). A measure of dispersion for variables that have been organized into frequency distributions. Chapter 4

Inferential statistics. The branch of statistics concerned with making generalizations from samples to populations. Chapter 1

Interaction. A multivariate relationship in which a bivariate relationship changes across the categories of the control variable. Chapter 17

Interpretation. *See* **intervening relationship.** Chapter 17

Interquartile range (Q). The distance from the third quartile to the first. Chapter 4

Interval estimate. *See* **confidence interval.** Chapter 7

Intervening relationship. A multivariate relationship in which the independent and dependent variables are linked primarily through the control variable. Chapter 17

Kendall's tau-*b*. A measure of association for ordinal variables organized in table format. Especially appropriate for square tables. Chapter 15

Keywords. Commands that SPSS/PC+ recognizes automatically. Appendix F.

Lambda (λ). A measure of association for nominal-level variables that have been organized into a bivariate table. Lambda is based on the logic of PRE. Chapter 14

Level of measurement. The mathematical characteristics of a variable as determined by the measurement process. A major criterion for selecting statistical techniques. See Table 1.3. Chapter 1

Linear relationship. A relationship between two variables in which the observation points (dots) in the scattergram can be approximated with a straight line. Chapter 16

Mann-Whitney *U*. A nonparametric test of significance for the two-sample case with ordinal-level variables. Chapter 10

Marginals. The row and column totals of a bivariate table. Chapter 11

Mean ($\bar{X}$). The arithmetic average of a set of scores. Chapter 3

Mean square. In the analysis of variance, an estimate of the variance calculated by dividing the sum of squares within (SSW) or the sum of squares between (SSB) by the appropriate degrees of freedom. Chapter 12

Measures of association. Statistics that summarize the strength and direction of the relationship between variables. Chapter 13

Measures of central tendency. Statistics that summarize a distribution of scores by reporting the most typical or representative value of the distribution. Chapter 3

Measures of dispersion. Statistics that indicate the amount of variety or heterogeneity in a distribution of scores. Chapter 4

Median (Md). The point in a distribution of scores above and below which exactly half of the cases fall. Chapter 3

Menu. A list of options in a statistical package. Appendix F

Midpoint. The point halfway between the upper and lower limits of a class interval. Chapter 2

Mode. The most common value in a distribution or the largest category of a variable. Chapter 3

Multiple correlation. A multivariate technique for examining the combined effects of more than one independent variable on a dependent variable. Chapter 18

Multiple correlation coefficient (R). A statistic that indicates the strength of the correlation between a dependent variable and two or more independent variables. Chapter 18

Multiple regression. A multivariate technique that breaks down the separate effects of the independent variables on the dependent variable. Chapter 18

Negative association. A bivariate relationship in which the variables vary in opposite directions. As one variable increases, the other decreases, and high scores on one variable are associated with low scores on the other. Chapter 13

Nonparametric test. A type of significance test in which no assumptions about the shape of the sampling distribution are made. Chapter 10

Normal curve. A theoretical distribution of scores that

is symmetrical, unimodal, and bell-shaped. The standard normal curve always has a mean of 0 and a standard deviation of 1. Chapter 5

Normal curve table. Appendix A; a detailed description of the area between a Z score and the mean of a standardized normal distribution. Chapter 5

Null hypothesis (H_0). A statement of "no difference." The specific form varies from test to test. Chapter 8

Observed frequency (f_0). The cell frequencies actually observed in a bivariate table. Chapter 11

One-tailed test. A type of hypothesis test that can be used when (1) the direction of the difference can be predicted or (2) concern is focused on only one tail of the sampling distribution. Chapter 8

One-way analysis of variance. An application of ANOVA in which the effect of a single variable on another is observed. Chapter 12

Partial correlation. A multivariate technique for examining a bivariate relationship while controlling for other variables. Chapter 18

Partial correlation coefficient. A statistic that shows the relationship between two variables while controlling for other variables; $r_{yx.z}$ is the symbol for the partial correlation coefficient when controlling for one variable. Chapter 18

Partial gamma (G_p). A statistic that indicates the strength of the association between two variables after the effects of a third variable have been removed. Chapter 17

Partial slopes. In a multiple regression equation, the slope of the relationship between a particular independent variable and the dependent variable while controlling for all other independents in the equation. Chapter 18

Partial tables. Tables produced when controlling for a third variable. Chapter 17

Pearson's r (r). A measure of association for variables that have been measured at the interval-ratio level. Chapter 16

Percentage (%). The number of cases in a category divided by the number of cases in all categories, the entire quantity multiplied by 100. Chapter 2

Percentile. The point in a distribution of scores below which a specific percentage of the cases fall. Chapter 3

Phi (ϕ). A chi square–based measure of association. Chapter 14

Pie chart. A graphic display device for nominal- and ordinal-level variables. Chapter 2

Point estimate. An estimate of a population value in which a single value is specified. Chapter 7

Population. The total collection of all cases in which the researcher is interested. Chapter 1

Positive association. A bivariate relationship in which the variables vary in the same direction. As one variable increases, the other also increases, and high scores on one variable are associated with high scores on the other. Chapter 13

Proportion (p). The number of cases in a category divided by the number of cases in all categories. Chapter 2

Proportional reduction in error (PRE). The logic that underlies the definition and computation of several different measures of association. Statistics are derived by comparing the number of errors made in predicting the dependent variable while ignoring the independent variable with the number of errors made while taking the independent variables into account. Chapter 14

Quartiles. The points in a distribution of scores that divide the distribution into quarters. Chapter 3

Random samples. *See* **EPSEM.** Chapter 6

Range (R). The highest score minus the lowest score. Chapter 4

Rate. The number of actual occurrences divided by the number of possible occurrences per some unit of time. Chapter 2

Ratio. The number of cases in one category divided by the number of cases in another category. Chapter 2

Real class limits. The limits of a class interval used when the distribution is conceptualized as a continuous series of categories. Chapter 2

Region of rejection. *See* **critical region.** Chapter 8

Regression line. The best-fitting straight line that summarizes the relationship between two variables. The regression line is fitted to the data points by the least squares criteria whereby the line touches all conditional means of Y or comes as close to doing so as possible. Chapter 16

Replication. *See* **direct relationship.** Chapter 17

Representative. A quality of a sample. If a sample reproduces the major characteristics of the population from which it was drawn, it is said to be representative of that population. Chapter 6

Research hypothesis (H_1). A statement that contradicts the null hypothesis. The specific form varies from test to test. Chapter 8

Row. The horizontal dimension of a table. Chapter 2

Runs test. A nonparametric test of significance for the two-sample case with ordinal-level variables. Chapter 10

Sample. A carefully chosen subset of a population. In inferential statistics, information is gathered from samples and then generalized to populations. Chapter 1

Sampling distribution. The distribution of all possible sample outcomes of a given statistic. Chapter 6

Scattergram. A graphic display device that depicts the relationship between two variables. Chapter 16

Simple random sample. A method for selecting a sample from a population by which every case has an equal chance of being included in the sample. Chapter 6

Skew. The extent to which a distribution of scores has a few cases that are extremely high (positive skew) or extremely low (negative skew). Chapter 3

Slope (b). The amount of change in a variable per unit change in the other variable. Chapter 16

Somer's d. A measure of association for ordinal variables organized in table format. Chapter 15

SPSS/PC+. A statistical package designed for the analysis of social science data. The version used in this text runs on personal computers. Appendix F

Spearman's rho (r_s). A measure of association for ordinal variables that are in "continuous" format. Chapter 15

Specification. *See* **interaction.** Chapter 17

Spurious relationship. A multivariate relationship in which both the independent and dependent variables are actually caused by the control variable. The independent and dependent are not causally related. Chapter 17

Standard deviation (s or σ). The square root of the squared deviations of the scores around the mean, divided by N. The most commonly used measure of dispersion; s represents the standard deviation of a sample, and σ represents the standard deviation of a population. Chapter 4

Standard error of the mean. The standard deviation of a sampling distribution of sample means. Chapter 6

Standardized partial slopes (beta-weights). The slope of the relationship between a particular independent variable and the dependent when all scores are expressed as Z scores. Chapter 18

Stated class limits. The limits of a class interval used in a frequency distribution. Chapter 2

Statistical package (statpak). A set of computer programs designed to manipulate and statistically analyze data. Appendix F

Statistics. A set of mathematical techniques for organizing and analyzing data. Chapter 1

Stratified sampling. A method of sampling by which cases are selected from sublists of the population in numbers proportional to their representation in the population. Chapter 6

Sum of squares between (SSB). The sum of the squared deviations of the sample means from the overall mean, weighted by sample size. Chapter 12

Sum of squares total (SST). The sum of the squared deviations of the scores from the overall mean. Chapter 12

Sum of squares within (SSW). The sum of the squared deviations from the category means. Chapter 12

System file. A special organizational format for SPSS/PC+ files. Appendix F

Systematic sampling. A method of sampling by which the first case from a list of the population is selected by a random process and, thereafter, every kth case is selected. Chapter 6

t (critical). The t score that marks the beginnings of the critical region of a t distribution. Chapter 8

t distribution. A distribution used to find the critical region for tests of sample means when N is small and σ is unknown. Chapter 8

t (obtained). The test statistic computed in step 4 of the five-step model for tests of sample means when N is small and σ is unknown. Chapter 8

Test statistic. The value computed in step 4 of the five-step model that places the sample outcome on the sampling distribution. Chapter 8

Theory. A generalized explanation of the relationship between two or more variables. Chapter 1

Total variation. The spread of the Y scores around the mean of Y. Chapter 16

Two-tailed test. A type of hypothesis test that can be used when (1) the direction of the difference cannot be predicted or (2) concern is focused on both tails of the sampling distribution. Chapter 8

Type I error (alpha error). The probability of rejecting a null hypothesis that is true. Chapter 8

Type II error (beta error). The probability of failing to reject a null hypothesis that is false. Chapter 8

Unexplained variation. The proportion of the total variation in Y that is not accounted for by X. Chapter 16

Variable. Any trait that can change values from case to case. Chapter 1

Variance (s^2 or σ^2). The squared deviations of the scores around the mean, divided by N. A measure of dispersion used in inferential statistics and in regression techniques; s^2 represents the variance of a sample, and σ^2 represents the variance of a population. Chapter 4

Warning. When SPSS/PC+ detects a minor difficulty, it will print a warning to the screen. Appendix F

Y intercept (a). The point where the regression line crosses the Y axis. Chapter 16

Z (critical). The Z score that marks the beginnings of the critical region of a Z distribution. Chapter 8

Z (obtained). The test statistic computed in step 4 of the five-step model. Chapter 8

Z scores. Standard scores; the way scores are expressed after they have been standardized to the theoretical normal curve. Chapter 5

Index

Glossary of Symbols

(The number in parentheses indicates the chapter in which the symbol is introduced.)

a	Point at which the regression line crosses the Y axis (16)
AD	Average deviation (4)
ANOVA	The analysis of variance (12)
b	Slope of the regression line (16)
b_i	Partial slope of the linear relationship between the ith independent variable and the dependent variable (18)
b^*_i	Standardized partial slope of the linear relationship between the ith independent variable and the dependent variable (18)
d	Somer's d (15)
df	Degrees of freedom (8)
f	Frequency (2)
F	The F ratio (12)
f_e	Expected frequency (11)
f_o	Observed frequency (11)
G	Gamma for a sample (15)
G_p	Partial gamma (17)
H_o	Null hypothesis (8)
H_1	Research or alternate hypothesis (8)
IQV	Index of qualitative variation (4)
Md	Median (3)
Mo	Mode (3)
N	Number of cases (2)

N_d	Number of pairs of cases ranked in different order on two variables (15)
N_s	Number of pairs of cases ranked in the same order on two variables (15)
%	Percentage (2)
P	Proportion (2)
P_s	A sample proportion (7)
P_u	A population proportion (7)
PRE	Proportional reduction in error (14)
Q	Interquartile range (4)
r	Pearson's correlation coefficient for a sample (16)
r^2	Coefficient of determination (16)
R	Range (4)
r_s	Spearman's rho for a sample (15)
$r_{xy \cdot z}$	Partial correlation coefficient (18)
R^2	Multiple correlation coefficient (18)
s	Sample standard deviation (4)
SSB	The sum of squares between (12)
SST	The total sum of squares (12)
SSW	The sum of squares within (12)
s^2	Sample variance (4)
t	Student's t score (8)
T_x	Number of pairs of cases tied on the independent variable (15)